New Polymeric Materials Based on Element-Blocks

Yoshiki Chujo
Editor

New Polymeric Materials Based on Element-Blocks

Springer

Editor
Yoshiki Chujo
Department of Polymer Chemistry
Kyoto University
Kyoto, Japan

ISBN 978-981-13-2888-6 ISBN 978-981-13-2889-3 (eBook)
https://doi.org/10.1007/978-981-13-2889-3

Library of Congress Control Number: 2018963836

This Springer imprint is published by the registered company Springer Nature Singapore Pte Ltd.
The registered company address is: 152 Beach Road, #21-01/04 Gateway East, Singapore 189721, Singapore

Preface

This book, *New Polymeric Materials Based on Element-Blocks*, aims to summarize the major development in the fruitful results produced by the project of MEXT Scientific Research on Innovative Areas "New Polymeric Materials Based on Element-Blocks," during 2012–2017. A structural unit consisting of various groups of elements is called an "element-block." The design and synthesis of new element-blocks, polymerization of these blocks, and development of methods of forming higher-order structures and achieving hierarchical interface control to yield the desired functions are expected to result in manifold advantages. These benefits will encourage the creation of new polymeric materials that share, at a high level, electronic, optical, and magnetic properties not achievable with conventional organic polymeric materials. The element-block materials are expected to show form properties of molding processability and flexible designability that inorganic materials lack. By pioneering innovative synthetic processes that exploit the reactivity of elements and the preparation techniques employed for inorganic element-blocks, the aim is (1) to create a new series of innovative polymers based on the novel concept of element-block polymers, in which the characteristics of elements are extensively combined and utilized, and (2) to formulate theories related to these polymers. This book demonstrates especially the design strategies and the resulting successful examples offering highly functional materials that utilize element-block polymers as a key unit. In other words, the element-block materials are one of the organic-inorganic hybrid materials but especially at the element level. In such a sense, the idea of "element-block" should be a new advanced concept of the organic-inorganic hybrid materials.

In each of the chapters, all of them written by internationally acclaimed experts, the book covers the whole spectrum of the concept, new materials, and some useful industrial applications such as optical, electronic, magnetic, and biomedical materials. I hope that the readers will enjoy this novel chemistry and new materials developed throughout the book.

Kyoto, Japan
August, 2018

Yoshiki Chujo

Contents

Part I New Materials Based on Novel Concept of Element-Blocks

1 **Element-Block Materials: New Concept for the Development of Advanced Hybrids and Inorganic Polymers** 3
Masayuki Gon, Kazuo Tanaka, and Yoshiki Chujo

2 **Si-, Ge-, and Sn-Bridged Biaryls as π-Conjugated Element Blocks** 27
Yohei Adachi and Joji Ohshita

3 **Element-Blocks π-Conjugated Polymers by Post-element-Transformation Technique** 49
Ikuyoshi Tomita

4 **Borylated Polystyrenes as Versatile Functional Materials** 59
Frieder Jäkle

5 **Element-Block Polymeric Materials Based on Cage Silsesquioxane Frameworks** 77
Kensuke Naka

6 **Silsesquioxane-Based Hierarchical and Hybrid Materials** 95
Fuping Dong and Chang-Sik Ha

7 **Nanocomposite Materials Properties of Aminophenylsilsesquioxanes** 121
R. M. Laine, K. Takahasi, R. Tamaki, J. Choi, S. G. Kim, C. Brick, M. Z. Asuncion, E. Chetioui, S. Sulaiman, and R. Basheer

8 **Necklace-Shaped Dimethylsiloxane Polymers Bearing Polyhedral Oligomeric Silsesquioxane Cages as a New Type of Organic–Inorganic Hybrid** 139
Masashi Kunitake

9 **Alternative Aspects of Polythiophenes** 153
Takashi Nishino, Takuya Matsumoto, and Atsunori Mori

10 **Polyphosphazenes as an Example of the Element-Blocks Approach to New Materials** 167
Harry R. Allcock

11 **Electronic Structure of Element-Block Material As_4S_6 with Cage Shape Toward "Soft Electride"** 189
Kazuyoshi Tanaka and Hiroyuki Fueno

12 **Metalloxane Cage Compounds as an Element-Block** 199
Takahiro Gunji and Satoru Tsukada

13 **Preparation of Element-Block Materials Using Inorganic Nanostructures and Their Applications** 219
Naokazu Idota and Yoshiyuki Sugahara

14 **Design of Element Blocks for Photoresponsive Organosiloxane-Based Materials** 243
Sufang Guo, Kazuyuki Kuroda, and Atsushi Shimojima

15 **How Can We Control the "Element-Blocks" in Transition Metal Oxide Crystals?** 253
Katsuhisa Tanaka and Koji Fujita

Part II Applications of Element-Blocks as New Functional Materials

16 **Organic-Inorganic Hybrid Material with Surface-Modified Zirconia Nanoparticles as Element Blocks** 275
Kimihiro Matsukawa, Koji Mitamura, and Seiji Watase

17 **Network Formation Conditions Control Water Drop Adhesion for VK100 and a Model Pt-Cured Silicone** 291
Jennie B. Lumen, Rebecca M. Jarrell, Sithara S. Nair, Chenyu Wang, Ashraf M. Kayesh, and Kenneth J. Wynne

18 **Block Copolymers with Element Blocks: The Metal-Bisterpyridine Linkage** 307
Andreas Winter and Ulrich S. Schubert

19 **Three-Dimensional Coordination Polymers Composed of Luminescent Lanthanide Element Blocks** 347
Yasuchika Hasegawa, Takayuki Nakanishi, and Yuichi Kitagawa

20 **Organic Hybrid Thermoelectric Materials Containing Nano-dispersed Poly(nickel 1,1,2,2-ethenetetrathiolate) as an Element-Block** 371
Naoki Toshima, Keisuke Oshima, and Yukihide Shiraishi

21 Crystallization Kinetics-Induced Self-Assembly of Inorganic Element Blocks and the Surface-Enhanced Raman Scattering Based on Ag Hierarchical Structures 385
Jinguang Cai and Akira Watanabe

22 Air-Stable Optoelectronic Devices with Metal Oxide Cathodes 413
Makoto Takada and Hiroyoshi Naito

23 Design of Multifunctional Soft Biomaterials: Based on the Intermediate Water Concept 423
Masaru Tanaka

24 Synthesis of Calcium-Phosphate-Based Nanoparticles as Biocompatible and Biofunctional Element Blocks 433
Ayako Oyane and Maki Nakamura

Part I
New Materials Based on Novel Concept of Element-Blocks

Chapter 1
Element-Block Materials: New Concept for the Development of Advanced Hybrids and Inorganic Polymers

Masayuki Gon, Kazuo Tanaka, and Yoshiki Chujo

Abstract By incorporating highly functional inorganic units in organic materials, the creation of advanced materials possessing both advantages of organic components such as designability and good processability and inorganic elements can be expected. However, there are several difficulties in combination with organic and inorganic components due to intrinsic low compatibility between organic and inorganic components. Although organic-inorganic hybrids have been developed, further strategies for material design such as for precise controls of nanostructures in the hybrids are strongly required. To meet these demands, we propose the new concept for material design based on an "element-block" which is defined as a minimum functional unit composed of heteroatoms. In this chapter, the basic idea of an "element-block" and the recent progresses in the development of "element-block materials" are mainly from our recent works. As a representative example, we illustrate the element-blocks involving specific steric structures such as polyhedral oligomeric silsesquioxane (POSS), modified boron dipyrromethenes (BODIPYs) having the cardo boron and [2.2]paracyclophanes as a chiral source and explain material properties originated from these element-blocks. The roles of these element-blocks in the materials are explained.

Keywords Hybrid · POSS · Boron · BODIPY · [2.2]Paracyclophane

---

M. Gon · K. Tanaka (✉) · Y. Chujo (✉)
Department of Polymer Chemistry, Graduate School of Engineering, Kyoto University, Kyoto, Japan
e-mail: tanaka@poly.synchem.kyoto-u.ac.jp; chujo@poly.synchem.kyoto-u.ac.jp

Y. Chujo (ed.), *New Polymeric Materials Based on Element-Blocks*,
https://doi.org/10.1007/978-981-13-2889-3_1

1.1 Introduction

1.1.1 New Concept for Material Design Based on "Element-Blocks"

Inorganic materials show high performances especially in photonics, electronics, and magnetics, and many researchers have continuously devoted huge efforts for exploring new substances. By incorporating these functional inorganic units in organic materials, it can be expected that advanced materials possessing both advantages of organic components such as designability and good processability as well as inorganic elements. However, there are several difficulties in combination with organic and inorganic components. Basically, due to distinct different compatibility between both elements, further technological advancements are commonly needed in all size scales. In the molecular scale, only small electronic interaction between both components was often obtained. Moreover, the introduction of heteroatoms often caused distortion of molecular skeletons, followed by loss of electronic conjugation. From the nano-sizes to bulk sizes, phase separation often occurred, resulting in losses of physical properties as a material. To overcome this problem, the idea of organic-inorganic hybrids was proposed, and indeed when homogeneous mixing states were able to be realized, synergetic effects were obtained [1]. However, in the conventional preparation methods for hybrid formation known as a sol-gel method, chemical modification to each component and/or severe condition controls are necessary to achieve homogeneous mixtures which are essential to express superior functions. Moreover, it is still challenging to form nanostructures according to the design. In view of this situation, we proposed the new concept "element-block materials" [2]. An element-block is defined as a minimum functional unit composed of heteroatoms, and it was expected that element-block materials can be fabricated by combination, assembly, and organization with these element-blocks (Fig. 1.1). Through discovery of new element-blocks and development of element-block materials, we desired not only to accumulate significant information on interaction and interfaces among element-blocks and matrices but also to establish robust and effective hybridization protocols between organic and inorganic components for producing advanced functional materials based on the preprogrammed design.

1.1.2 Overview

We describe here several examples from our recent works based on the idea of "element-block materials." As a representative research, we selected the element-blocks involving specific steric structures such as polyhedral oligomeric silsesquioxane (POSS), modified boron dipyrromethenes (BODIPYs) having the cardo boron and [2.2]paracyclophanes as a chiral source and explain material

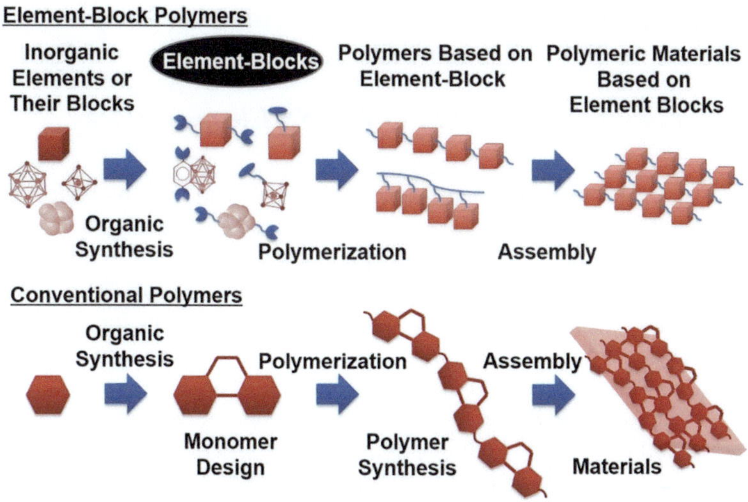

Fig. 1.1 Schematic models of the development of element-block materials

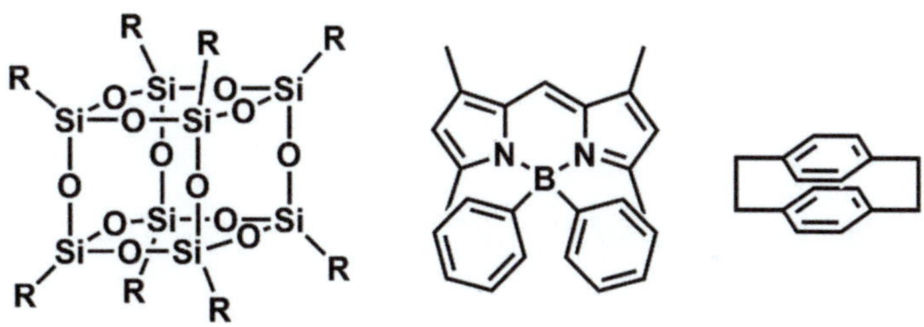

Fig. 1.2 Element-blocks introduced in this section

properties originated from these element-blocks (Fig. 1.2). Initially, the recent progresses in organic-inorganic hybrids based on POSS element-blocks with high thermal stability and extra functions are demonstrated. By employing POSS, various optical materials can be transformed to the hybrid materials. As a result, thermally stable optical materials were obtained. Furthermore, multiple functions including the trade-off relationships were observed. The mechanisms are explained here. Next topics are on the construction of conjugated polymers involving the cardo boron. The series of phenyl-bearing modified BODIPYs with the cardo structure were built, and their luminescent properties were evaluated. Characteristic behaviors in solid-state emission and energy transferring through the cardo boron are shown here. The roles of [2.2]paracyclophane element-blocks in the optically active materials are also described as the last topic. By incorporating the [2.2]paracyclophane unit into electronic conjugation, significant optical behaviors can be induced. Unique functions originated from structural features of these element-blocks are presented here.

1.2 Unique Thermal Properties of "Designable Hybrids" Based on POSS Element-Blocks

1.2.1 Fundamental Properties of POSS Element-Blocks

Initially, several POSS-based element-block materials and their multiple functions including unique thermal properties are demonstrated. Generally, by linking or interacting to the cubic core of POSS, [3–5] thermal stability of the connected molecules can be improved [6–11]. Suppression of molecular motions would efficiently occur because of the rigid cubic silica, followed by prohibition of degradation processes. This effect is versatile for enhancing thermal stability with various matrices which have been never used in a hybrid material. Therefore, it can be said that simply by employing POSS element-blocks, hybrid materials possessing thermal stability would be obtained. Additionally, the highly symmetric structure can play a significant role in elevating transition temperatures in the thermal processes by decreasing transition entropy changes [12–15]. This character has potential to be applicable for maintaining the compensate relationship to thermal property. Thus, unique behaviors were often observed from the POSS-based element-block materials. As a representative example, extension of the temperature range of the liquid crystal phase is explained according to the researches with the POSS-tethering ionic salts.

1.2.2 POSS Ionic Liquid Crystal

Ionic liquid crystals have attracted much attention as a platform for preparing multiple functional materials having optical property and anisotropic ion-carrier ability. Therefore, enhancement to thermal stability of liquid crystal phases as well as durability of ionic components is strongly required. One of the strategies to meet these demands is to introduce POSS [12–15]. According to POSS-containing ionic liquids, by connecting the ion pairs to POSS, molecular tumbling would be highly restricted. As a result, pyrolysis was able to be suppressed. Moreover, the POSS core played a critical role in isolation of ion pairs, followed by lowering melting temperatures of the ion pairs. Finally, it was shown that ionic liquid phases can be greatly stabilized by the POSS element-block. From these data, we presumed that similar effects on thermal properties were able to be expected with ionic liquid crystals. To examine the validity of this idea, the ion pairs composed of carboxylate and imidazolium with variable alkyl chains were prepared, and their properties were examined (Fig. 1.3) [16]. Initially, it was found from the X-ray analyses and polarized optical microscopic observation that the ion salts with the octadecyl alkyl chains at the imidazolium moiety formed the liquid crystal phases. Interestingly, the liquid crystal phases were maintained until the thermal decomposition occurred, while the pristine salt without POSS showed typical phase transition to the isotropic phase by heating. Furthermore, comparing to the pristine ionic salt, the

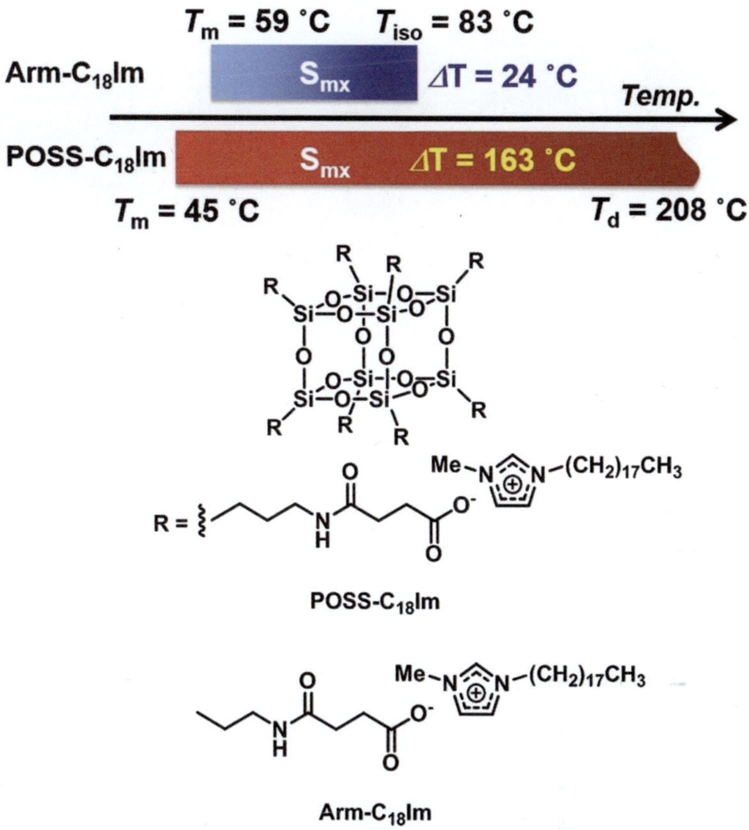

Fig. 1.3 Chemical structures and thermal properties of ionic liquid crystals

POSS-tethered ion pairs showed lower melting temperature. These data indicated that the temperature range of the liquid crystal phase was extended by the POSS core. The highly symmetric structure of POSS should contribute not only to the suppression of the molecular motions of the ion salts but also to the formation of the regular structures. This is one of the typical examples to offer the significant stabilization for nanostructures by the structural feature of the POSS element-block.

1.2.3 Thermally Stable Luminescent Materials

Next, it was demonstrated that the inorganic cubic core is an advantageous scaffold for realizing solid-state emissive materials with high thermal stability. Recently, higher affinity of POSS with conjugated polymers was demonstrated [17]. This fact implied that modified POSS with conjugated molecules can be thermally stable luminescent materials. The series of the modified POSS with π-conjugated

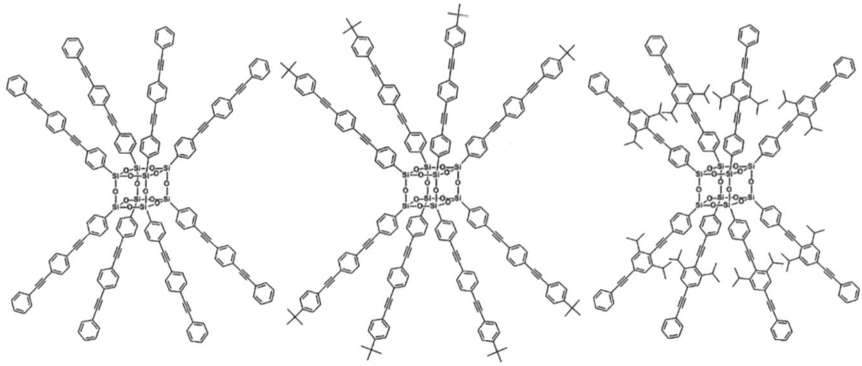

Fig. 1.4 Chemical structures of thermally stable luminescent molecules based on POSS

luminophores connected to eight vertices of the cubic were synthesized (Fig. 1.4) [18]. From the optical measurements in a dilute solution, the excimer formation was obtained. It was proposed that the π-conjugated luminophores around the POSS scaffold interacted with the neighboring π-conjugated luminophores. Indeed, the intrinsic luminescent property of the luminophore moieties was readily recovered by modification with bulky alkyl chains. In the solid state, intriguing behaviors were observed. The π-conjugated luminophores linked to the POSS scaffold presented similar optical properties to the intrinsic emission even in the absence of the bulky substituents. This result represents that POSS played a critical role in inhibition of intermolecular interaction, which often brings unfavorable alteration in optical properties such as aggregation-caused quenching (ACQ) and peak broadening in the condensed state. Furthermore, improvement of thermal stabilities of the π-conjugated luminophores by POSS was exhibited. Finally, it was demonstrated that the POSS materials presented bright blue emission even beyond 200 °C in the open air; meanwhile the π-conjugated luminophore no longer showed emission. Owing to the cubic structure of the POSS element-block, thermally durable optical materials were able to be produced.

1.2.4 Thermally Durable Mechanochromic Luminescent Material

Solid-state emissive molecules occasionally showed luminescent chromism by physical stresses such as pressing, crashing, and grinding [19–22]. These mechanochromic luminescent behaviors are especially beneficial for constructing pressure sensors as well as for fabricating optical recording/memory devices. Therefore, many researchers still have devoted their much efforts not only for exploring new materials but also for regulating chromic properties. The solid-state emissive boron complexes [23–33] were designed and connected to each vertex of POSS

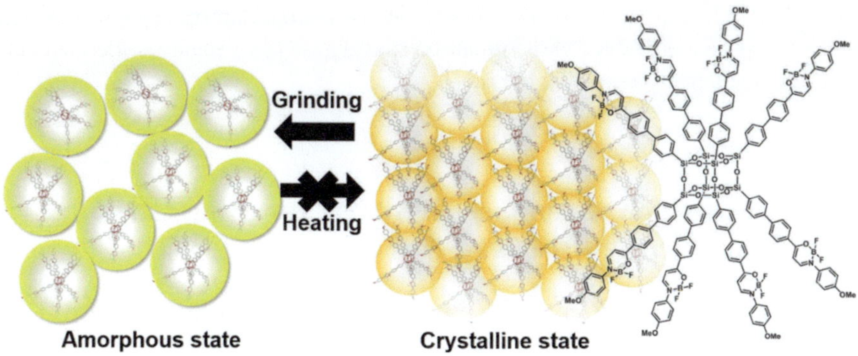

Fig. 1.5 Chemical structure of the dye-modified POSS with thermally durable mechanochromic luminescent property

(Fig. 1.5) [34]. The crystalline sample of the modified POSS presented the bathochromic emission band than that from the single molecule. By adding mechanical stresses to the crystalline sample, regular structures disappeared, followed by luminescent chromism from orange to yellow. This result indicated that each boron complex should be isolated from electronic interaction with other molecules even in the amorphous state after the grinding treatment. Interestingly, the emission color can be maintained by heating until melting. Commodity mechanochromic luminescent dyes often showed reversible changes by heating; meanwhile the POSS-tethered boron complex showed thermally durable luminescent chromism. Similar to the above results, the amorphous state can be stabilized by POSS. As a result, phase transition should be highly restricted. This thermally durable mechanochromic luminescent material is strongly required in the industrial applications as a pressure-sensing paint in the wind tunnel test. It is likely that luminescent chromism of commodity dyes induced by the air pressure could be spoiled by recovering with fraction heat which is inevitably generated by the air flow. On the other hand, the POSS-based "hybrid" dye would maintain the significant changes in luminescence color. Thus, precise detection system could be obtained.

1.2.5 Summary

Intrinsic lower stabilities of organic molecules than those of inorganics are still limitation to practical applications such as for organic optoelectronic devices. Although hybrid formation is one of the valid strategies for improving the durability of organic materials, distinctive inventions for obtaining homogeneous dispersion state at the nano-level such as chemical modification to organic units by polar groups are commonly necessary to receive reinforcement originated from the inorganic component. In contrast, simply by employing POSS, these demands can be readily satisfied without losses of optical and electrical properties of organic components in the hybrid material. In particular, based on a preprogrammed design, intended functions are conquerable. Thus, it can be said that POSS should be the element-block for realizing "designable hybrids."

1.3 Construction of Conjugated System Involving Cardo Boron Element-Blocks

1.3.1 Optical and Electronic Properties of Cardo Structures

Next, optical properties of the conjugated polymers containing cardo boron element-blocks are illustrated. From the series of researches on the electronic properties of conjugated polymers involving the cardo fluorene, it was demonstrated that the cardo structure should be a versatile scaffold for obtaining multiple functional optical materials by assembling dye molecules [35–39]. For example, significant emission from the main-chain conjugation was observed in the film state by suppressing the ACQ owing to the substituents connected via the cardo carbon. Moreover, it was shown that the main-chain conjugation can be efficiently preserved even in the presence of the electron-donating and/or electron-withdrawing units at the end of the side groups via the cardo carbons [35, 36]. Furthermore, the energy transfer between the fluorescent dyes at the side chains and the main-chain conjugation was controllable in the dye-modified cardo fluorene polymers where several kinds of dyes were placed with orthogonal directions of each transition moment [37–39]. Finally, by choosing the type of the fluorescent dyes at the side chains and the copolymers in the main chain, not only the light-harvesting materials but also the multi-emissive polymer films were accomplished [37–39]. From these data, we constructed the cardo structure in the boron atom of BODIPY, which is known as a versatile luminescent dye because of various advantages, [40–46] and built up conjugated polymers containing the cardo boron structure.

1.3.2 Solid-State Luminescent Properties of Cardo BODIPYs

The BODIPY derivatives having single or dual phenyl groups into the boron center were prepared (Fig. 1.6) [47]. From optical measurements, it was observed that the pristine BODIPY showed the sharp intense emission band in the diluted solution, meanwhile critical ACQ and unexpected bathochromic peak shift occurred in the solid state. In contrast, it was revealed that the phenyl substituents played a significant role in suppression of critical losses of emission efficiencies in the solid state. Additionally, it should be noted that the single- and dual-phenyl substituted BODIPYs showed sharp emission bands both in the solution and solid states. It

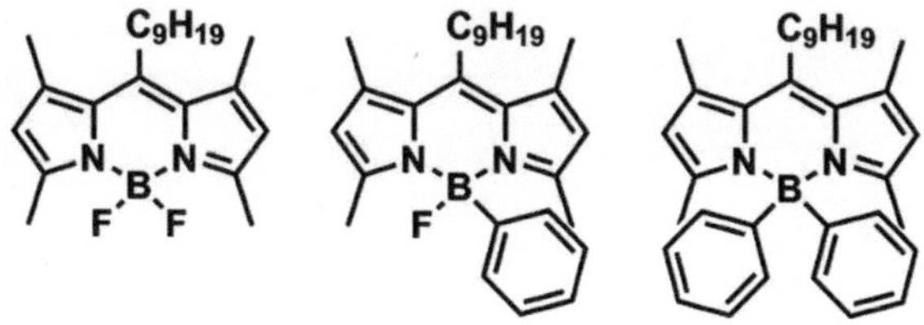

	B0	B1	B2
$\lambda_{em,solution}$	511 nm	511 nm	539 nm
$\Phi_{solution}$	0.81	0.79	0.34
$\lambda_{em,solid}$	614 nm	569 nm	577 nm
Φ_{solid}	0.09	0.25	0.30

Fig. 1.6 Chemical structures and optical properties of the pristine and cardo BODIPYs

was suggested that the steric hindrance by the phenyl groups can inhibit ACQ in the condensed state.

1.3.3 Conjugated Polymers Involving Cardo BODIPYs

Next, the conjugated polymers containing the cardo boron in BODIPY were synthesized, and influence of the substituent on electronic properties of the boron center was examined (Fig. 1.7) [48]. The electron-donating and/or electron-withdrawing groups were introduced into the phenyl substituents connected to the boron center. Similarly to the dye-modified cardo fluorene polymers, it was indicated that electronic structures of the main-chain conjugation can be preserved from the introduction of substituent groups. Both in the UV-vis absorption and photoluminescence spectra of the polymers, similar shapes were obtained. These data represent isolation of the main-chain conjugation in the conjugated polymers from the electronic interaction with the side chains by the cardo boron. Thus, it can be expected that the cardo boron element-block could be a potential platform for constructing multifunctional optoelectronic polymers by assembling the functional units according to the preprogrammed designs without losses of properties of each component.

1.4 [2.2]Paracyclophane as a Chiral Element-Block

1.4.1 General Evaluation of Chiroptical Properties

Final topics are concerned to the chiral element-blocks and their unique optical properties. When two mirror-image molecules are non-superimposable like a right

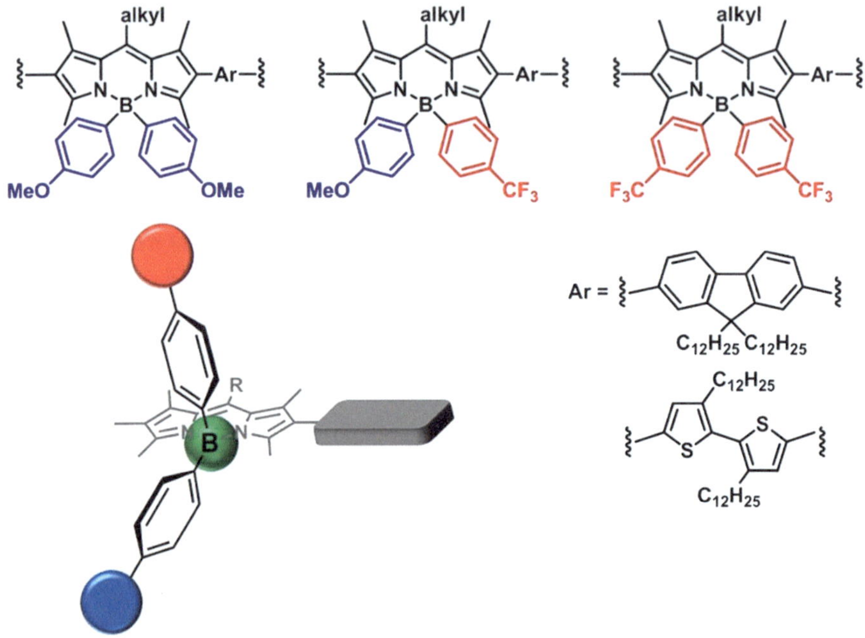

Fig. 1.7 Chemical structures of the conjugated polymers containing cardo BODIPYs

hand and a left hand, we call that the molecule is chiral. Chirality exists in various molecules including amino acids and sugars which make up natural world, and it is an essential for understanding the intrinsic properties of substances. Control of chirality is widely studied such as selective asymmetric synthesis with a chiral catalyst in the field of organic synthetic chemistry and biochemistry. Recognition of chirality is effective tool for discriminating the optically active compounds. Those methods definitely have developed the recent chiral chemistry. In addition, absorption and luminescent properties based on chirality are widely known as circularly dichroism (CD) and circularly polarized luminescence (CPL), respectively. The chiroptical data obtained from CD and CPL spectra measurements are useful for understanding the structural information in the ground state and in the excited state, respectively. Especially, CPL has received much attention as significant tools for potential application to cryptography, memory, or light source of a display. Therefore, chirality is important in the field of chemistry, and it is effective in creating functional materials. From those backgrounds, in this section, we suggest that functional units based on chiral molecules are called "chiral element-blocks," and combination of the element-blocks should create versatile functional materials. As examples of the chiral element-blocks, planar chiral [2.2]paracyclophanes were selected. Planar chiral [2.2]paracyclophanes [49, 50] are designed for constructing chiral π-conjugated systems. As opposed to conformationally flexible chiral molecules, which have an axis of rotation as well as a rotatable bond around a stereogenic center, substituted [2.2]paracyclophanes are conformationally stable and show planar chirality because of the suppressed rotation of the aromatic rings [51–53]. In addition, the planarity of the [2.2]paracyclophanes takes great advantage for expanding π-conjugated systems via through-space conjugation and through-bond conjugation [41]. That is effective in obtaining good chiroptical performance such as CD and CPL [54–66]. To evaluate the chirality on CD and CPL spectra, dissymmetry factors, g_{abs} and g_{lum} were used. They were defined as $g_{abs} = \Delta\varepsilon/\varepsilon$ ($\Delta\varepsilon = \varepsilon_{left} - \varepsilon_{right}$), where ε_{left} and ε_{right} indicate absorbances of left- and right-handed circularly polarized light, respectively, and $g_{lum} = \Delta I/I$ ($\Delta I = I_{left} - I_{right}$), where I_{left} and I_{right} indicate luminescence intensities of left- and right-handed CPL, respectively. In this section, our recent researches based on the concept of chiral element-blocks are introduced briefly.

1.4.2 [2.2]Paracyclophane as a Chiral Element-Block

[2.2]Paracyclophane consists of face-to-face-oriented two benzene rings combined by two ethylene chains at *para* positions [49, 50]. It was known that several groups of substituted [2.2]paracyclophanes at the aromatic rings show planar chirality because of the restricted rotation of the aromatic rings. Therefore, the planar chirality can be handled as rigid and stable chirality [51–53]. In addition, the rigidity and planarity should be a great advantage for the expansion of π-conjugated systems, and that can be a different character from the other representative chiral sources such as chiral center, axial chirality, and helicity. However, planar chiral [2.2]paracyclophanes have not been studied in the field of material chemistry such as polymer, chemistry, or optoelectronics but in the field of organic and organometallic chemistry. From the above reasons, we consider that the planar chiral [2.2]paracyclophane can be effective chiral building blocks to construct π-conjugated system for the material chemistry, and the functional units can be treated as chiral element-blocks. Concretely, we focused on pseudo-*ortho*-disubstituted [2.2]paracyclophane and 4,7,12,15-terasubstituted [2.2]paracyclophane as planar chiral element-blocks and synthesized construct optically active second-ordered structures, such as V-, N-, M- [57], triangle- [56, 57], propeller- [59, 60, 63], and X-shaped [61, 62, 65, 66], one-handed double-helical [64] and self-assembled structures [65]. All compounds exhibited intense CPL [67, 68] with large molar extinction coefficients (ε), good photoluminescence (PL) quantum efficiencies (Φ_{PL}), and excellent CPL dissymmetry factors (g_{lum}). In this term, recently our works are demonstrated briefly.

1.4.3 *Planar Chiral Pseudo-*ortho*-disubstituted [2.2] Paracyclophanes*

At first, we started the development of the practical optical resolution method of planar chiral pseudo-*ortho*-disubstituted [2.2]paracyclophanes [55]. There are several reports about optical resolution method of them; however, the diversity for the functionality after the optical resolution was limited. We selected a diastereomer method as a practical optical resolution way because it is possible to separate enantiomers in large scale with SiO_2 column chromatography and the enantiomers were obtained with the same amount, respectively. The detail synthetic method is shown in Scheme 1.1. Treatment of commercially available *rac*-pseudo-*ortho*-dibromo[2.2]paracyclophane *rac*-**1** with 1.1 equivalent of *n*-BuLi and (1*R*,2*S*,5*R*)-(−)-menthyl (*S*)-*p*-toluenesulfinate converted one of the bromines into a sulfinyl group. The resulting mixture of the diastereomers was readily separated by conventional SiO_2 column chromatography to obtain (R_p,*S*)-**2** and (S_p,*S*)-**2** (each in 39% isolated yield). The reaction of (R_p,*S*)-**2** with 4 equiv. *t*-BuLi afforded the (R_p)-pseudo-*ortho*-dilithio[2.2]paracyclophane (R_p)-**3**, which was able to react with

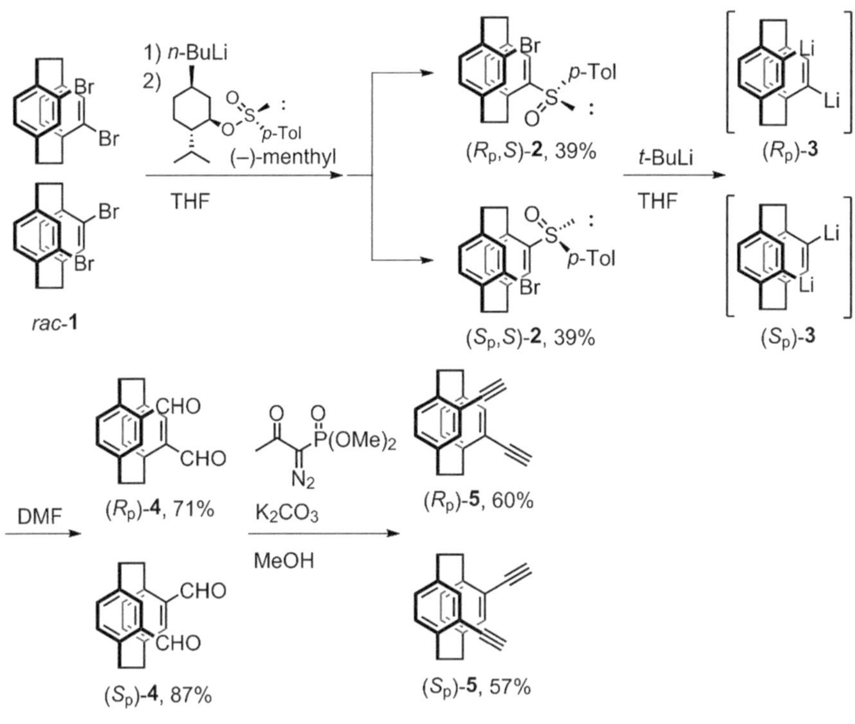

Scheme 1.1 Optical resolution of *rac*-**1** and synthesis of (R_p) and (S_p)-**5**

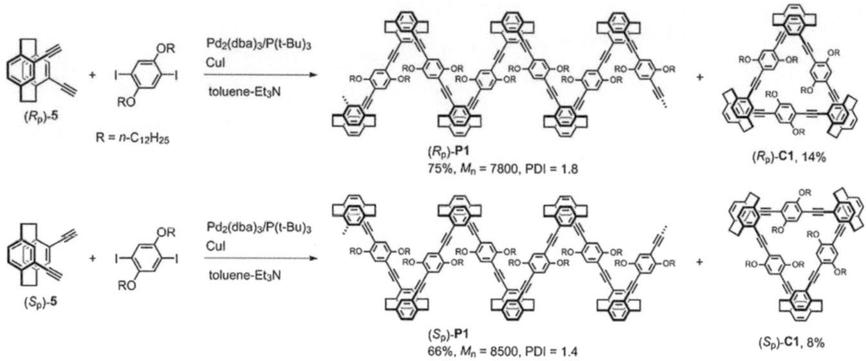

Scheme 1.2 Synthesis of polymers (R_p)- and (S_p)-**P1** and cyclic compounds (R_p)- and (S_p)-**C1**

various electrophiles [55–58]. For example, the (R_p)-**3** was allowed to react with DMF to afford (R_p)-pseudo-*ortho*-diformyl[2.2]paracyclophane (R_p)-**4**. Formyl groups of (R_p)-**4** (or (S_p)-**4**) were converted into ethynyl groups to afford (R_p)-pseudo-*ortho*-diethynyl[2.2]paracyclophane (R_p)-**5**, which is readily used for the chiral element-blocks to construct π-conjugated polymer. Through-space-conjugated polymers (R_p)-**P1** and (S_p)-**P1** were synthesized from (R_p)-**5** and (S_p)-**5**, respectively, by the Sonogashira-Hagihara coupling polymerization (Scheme 1.2). The π-electron systems of (R_p)-**P1** and (S_p)-**P1** partially overlapped to form chiral V-shaped zigzag structures, and the resulting polymers were mirror images of each other. At the same time, optically active cyclic trimers (R_p)-**C1** and (S_p)-**C1** were isolated in this reaction and separated by size exclusion column chromatography. The CD and CPL spectra of (R_p)-**P1** and (S_p)-**P1** in dilute $CHCl_3$ solution (1.0×10^{-5} M, excitation wavelength: 320 nm) exhibited mirror-image signals in the observed PL range (Fig. 1.8), and the g_{lum} of (R_p)-**P1** and (S_p)-**P1** at 415 nm were estimated to be $+2.3 \times 10^{-3}$ and -2.0×10^{-3}, respectively. (R_p)-**P1** and (S_p)-**P1** showed intense CPL with a large g_{lum} in dilute solutions (1.0×10^{-5} M) because of the rigid structure with planar chirality and the stacked π-electron systems. To elucidate the feature of the cyclic trimer (R_p)-**C1** and (S_p)-**C1**, V-shaped zigzag dimer, trimer, and tetramer were prepared from the oligomeric approach [57]. As a result, in the ground state, observed similarities in the CD profiles of the oligomers were attributed to the equivalent orientations of two adjacent chromophores. On the other hand, in the excited state, the oligomers were folded into a form analogous to a one-handed helix by photoexcitation. We successfully predicted the structure both in the ground state and in the excited state in comparing the cyclic trimer and the corresponding oligomers (Fig. 1.9). All the compounds in dispersed solution exhibited intense circularly polarized luminescence with relatively large dissymmetry factors on the order of 10^{-3}.

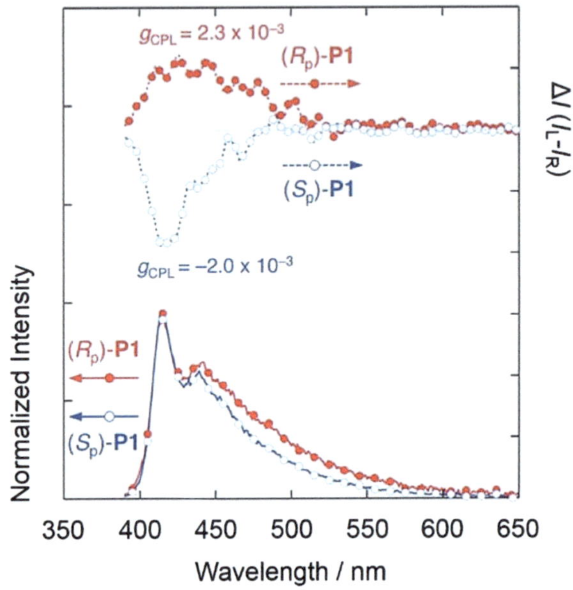

Fig. 1.8 Optical properties of (R_p)-**P1** and (S_p)-**P1**. (**a**) UV-vis absorption spectra (1.0×10^{-5} M) and CD spectra (1.0×10^{-5} M). (**b**) Photoluminescence spectra (1.0×10^{-6} M, excited at 370 nm) and CPL spectra (1.0×10^{-5} M, excited at 320 nm). (Reproduced from Ref. [56] with permission from the Royal Society of Chemistry)

1.4.4 Planar Chiral Tetrasubstituted [2.2]Paracyclophanes

Next, as the other chiral element-blocks based on a planar chirality, we focused on 4,7,12,15-tetrasubstituted [2.2]paracyclophane scaffold. 4,7,12,15-Tetrasubstituted [2.2]paracyclophane has highly symmetrical structure, and the π-conjugated system was expanded whole of the molecule via through-space conjugation because the through-space conjugation occurs at the center of the structure [54]. However, there are no reports about the optical resolution method of planar chiral 4,7,12,15-tetrasubstituted [2.2]paracyclophanes. Therefore, we developed the optical resolution method and induction way to the ethynyl moiety, planar chiral 4,7,12,15-tetraethynyl [2.2]paracyclophane, which is key component for constructing π-conjugated systems. Optical resolution of tetrasubstituted [2.2]paracyclophane was carried out by a diastereomer method with (−)-(1*S*,4*R*)-camphanoyl chloride (Scheme 1.3). One of bromines of 4,7,12,15-tetrabromo[2.2]paracyclophanes [69] *rac*-**6**, was converted to a hydroxyl group to obtain *rac*-**7** in 69% isolated yield, which was reacted with (−)-(1*S*,4*R*)-camphanoyl chloride to obtain a mixture of diastereomers. These were readily separated by SiO_2 column chromatography and purified by recrystallization to obtain (S_p,1*S*,4*R*)-**8** and (R_p,1*S*,4*R*)-**8** in 38% and 34% isolated yield, respectively. Subsequent transformation of (S_p,1*S*,4*R*)-**8** into ethynyl moiety is shown in Scheme 1.3. The reactions were proceeded in high yields, and finally, we succeeded in synthesizing enantiopure 4,7,12,15-tetra-ethynyl[2.2]paracyclophane (S_p)-**12**. The enantiomer (R_p)-**12** was synthesized as the same method from (R_p,1*S*,4*R*)-**8**. They can be employed as conformationally stable chiral building blocks for various optically active π-conjugated compounds. To investigate

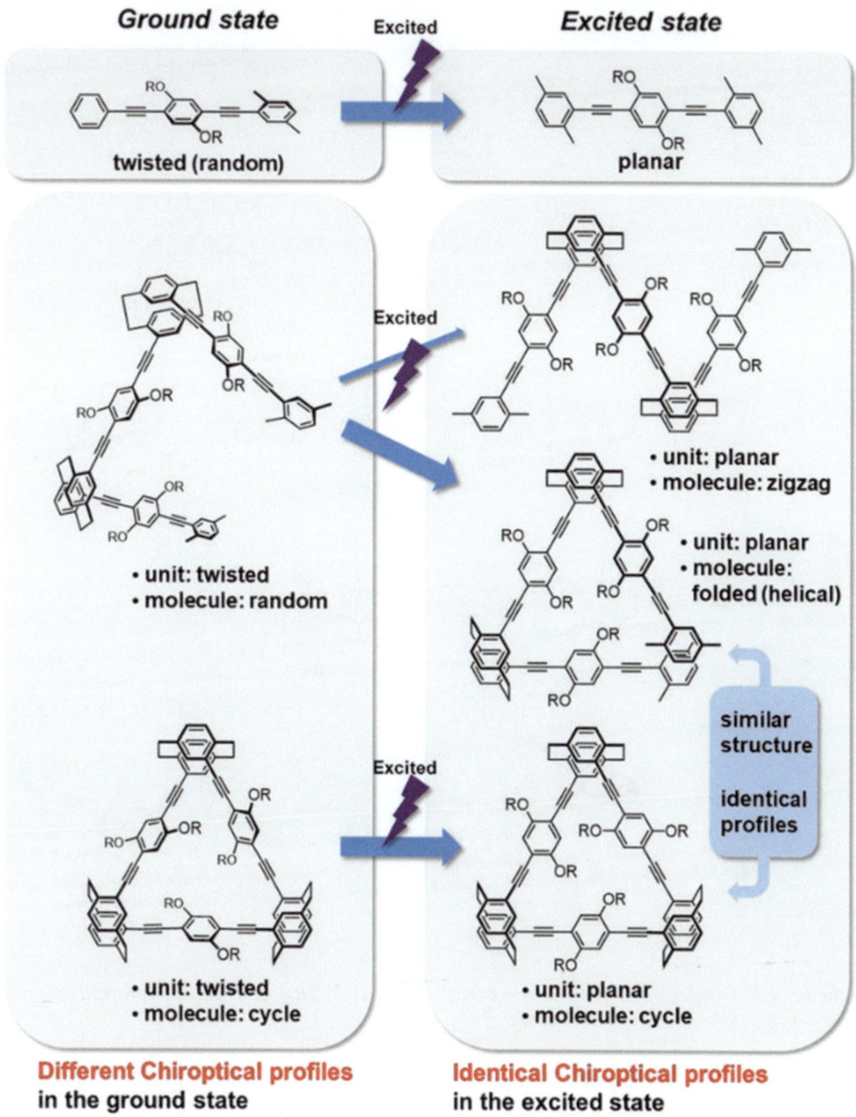

Fig. 1.9 Plausible structures of the oligomers: the linear trimer versus the cyclic trimer in the ground and excited states

the chiroptical properties of planar chirality based on the tetrasubstituted [2.2] paracyclophane, optically active propeller-shaped macrocycles were synthesized. This structure has previously been synthesized by Hopf, Haley, and co-workers as a racemic compound [70], although the positions of the *t*-butyl groups were different. Surprisingly, the propeller-shaped macrocycle showed excellent CD and CPL performance ($g_{abs} = 0.88 \times 10^{-2}$, $\varepsilon_{abs} = 129{,}000\ M^{-1}\ cm^{-1}$, $g_{lum} = 1.3 \times 10^{-2}$, and

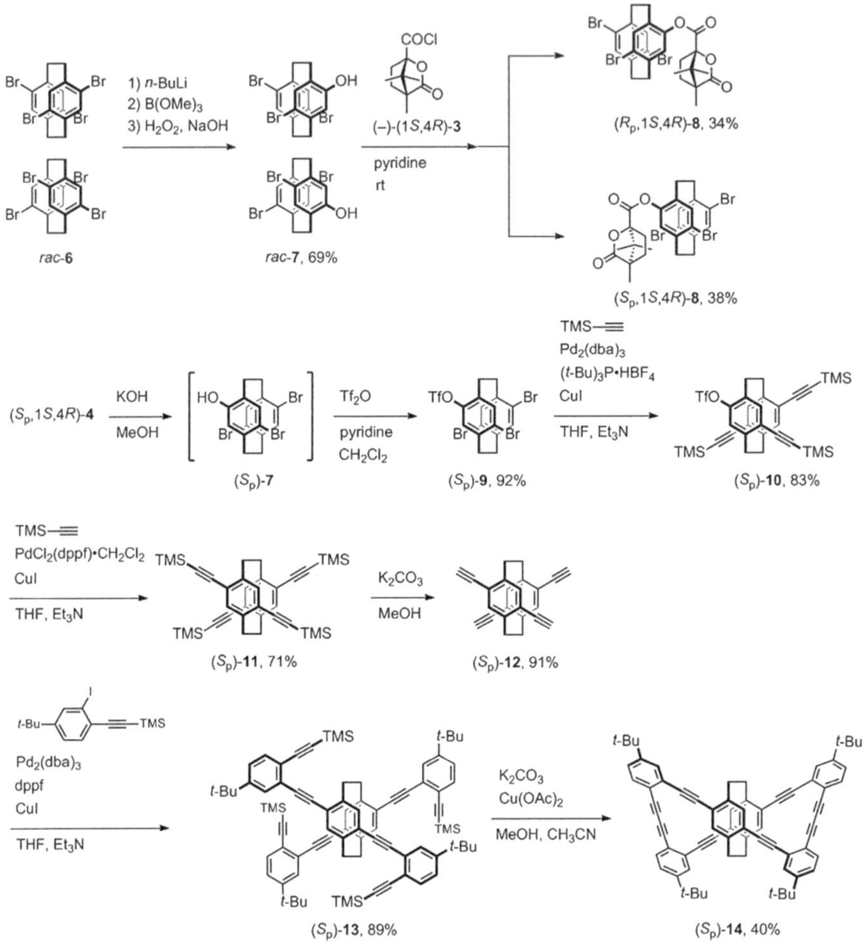

Scheme 1.3 Optical resolution of *rac*-**6** and synthesis of (R_p) and (S_p)-**12**, propeller-shaped macrocycles (S_p)-**14**

$\Phi_{lum} = 0.41$) in the dilute solution condition (Fig. 1.10). It is the first report that planar chirality based on the tetrasubstituted [2.2]paracyclophane has excellent potential to be CPL emitter. Several compounds showing great CPL performances in the diluted solution were reported in these days; however, the compounds showing good Φ_{lum} and g_{lum} with 10^{-2} order were still limited. Following those results, we prepared several derivatives based on a propeller-shaped macrocyclic structure [50, 63], and at the present stage, the CD and CPL performances were modified to be $g_{abs} = 1.0 \times 10^{-2}$, $\varepsilon_{abs} = 216{,}000\ M^{-1}\ cm^{-1}$, $g_{lum} = 1.0 \times 10^{-2}$, and $\Phi_{lum} = 0.60$ (Fig. 1.11) [60].

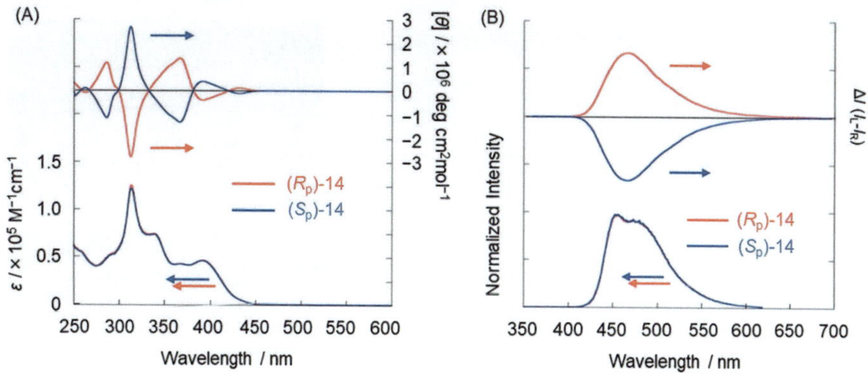

Fig. 1.10 (**a**) UV-vis absorption and CD spectra of (R_p)- and (S_p)-**14** in $CHCl_3$ (1.0×10^{-5} M) at room temperature. (**b**) PL and CPL spectra of (R_p)- and (S_p)-**14** in $CHCl_3$ (1.0×10^{-6} M for PL and 1.0×10^{-5} for CPL) at room temperature, excited at 314 nm. (Reprinted with permission from Ref. [59]. Copyright 2014 American Chemical Society)

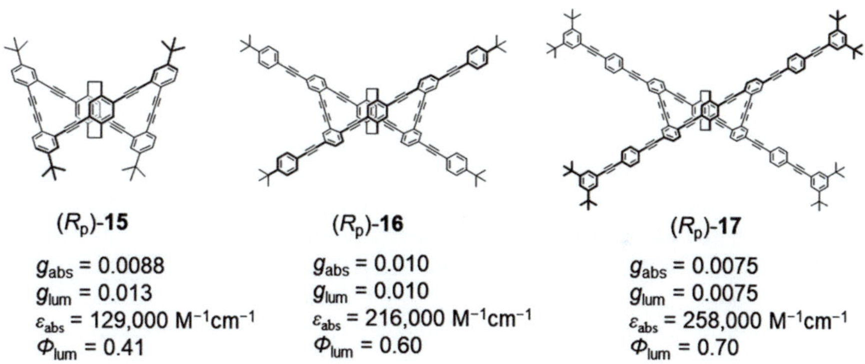

Fig. 1.11 The series of macrocycles and the typical chiroptical properties

1.4.5 Development of Chiral Dendrimers

Taking advantage of the unique structure of the tetrasubstituted [2.2]paracyclophane, chiral dendrimers [62] and chiral dimers [61, 65, 66] were synthesized. In addition, self-assembly of the chiral dimers [65] was prepared and investigated the properties. Chiral dendrimer consisted of planar chiral [2.2]paracyclophane core and Fréchet-type dendrimer which is famous as light-harvesting antenna. As a result, the energy absorbed by Fréchet-type dendrimer was transferred to the [2.2]paracyclophane core, and the core emitted high-intense CPL with preservation of chiroptical parameters such as g_{lum}. In addition, the CPL film was easily prepared because of the steric hindrance of the dendrimer (Fig. 1.12).

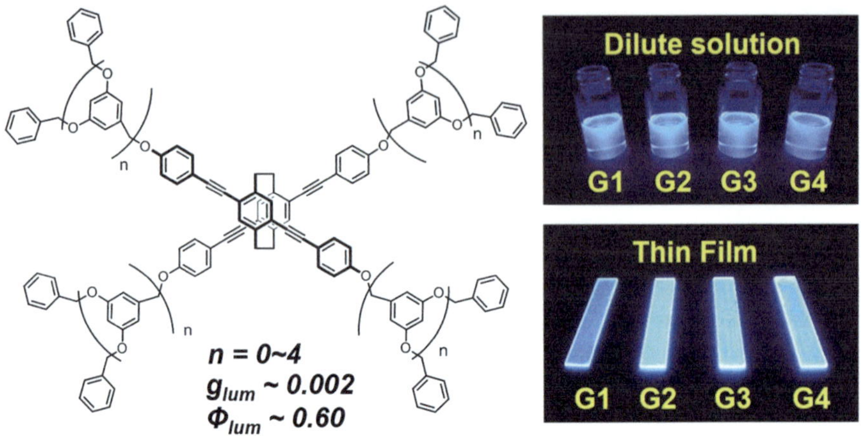

Fig. 1.12 Chemical structures of the dendrimers and photography-irradiated UV lamp (365 nm). (Reprinted with permission from Ref. [62]. Copyright 2016 Wiley-VCH Verlag GmbH & Co. KGaA)

1.4.6 Enhancement of Chiroptical Properties by Molecular Assembly

When the chiral molecules are self-assembled, it is known that the g_{lum} performances are often enhanced [71]. In our case, the self-assembly with oligo-phenylene-ethynylene-based planar chiral tetrasubstituted [2.2]paracyclophane with dodecyl alkyl chains showed dramatic enhancement of g_{lum}; finally the g_{lum} increased in 10^{-1} order after annealing ($g_{\mathrm{lum}} = -0.25$ for (R_{p})-**CP5** and $g_{\mathrm{lum}} = +0.27$ for (S_{p})-**CP5**). Those are one of the largest g_{lum} in the assembled systems with organic compounds. In addition, sign inversion was observed corresponding to the difference of preparation method of the films. The high CPL performance and sign inversion should be caused by the easily assembled planar structure of the planar chiral [2.2] paracyclophanes. The proposal mechanism of sign inversion is shown in Fig. 1.13. When the stacking formation of the chromophores changes, inversion of the CD and CPL signs can occur, and the phenomenon was estimated by exciton coupling theory.

1.4.7 Summary

A series of our research based on planar chiral [2.2]paracyclophanes indicated that the combination of the functional units such as planar chiral pseudo-*ortho*-disubstituted [2.2]paracyclophane and 4,7,12,15-tetrasubstituted [2.2]paracyclophane creates unique characteristics and the concept was matched with element-blocks. Recently, we successfully prepared the oligomers showing unidirectional energy

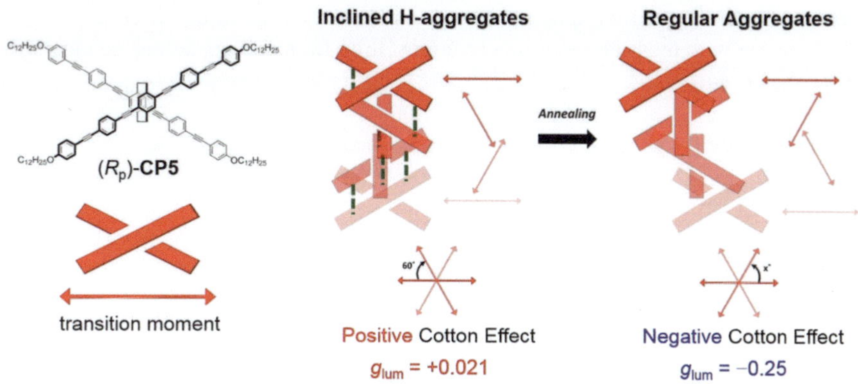

Fig. 1.13 Proposal mechanism of self-assembly of (R_p)-**CP5**. The direction of transition moment was estimated by TD-DFT (B3LYP/6-31G(d,p)//B3LYP/6-31G(d,p))

transfer between chromophores with pseudo-*para*-disubstituted [2.2] paracyclophane as a linker [72–75]. The molecular design was a combination of the functional units, and this is also one of the examples applied with the concept of element-blocks.

1.5 Conclusion

The series of element-blocks involving unique steric structures are reviewed. By employing POSS, thermally stable materials similar to organic-inorganic hybrids can be readily prepared via conventional organic reactions. In particular, by designing the side chains of POSS, various functions can be loaded. Finally, multifunctional element-block materials having combinational properties originated from both organic molecules and POSS were obtained. From these data, we can say that POSS element-blocks contributed to extending applicability of hybrids to various matrices. Finally, the concept of "designable hybrids" was proposed. The possibility of the cardo boron element-blocks as a promising platform for constructing conjugated materials consisting of electronic functional units was demonstrated from the recent studies. By modulating distribution and direction of these functional units at the cardo boron center, energy transfer efficiency and degree of electronic interaction were regulated. As a result, useful optical properties such as solid-state luminescence were obtained. It was demonstrated that [2.2] paracyclophane element-blocks work as the origin for expressing superior CD and CPL signals, and by introducing conjugated system, these optical properties were enhanced and modulated. These results indicate that chiral element-blocks can be a foundation stone for constructing next generation of chiral optical materials. Except for these examples, there are still a numerous number of heteroatom-based steric structures, and these structural units are promised to be individually as a new element-block. Further exploration of new element-blocks and the discovery of the combination with these element-blocks are exciting for receiving advanced element-block materials.

Acknowledgments This work was supported by a Grant-in-Aid for Scientific Research on Innovative Areas "New Polymeric Materials Based on Element-Blocks (No.2401)" (JSPS KAKENHI Grant Number JP24102013).

References

1. Gon M, Tanaka K, Chujo Y (2017) Creative synthesis of organic–inorganic molecular hybrid materials. Bull Chem Soc Jpn 90:463–474
2. Chujo Y, Tanaka K (2015) New polymeric materials based on element-blocks. Bull Chem Soc Jpn 88:633–643
3. Tanaka K, Chujo Y (2012) Advanced functional materials based on polyhedral oligomeric silsesquioxane (POSS). J Mater Chem 22:1733–1746
4. Tanaka K, Chujo Y (2013) Chemicals-inspired biomaterials; developing biomaterials inspired by material science based on POSS. Bull Chem Soc Jpn 86:1231–1239
5. Tanaka K, Chujo Y (2013) Unique properties of amphiphilic POSS and their applications. Polym J 45:247–254
6. Tanaka K, Kozuka H, Ueda K, Jeon JH, Chujo Y (2017) POSS-based molecular fillers for simultaneously enhancing thermal and viscoelasticity of poly(methyl methacrylate) films. Mater Lett 203:62–67
7. Ueda K, Tanaka K, Chujo Y (2017) Synthesis of POSS derivatives having dual types of alkyl substituents via in situ sol−gel reactions and their application as a molecular filler for low-refractive and highly-durable materials. Bull Chem Soc Jpn 90:205–209
8. Tanaka K, Yamane H, Mitamura K, Watase S, Matsukawa K, Chujo Y (2014) Transformation of sulfur to organic−inorganic hybrids employed by POSS networks and their application for the modulation of refractive indices. J Polym Sci Part A Polym Chem 52:2588–2595
9. Jeon JH, Tanaka K, Chujo Y (2013) Rational design of POSS fillers for simultaneous improvements of thermomechanical properties and lowering refractive indices of polymer films. J Polym Sci Part A Polym Chem 51:3583–3589
10. Tanaka K, Adachi S, Chujo Y (2010) Side-chain effect of octa-substituted POSS fillers on refraction in polymer composites. J Polym Sci Part A Polym Chem 48:5712–5717
11. Tanaka K, Adachi S, Chujo Y (2009) Structure-property relationship of octa-substituted POSS in thermal and mechanical reinforcements of conventional polymers. J Polym Sci Part A Polym Chem 47:5690–5697
12. Tanaka K, Ishiguro F, Chujo Y (2010) POSS ionic liquid. J Am Chem Soc 132:17649–17651
13. Tanaka K, Ishiguro F, Chujo Y (2011) Thermodynamic study of POSS-based ionic liquids with various numbers of ion pairs. Polym J 43:708–713
14. Jeon JH, Tanaka K, Chujo Y (2013) POSS fillers for modulating thermal properties of ionic liquids. RSC Adv 3:2422–2427
15. Jeon JH, Tanaka K, Chujo Y (2014) Synthesis of sulfonic acid-containing POSS and its filler effects for enhancing thermal stabilities and lowering melting temperatures of ionic liquids. J Mater Chem A 2:624–630
16. Tanaka K, Ishiguro F, Jeon JH, Hiraoka T, Chujo Y (2015) POSS ionic liquid crystals. NPG Asia Mater 7:e174
17. Ueda K, Tanaka K, Chujo Y (2016) Remarkably high miscibility of octa-substituted POSS with commodity conjugated polymers and molecular fillers for the improvement of homogeneities of polymer matrices. Polym J 48:1133–1139

18. Gon M, Sato K, Tanaka K, Chujo Y (2016) Controllable intramolecular interaction of 3D arranged π-conjugated luminophores based on a POSS scaffold, leading to highly thermostable and emissive materials. RSC Adv 6:78652–78660
19. Suenaga K, Tanaka K, Chujo Y (2017) Design and luminescent chromism of fused boron complexes having constant emission efficiencies in solution, amorphous and crystalline states. Eur J Org Chem. https://doi.org/10.1002/ejoc.201700704
20. Ohtani S, Gon M, Tanaka K, Chujo Y (2017) Flexible fused azomethine–boron complex: thermally-induced switching of crystalline-state luminescent property and thermosalient behaviors based on phase transition between polymorphs. Chem Eur J. https://doi.org/10.1002/chem.201702309
21. Yamaguchi M, Ito S, Hirose A, Tanaka K, Chujo Y (2016) Modulation of sensitivity to mechanical stimulus in mechanofluorochromic properties by altering substituent positions in solid-state emissive diiodo boron diiminates. J Mater Chem C 3:5314–5319
22. Yoshii R, Suenaga K, Tanaka K, Chujo Y (2015) Mechanofluorochromic materials based on aggregation-induced emission-active boron ketoiminates: regulation of the direction of the emission color changes. Chem Eur J 21:7231–7237
23. Tanaka K, Chujo Y (2015) Recent progress of optical functional nanomaterials based on organoboron complexes with β-diketonate, ketoiminate and diiminate. NPG Asia Mater 7:e223
24. Yoshii R, Hirose A, Tanaka K, Chujo Y (2014) Functionalization of boron diiminates with unique optical properties: multicolor tuning of crystallization-induced emission and introduction into the main-chain of conjugated polymers. J Am Chem Soc 136:18131–18139
25. Tanaka K, Yanagida T, Hirose A, Yamane H, Yoshii R, Chujo Y (2015) Synthesis and color tuning of boron diiminate conjugated polymers with aggregation-induced scintillation properties. RSC Adv 5:96653–96659
26. Yamaguchi M, Ito S, Hirose A, Tanaka K, Chujo Y (2017) Control of aggregation-induced emission versus fluorescence aggregation-caused quenching by the bond existence at the single site in boron pyridinoiminate complexes. Mater Chem Front. https://doi.org/10.1039/c7qm00076f
27. Suenaga K, Yoshii R, Tanaka K, Chujo Y (2016) Sponge-type emissive chemosensors for the protein detection based on boron ketoiminate-modifying hydrogels with aggregation-induced blue shift emission property. Macromol Chem Phys 217:414–417
28. Hirose A, Tanaka K, Yoshii R, Chujo Y (2015) Film-type chemosensors based on boron diminate polymers having oxidation-induced emission properties. Polym Chem 6:5590–5595
29. Kajiwara Y, Nagai A, Tanaka K, Chujo Y (2013) Efficient simultaneous emission from RGB-emitting organoboron dyes incorporated into organic-inorganic hybrids and preparation of white light-emitting materials. J Mater Chem C 1:4437–4444
30. Yoshii R, Tanaka K, Chujo Y (2014) Conjugated polymers based on tautomeric units: regulation of main-chain conjugation and expression of aggregation induced emission property via boron-complexation. Macromolecules 47:2268–2278
31. Yoshii R, Nagai A, Tanaka K, Chujo Y (2014) Boron ketoiminate-based polymers: fine-tuning of the emission color and expression of strong emission both in the solution and film state. Macromol Rapid Commun 35:1315–1319
32. Yoshii R, Hirose A, Tanaka K, Chujo Y (2014) Boron diiminate with aggregation-induced emission and crystallization-induced emission enhancement characteristics. Chem Eur J 20:8320–8324
33. Yoshii R, Nagai A, Tanaka K, Chujo Y (2013) Highly emissive boron ketoiminate derivatives as new class of aggregation-induced emission fluorophores. Chem Eur J 19:4506–4512
34. Suenaga K, Tanaka K, Chujo Y (2017) Heat-resistant mechanoluminescent chromism of the hybrid molecule based on boron ketoiminate-modified octa-substituted polyhedral oligomeric silsesquioxane. Chem Eur J 23:1409–1414
35. Yeo H, Tanaka K, Chujo Y (2012) Isolation of π-conjugated system through polyfluorene from electronic coupling with side-chain substituents by cardo structures. J Polym Sci Part A Polym Chem 50:4433–4442

36. Yeo H, Tanaka K, Chujo Y (2015) Synthesis of dual-emissive polymers based on ineffective energy transfer through cardo fluorene-containing conjugated polymers. Polymer 60:228–233
37. Yeo H, Tanaka K, Chujo Y (2015) Synthesis and energy transfer through heterogeneous dyes-substituted fluorene-containing alternating copolymers and their dual-emission properties. J Polym Sci Part A Polym Chem 53:2026–2035
38. Yeo H, Tanaka K, Chujo Y (2016) Tunable optical property between pure red luminescence and dual-emission depended on the length of light-harvesting antennae in the dyads containing the cardo structure of BODIPY and oligofluorene. Macromolecules 49:8899–8904
39. Yeo H, Tanaka K, Chujo Y (2013) Effective light-harvesting antennae based on BODIPY-tethered cardo polyfluorenes via rapid energy transferring and low concentration quenching. Macromolecules 46:2599–2605
40. Tanaka K, Chujo Y (2012) Advanced luminescent materials based on organoboron polymers. Macromol Rapid Commun 33:1235–1255
41. Yoshii R, Nagai A, Tanaka K, Chujo Y (2013) Highly NIR emissive boron di(iso)indomethene (BODIN)-based polymer: drastic change from deep-red to NIR emission via quantitative polymer reaction. J Polym Sci Part A Polym Chem 51:1726–1733
42. Tanaka K, Yamane H, Yoshii R, Chujo Y (2013) Efficient light absorbers based on thiophene-fused boron dipyrromethene (BODIPY) dyes. Bioorg Med Chem 21:2715–2719
43. Yoshii R, Yamane H, Nagai A, Tanaka K, Taka H, Kita H, Chujo Y (2014) π-Conjugated polymers composed of BODIPY or Aza-BODIPY derivatives exhibiting high electron mobility and low threshold voltage in electron-only devices. Macromolecules 47:2316–2323
44. Yoshii R, Yamane H, Tanaka K, Chujo Y (2014) Synthetic strategy for low-band gap oligomers and homopolymers using characteristics of thiophene-fused boron dipyrromethene. Macromolecules 47:3755–3760
45. Tanaka K, Yanagida T, Yamane H, Hirose A, Yoshii R, Chujo Y (2015) Liquid scintillators with near infrared emission based on organoboron conjugated polymers. Bioorg Med Chem Lett 25:5331–5334
46. Yamane H, Ohtani S, Tanaka K, Chujo Y (2017) Synthesis of furan-substituted Aza-BODIPYs having strong near-infrared emission. Tetrahedron Lett. https://doi.org/10.1016/j.tetlet.2017.06.054
47. Yamane H, Tanaka K, Chujo Y (2015) Simple and valid strategy for the enhancement of the solid-emissive property based on boron dipyrromethene. Tetrahedron Lett 56:6786–6790
48. Yamane H, Ito S, Tanaka K, Chujo Y (2016) Preservation of main-chain conjugation through BODIPY-containing alternating polymers from electronic interactions with side-chain substituents by cardo boron structures. Polym Chem 7:2799–2807
49. Vögtle F (1993) Cyclophane chemistry: synthesis, structures and reactions. Wiley, Chichester
50. Gleiter R, Roers R (2004) Modern cyclophane chemistry. In: Gleiter R, Hopf H (eds) Wiley-VCH. Weinheim, Germany
51. Cram DJ, Allinger NL (1955) Macro rings. XII. Stereochemical consequences of steric compression in the smallest paracyclophane. J Am Chem Soc 77:6289–6294
52. Rowlands GJ (2008) The synthesis of enantiomerically pure [2.2]paracyclophane derivatives. Org Biomol Chem 6:1527–1534
53. Vorontsova NV, Rozenberg VI, Sergeeva EV, Vorontsov EV, Starikova ZA, Lyssenko KA, Hopf H (2008) Symmetrically tetrasubstituted [2.2]paracyclophanes: their systematization and regioselective synthesis of several types of bis-bifunctional derivatives by double electrophilic substitution. Chem Eur J 14:4600–4617
54. Bazan GC (2007) Novel organic materials through control of multichromophore interactions. J Organomet Chem 72:8615–8635
55. Morisaki Y, Hifumi R, Lin L, Inoshita K, Chujo Y (2012) Practical optical resolution of planar chiral pseudo-ortho-disubstituted [2.2]paracyclophane. Chem Lett 41:990–992
56. Morisaki Y, Hifumi R, Lin L, Inoshita K, Chujo Y (2012) Through-space conjugated polymers consisting of planar chiral pseudo-*ortho*-linked [2.2]paracyclophane. Polym Chem 3:2727–2730

57. Morisaki Y, Inoshita K, Chujo Y (2014) Planar-chiral through-space conjugated oligomers: synthesis and characterization of chiroptical properties. Chem Eur J 20:8386–8390
58. Morisaki Y, Inoshita K, Shibata S, Chujo Y (2015) Synthesis of optically active through-space conjugated polymers consisting of planar chiral [2.2]paracyclophane and quaterthiophene. Polym J 47:278–281
59. Morisaki Y, Gon M, Sasamori T, Tokitoh N, Chujo Y (2014) Planar chiral tetrasubstituted [2.2] paracyclophane: optical resolution and functionalization. J Am Chem Soc 136:3350–3353
60. Gon M, Morisaki Y, Chujo Y (2015) Optically active cyclic compounds based on planar chiral [2.2]paracyclophane: extension of the conjugated systems and chiroptical properties. J Mater Chem C 3:521–529
61. Gon M, Morisaki Y, Chujo Y (2015) Highly emissive optically active conjugated dimers consisting of a planar chiral [2.2]paracyclophane showing circularly polarized luminescence. Eur J Org Chem 2015:7756–7762
62. Gon M, Morisaki Y, Sawada R, Chujo Y (2016) Synthesis of optically active, x-shaped, conjugated compounds and dendrimers based on planar chiral [2.2]paracyclophane, leading to highly emissive circularly polarized luminescence. Chem Eur J 22:2291–2298
63. Gon M, Kozuka H, Morisaki Y, Chujo Y (2016) Optically active cyclic compounds based on planar chiral [2.2]paracyclophane with naphthalene units. Asian J Org Chem 5:353–359
64. Morisaki Y, Sawada R, Gon M, Chujo Y (2016) New types of planar chiral [2.2] paracyclophanes and construction of one-handed double helices. Chem Asian J 11:2524–2527
65. Gon M, Sawada R, Morisaki Y, Chujo Y (2017) Enhancement and controlling the signal of circularly polarized luminescence based on a planar chiral tetrasubstituted [2.2]paracyclophane framework in aggregation system. Macromolecules 50:1790–1802
66. Gon M, Morisaki Y, Chujo Y (2017) Optically active phenylethene dimers based on planar chiral tetrasubstituted [2.2]paracyclophane. Chem Eur J 23:6323–6329
67. Riehl JP, Richardson FS (1986) Circularly polarized luminescence spectroscopy. Chem Rev 86:1–16
68. Riehl JP, Muller F (2012) Comprehensive Chiroptical spectroscopy. Wiley, New York
69. Chow HF, Low KH, Wong KY (2005) An improved method for the regiospecific synthesis of polysubstituted [2.2]paracyclophanes. Synlett 2005:2130–2134
70. Hinrichs H, Boydston AJ, Jones PG, Hess K, Herges R, Haley MM, Hopf H (2006) Phane properties of [2.2]paracyclophane/dehydrobenzoannulene hybrids. Chem Eur J 12:7103–7115
71. Kumar J, Nakashima T, Kawai T (2015) Circularly polarized luminescence in chiral molecules and supramolecular assemblies. J Phys Chem Lett 6:3445–3452
72. Morisaki Y, Kawakami N, Nakano T, Chujo Y (2013) Energy-transfer properties of a [2.2] paracyclophane-based through-space dimer. Chem Eur J 19:17715–17718
73. Morisaki Y, Kawakami N, Nakano T, Chujo Y (2013) Synthesis and properties of a through-space-conjugated dimer. Chem Lett 43:426–428
74. Morisaki Y, Kawakami N, Shibata S, Chujo Y (2014) Through-space conjugated molecular wire comprising three π-electron systems. Chem Asian J 9:2891–2895
75. Morisaki Y, Shibata S, Chujo Y (2016) [2.2]Paracyclophane-based single molecular wire consisting of four π-electron systems. Can J Chem 95:424–431

Chapter 2
Si-, Ge-, and Sn-Bridged Biaryls as π-Conjugated Element Blocks

Yohei Adachi and Joji Ohshita

Abstract Biaryls, such as bithiophene and bipyridyl, are intramolecularly bridged by group 14 heavy elements, silicon, germanium, and tin, to produce element blocks having condensed tricyclic systems with silole, germole, and stannole as the center ring. These bridged biaryls exhibit interesting properties arising from the characteristics of both the biaryl units and the bridging elements. In this chapter, recent development in functional organic materials containing the bridged biaryl systems as the core structure is described. The electronic states and properties of the bridged biaryls are discussed on the basis of the results of optical and electrochemical measurements and quantum chemical calculations, to explore how the element bridges contribute to the improvement of desired properties, such as photoluminescent, carrier transport, photovoltaic, and chromic properties.

Keywords Dithienometallole · Dipyridinometallole · Group 14 element · Conjugated polymer · Optoelectronic application

2.1 Introduction

There has been a surge of interest in silole (silacyclopentadiene) derivatives as functional materials. The effective conjugation in these compounds is due primarily to the high planarity of the cyclic system. Bonding interaction between the silicon σ*-orbital and the butadiene π*-orbital, the so-called σ*-π* conjugation, is also involved, which lowers LUMO levels to further enhance the conjugation (Chart 2.1) [1–3]. In addition, highly photo- and electroluminescent (PL and EL) properties have been reported in silole compounds. These interesting electronic states and properties of the silole system have resulted in the application of silole-

Y. Adachi · J. Ohshita (✉)
Department of Applied Chemistry, Graduate School of Engineering, Hiroshima University, Hiroshima, Japan
e-mail: jo@hiroshima-u.ac.jp

Y. Chujo (ed.), *New Polymeric Materials Based on Element-Blocks*,
https://doi.org/10.1007/978-981-13-2889-3_2

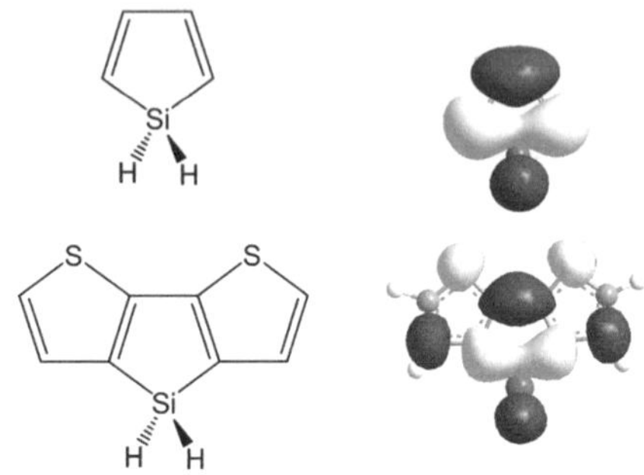

Chart 2.1 σ^*-π^* Conjugation in LUMOs of silole and DTS

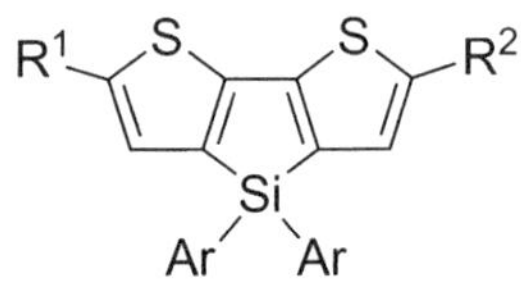

Chart 2.2 DTS derivatives with electron-transporting properties (**a**) and highly PL properties with quantum yield in the solid state (**b**)

(a) Electron-transporting material
$R^1 = R^2$ = 2-(6-Me_3Si)pyridyl, Ar = Ph

(b) Solid state PL material
$R^1 = R^2 = B(Mes)_2$, Ar = Ph Φ= 0.85
$R^1 = B(Mes)_2$, R^2 = SMe, Ar = Ph Φ= 0.58
$R^1 = R^2$ = *p*-$CF_3(C_6F_4)$, Ar = *p*-*n*-Bu(C_6H_4) Φ= 0.60
$R^1 = R^2 = PPh_2(AuCl)$, Ar = *p*-*n*-Bu(C_6H_4) Φ= 0.33
(Mes = 2,4,6-trimethylphenyl)

based compounds as functional materials, such as semiconductors and light emitters for organic optoelectronic devices [1–7]. In the course of our studies of the opto-electronic applications of silicon-bridged π-conjugated systems, we synthesized dithienosiloles (DTSs) as the first example of siloles fused with heteroaromatic units [8]. Like simple siloles, DTS exhibits extended conjugation arising from the high planarity and the σ^*-π^* conjugation [9, 10] (Chart 2.1) and is widely used as the core fragment of conjugated organic functional materials. For example, pyridyl-substituted DTS (**DTSPy**), with its excellent electron-transporting properties, has been applied to multilayered organic light-emitting diodes (OLEDs) (Chart 2.2a) [11]. The rigid tricyclic systems often lead to highly PL properties of the DTS-based compounds. As illustrated in Chart 2.2b), some DTS compounds possess high PL efficiencies even in the solid state [12–14].

It has been also demonstrated that DTS units can be used as building blocks of functional π-conjugated polymers. In 2006, Marks et al. reported the first example of semiconducting DTS polymers (**pDTS-T** and **pDTS-2T** in Chart 2.3) that could be

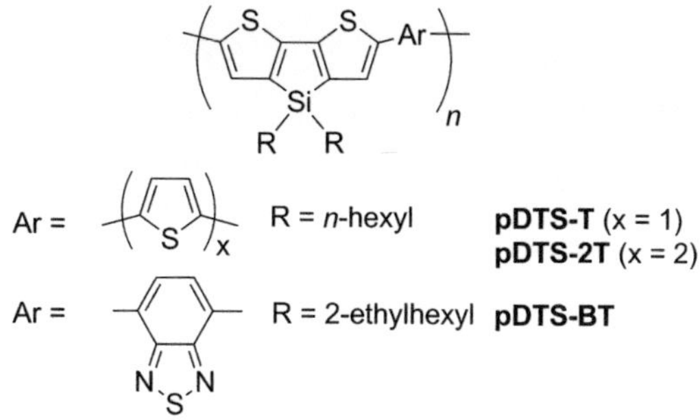

Chart 2.3 DTS-containing conjugated polymers

used as the active layers of p-type organic thin-film transistors (OTFTs) [15, 16]. In 2008, Yang and coworkers reported the application of a DTS-containing polymer as the donor material of bulk heterojunction polymer solar cells (BHJ-PSCs). The BHJ-PSC with a blend film composed of **pDTS-BT** and $PC_{71}BM$ as the donor and acceptor materials, respectively, provided a power conversion efficiency (PCE) of 5.1%, which was excellent at that time (Chart 2.3) [17]. In **pDTS-BT**, DTS and benzothiadiazole (BT) units served as electron donor and acceptor units, respectively, giving rise to donor-acceptor (D-A) interaction in the polymeric chain and realizing the low band gap of the polymer. Following this intriguing report, many papers describing the synthesis and photovoltaic applications of D-A polymers with DTS as the donor appeared [4–6, 10, 18]. Dithienogermole (DTG), the germanium congener of DTS, was first prepared in 2011 independently by us and two other research groups [19–21], as the donor unit of photovoltaic D-A polymers (Chart 2.4). Since then, DTG-containing D-A oligomers and polymers with high photovoltaic functionalities have been prepared [22–26]. DTS and DTG possess similar electronic states and similar HOMO and LUMO energy levels, although the σ^*-π^* conjugation of DTG is slightly suppressed due to the longer end cyclic Ge-C bonds that minimize the orbital overlap. DTG-containing polymers often show larger van der Waals interchain interaction in the solid state than their DTS analogues, which facilitates the hopping-type carrier conduction, thereby improving the current density of BHJ-PSC [27]. Highly condensed DTG moieties have been also investigated as the donor component of D-A polymers (Chart 2.4) [28–30].

In this chapter, recent achievements in the synthesis and properties of group 14 heavy element-bridged biaryls as π-conjugated element blocks are described. Their potential applications as building units of functional materials, such as photovoltaic, semiconducting, luminescent, chromic, and sensing materials, are also discussed.

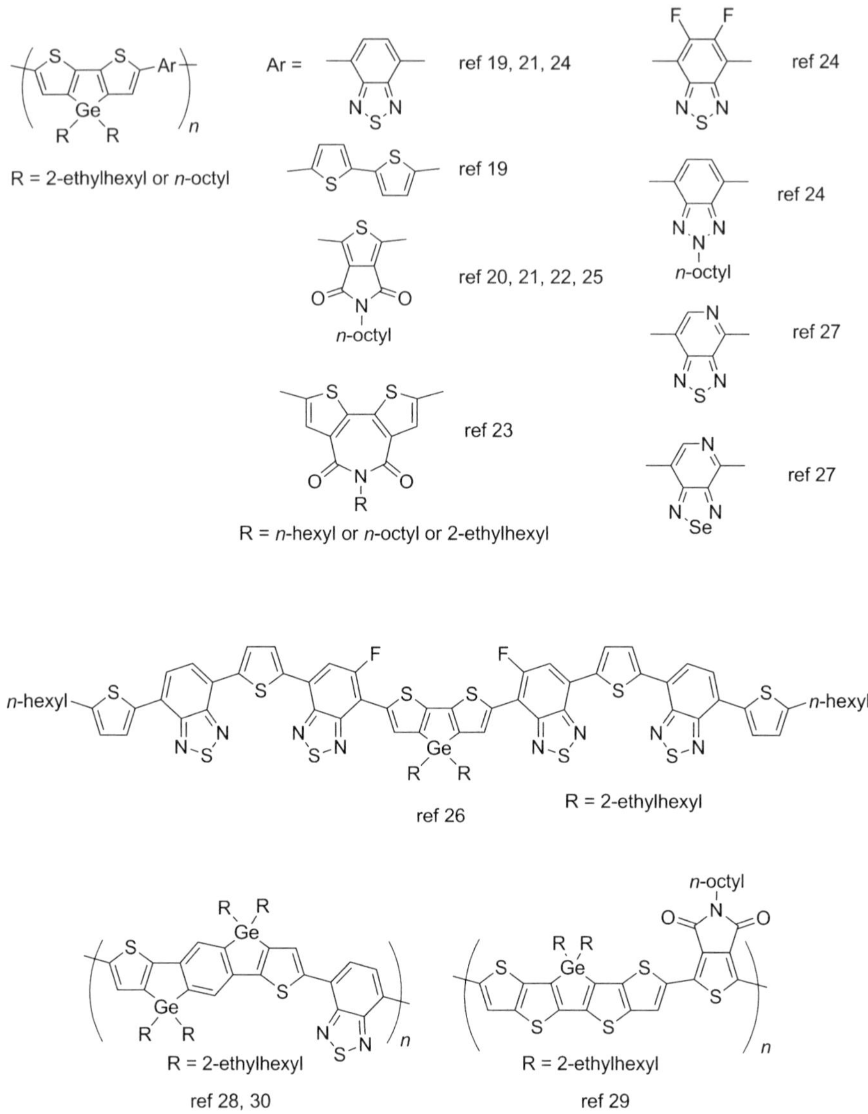

Chart 2.4 Examples of DTG-containing photovoltaic polymers

2.2 DSBT-Containing Conjugated Polymers

Several years ago, we demonstrated that disilanobithiophene (DSBT) should have lower-lying HOMO and LUMO with a similar energy gap compared with those of DTS by quantum chemical calculations at the B3LYP/6-31G(d,p) level of theory [31]. The HOMO and LUMO levels of model compounds **DTS1** and **DSBT1**, which

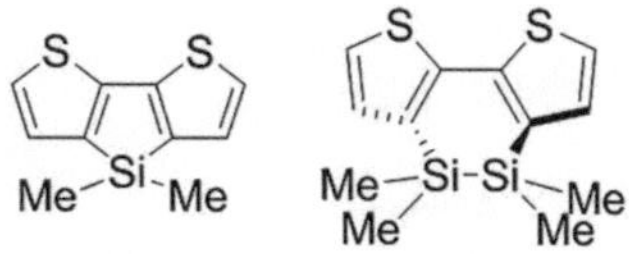

	DTS1	DSBT1
LUMO/eV	-1.29	-1.33
HOMO/eV	-5.30	-5.38
Δ(LUMO-HOMO)/eV	4.01	4.05
dihedral angle (S-C-C-S)/°	0	18.2

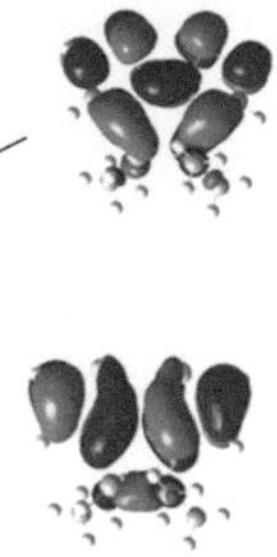

Chart 2.5 Relative HOMO and LUMO energy levels of **DTS1** and **DSBT1** and orbital profiles of **DSBT1**, derived from DFT calculations at B3LYP/6-31G(d,p) level

were derived from the calculations, are listed in Chart. 2.5. The introduction of a disilane bridge in place of the monosilane unit in DTS results in a twisted conformation that suppresses the π-conjugation. However, the LUMO of **DSBT1** is at a lower energy than that of **DTS1**. This is ascribable to the enhanced σ*-π* conjugation in DSBT by utilizing the Si-Si σ*-orbital. At the same time, the HOMO is also lowered, likely due to the suppressed π-conjugation of the twisted bithiophene. In the HOMO profile presented in Chart 2.5, however, the σ-π interaction is clearly seen, which may compensate the suppressed π-conjugation by twisting to a certain extent. As a consequence, **DTS1** and **DSBT1** have similar HOMO-LUMO energy gaps.

As the lower-lying HOMO of the donor materials usually leads to the higher voltage from BHJ-PSCs, we prepared D-A polymers with DSBT as the donor component [32]. As expected, the BHJ-PSC based on a blend film of **pDSBT-BT** and $PC_{71}BM$ as the active layer provided a good PCE of 6.38% with a high open-circuit voltage ($V_{oc} = 0.82$ V). Crystal orbital calculations of polymer **pDSBT-BT**, however, indicated little contribution of the disilane unit to the HOCO and the LUCO. Presumably, the contribution of silicon σ-orbital is negligible in highly extended π-conjugated systems. A similar tendency to suppress the σ*-π* conjugation by expanding the π-conjugation is also seen in DTS-containing polymers. Attempts to optimize the DSBT-based polymer structures to improve cell performance have been unsuccessful, and **pDSBT-BT** shows the best photovoltaic performance in the cells among the DSBT-containing D-A polymers that we have prepared (Chart 2.6) [33–35].

Polymers composed of alternating π-conjugated units and Si-Si bonds are known to react with hydroxyl groups (M-OH) on the inorganic oxide surface to form M-OSi bonds [36] (Chart 2.7a). DSBT-containing polymers react also with TiO_2 surface. TiO_2 modified with DSBT-containing polymers can be used as photoelectrodes of dye-sensitized solar cells (DSSCs) (Chart 2.7b) [37]. DSSCs are of current interest because of their low cost and easy fabrication and their potential application as flexible and bendable cells. However, the maximal PCE of the DSSCs using the

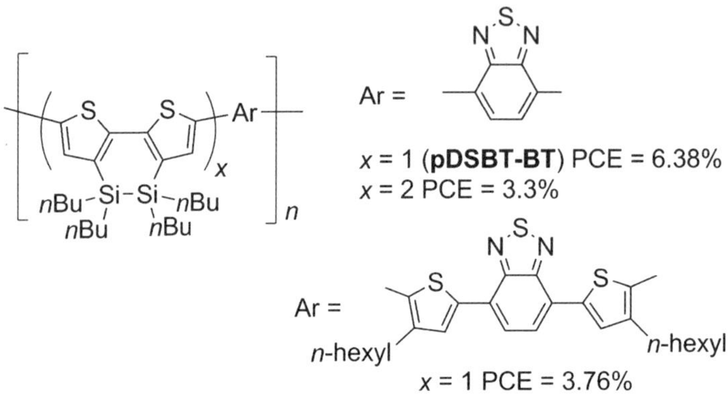

Chart 2.6 DSBT-containing photovoltaic polymers and PCEs of BHJ-PSCs with polymer/$PC_{71}BM$ blend films as active layers

present DSBT-polymer-attached TiO_2 is not very high yet (0.89%). The present modification of TiO_2 surface to suppress production of such by-products as water seems to be potentially useful because the by-products may affect cell performance. In addition, it should be mentioned that Ti-O-Si linkages are usually stable toward hydrolysis [38]. This is in contrast to the conventional TiO_2 photoelectrodes to which sensitizing dyes are attached via an ester linkage. The ester linkage (Ti-OOC-) is rather unstable toward hydrolysis with atmospheric moisture, resulting in the liberation of the dye fragments from the TiO_2 surface (Chart 2.7c).

2.3 Applications of DTG as PL and Sensing Materials and Photosensitizers

In general, DTS and DTG systems are readily prepared by the reactions of dilithiobithiophenes with dihalosilanes and dihalogermanes, respectively (Scheme 2.1) [10]. However, the synthetic strategies are still limited. Recently, we prepared 4,4-dichlorodithienogermoles by the reactions of dilithiobithiophenes with tetrachlorogermane [39]. The DTG dichlorides readily undergo direct derivatization at the germanium atom, as shown in Scheme 2.2. The reactions with lithium aluminum hydride and organometallic reagents proceed smoothly to yield the corresponding germanium-substituted products. Hydrolysis of the dichlorides provides cyclic tetragermoxanes. The cyclotetragermoxane with trimethylsilyl substituents (**cDTGO-Si**) shows good PL properties even in the solid state with quantum efficiencies of $\Phi = 0.80$ in THF and 0.49 as a solid. Interestingly, the **cDTGO-Si** film responds to nitrobenzene vapor, thereby decreasing the PL intensity. The detection of nitroaromatic compounds is of importance in sensing technology. In particular, solid-state sensing of vapor is desired for convenient detection. However, organic PL materials often show low PL efficiencies in the solid state by concentration quenching

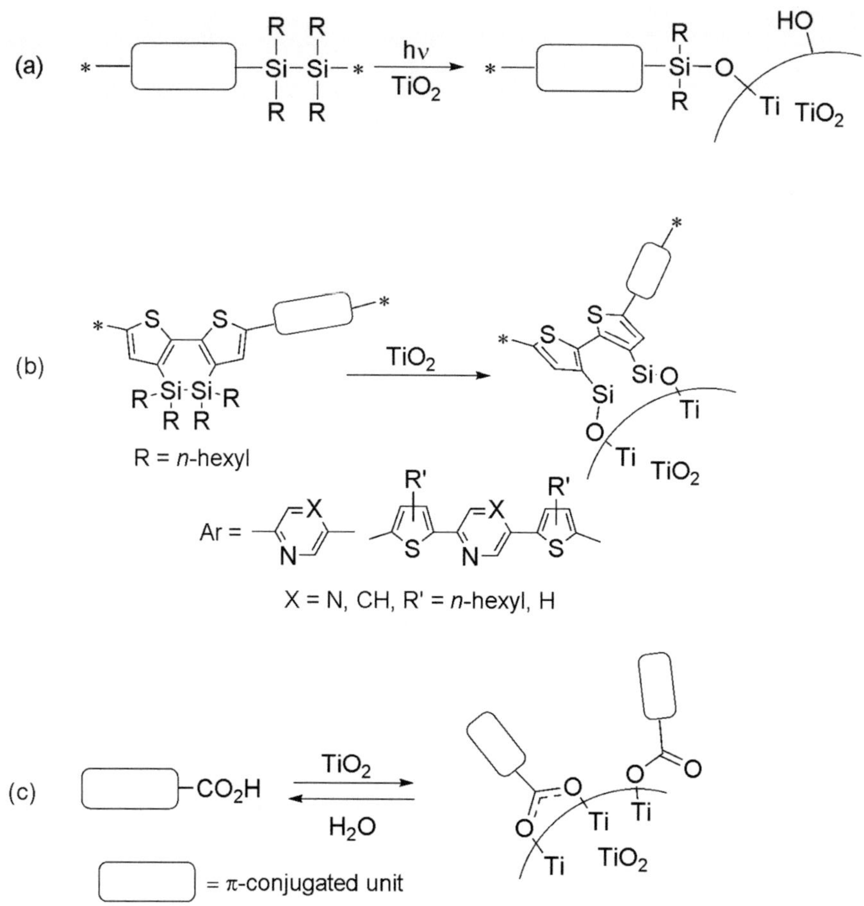

Chart 2.7 Reactions of disilane (**a**), DSBT (**b**), and carboxylic acid (**c**) with TiO_2 surface

and are difficult to utilize as turn-off sensors. In this regard, **cDTGO-Si** with high solid-state PL efficiency is potentially useful.

Polygermanes bearing perpendicularly arranged DTG rings (**pDTG**) are prepared by the reductive coupling of DTG dichlorides with sodium [40]. The polygermanes show both σ- and σ-π conjugations with respect to the UV-vis absorption spectra (Scheme 2.2). Strong through-space π-π interaction is also suggested by the low PL efficiency, likely due to the intramolecular concentration quenching. In contrast to DTG dichlorides, attempted preparation of the silicon congeners has been unsuccessful. The silicon-halogen bond is more polar than the germanium-halogen bond and therefore more reactive; thus, the reactions are less selective and more difficult to control. In fact, spirodi(dithienosilole)s are always obtained as the major products from the reactions of dilithiobithiophenes with tetrachlorosilane under several conditions.

Scheme 2.1 Synthesis of DTS and DTG derivatives

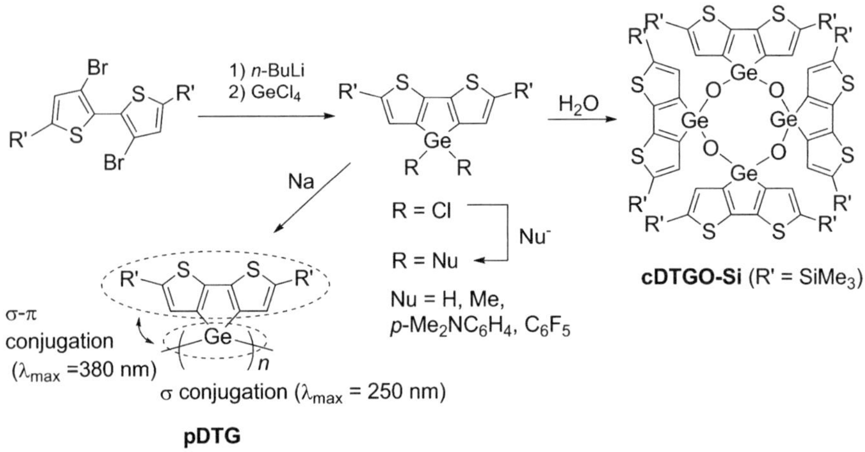

Scheme 2.2 Synthesis and reactions of DTG dichlorides

DTG is chemically more stable than DTS and can be polymerized under alkali-catalyzed sol-gel conditions [41]. DTS derivatives undergo hydrolysis of the endocyclic Si-C bond to form bithiophenes and silanols under the same conditions. The sol-gel reaction of bis(trimethoxysilyl)-substituted DTG provides a self-standing polysilsesquioxane (PSQ) film (Fig. 2.1a). The homopolymer film is non-emissive, but copolymerization with tri(methoxy)methylsilane gives blue-PL PSQ films (Fig. 2.1b). Copolymerization with a carbazole-containing monomer gives also PSQ films that show smooth energy transfer from the photoexcited carbazole units to the DTG units, i.e., only PL from DTG is observed even when carbazole is excited. The PSQ films are applicable also as hole-transporting EL materials in OLED. The sol-gel reactions of DTG dichlorides with bis(triethoxysilyl)ethane (BTESE) provide porous PSQ films with good PL properties, which respond to the vapors of several nitroaromatic compounds, including TNT, much more sensitively than the **cDTGO-Si** films (Fig. 2.2 and Table 2.1) [42].

DTS has been used also as building units of sensitizing dyes of DSSCs [43]. In contrast, no attention had been given to DTG-based sensitizing dyes until recently. We prepared D-π1-A1-π2-A2 dyes with a DTG unit as the π1 linker, expecting that the extended conjugation of DTG arising from the high planarity would facilitate communication between the D and A1 units (Fig. 2.3) [44]. In addition, two

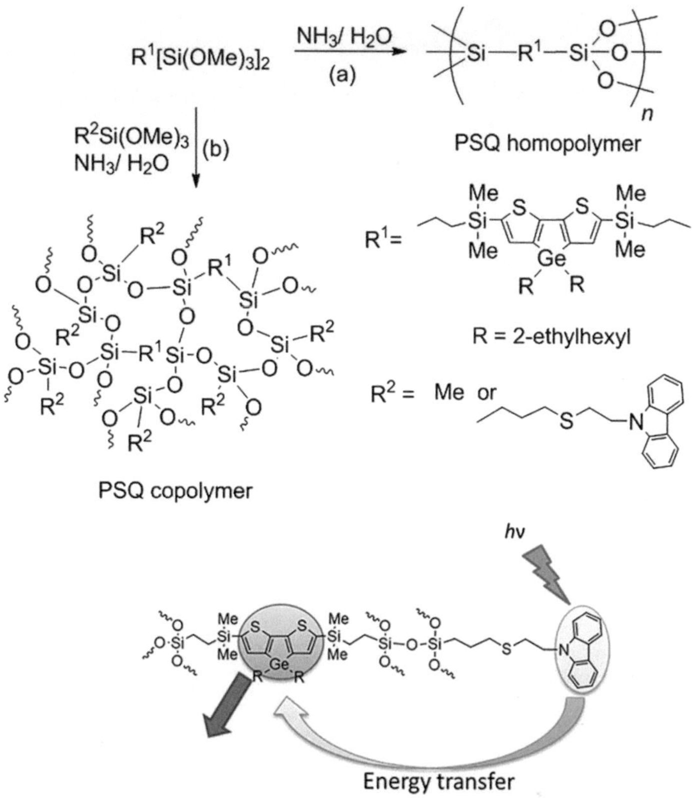

Fig. 2.1 Synthesis of DTG-containing PSQ homopolymer (**a**) and copolymer (**b**) and schematic representation of photo-energy transfer of the carbazole-containing copolymer

substituents on the germanium bridging atom suppress the intermolecular interaction when attached to the TiO_2 electrode surface. Intermolecular interaction of the dye molecules is known to cause energy migration to decrease cell efficiencies. Among the dyes that we prepared, the one with phenylene and bis(hexyloxyphenyl) aminophenyl fragments as the second linker (π2) and donor (D) units, respectively, exhibits the highest performance with a PCE of 6.09%, indicating the potential of DTG as the π-linker of sensitizing dyes.

We also prepared DTG-containing polythiophene derivatives [45, 46] and compared their optical and electrochemical properties as well as applications as p-type semiconductors in OTFTs with those of DTS and carbon-bridged bithiophene (DTC)-containing analogues [47] (Fig. 2.4). The substituents on the bridging elements and the molecular weights of the polymers exert a significant influence on the semiconducting properties. Regardless of the bridging elements, the polymers with R = *n*-octyl always provide higher carrier mobility than those with R = 2-ethylhexyl, and higher molecular weights lead to higher mobility. These seem to reflect the enhanced interchain interaction in the films. It is also suggested that the

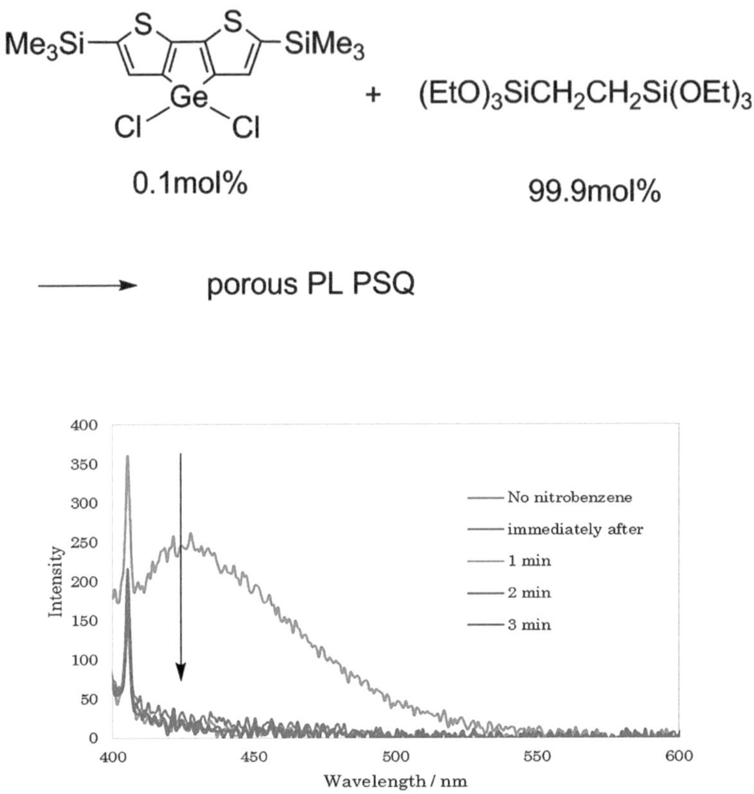

Fig. 2.2 Synthesis of PSQ from DTG dichloride and BTESE and PL spectral changes upon exposure of film to nitrobenzene vapor

Table 2.1 PL quenching of PSQ film prepared from DTG dichloride and BTESE upon exposure to vapors of nitroaromatic compounds

	Relative PL intensity/%[a]	
Compound	10 min	30 min
Nitrobenzene	1	1
4-Nitrotoluene	14	12
4-Chloronitrobenzene	6	5
1,3-Dinitrobenzene	55	33
Trinitrotoluene	92	86

[a]Relative to PL intensity before exposure

interchain interaction in the films is enhanced in the order of DTC < DTS < DTG-containing polymers. The OTFT activity of the DTG polymers is similar to that of the corresponding DTS polymers but superior to that of the DTC polymer, likely because of the inferior interchain interaction of the DTC polymer. It is also

Fig. 2.3 DTG-containing sensitizing dye of DSSC

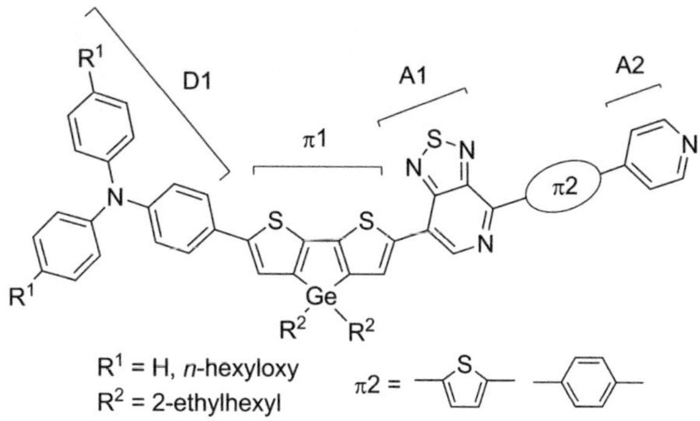

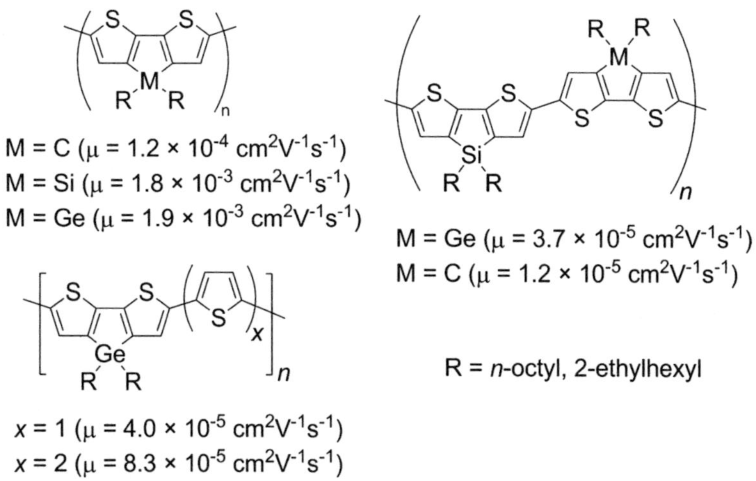

Fig. 2.4 DTG-containing polythiophenes with p-type OTFT mobility for R = *n*-octyl

speculated that less flexible properties of DTC substituents arising from the short C-C bonds compared with Si-C and Ge-C bonds result in the less soluble properties of DTC polymers, being responsible for the film-forming properties. The highest OTFT mobility of $\mu = 1.9 \times 10^{-3}$ $cm^2\ V^{-1}\ s^{-1}$ with on/off = 10^4 was achieved with the DTG homopolymer with R = *n*-octyl.

2.4 Dithienostannole

Dithienostannole (DTSn) is prepared by reacting dilithiobithiophene with dichlorostannane, similar to DTS and DTG (Scheme 2.3) [48]. DTSn derivatives are more reactive and unstable than their DTS and DTG analogues [49] but can be

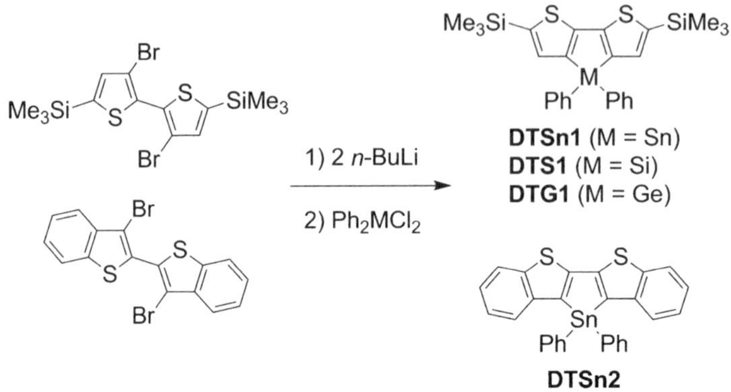

Scheme 2.3 Synthesis of DTSn derivatives

Table 2.2 Optical properties of dithienometalloles

Comp	Abs λ_{max}/nm	PL λ_{max}/nm	Φ_f
DTSn1	350	410	<0.02
DTS1	356	425	0.69
DTG1	350	409	0.71
DTSn2	369	431	0.30

stabilized by introducing sterically bulky phenyl groups on the bridging tin atom, thereby enabling the detailed investigation of their properties. The UV absorption and PL bands of **DTSn1** appear at nearly the same wavelengths as those of **DTS1** and **DTG1** with the same substituents, suggesting similarity of the electronic states of these compounds regardless of the bridging elements (Table 2.2). However, **DTSn1** exhibits a much lower PL quantum yield than **DTS1** and **DTG1**, probably because the Sn-C bonds are longer than the Si-C and Ge-C bonds. Enhanced vibration with respect to the distant Sn-Ph bonds accelerates the non-radiative decay. One may consider the possibility of the heavy atom effects of tin, facilitating the intersystem crossing to form triplet states from the singlet photoexcited states, resulting in fluorescence quenching. In contrast, bis(benzothieno)stannole **DTSn2** shows moderate PL efficiency, indicating that the heavy atom effects are not essential to decrease the PL quantum efficiency. Benzo-annulation in **DTSn2** would lead to restricted vibration of the phenyl rings arising from the steric repulsion between the phenyl ring and the benzo-annulated units. **DTSn1** is nearly non-emissive also in the amorphous condensed phase but is emissive in the crystal phase (Table 2.2, Fig. 2.5). Single-crystal X-ray diffraction study of **DTSn1** reveals intermolecular Ph-Ph interaction that suppresses the vibration. Such crystallization-enhanced and crystallization-induced emissions (CEE and CIE) are interesting phenomena, and this is the first example of element-based CEE.

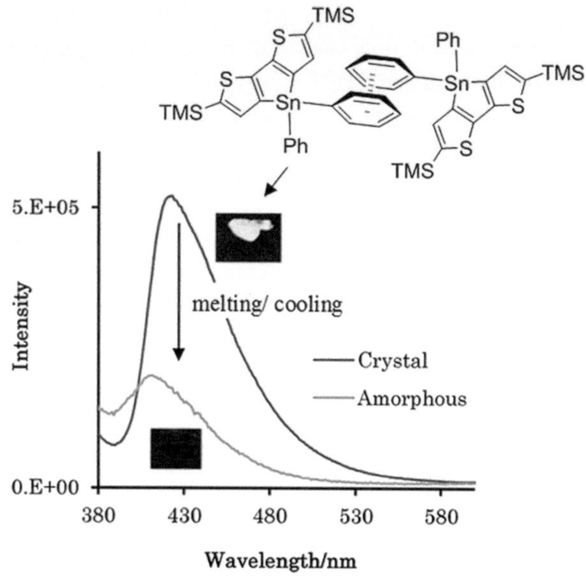

Fig. 2.5 PL spectra of **DTSn1**. Insets are photographs of **DTSn1** as crystals and amorphous solid under UV irradiation (365 nm). (Reprinted with permission from Ref. [48]. Copyright (2013) American Chemical Society)

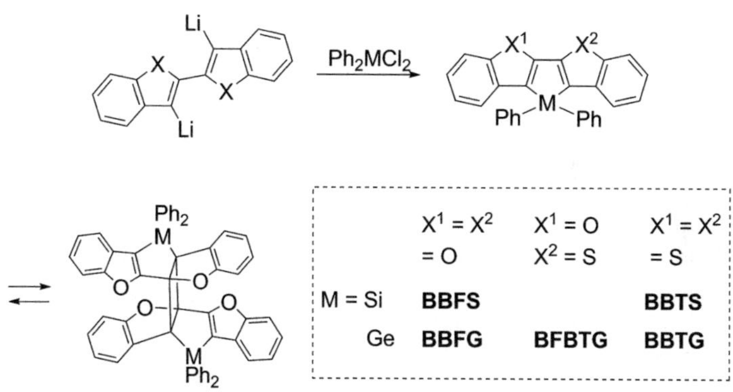

Scheme 2.4 Synthesis and dimerization of benzofuran-fused siloles and germoles

2.5 Bis(benzofurano)silole and Bis(benzofurano)germole

Bridged dibenzofurans, bis(benzofurano)siloles and bis(benzofurano)germoles (**BBFS** and **BBFG**) were prepared as shown in Scheme 2.4 [50]. Unsymmetrical (benzofurano)(benzothieno)germole **BFBTG** was also prepared [51]. These compounds show similar optical properties to analogous bis(benzothieno)metalloles (**BBTS** and **BBTG**), suggesting similarity in the electronic states. DFT calculations and CV analyses, however, indicate that both HOMO and LUMO energy levels are gradually elevated by replacing the sulfur atoms one by one by oxygen. Actually, the HOMO levels estimated from the CV anodic onset potentials are −5.50 eV,

−5.47 eV, and −5.40 eV for **BBTG**, **BFBTG**, and **BBFG**, respectively. This clearly indicates possible fine-tuning of the electronic states by changing the chalcogen atoms. Interestingly, **BBFS** and **BBFG** undergo crystallization-induced [2 + 2] dimerization. They exist as the monomers in solution but yield dimers when recrystallized from hexane/dichloromethane. In marked contrast, recrystallization from diethyl ether does not promote the dimerization but yields the monomer crystals. Similarly, **BBTS**, **BBTG**, and **BFBTG** do not produce dimers at all. It is speculated therefore that the high-lying HOMO is responsible for the dimerization, although the calculated heats of reaction for the dimerization are not affected by the nature of the chalcogen atoms, i.e., sulfur or oxygen. As other bridged bichalcogenophenes, bisselenophenosilole and -germole were also synthesized [52].

2.6 Dipyridinosilole and Dipyridinogermole

Bridging electron-deficient bipyridyl with a silicon or germanium atom results in even more electron-deficient dipyridinosilole and dipyridinogermole (DPyS and DPyG in Scheme 2.5) [53]. This is confirmed by the more positive cathodic signals of **DPyS1**, **DPyS2**, and **DPyG** than that of bridge-free bipyridyl (**BPy**) itself. No evident dependence of the potentials on the nature of the bridging element (silicon or germanium) is seen. The enhanced electron deficiency of DPyS and DPyG is likely ascribed to the planar structures and the σ*-π* conjugation in the LUMOs, as shown in Fig. 2.6. Quantum chemical calculations of the bridged and non-bridged bipyridyls indicate that the optimized geometry of **BPy** is twisted and fixing the pyridyl rings to planar results in the lowering of the LUMOs. However, the LUMOs of **DPyS0** and **DPyG0** are still lower than that of planar **BPy**. These are indicative of the contribution of σ*-π* conjugation to the LUMOs of **DPyS0** and **DPyG0**. However, the σ*-π* conjugation of the dipyridinometallole systems seems to be less pronounced than that of DTS, DTG, and other metalloles fused with electron-rich heteroaromatics.

Durben and Baumgartner reported the preparation of dipyridinophosphole and its oxide and examined the electron deficiency of the oxide (**DPyPO**) (Fig. 2.7) [54]. Based on the CV data, it is concluded that the electron deficiency is enhanced as **BPy** < DPyS and DPyG < **DPyPO**, again indicating the possible tuning of the electronic states of bipyridyls by introducing element bridges.

The dipyridinometalloles are hardly emissive at room temperature with PL quantum efficiencies (Φ) of less than 0.02 in solution but exhibit aggregation-induced emission. In fact, the PL efficiency of solid **DPyG** is approximately $\Phi = 0.07$. Measuring the spectra of the solids at low temperature provides broad phosphorescence bands in the range of 450–600 nm (Fig. 2.8) with the lifetime of $\tau = 6$ μs. The phosphorescence quantum efficiencies at 77 K are $\Phi = 0.22$ and <0.01 for **DPyG** and **DPyS1**, respectively. The higher efficiency of **DPyG** is likely ascribed to the germanium heavy atom effect, facilitating the intersystem crossing. Blue phosphorescence is clearly visible from solid **DPyG** at 77 K.

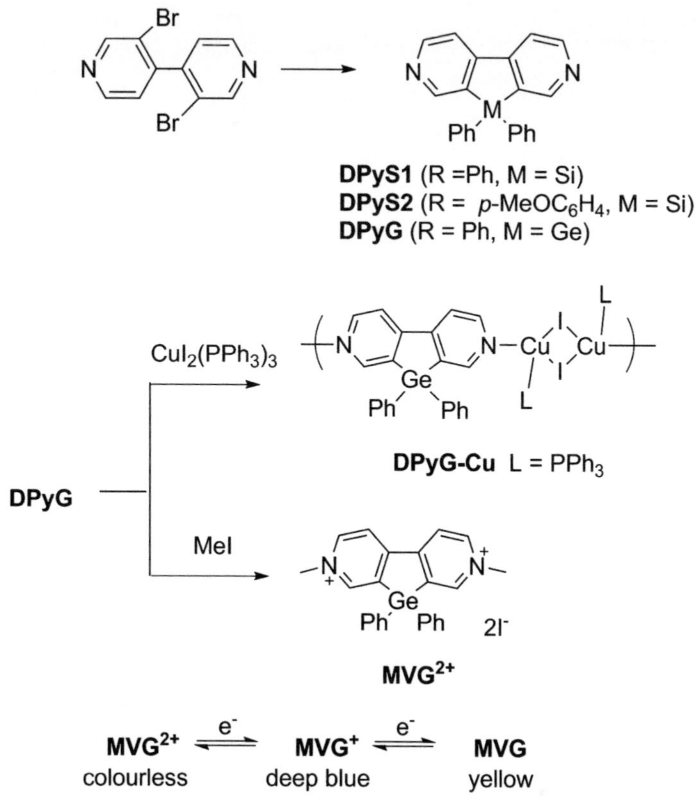

Scheme 2.5 Preparation of dipyridinometallole derivatives

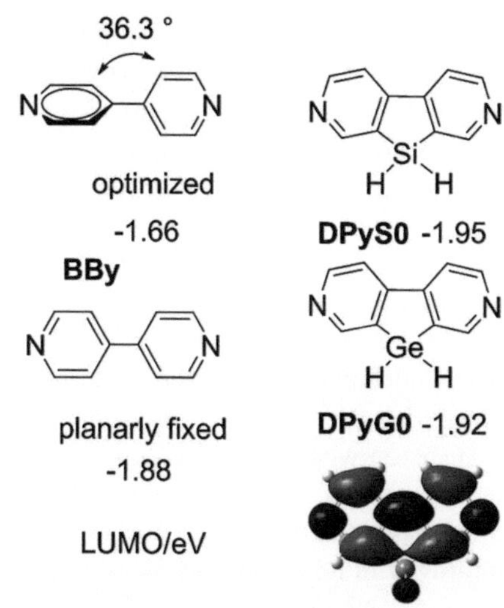

Fig. 2.6 LUMO energy levels of bipyridyl derivatives and LUMO profile of **DPyG0**, derived from computation at B3LYP/6-31G(d,p) level

Fig. 2.7 Structures of **DPyPO** and **MVPO**

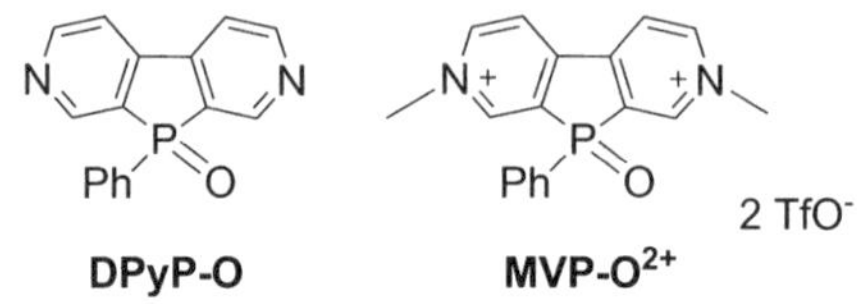

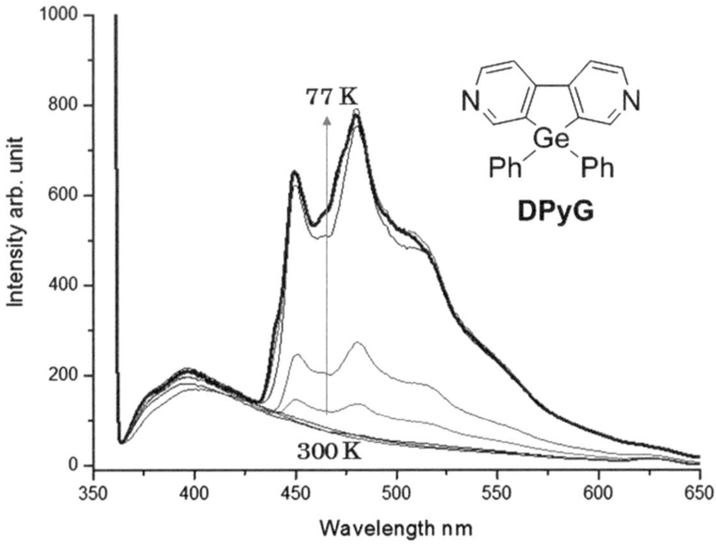

Fig. 2.8 PL spectra of **DPyG1** at temperatures ranging from 300 to 77 K

The generation of singlet oxygen is of current interest because of the possible applications to photocatalysts. In particular, photodynamic therapy in which cancer cells are killed by singlet oxygen generated by photocatalysts has attracted a great deal of attention. The DPyG system that exhibits efficient intersystem crossing can be used as the core fragment of photosensitizers for singlet oxygen generation [55]. Among the DPyG compounds examined, the compound composed of spiro-condensed DPyG and bis(dithiophenyl)-substituted DTG units (**sDPyDTG2T** in Fig. 2.9) shows high sensitizing ability with a quantum yield of $\Phi = 0.72$. Although the efficiency leaves much to be desired, the spiro-condensed sDPyDTG system is a new core structure of photocatalysts utilizing the germanium heavy atom effect.

The UV-vis absorption spectra of sDPyDTG derivatives show two independent bands attributable to perpendicularly arranged DPyG and DTG [56]. On the other hand, their PL spectra revealed only one broad band assignable to the emission from DTG, even when the DPyG unit is irradiated. No emission from DPyG is observed, indicating efficient photo-energy transfer from DPyG to DTS. The PL quantum efficiency of **sDPyDTG-Si** (Fig. 2.9) is very low ($\leq$0.03), although it is slightly improved in nonpolar solvent, suggesting the photo-induced electron transfer from DTG to DPyG. The introduction of bithiophenyl groups on the DTG unit in

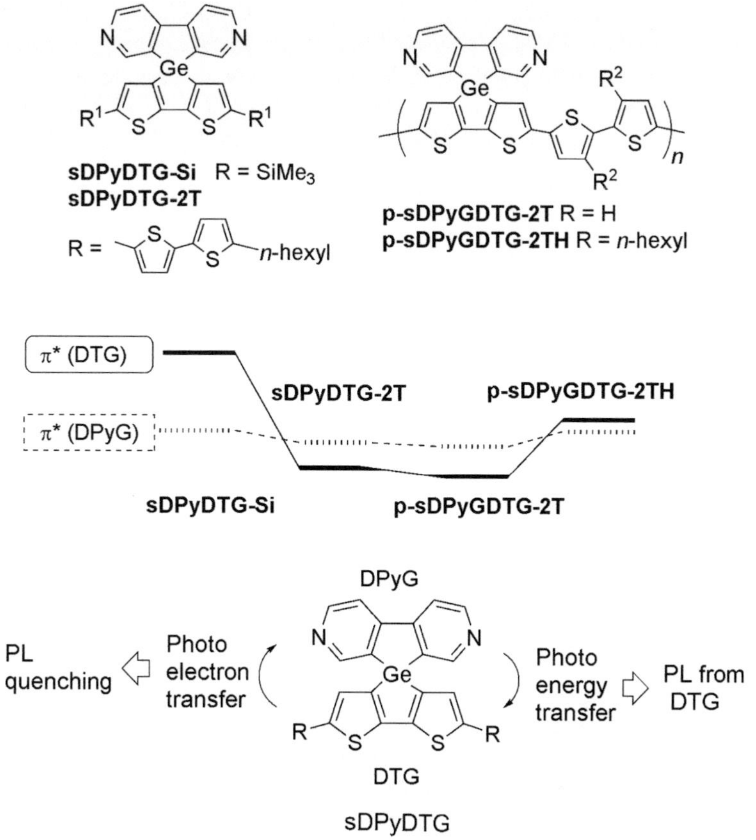

Fig. 2.9 Energy diagram of π* orbitals of DPyG and DTG units and schematic representation of energy and electron transfer of sDPyDTG derivatives

sDPyDTG-2T, however, increases the PL efficiency drastically to 0.13–0.22. The extended conjugation by the bithiophenyl substituents would lower the LUMO level of the DTG unit, making the electron transfer difficult. The sDPyDTG unit is introduced to polymeric conjugated systems, as shown in Fig. 2.9. Polymer **p-sDPyDTG-2T** shows relatively high PL efficiencies in the range of $\Phi = 0.26$–0.39 depending on the solvent, whereas **p-sDPyGDTG-2TH** is much less emissive with $\Phi = 0.03$–0.15. The hexyl groups that are introduced to improve the solubility cause steric repulsion, which results in loss of planarity of the polythiophene backbone. Suppressing conjugation by twisting the backbone raises the DTG LUMO level to facilitate the electron transfer.

DPyG1 can be also used as the ligand; in this case, a complex with copper (**DPyG-Cu**) is obtained, which shows phosphorescence in the solid state at room temperature (Scheme 2.5) [57]. Complex **DPyG-Cu** is soluble in organic solvents, including THF, chloroform, and toluene, in contrast to a similar bridge-free complex that is insoluble in common organic solvents [58]. **DPyG-Cu** can be dispersed in such polymer films as PMMA (poly(methyl methacrylate)) to form phosphorescent self-standing films. Treatment of **DPyG1** with methyl iodide provides a germanium-bridged methyl viologen derivative (**MVG**$^{2+}$ in Scheme 2.5) [59]. Like methyl viologen (**MV**$^{2+}$), the bridged **MVG**$^{2+}$ undergoes reversible two-step cathodic reduction with reversible color changes. The reduction potentials of **MVG**$^{2+}$ are slightly more positive than those of **MV**$^{2+}$, again reflecting the σ^*-π^* conjugation that enhances the electron deficiency. A similar viologen derivative of **MVP-O**$^{2+}$ is also reported, which shows more pronounced electron deficiency than **MVG**$^{2+}$ (Fig. 2.7) [54].

2.7 Conclusions

A variety of group 14 element-bridged biaryls have been synthesized. Detailed investigation through experimental observations and theoretical calculations reveals that these bridged biaryls show electronic states characteristic of the bridging elements as well as the nature of the biaryls. Compared to DTS, DTG exhibits rather limited σ^*-π^* conjugation due to the long endocyclic Ge-C bonds that sterically separate the germanium σ^* from the bithiophene π^* orbitals. Enhancement of phosphorescence properties by the introduction of a germanium bridge is observed in DPyG. This is ascribable to the germanium heavy atom effect. The long tin bond influences the molecular vibration, thereby suppressing the PL properties of DTSn. In polymeric systems, it is likely that the introduction of heavier germanium atom leads to a higher degree of interchain interaction in films and improves carrier mobility, compared with the silicon congener. The flexibility of substituents on silicon and germanium arising from distant Si-C and Ge-C bonds seems to improve polymer solubility and processibility. Other combinations of biaryl structures and bridging elements are awaiting discovery, and further studies aimed at the development of new functional materials with bridged biaryl systems are in progress.

Acknowledgments The work presented in this chapter was partly supported by a Grant-in-Aid for Scientific Research on Innovative Areas, "New Polymeric Materials Based on Element-Blocks (No. 2401)" (JSPS KAKENHI Grant Number JP24102005).

References

1. Tamao K, Uchida M, Izumizawa T, Furukawa K, Yamaguchi S (1996) Silole derivatives as efficient electron transporting materials. J Am Chem Soc 118(47):11974–11975. https://doi.org/10.1021/ja962829c
2. Yamaguchi S, Tamao K (1998) Silole-containing σ- and π-conjugated compounds. J Chem Soc Dalton Trans 22:3693–3702. https://doi.org/10.1039/A804491K

3. Yamaguchi S, Tamao K (2005) A key role of orbital interaction in the main group element-containing π-electron systems. Chem Lett 34(1):2–7. https://doi.org/10.1246/cl.2005.2
4. Ponomarenko SA, Kirchmeyer S (2011) Conjugated organosilicon materials for organic electronics and photonics. Adv Polym Sci 235:33–110. https://doi.org/10.1007/12_2009_48
5. Liu J, Lam JWY, Tang BZ (2009) Aggregation-induced emission of silole molecules and polymers: fundamental and applications. J Inorg Organomet Polym 19(3):249–285. https://doi.org/10.1007/s10904-009-9282-8
6. Chen J, Cao Y (2007) Silole-containing polymers: chemistry and optoelectronic properties. Macromol Rapid Commun 28(17):1714–1742. https://doi.org/10.1002/marc.200700326
7. Zhao Z, He B, Tang BZ (2015) Aggregation-induced emission of siloles. Chem Sci 6(10):5347–5365. https://doi.org/10.1039/C5SC01946J
8. Ohshita J, Nodono M, Watanabe T, Ueno Y, Kunai A, Harima Y, Yamashita K, Ishikawa M (1998) Synthesis and properties of dithienosiloles. J Organomet Chem 553(1–2):487–491. https://doi.org/10.1016/S0022-328X(97)00643-8
9. Ohshita J, Nodono M, Kai H, Watanabe T, Kunai A, Komaguchi K, Shiotani M, Adachi A, Okita K, Harima Y, Yamashita K, Ishikawa M (1999) Synthesis and optical, electrochemical, and electron-transporting properties of silicon-bridged bithiophenes. Organometallics 18(8):1453–1459. https://doi.org/10.1021/om980918n
10. Ohshita J (2009) Conjugated oligomers and polymers containing dithienosilole units. Macromol Chem Phys 210(17):1360–1370. https://doi.org/10.1002/macp.200900180
11. Ohshita J, Kai H, Takata A, Iida T, Kunai A, Ohta N, Komaguchi K, Shiotani M, Adachi A, Sakamaki K (2001) Effects of conjugated substituents on the optical, electrochemical, and electron-transporting properties of dithienosiloles. Organometallics 20(23):4800–4805. https://doi.org/10.1021/om0103254
12. Ohshita J, Kurushima Y, Lee KH, Kunai A, Ooyama Y, Harima Y (2007) Synthesis of bis (diaryl phosphino)dithienosilole derivatives as novel photo- and electroluminescence materials. Organometallics 26(26):6591–6595. https://doi.org/10.1021/om700765w
13. Ohshita J, Tominaga Y, Mizumo T, Kuramochi Y, Higashimura H (2011) Synthesis and optical properties of a bis(diphenylphosphino)dithienosilole-digold(I) complex. Heteroat Chem 22(3–4):514–517. https://doi.org/10.1002/hc.20715
14. Ohshita J, Tominaga Y, Tanaka D, Ooyama Y, Mizumo T, Kobayashi N, Higashimura H (2013) Synthesis of dithienosilole-based highly photoluminescent donor-acceptor type compounds. Dalton Trans 42(10):3646–3652. https://doi.org/10.1039/C2DT32738D
15. Usta H, Lu G, Facchetti A, Marks TJ (2006) Dithienosilole− and dibenzosilole−thiophene copolymers as semiconductors for organic thin-film transistors. J Am Chem Soc 128(28):9034–9035. https://doi.org/10.1021/ja062908g
16. Lu G, Usta H, Risko C, Wang L, Facchetti A, Ratner MA, Marks TJ (2008) Synthesis, characterization, and transistor response of semiconducting silole polymers with substantial hole mobility and air stability. Experiment and theory. J Am Chem Soc 130(24):7670–7685. https://doi.org/10.1021/ja800424m
17. Hou J, Chen HY, Zhang S, Li G, Yang Y (2008) Synthesis, characterization, and photovoltaic properties of a low band gap polymer based on silole-containing polythiophenes and 2,1,3-benzothiadiazole. J Am Chem Soc 130(48):16144–16145. https://doi.org/10.1021/ja806687u
18. Chu TY, Lu J, Beaupré S, Zhang Y, Pouliot JR, Zhou J, Najari A, Leclerc M, Tao Y (2012) Effects of the molecular weight and the side-chain length on the photovoltaic performance of dithienosilole/thienopyrrolodione copolymers. Adv Funct Mater 22(11):2345–2351. https://doi.org/10.1002/adfm.201102623
19. Ohshita J, Hwang YM, Mizumo T, Yoshida H, Ooyama Y, Harima Y, Kunugi Y (2011) Synthesis of dithienogermole-containing π-conjugated polymers and applications to photovoltaic cells. Organometallics 30(12):3233–3236. https://doi.org/10.1021/om200081b
20. Amb CM, Chen S, Graham KR, Subbiah J, Small CE, So F, Reynolds JR (2011) Dithienogermole as a fused electron donor in bulk heterojunction solar cells. J Am Chem Soc 133(26):10062–10065. https://doi.org/10.1021/ja204056m

21. Gendron D, Morin PO, Berrouard P, Allard N, Aïch BR, Garon CN, Tao Y, Leclerc M (2011) Synthesis and photovoltaic properties of poly(dithieno[3,2-b:2′,3′-d]germole) derivatives. Macromolecules 44(18):7188–7193. https://doi.org/10.1021/ma2013496
22. Small CE, Chen S, Subbiah J, Amb CM, Tsang SW, Lai TH, Reynolds JR, So F (2011) High-efficiency inverted dithienogermole-thienopyrrolodione-based polymer solar cells. Nat Photon 6:115–120. https://doi.org/10.1038/NPHOTON.2011.317
23. Guo X, Zhou N, Lou SJ, Hennek JW, Ortiz RP, Butler MR, Boudreault PLT, Strzalka J, Morin PO, Leclerc M, Navarrete JTL, Ratner MA, Chen LX, Chang PH, Facchetti A, Marks TJ (2012) Bithiopheneimide–dithienosilole/dithienogermole copolymers for efficient solar cells: information from structure–property–device performance correlations and comparison to thieno[3,4-*c*] pyrrole-4,6-dione analogues. J Am Chem Soc 134:18427–18439. https://doi.org/10.1021/ja3081583
24. Yau CP, Fei Z, Ashraf RS, Shahid M, Watkins SE, Pattanasattayavong P, Anthopoulos TD, Gregoriou VG, Chohos CL, Heeney M (2014) Influence of the electron deficient co-monomer on the optoelectronic properties and photovoltaic performance of dithienogermole-based co-polymers. Adv Funct Mater 24(5):678–687. https://doi.org/10.1002/adfm.201302270
25. Constantinou I, Lai TH, Zhao D, Klump ED, Deininger JJ, Lo CK, Reynolds JR, So F (2016) High efficiency air-processed dithienogermole-based polymer solar cells. ACS Appl Mater Interfaces 7(8):4826–4832. https://doi.org/10.1021/am5087566
26. Gupta V, Lai LF, Datt R, Chand S, Heeger AJ, Bazan GC, Singh SP (2016) Dithienogermole-based solution-processed molecular solar cells with efficiency over 9%. Chem Commun 52:8596–8599. https://doi.org/10.1039/C6CC03998G
27. Ohshita J, Miyazaki M, Zhang FB, Tanaka D, Morihara Y (2013) Synthesis and properties of dithienometallole-pyridinochalcogenadiazole alternate polymers. Polym J 45(9):979–984. https://doi.org/10.1038/pj.2013.13
28. Fei Z, Ashraf RS, Huang Z, Smith J, Kline RJ, Angelo PD, Anthopoulos TD, Durrant JR, McCulloch I, Heeney M (2012) Germaindacenodithiophene based low band gap polymers for organic solar cells. Chem Commun 48:2955–2957. https://doi.org/10.1039/C2CC17996B
29. Zhong H, Li Z, Deledalle F, Fregoso EC, Shahid M, Fei Z, Nielsen CB, Yaacobi-Gross N, Rossbauer S, Anthopoulos TD, Durrant JR, Heeney M (2013) Fused dithienogermolodithiophene low band gap polymers for high-performance organic solar cells without processing additives. J Am Chem Soc 135(6):2040–2043. https://doi.org/10.1021/ja311700u
30. Ashraf RS, Schroeder BC, Bronstein HA, Huang Z, Thomas S, Kline RJ, Brabec CJ, Rannou P, Anthopoulos TD, Durrant JR, McCulloch I (2013) The influence of polymer purification on photovoltaic device performance of a series of indacenodithiophene donor polymers. Adv Mater 25(14):2029–2034. https://doi.org/10.1002/adma.201300027
31. Kai H, Ohshita J, Ohara S, Nakayama N, Kunai A, Lee IS, Kwak YW (2008) Disilane- and siloxane-bridged biphenyl and bithiophene derivatives as electron-transporting materials in OLEDs. J Organomet Chem 693(1):3490–3494. https://doi.org/10.1016/j.jorganchem.2008.08.018
32. Ohshita J, Nakashima M, Tanaka D, Morihara Y, Fueno H, Tanaka K (2014) Preparation of a D–A polymer with disilanobithiophene as a new donor component and application to high-voltage bulk heterojunction polymer solar cells. Polym Chem 5(2):346–349. https://doi.org/10.1039/C3PY01157G
33. Nakashima M, Murata N, Suenaga Y, Naito H, Sasaki T, Kunugi Y, Ohshita J (2016) Disilanobithiophene-dithienylbenzothiadiazole alternating polymer as donor material of bulk heterojunction polymer solar cells. Synth Met 215:116. https://doi.org/10.1016/j.synthmet.2016.02.012
34. Nakashima M, Otsura T, Naito H, Ohshita J (2015) Synthesis of new D-A polymers containing disilanobithiophene donor and application to bulk heterojunction polymer solar cells. Polym J 47(11):733–738. https://doi.org/10.1038/pj.2015.61

35. Nakashima M, Ooyama Y, Sugiyama T, Naito H, Ohshita J (2016) Synthesis of a conjugated D-A polymer with bi(disilanobithiophene) as a new donor component. Molecules 21(6):789–795. https://doi.org/10.3390/molecules21060789
36. Ohshita J, Matsukawa J, Hara M, Kunai A, Kajiwara S, Ooyama Y, Harima Y, Kakimoto M (2008) Chem Lett 37(11):316–317. https://doi.org/10.1246/cl.2008.316
37. Ohshita J, Adachi Y, Tanaka D, Nakashima M, Ooyama Y (2015) Synthesis of D–A polymers with a disilanobithiophene donor and a pyridine or pyrazine acceptor and their applications to dye-sensitized solar cells. RSC Adv 5(46):36673–36679. https://doi.org/10.1039/c5ra01055a
38. Unno M, Kakiage K, Yamamura M, Kogure T, Kyomen T, Hanaya M (2010) Silanol dyes for solar cells: higher efficiency and significant durability. Appl Organomet Chem 24:247–250. https://doi.org/10.1002/aoc.1612
39. Ohshita J, Nakamura M, Ooyama Y (2015) Preparation and reactions of dichlorodithienogermoles. Organometallics 34(23):5609–5614. https://doi.org/10.1021/acs.organomet.5b00832
40. Nakamura M, Ooyama Y, Hayakawa S, Nishino M, Ohshita J (2016) Synthesis of poly (dithienogermole)s. Organometallics 35(14):2333. https://doi.org/10.1021/acs.organomet.6b00263
41. Ohshita J, Nakamura M, Yamamoto K, Watase S, Matsukawa K (2015) Synthesis of dithienogermole-containing oligo- and polysilsesquioxanes as luminescent materials. Dalton Trans 44(17):8214–8220. https://doi.org/10.1039/C5DT00777A
42. Nakamura M, Shigeoka K, Adachi Y, Ooyama Y, Watase S, Ohshita J (2017) Preparation of dithienogermole-containing polysilsesquioxane films for sensing nitroaromatics. Chem Lett 46(4):438–441. https://doi.org/10.1246/cl.161119
43. Zeng W, Cao Y, Bai Y, Wang Y, Shi Y, Zhang M, Wang F, Pan C, Wang P (2010) Efficient dye-sensitized solar cells with an organic photosensitizer featuring orderly conjugated ethylenedioxythiophene and dithienosilole blocks. Chem Mater 22(5):1915–1925. https://doi.org/10.1021/cm9036988
44. Adachi Y, Ooyama Y, Shibayama N, Ohshita J (2017) Dithienogermole-containing D-π-A-π-A photosensitizers for dye-sensitized solar cells. Chem Lett 46(3):310–312. https://doi.org/10.1246/cl.161034
45. Ohshita J, Miyazaki M, Tanaka D, Morihara Y, Fujita Y, Kunugi Y (2013) Synthesis of poly (dithienogermole-2,6-diyl)s. Polym Chem 4(10):3116–3122. https://doi.org/10.1039/c3py00253e
46. Ohshita J, Adachi Y, Sagisaka R, Nakashima M, Ooyama Y, Kunugi Y (2017) Synthesis of dithienogermole-containing polythiophenes. Synth Met 227:87–92. https://doi.org/10.1016/j.synthmet.2017.03.009
47. Nakashima M, Miyazaki M, Ooyama Y, Fujita Y, Murata S, Kunugi Y, Ohshita J (2016) Synthesis of silicon- or carbon-bridged polythiophenes and application to organic thin-film transistors. Polym J 48(5):645–651. https://doi.org/10.1038/pj.2015.121
48. Tanaka D, Ohshita J, Ooyama Y, Kobayashi N, Higashimura H, Nakanishi T, Hasegawa Y (2013) Synthesis, optical properties, and crystal structures of dithienostannoles. Organometallics 32(15):4136–4141. https://doi.org/10.1021/om400213q
49. Nagao I, Shimizu M, Hiyama T (2009) 9-Stannafluorenes: 1,4-dimetal equivalents for aromatic annulation by double cross-coupling. Angew Chem Int Ed 48(41):7573–7576. https://doi.org/10.1002/anie.200903779
50. Zhang FB, Adachi Y, Ooyama Y, Ohshita J (2016) Synthesis and properties of benzofuran-fused silole and germole derivatives: reversible dimerization and crystal structures of monomers and dimers. Organometallics 35(14):2327–2332. https://doi.org/10.1021/acs.organomet.6b00222
51. Zhang FB, Ooyama Y, Ohshita J (2017) Synthesis of (benzofurano)(benzothieno)germole. Chem Sel 2(10):3106–3109. https://doi.org/10.1002/slct.201700597

52. Pao YC, Chen YL, Chen YT, Cheng SW, Lai YY, Haung WC, Cheng YJ (2014) Synthesis and molecular properties of tricyclic biselenophene-based derivatives with nitrogen, silicon, germanium, vinylidene, and ethylene bridges. Org Lett 16(21):5724–5727
53. Ohshita J, Murakami K, Tanaka D, Ooyama Y, Mizumo T, Kobayashi N, Higashimura H, Nakanishi T, Hasegawa Y (2014) Synthesis of group 14 dipyridinometalloles with enhanced electron-deficient properties and solid-state phosphorescence. Organometallics 33(2):517–521. doi:https://doi.org/10.1021/om401019b
54. Durben S, Baumgartner T (2011) 3,7-Diazadibenzophosphole oxide: a phosphorus-bridged viologen analogue with significantly lowered reduction threshold. Angew Chem Int Ed 50(34):7948–4952. https://doi.org/10.1002/anie.201102453
55. Ohshita J, Hayashi Y, Murakami K, Enoki T, Ooyama Y (2016) Single oxygen generation sensitized by spiro(dipyridinogermole)(dithienogermole)s. Dalton Trans 45(39):15679–15683. https://doi.org/10.1039/c6dt02767a
56. Murakami K, Ooyama Y, Higashimura H, Ohshita J (2016) Synthesis, properties, and polymerization of spiro[(dipyridinogermole)(dithienogermole)]. Organometallics 35(1):20–26. https://doi.org/10.1021/acs.organomet.5b00817
57. Murakami K, Ooyama Y, Watase S, Matsukawa K, Omagari S, Nakanishi T, Hasegawa Y, Inumaru K, Ohshita J (2016) Synthesis of dipyridinogermole-copper complex as soluble phosphorescent material. Chem Lett 45(5):502–504. https://doi.org/10.1246/cl.160036
58. Araki H, Tsuge K, Sasaki Y, Ishizaka S, Kitamura N (2005) Luminescence ranging from red to blue: a series of copper(I)−halide complexes having rhombic $\{Cu_2(\mu\text{-}X)_2\}$ (X = Br and I) units with N-Heteroaromatic ligands. Inorg Chem 44(26):9667–9675. https://doi.org/10.1021/ic0510359
59. Murakami K, Ohshita J, Inagi S, Tomita I (2015) Synthesis, and optical and electrochemical properties of germanium-bridged viologen. Electrochemistry 83(8):605–608. https://doi.org/10.5796/electrochemistry.83.605

Chapter 3
Element-Blocks π-Conjugated Polymers by Post-element-Transformation Technique

Ikuyoshi Tomita

Abstract The synthesis of a new class of π-conjugated polymers possessing versatile element-blocks is described. That is, regioregular organotitanium polymers having titanacyclopentadiene-2,5-diyl units, which are obtained from a low-valent titanium complex and diynes, are subjected to the polymer reactions with elements-containing electrophiles to produce π-conjugated polymers possessing unique element-blocks. Optoelectronic features of the resulting element-blocks π-conjugated polymers are also described.

Keywords Polymer reactions · π-Conjugated polymers · Organometallic polymers · Heteroles · Chemosensors

3.1 Introduction

The precision control of electronic features of π-conjugated polymers is an important subject for many applications such as field-effect transistors, dyes for solar cells, and organic light-emitting devices [1–4]. In accordance with the idea that incorporation of appropriate element-blocks into polymers is effective to build up advanced functional materials [5], it has also been proposed that the addition of heteroatoms into the π-electronic systems is quite effective to tune up their electronic properties such as HOMO and LUMO energy levels and their bandgaps based on the theoretical calculations [6]. Further, the theoretical studies suggest that the incorporation of plural elements often affects dramatically on their electronic properties [7]. Nevertheless, the elements that have successfully been introduced into π-conjugated polymers are still limited so far, and many elements-containing π-conjugated polymers (i.e., element-blocks π-conjugated polymers) still remained untouched.

I. Tomita (✉)
Department of Chemical Science and Engineering, School of Materials and Chemical Technology, Tokyo Institute of Technology, Yokohama, Japan
e-mail: tomita@echem.titech.ac.jp

Y. Chujo (ed.), *New Polymeric Materials Based on Element-Blocks*,
https://doi.org/10.1007/978-981-13-2889-3_3

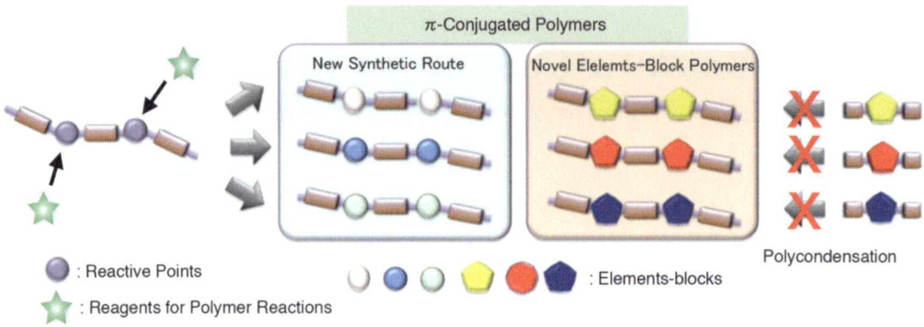

Fig. 3.1 Synthesis of versatile element-blocks π-conjugated polymers by means of polymer reactions accompanying transformation of main chain structures

In general, π-conjugated polymers are prepared by the polycondensation of the monomers possessing the corresponding building blocks, and the transition metal-catalyzed coupling reactions are often employed for the polycondensation processes. However, π-conjugated polymers possessing the corresponding units are hardly accessible by the conventional polycondensation approach, because most of the elements-carbon and/or inter-element bonds in the elements-containing monomers are unstable under the polymerization conditions. Accordingly, new synthetic methods are required to produce the element-blocks π-conjugated polymers.

This chapter deals with a new synthetic approach based on the polymer reactions that enable the synthesis of versatile element-blocks π-conjugated polymers by means of the transformation of the transition metals in the main chain into various elements (Fig. 3.1). That is, regioregular organotitanium polymers having titanacyclopentadiene-2,5-diyl units were prepared from a low-valent titanium complex and diynes, and their polymer reactions with elements-containing electrophiles were carried out as new synthetic methods to produce π-conjugated polymers possessing unique element-blocks. Optoelectronic features of the resulting element-block π-conjugated polymers are also described.

3.2 Synthesis of Element-Blocks π-Conjugated Polymers via Organotitanium Polymers

Based on the regioregular metallacyclization of a low-valent titanium (**2**) and terminal alkynes established by Sato and his coworkers [8, 9], organotitanium polymers having titanacyclopentadiene-2,5-diyl units (**3**) were prepared from aromatic terminal diynes (**1**) [10]. The organotitanium polymers (**3**) are air-unstable and thermally unstable. However, they exhibit excellent reactivity toward many electrophilic reagents. For example, the treatment of the polymers (**3**) with hydrochloric acid and iodine gave polymers having 1,3-butadiene-1,4-diyl and 1,4-diiodo-1,3-butadiene-1,4-diyl units (**4** and **5**), respectively (Scheme 3.1).

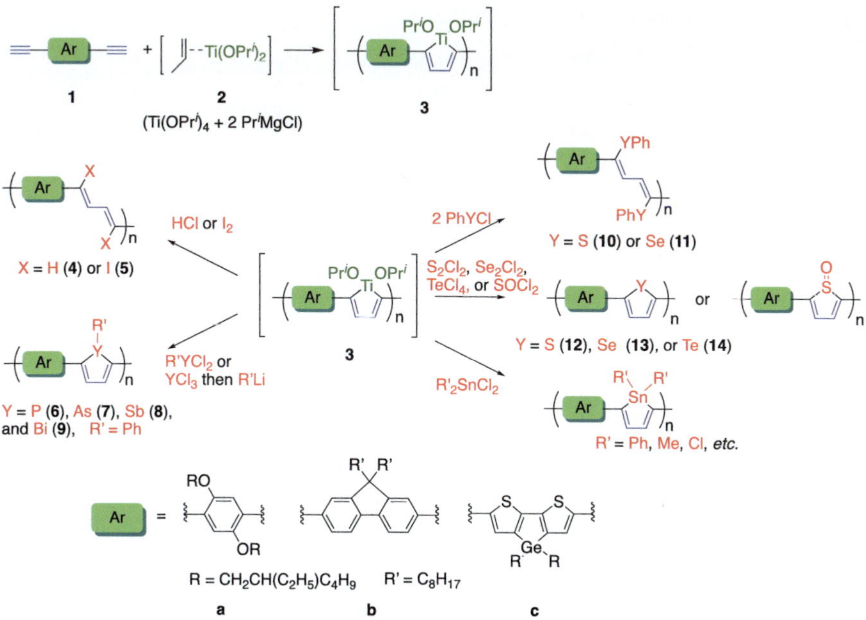

Scheme 3.1 Synthesis of element-blocks π-conjugated polymers by post-element-transformation processes

Likewise, reactions of the organotitanium polymers (**3**) with electrophiles possessing the group 14, 15, and 16 elements were performed to produce π-conjugated polymers possessing heteroles of the group 14, 15, and 16 elements. For example, the reaction of the organotitanium polymer (**3a**) with dichlorophenylphosphine produced a red-colored phosphole-containing π-conjugated polymer (**6a**) [11]. Compared to that of the corresponding polymer with thiophene moieties (**12a**, vide infra), the phosphole-containing polymer thus obtained (**6a**) proved to have low-lying LUMO energy level due to the σ^*-π^* orbital interaction, [12] while the HOMO energy levels for both the polymers were comparably high, as predicted from the theoretical calculations [6] and also from the preceding report of the analogous phosphole-containing polymers [13–15]. The phosphole-containing π-conjugated polymer (**6a**) exhibits yellow and orange photoluminescence in solution and in film, respectively.

From the theoretical calculations of the heteroles of the group 15 elements, it was supposed that the stibole has the lowest LUMO energy level and the LUMO decreases in the order of P ~ As > Bi > Sb. In accordance with the calculations, a stibole-containing polymer (**8a**), which could be prepared by the transformation reaction of **3a**, was found to have the narrowest bandgap as a result of the lowering of the LUMO energy level (Table 3.1). An arsole-containing polymer (**7a**) could successfully be obtained by the collaborative research with Naka and his coworkers by the use of diiodophenylarsine generated in situ from a less volatile and safer arsenic precursor [16, 17]. The optical and electronic properties (i.e., HOMO and

Table 3.1 Optical properties and bandgap of **6a–9a**

	Polymer	E	λ_{max} (nm)[a]	λ_{onset} (nm)[a]	Bandgap (eV)[b]
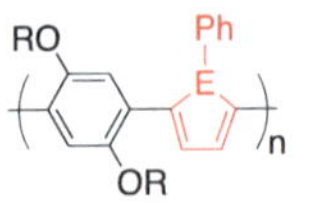 $R = CH_2CH(C_2H_5)C_4H_9$	**6a**	**P**	510	622	1.99
	7a	**As**	517	612	2.03
	8a	**Sb**	550	685	1.81
	9a	**Bi**	517	648	1.93

[a]Measured in THF
[b]Optical bandgap, estimated from their λ_{onset}

LUMO energy levels, UV-vis absorptions, and photoluminescence spectra) of the arsole-containing polymer (**7a**) were found to be very close to those of the phosphole-containing polymer (**6a**). However, the arsole-containing polymer (**7a**) exhibits remarkably higher oxidation-tolerant nature compared to that of the phosphole-containing polymer (**6a**), which readily undergoes oxidation to produce a phosphole oxide-containing polymer.

π-Conjugated polymers having 1,4-phenylthio-1,3-butadiene-1,4-diyl and 1,4-phenylseleno-1,3-butadiene-1,4-diyl units (**10** and **11**) were prepared by the reactions with PhSCl and PhSeCl, respectively [18]. The transformation of the organotitanium polymers (**3**) into π-conjugated polymers having heteroles of the group 16 elements was likewise attainable by the use of appropriate electrophilic reagents. That is, thiophene-, selenophene-, and tellurophene-containing π-conjugated polymers (**12**, **13**, and **14**) were obtained by the reactions of **3** with sulfur monochloride, selenium monochloride, and tellurium tetrachloride, respectively [19]. The π-conjugated polymers having the group 16 elements thus prepared were found to have high HOMO and LUMO energy levels, where the LUMO energy levels tend to decrease by the use of heavier elements, resulting in the bathochromic shift of the UV-vis spectra.

As an electrophilic reagent, thionyl chloride works effectively to give thiophene-1-oxide-containing polymers (**15**) from **3**. The UV-vis spectrum of the **15a** exhibited a significant bathochromic shift compared to that of the corresponding thiophene-containing polymer (**12a**), which indicated the excellent effect of the plural elements on the optical and electronic properties of the π-conjugated materials.

Concerning the transformation of **3** into π-conjugated polymers having heteroles of the group 14 elements, the reaction with organotin dichloride or tin tetrachloride takes place smoothly to produce stannole-containing π-conjugated polymers (**16**), while the reactions with germanium halides or silicon halides did not proceed sufficiently under analogous conditions.

As indicated above, π-conjugated polymers having the group 14–16 elements (**6–16**) were synthesized by the polymer reactions of the organotitanium intermediates (**3**). Their electronic and optical properties were found to be affected largely by the nature of the element-blocks. As shown in Fig. 3.2, the UV-vis absorptions of the π-conjugated polymers varied in a wide range. Some of the polymers are photoluminescence-active, and the color of the emission also altered in a wide range, depending on the elements incorporated in the π-conjugated system.

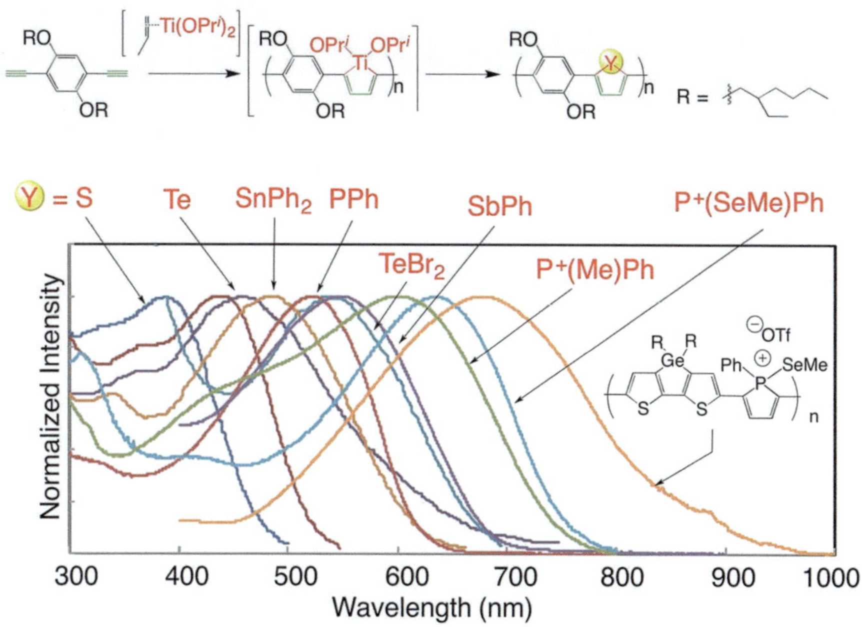

Fig. 3.2 UV-vis spectra of **12a**, **14a**, **16a**, **6a**, Br_2 adduct of **14a**, onium salts of **6a** and **6c**

3.3 Synthesis of π-Conjugated Polymers Possessing Plural Elements by Further Reactions of Element-Blocks π-Conjugated Polymers

Some of the elements on the π-conjugated polymers thus obtained can be modified or transformed into other elements by the further polymer reactions. For example, the phosphole-containing π-conjugated polymer (**6a**) was subjected to oxidation with sulfur and selenium to produce polymers having pentavalent phosphorous atoms (Scheme 3.2). The sulfur and selenium atoms can be alkylated further to give the corresponding phosphonium salt-containing polymers. The reactions of **6a** by the alkylation and complexation with boron trifluoride and gold(I) also produced the corresponding phosphorous-containing π-conjugated polymers. All the transformation processes proceed quantitatively as monitored by the ^{31}P NMR spectra. Through these chemical modifications, the color of the polymers turned from red to blue or purple, and the UV-vis absorption spectra exhibited remarkable bathochromic shift (Fig. 3.3). For example, the λ_{max} of the polymer shifted to longer wavelength by 130 nm after the selenation followed by the alkylation. Chemical modifications of the arsole-containing π-conjugated polymer (**7a**) can also be performed which also resulted in the change of the optical and electronic properties. From the theoretical studies performed by Tanaka, the lowering of the energy level of the σ^* orbital in the arsine-carbon or arsine-elements bond effectively lowered the LUMO energy level through the σ^*-π^* molecular orbital interaction [17].

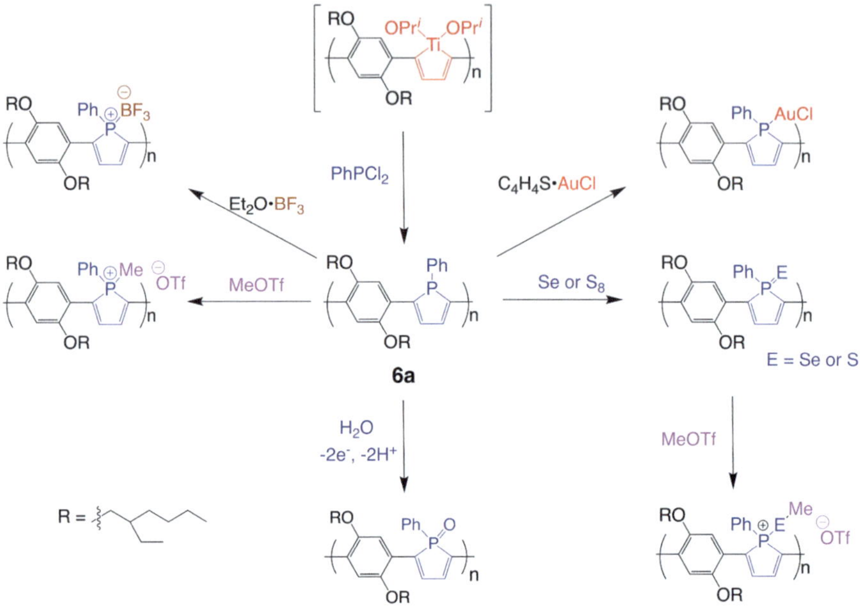

Scheme 3.2 Chemical modifications of **6a**

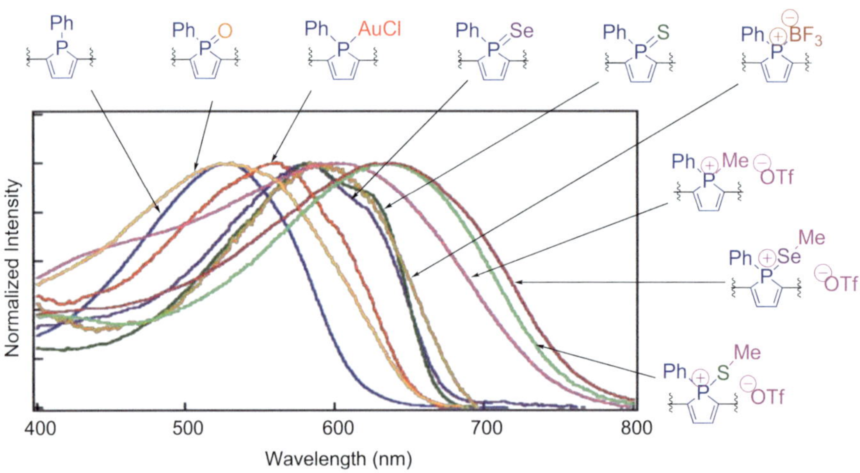

Fig. 3.3 UV-vis spectra of polymers derived from **6a**

As mentioned above, the phosphole-containing π-conjugated polymers (**6**) exhibit yellow and orange photoluminescence in solution and in film, respectively. It was found that the photoluminescence was totally quenched through these chemical modifications, which may suggest the application of the polymers as the turnoff photoluminescence chemosensors.

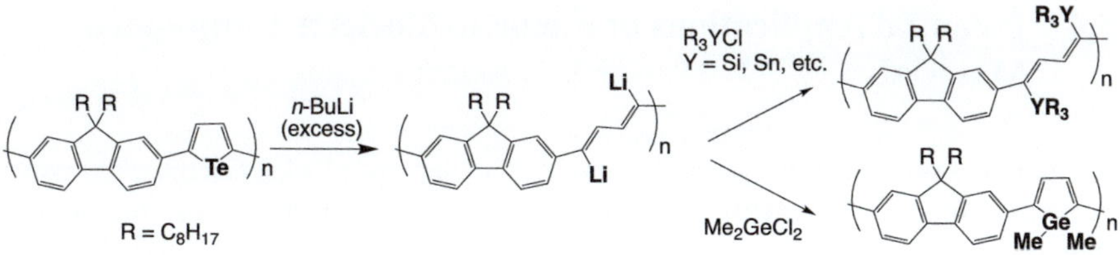

Scheme 3.3 Synthesis of element-blocks π-conjugated polymers via the lithiation of the tellurophene-containing polymer (**14b**)

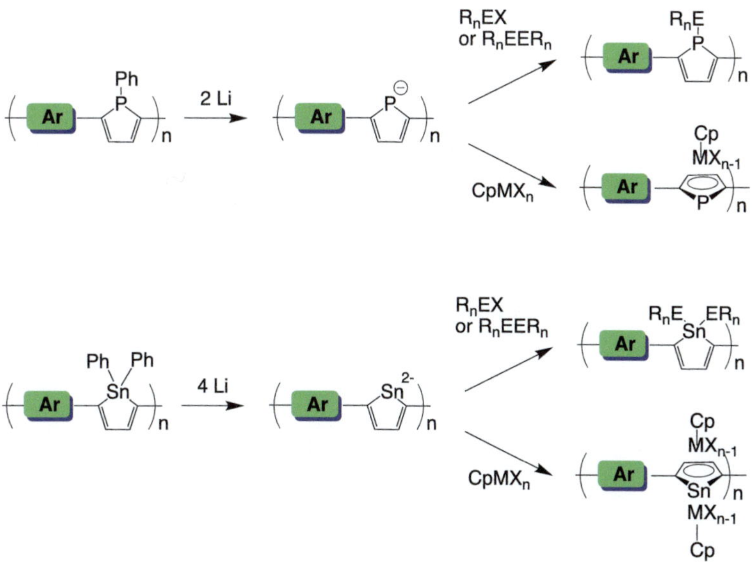

Scheme 3.4 Transformation of phosphole- and stannole-containing polymers

The tellurophene-containing polymers **(14)** can also be reacted further to produce π-conjugated polymers that have not been prepared by the direct transformation of the organotitanium intermediates (**3**). That is, the tellurophene-containing π-conjugated polymer (**14b**) was treated with *n*-butyllithium to generate a highly reactive lithiated polymer. Through the lithiated form, reagents having lower electrophilicity such as germanium halides, silicon halides, and others can be reacted smoothly to produce the corresponding element-blocks π-conjugated polymers (Scheme 3.3).

Reduction of the phosphole- and stannole-containing polymers (**6** and **16**) takes place quantitatively to give phospholyl anion-containing and stannole dianion-containing polymers, respectively. These intermediates not only enable the exchange reactions of the substituents on the phosphorus and tin elements but also provide the chances to prepare polymers having unique heterocyclopentadienyl transition metal complexes (Scheme 3.4).

3.4 Potential Applications of Element-Blocks π-Conjugated Materials

The unique optical and electronic features of the element-blocks π-conjugated polymers are attractive for many applications. As mentioned above, the fluorescent chemosensors can be constructed based on the specific interaction of elements attached to the π-conjugated polymers. Not only the phosphole-containing polymers but also arsole-, thiophene-, selenophene-, tellurophene, stannole-, and 1,4-dithio-1,3-butadiene-containing polymers exhibit stimuli-responsive luminescence and/or UV-vis absorption changes which may also be attractive for chemosensor applications.

On the basis of their low LUMO and high HOMO energy levels, the phosphole-containing polymers serve as both n- and p-type semiconductors, depending on the nature of the silicon substrate in the fabricated diode device (Fig. 3.4) [11].

Because the polymer reaction routes involving the post-element-transformation processes are suitable to carry out the parallel synthesis of a series of π-conjugated polymers, it is possible to perform the rational approach to obtain functional materials having desired features. For example, photoluminescent polymers with various colors can be designed by the use of a designed diyne (**1c**) [20].

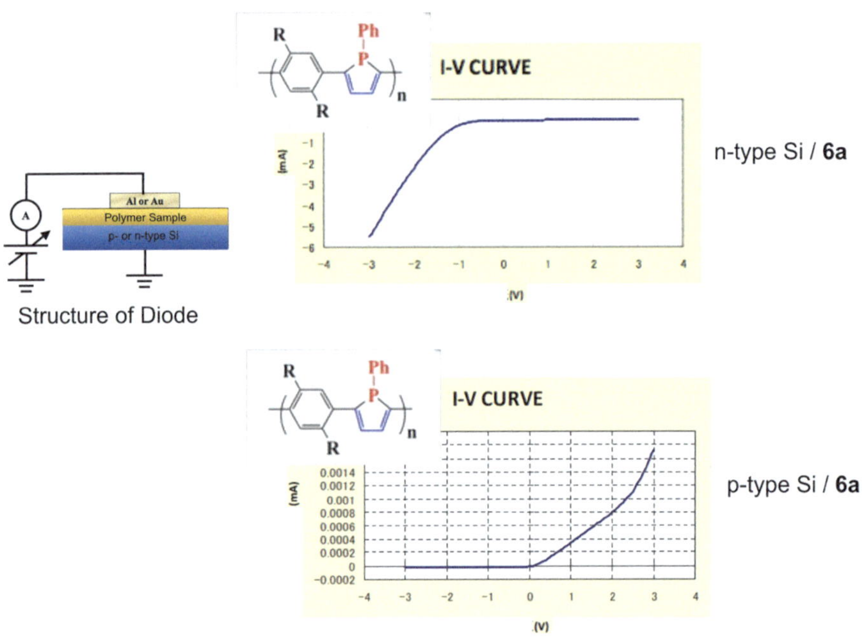

Fig. 3.4 Semiconducting properties of **6a**

Low-bandgap intramolecular charge transfer π-conjugated polymers with varied LUMO energy levels such as **6d** and its derivatives can also be designed by using a diyne having a high HOMO energy level (**1d**) through collaborative research with Ohshita and his coworkers.

3.5 Summary

By the progress of the synthetic methods for π-conjugated polymers possessing versatile element-blocks, unprecedented materials with unique optical and electronic features could be created which potentially exhibit excellent functions for optical and electronic devices and chemosensors. The great potential of the element-blocks to control the electronic state of π-electron systems of organic molecules and macromolecules could be convinced at least in part by the present studies. The author strongly wishes the further progress of the related synthetic researches as well as the applications of the new materials.

References

1. Grimsdale AC, Chan KL, Martin RE, Jokisz PG, Holmes AB (2009) Synthesis of light-emitting conjugated polymers for applications in electroluminescent devices. Chem Rev 109:897
2. Günes S, Neugebauer H, Sariciftci NS (2007) Conjugated polymer-based organic solar cells. Chem Rev 107:1324
3. McQuade DT, Pullen AE, Swager TM (2000) Conjugated polymer-based chemical sensors. Chem Rev 100:2537
4. Kularatne RS, Magurudeniya HD, Sista P, Biewer MC, Stefan MC (2013) Donor–acceptor semiconducting polymers for organic solar cells. J Polym Sci, Part A: Polym Chem 51:743
5. Chujo Y, Tanaka K (2015) New Polymeric Materials Based on Element-Blocks. *Bull. Chem. Soc. Jpn*. 88:633
6. Salzner U, Lagowski JB, Pickup PG, Poirier RA (1998) Comparison of geometries and electronic structures of polyacetylene, polyborole, polycyclopentadiene, polypyrrole, polyfuran, polysilole, polyphosphole, polythiophene, polyselenophene and polytellurophene. Synth Met 96:177
7. Tanaka K, Wang S, Yamabe T (1989) A study of the electroactivity decay of polythiophene film electrodes. Synth Met 30:57
8. Yamaguchi S, Jin RZ, Tamao K, Sato F (1998) A convenient one-pot synthesis of 1,4-Dihalobutadienes from alkynes via Titanacyclopentadienes and their transformation to a series of Silole derivatives. J Org Chem 63:10060
9. Hamada T, Suzuki D, Urabe H, Sato F (1999) J Am Chem Soc 121:7342
10. Atami K, Kino T, Zhou W-M, Nishiyama H, Tomita I (2009) Synthesis of π-conjugated polymers possessing 1,3-butadiene-1,4-diyl units by reactions of regioregular organometallic polymer having titanacyclopentadiene moieties in the main chain. Synth Met 159:949
11. Matsumura Y, Ueda M, Fukuda K, Fukui K, Takase I, Nishiyama H, Inagi S, Tomita I (2015) Synthesis of π-conjugated polymers containing Phosphole units in the Main chain by reaction of an organometallic polymer having a Titanacyclopentadiene unit. ACS Macro Lett 4:124

12. Baumgartner T, Réau R (2006) Organophosphorus π-conjugated materials. Chem Rev 106:4681
13. Mao SSH, Tilley TD, Versatile A (1997) Transition-metal mediated route to blue-light-emitting polymers with chemically Tunable luminescent properties. Macromolecules 30:5566
14. Morisaki Y, Aiki Y, Chujo Y (2003) Synthesis and properties of first well-defined Phosphole-containing π-conjugated polymers. Macromolecules 36:2594
15. Talancé VL, Hissler M, Zhang LZ, Kárpáti T, Nyulászi L, Quintero DC, Bäuerle P, Réau R (2008) Synthesis, electronic properties and electropolymerisation of EDOT-capped σ^3-phospholes. Chem Commun:2200
16. Ishidoshiro M, Matsumura Y, Imoto H, Irie Y, Kato T, Watase S, Matsukawa K, Inagi S, Tomita I, Naka K (2015) Practical synthesis and properties of 2,5-Diarylarsoles. Org Lett 17:4854
17. Matsumura Y, Ishidoshiro M, Irie Y, Imoto H, Naka K, Tanaka K, Inagi S, Tomita I (2016) Arsole-containing π-conjugated polymer by the post-element-transformation technique. Angew Chem Int Ed 55:15040
18. Nishiyama H, Tomita I (2010) Synthesis of π-conjugated polymer possessing Mercapto-substituted 1,3-Butadiene-1,4-diyl units by reaction of Regioregular organometallic polymer having Titanacyclopentadiene moieties in the Main chain. Macromol Chem Phys 20:2248
19. Nishiyama H, Kino T, Tomita I (2012) Transformation of Regioregular Organotitanium polymers into group 16 Heterole-containing π-conjugated materials. Macromol Rapid Commun 33:545
20. Matsumura Y, Fukuda K, Inagi S, Tomita I (2015) Parallel synthesis of Photoluminescent π-conjugated polymers by polymer reactions of an Organotitanium polymer with a Titanacyclopentadiene unit. Macromol Rapid Commun 36:660

Chapter 4
Borylated Polystyrenes as Versatile Functional Materials

Frieder Jäkle

Abstract The use of inorganic and organometallic compounds as building blocks of functional polymeric materials has developed into a flourishing research direction. Boron-containing polymers in particular are attracting tremendous attention for applications ranging from precursors to high-performance ceramics, flame retardants, electrolytes for lithium-ion batteries, supramolecular materials, optical and electronic materials, sensors and bioimaging agents, and polymer-supported reagents and catalysts. New synthetic approaches have been introduced in recent years that allow for controlled functionalization of polymeric materials with structurally diverse organoborane moieties, utilizing either polymer modification procedures or the direct polymerization of borane monomers. In this chapter we summarize our efforts toward the discovery of new functional materials derived from organoborane hybrid polymers, focusing on polystyrene-based systems.

Keywords Boron · Organoborane polymers · Polystyrene · Self-assembly

4.1 Introduction

Organometallic compounds are highly attractive "element blocks" in the development of new functional materials [1]. Those containing organoborane moieties are of key interest because of several characteristics that are unique to the element boron, including (i) the electron-deficient character of boron; (ii) the ability of organoboranes to (reversibly) form Lewis acid-base complexes; (iii) the tendency of organoboranes to display redshifted absorptions and emissions into the visible and NIR region due to π-bonding with or planarization of conjugated organic substituents; (iv) the capacity of boranes to be converted to boron oxide, boron carbide, and boron nitride ceramics; and (v) the large cross section of the ^{10}B isotope for neutrons.

F. Jäkle (✉)
Department of Chemistry, Rutgers University-Newark, Newark, NJ, USA
e-mail: fjaekle@rutgers.edu

Y. Chujo (ed.), *New Polymeric Materials Based on Element-Blocks*,
https://doi.org/10.1007/978-981-13-2889-3_4

Due to these unusual chemical and electronic characteristics, boron-containing polymers are promising candidates for a variety of applications, such as precursors for high-performance ceramics, flame retardants based on boron oxide char formation, electrolytes for lithium ion batteries, supramolecular and stimuli-responsive materials, optoelectronic materials, sensors for anions and toxic small molecules, materials for biomedical imaging and therapeutics, and polymer-supported reagents and catalysts for organic transformations [2–9].

The use of polyolefins as "scaffolds" for attachment of borane functional groups is attractive because chain growth polymerization methods can be applied in their synthesis [10]. Thus, "controlled/living" polymerization techniques provide access to polymers of well-defined molecular weight and diverse architectures, including random, gradient and block copolymers, brush polymers, and star-shaped polymers [11]. The self-assembly of these polymers, in turn, allows for incorporation of borane moieties into various nanostructured materials.

Two different approaches have been pursued: the modification of preformed organic or organometallic polymers with borane moieties and the direct polymerization of borane-functionalized monomers [10]. An advantage of polymer modification reactions is that a range of different functional materials can be readily accessed from a single precursor polymer. Most of the early polymer modification approaches involved lithiation-borylation of bromo-functionalized polymers or the hydroboration of vinyl-functionalized polymers [10]. These methods were plagued by limited selectivity and not so broadly applicable. In 2002, our group introduced a versatile new method for the functionalization of polystyrene with borane moieties. As illustrated in Fig. 4.1, selective attachment of electron-deficient dibromoboryl groups to polystyrene is accomplished by electrophilic aromatic substitution of trimethylsilyl groups with boron tribromide, whereas subsequent nucleophilic replacement of the Br substituents gives rise to diverse classes of functional borane polymers [12, 13]. This approach afforded well-defined soluble polymers of controlled molecular weight and degree of functionalization while also offering access to other polymer architectures such as a telechelic and block copolymers [14–16].

One of the drawbacks of this approach remains that the borylation with BBr_3 tends to be incompatible with more polar functional groups that may be present in the polymer. Thus, recent studies have increasingly focused on the development of functional boron monomers as versatile "element blocks" in the synthesis of borane polymers. In this case, a major challenge is that the catalysts or initiators have to be well compatible with the particular borane functional groups.

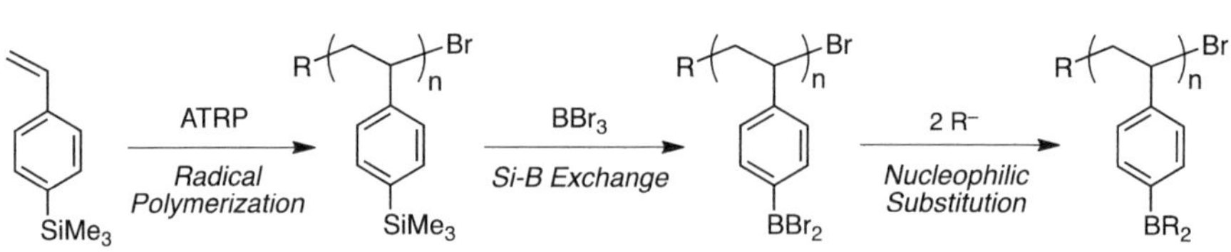

Fig. 4.1 Modification of polystyrene with borane functional groups by post-polymerization modification of silylated polystyrene

In this chapter, we discuss some of our efforts toward the development of new materials based on the borylation of styrenic polymers, using either polymer modification or direct polymerization approaches. We will introduce highly electron-deficient polymers that are promising as sensory and optoelectronic materials, polymers in which "CC" units are replaced with isoelectronic "BN" units, the attachment of boronic and borinic acid moieties and its implications on self-assembly and sensing applications, and the functionalization with anionic borate moieties that enable electrostatic or coordinative attachment of transition metal complexes.

4.2 Fluorescent Triarylborane Polymers as Anion Sensors

The utility of luminescent triorganoboranes as chemosensors has first been recognized by Yamaguchi and coworkers [17] and has since been extensively exploited for the detection of anions, toxic amines, and other Lewis basic substrates [5, 18]. In 2002, Chujo and coworkers reported that the luminescence of a boron-containing conjugated polymers, a poly(vinylene-phenylene-vinylene borane), is quenched upon addition of only small amounts of fluoride anions, suggesting an amplified detection mechanism [19]. Later studies by our group on conjugated oligomers and macrocycles suggested that this is a general phenomenon that likely results from the generation of a low-energy charge transfer state upon binding of the anion to one of the Lewis acidic borane moieties [20, 21].

For the installation of highly Lewis acidic borane moieties onto polyolefins, polymer modification reactions are usually used to avoid incompatibilities between (reactive) monomers and catalysts. Dibromoboryl-functionalized polymers are prepared from silicon precursors as discussed above, and, in a subsequent functionalization step, organotin or organocopper reagents are effectively used to diversely decorate the borane moieties. For instance, fine-tuning of the absorption and emission color of polymer **1** was achieved by variation of the pendent aryl groups on boron (Fig. 4.2a) [22]. Binding studies indicated that the Lewis acidic tricoordinate borane moieties can be exploited for the selective recognition of small

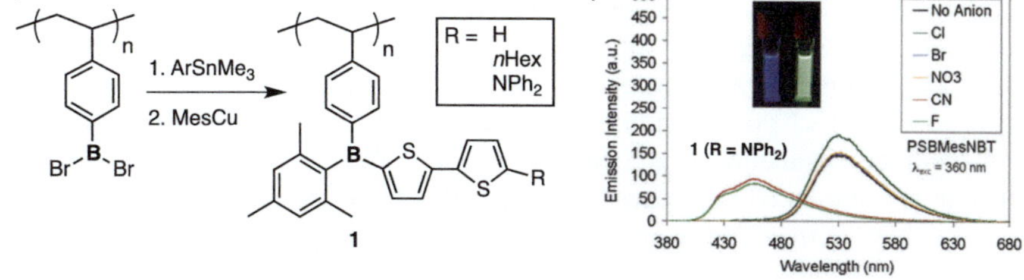

Fig. 4.2 (**a**) Conversion of borylated polystyrene to luminescent polymers; (**b**) illustration of changes in the emission of polymer **1** (R = NPh_2) upon addition of F^- or CN^- anions. (Adapted with permission from Ref. [22]. Copyright © 2006 American Chemical Society)

anions, based on changes in the absorption and emission characteristics. This phenomenon is illustrated for the binding of fluoride and cyanide, but not other larger anions, to **1** (R = NPh_2) in Fig. 4.2b. Further, a dramatic sensitivity enhancement for **1** (R = n-hexyl) in comparison to its monomeric model system suggested that highly functionalized polyolefins offer an interesting alternative to conjugated polymers as efficient sensory materials.

Systems with a fluorene or carbazole chromophore in combination with bulkier tri-*iso*-propylphenyl (Tip) substituents on boron were also prepared, and, given their enhanced stability, these strongly emissive polymers could be of interest in display or imaging applications [23]. Moreover, attachment of ferrocene moieties to boron led to deep red-colored triarylborane polymers that underwent reversible oxidation at the pendent ferrocenyl moieties [24]. Anion complexation studies suggested that these polymers bind fluoride less effectively than the corresponding molecular systems, and this phenomenon was attributed to neighboring group effects.

In a different approach, Do and Lee and coworkers prepared borane Lewis acid-functionalized polymers via Ziegler-Natta-type (co)polymerization of preformed vinyl-functionalized dimesitylborane monomers with ethylene or norbornene [25, 26]. They showed that the resulting polymers emit in the ultraviolet (UV) region with a maximum emission at 382 nm, which is quenched upon addition of fluoride anions. More recently, we reported the synthesis of a luminescent triarylborane polymer (**2**) by controlled free radical polymerization, using a reversible addition-fragmentation chain transfer (RAFT) protocol (Fig. 4.3a) [27]. The RAFT polymerization protocol also lends itself to the preparation of boron-containing block copolymers (**3**). GPC analyses vs PS standards gave molecular weights of M_n = 9630 Da ($Đ$ = 1.09) for homopolymer **2** and M_n = 57,460 Da ($Đ$ = 1.20) for block copolymer **3**, demonstrating excellent control.

Similar to the previously discussed tricoordinate boron polymers, **2** and **3** showed excellent sensing properties toward fluoride ions. Furthermore, the amphiphilic block copolymer **3** gave rise to interesting anion-responsive self-assembly behavior. Addition of a THF solution of **3** to a large excess of DMF led to aggregates with an average hydrodynamic diameter of $D_h = 93 \pm 24$ according to dynamic light scattering (DLS) analysis. Interestingly, an addition of only ca. 0.2 equivs of TBAF resulted in a clear solution, suggesting that (partial) conversion of neutral borane to charged fluoroborate moieties led to micelle dissociation into single chains (Fig. 4.3b, c). To achieve anion binding in aqueous solution, a random block copolymer (**4**) was prepared with cationic pyridinium moieties interspersed in the functional boron polymer block. This polymer responded to fluoride anions at a concentration of <1 ppm even in DMF-water mixtures, effectively overcoming the high hydration enthalpy of the fluoride anions. The greatly enhanced binding strength was attributed to electrostatic effects that favor the fluoroborate formation in the vicinity of the pyridinium moieties.

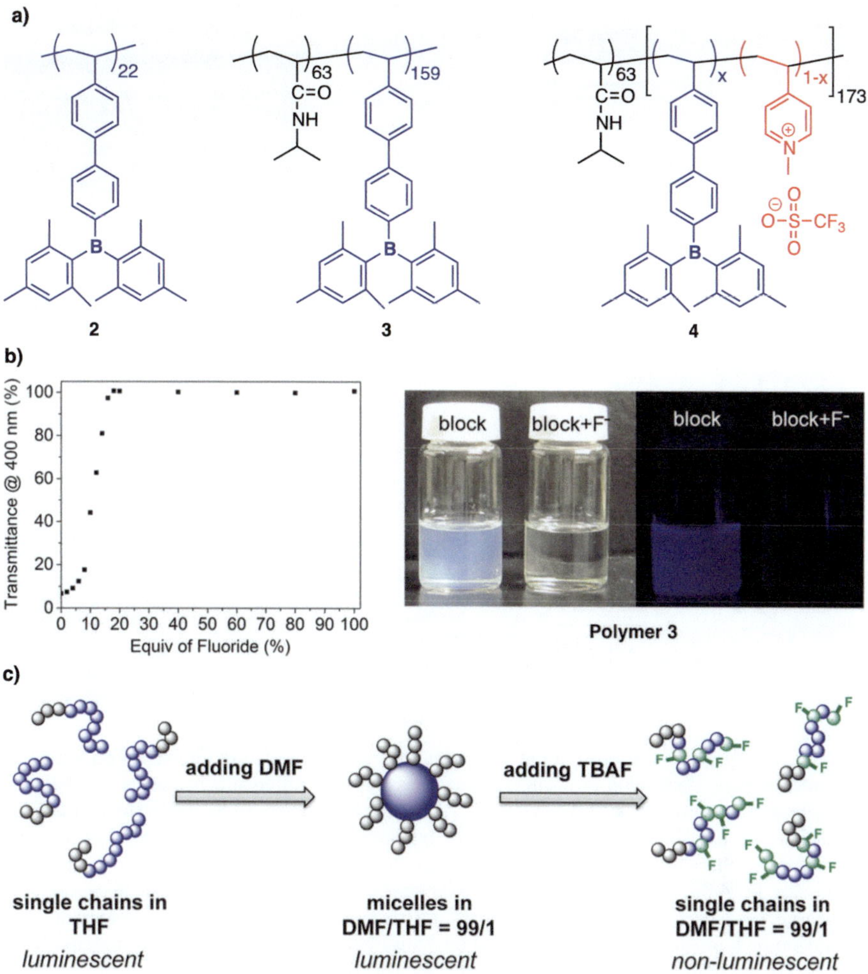

Fig. 4.3 (**a**) Structures of triarylborane-functionalized homo, block, and random block copolymers synthesized by RAFT; (**b**) illustration of changes in the transparency and emission of a solution of block copolymer **3** in DMF upon addition of F^- anions; (**c**) schematic illustration of the proposed mechanism. (Adapted with permission from Ref. [27]. Copyright © 2013 American Chemical Society)

4.3 Fluorescent Polymers Containing Boron Chelate Complexes

Not only conjugated triorganoboranes but also organoboron chelate complexes are attractive as fluorescent materials for imaging and device applications. In this case, the boron center is tetracoordinate, which frequently results in robust materials that

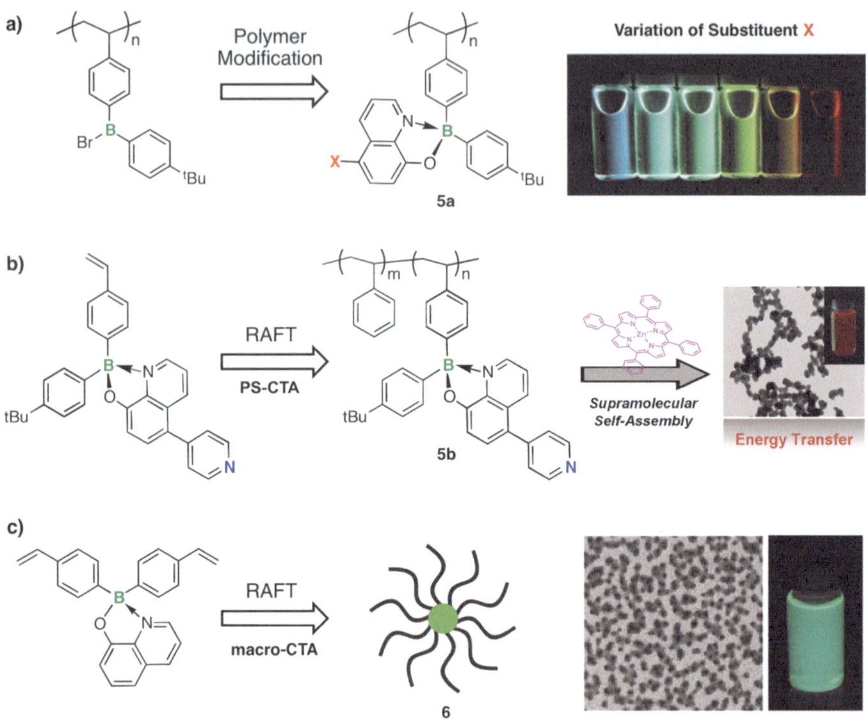

Fig. 4.4 (**a**) Synthesis of organoboron 8-hydroxyquinolate-functionalized polymers **5a** via polymer modification; illustration of changes in the emission with different substituents X. (Adapted with permission from Ref. [31]. Copyright © 2006 American Chemical Society); (**b**) synthesis of organoboron 8-hydroxyquinolate-functionalized polymers **5b** via RAFT polymerization; illustration of red luminescent particles upon coordination to zinc tetraphenylporphyrin. (Adapted with permission from Ref. [35]. Copyright © 2013 American Chemical Society); (**c**) synthesis of star-shaped copolymers **6** from a bifunctional distyryl monomer; TEM image and photograph of a dispersion of **6** in water (macro-CTA based on PS-b-PNIPAM). (Adapted with permission from Ref. [33]. Copyright © 2012 Royal Society of Chemistry)

are highly stable toward air and moisture. While boron dipyrromethenes (BODIPYs) are the most well known, many different boron chelates have been introduced in recent years [28, 29]. Using a post-polymerization modification approach similar to that described in Fig. 4.1, we prepared the first luminescent polymers containing 8-hydroxyquinolate-complexed borane chromophores [30, 31]. High-molecular-weight polymers were obtained with essentially quantitative chromophore functionalization. While the parent polymer **5a** (X = H) emits green light with quantum yields in the range of ca. 20–30%, the emission could be fine-tuned over almost the entire visible spectrum by simply varying the electronic structure of the substituent (X) at the 5-position of the ligand (Fig. 4.4a).

Again, the direct polymerization of luminescent 8-hydroxyquinolate-based organoboron monomers using the RAFT method offered advantages when pursuing more complex polymer architectures with diverse functionality [32–35]. For example, a pyridyl-substituted boron 8-hydroxyquinolate monomer was utilized to generate block copolymers **5b** that not only self-assemble in selective solvents but also bind to metal ions (Fig. 4.4b) [35]. Complexation with Zn tetraphenylporphyrin resulted in assemblies where efficient energy transfer from the boron quinolate chromophore to the porphyrin occurs. In addition, when using a PEO-based macro-CTA, amphiphilic block copolymers were obtained that formed stable luminescent micellar solutions in water, enabling potential applications as nano-sized fluorophores in biological environments [32]. Luminescent star-shaped copolymers **6** were obtained when using a bis-styryl-functionalized monomer as a cross-linking agent (Fig. 4.4c) [33].

We note that direct polymerization methods have also been widely used to functionalize polyolefins with BODIPY dyes, especially acrylate-type random and block copolymers [36–42]. For example, Chujo and coworkers used RAFT polymerization to prepare BODIPY random copolymers with dimethylaminoethyl methacrylate (DMAEMA). When an aqueous solution of the polymer was heated above the lower critical solution temperature (LCST), the absorption changed, and the fluorescence intensity increased dramatically. This process is reversible, allowing for fluorescence switching of the polymer solution. Gilroy and coworkers recently introduced a new class of boron formazanate dyes and embedded these chromophores into polymers by ring-opening metathesis polymerization (ROMP) [43].

4.4 "BN" for "CC" Substitution in Polystyrenes

The concept of replacing "CC" with "BN" units in aromatic systems presents yet another approach to boron heterocycles with interesting electronic structures [44–46]. In the simplest case, replacement of two carbons in benzene with a boron and a nitrogen results in 1,2-, 1,3-, or 1,4-dihydroazaborines that are isosteric and isoelectronic to benzene [47–49]. Significant aromatic character has been deduced for these BN heterocycles. This concept has also been extended to larger polycyclic aromatic hydrocarbons [50–53]. Much of this research is motivated by the question, how does such an isoelectronic replacement affect the electronic structure, photophysical properties, and the reactivity of these molecules? In the case of polymers, additional questions arise in respect to possible effects on material properties.

Although aromatic groups are abundant in many polymers, both conjugated materials and polyolefins, the "BN" for "CC" replacement remains relatively little explored [9]. In collaboration with the Liu group, we recently introduced the first example of a conjugated polymer that contains the parent 1,2-azaborine, a "BN" analog of polyphenylene [54], while the Pei group [55] reported on polymers derived from thiophene-fused azaborine systems and their use in field-effect transistors. A few early studies on the attachment of borazine heterocycles to polyolefins had been motivated primarily by their potential use as precursors to boron nitride and boron carbonitride ceramics. Sneddon and coworkers demonstrated the free radical

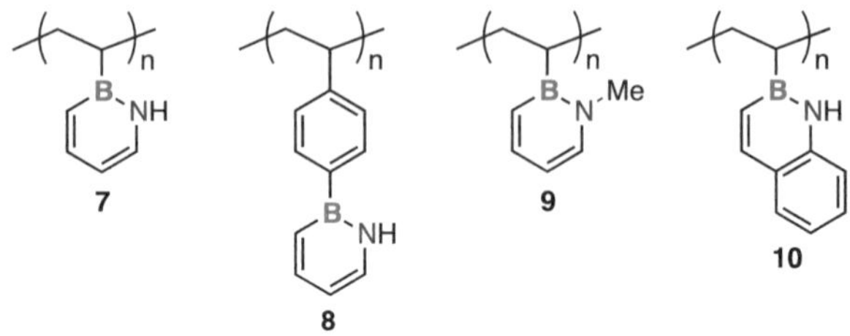

Fig. 4.5 Structures of BN-polystyrene, BN-poly(vinylbiphenyl), and BN-poly(vinylnaphthalene) derivatives obtained by free radical polymerization of azaborine monomers

polymerization of vinylborazines, and Allen and coworkers reported on the synthesis of borazine-substituted polystyrene [56, 57].

In collaboration with the Liu group, we recently succeeded in the free radical polymerization of B-vinyl- and B-styryl-functionalized azaborine monomers to give polymers **7** and **8** (Fig. 4.5) [58]. Higher molecular weights were achieved for **8** compared to **7**. This difference was tentatively attributed to the direct attachment of the vinyl group to boron in **7**, which may destabilize the propagating radical in the "benzylic" position. We found that these polymers exhibit enhanced solubility in polar solvents in comparison to the all-carbon analogs, which we attributed to the increased polarity of the side group and the presence of NH moieties capable of hydrogen bonding. Related polymers containing a methyl group on nitrogen (**9**) were reported by Staubitz and coworkers, and those polymers showed high molecular weights but low solubility in polar media (Fig. 4.5) [59]. Klausen and coworkers prepared corresponding polymers (**10**) derived from a "BN"-naphthalene building block [60]. Going forward, it will be interesting to see whether the "BN" for "CC" replacement will enable the use of polystyrene derivatives in application fields not previously accessible for the all-carbon polymer systems.

4.5 Boronic and Borinic Acid-Functionalized Polystyrene

The attachment of boronic acid functionalities to polystyrene has been accomplished many decades ago, using standard free radical polymerization of styrene boronic acid [61, 62]. Boronic acid polymers have since been studied extensively due to their ability to selectively bind sugars and other 1,2- or 1,3-diols. Thus, the development of boronic acid-functionalized polymers has implications in the biomedical field, and the interested reader is referred to a recent review on the subject by Sumerlin and coworkers [8].

When introducing boronic acid groups to block copolymers, the effects of pH and diol binding on the self-assembly behavior offer additional opportunities. We first demonstrated the formation of boronic acid-functionalized styrene block copolymers by simple hydrolysis of a BBr_2-functionalized polymer in 2005 [14] and subsequently explored the self-assembly of polymer **11** in aqueous and mixed solvents in dependence on the pH (Fig. 4.6a) [63]. We found that very regular micelles are generated at high pH in water. At neutral pH more complex morphologies were

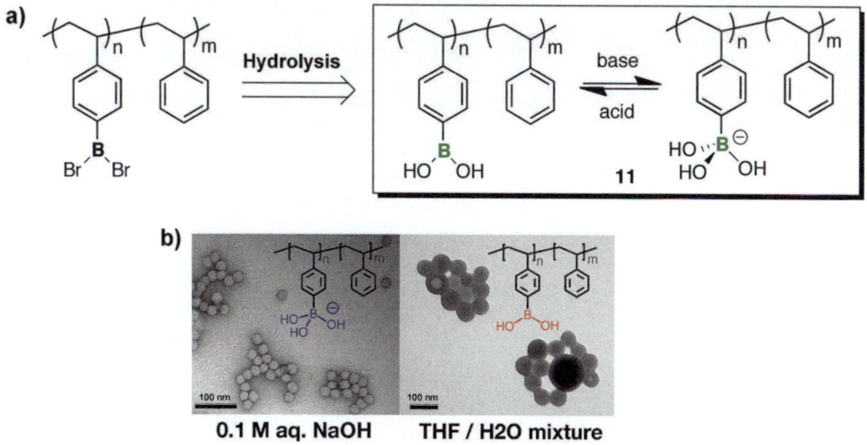

Fig. 4.6 (**a**) Synthesis of boronic acid-functionalized polystyrene block copolymer **11**; (**b**) TEM micrographs of micelles obtained from aqueous solution at high pH and from THF/water mixture at neutral pH. (Adapted with permission from Ref. [63]. Copyright © 2010 Wiley VCH)

detected. As illustrated in Fig. 4.6b, assembly in THF/water mixtures gave rise to vesicles, whereas assembly in an acetone/water mixture generated worm-like features. In related work, Kim and van Hest used block copolymers of polystyrene boronic acid with poly(ethylene glycol) (PEG-*b*-PSBA) and unfunctionalized polystyrene with poly(ethylene glycol) (PEG-*b*-PS) to generate vesicles that serve as responsive drug delivery vehicles [64]. Many examples of other polyolefin scaffolds, most notably polyacrylamides, have been utilized to prepare stimuli-responsive boronic acid polymers in recent years [65, 66].

Borinic acid polymers represent another emerging class of polymers that differ from the previously discussed boronic acid polymers in that only one B-OH functionality is present at boron. This tends to result in higher Lewis acidity of the boron center, while the additional R group offers opportunities of tuning of the steric and electronic structure [67]. In 2014, we introduced a first example of a borinic acid polymer (PBA, **13**), which was obtained by RAFT polymerization of the styryl-type monomer **12** (Fig. 4.7a). This polymer shows interesting thermo-responsive properties in that the upper critical solution temperature (UCST) in polar solvents increases linearly as the amount of added water increases, making it continuously tunable over a wide temperature range from 20 to 100 °C. The changes in UCST in different polar solvents in dependence on the percentage of water present is illustrated in Fig. 4.7b [68, 69].

We also prepared corresponding block copolymers with polystyrene and poly (N-isopropylacrylamide) (PNIPAM) [70]. Taking advantage of the hydrogen bond donating ability of the borinic acid groups, supramolecular aggregates of PNIPAM-*b*-PBA were generated by complexation with P4VP. Furthermore, in collaboration with the Wan group, we explored the processing of block copolymer PS-*b*-PBA (**14**) into porous films with tunable pore size by drop-casting from THF in the presence of

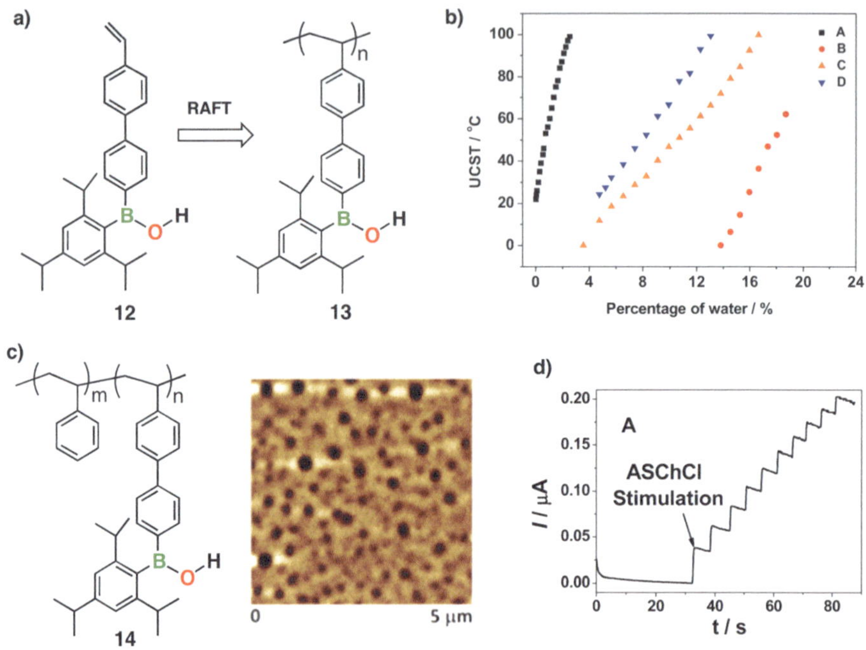

Fig. 4.7 (**a**) Synthesis of borinic acid polymer **13** by RAFT polymerization of **12**; (**b**) dependence of upper critical solution temperature (UCST) of **13** on the percentage of water present in polar organic solvents (A, DMSO; B, THF; C, DMF; D, PNIPAM-*b*-PBA in DMSO). (Adapted with permission from Ref. [69]. Copyright © 2015 Royal Society of Chemistry); (**c**) AFM image of a porous thin film of block copolymer **14**; (**d**) illustration of current response to the exposure of a porous film of **14** on an electrode that was treated with acetylcholine esterase to acetylthiocholine chloride (ASChCl). (Adapted with permission from Ref. [70]. Copyright © 2015 Royal Society of Chemistry)

trace amounts of water (Fig. 4.7c). A porous film that was deposited onto an electrode was then utilized to immobilize the enzyme acetylcholine esterase. The resulting biosensor allowed us to electrochemically detect acetylthiocholine chloride (Fig. 4.7d).

4.6 Weakly and Strongly Coordinating Organoborate Polymers

Organoborates are widely used as weakly coordinating anions to stabilize main group and transition metal complexes; they also serve important roles as electrolytes in lithium ion batteries, electrochemical redox processes, and as membrane materials [71–73]. Capitalizing on our post-polymerization modification approach to boron-functionalized polystyrenes, we devised a general synthetic protocol for the preparation of organoborate polymers where dibromoborylated polystyrene homo or

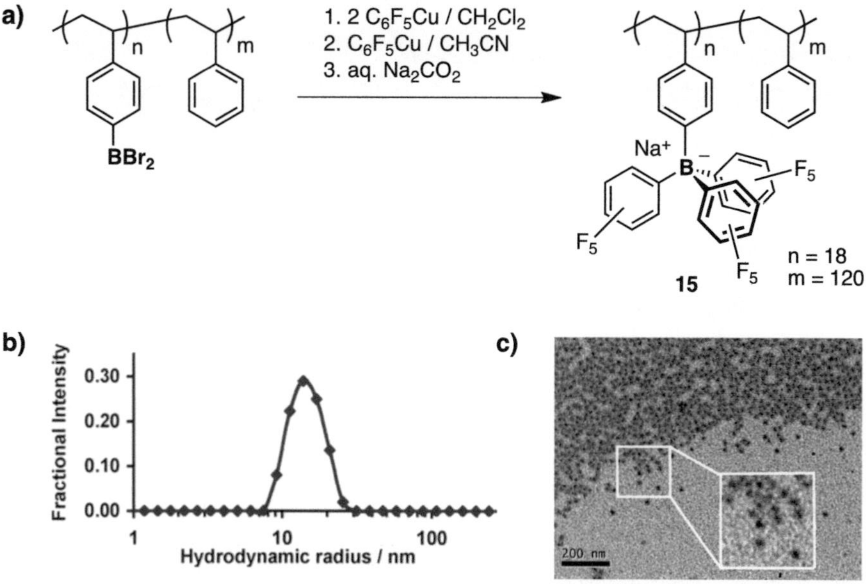

Fig. 4.8 (**a**) Synthesis of organoborate-functionalized block copolymer **15**; (**b**) size distribution histogram of a solution of the sodium salt of PSBPf$_3$-*b*-PS in toluene; (**c**) TEM image of block copolymer micelles of **15** from toluene, after loading with $[Rh(COD)(dppb)]^+$. (Adapted with permission from Ref. [74]. Copyright © 2010 American Chemical Society)

block copolymers are treated with an organometallic compound such as an aryl Grignard or an aryl copper reagent [74]. The process is illustrated in Fig. 4.8a for the synthesis of block copolymer **15** containing weakly coordinating fluorinated aryl borate moieties in the side chain. Multinuclear NMR (^{1}H, ^{13}C, ^{11}B, ^{19}F) and GPC analyses demonstrated high selectivity of the polymer modification process. The resulting ionic homopolymers were water-soluble, and the amphiphilic block copolymers assembled into micelles in selective solvents. For example, polymer **15** formed micelles with an average apparent hydrodynamic radius of $R_{\mathrm{h,app}} = 15$ nm in toluene as a selective solvent for the PS block (Fig. 4.8b). These weakly coordinating polyions are promising as supports for reactive cationic metal complexes. The loading of micelles of **15** in toluene with the transition metal catalyst, [Rh(COD) (dppb)]$^+$ (COD = 1,5-cyclooctadiene, dppb = bis(diphenylphosphino)butane), was demonstrated and verified by TEM imaging. The micrograph in Fig. 4.8c shows particles with a dark core region surrounded by a lighter corona. The higher contrast of the core region is attributed to the presence of the Rh heavy atoms.

Charge-reverse, cationic boronium-functionalized polymers were also prepared by reaction of BBr_2-functionalized polystyrene with 2,2′-biypridine [75, 76]. The boronium functionalities are isoelectronic to ammonium groups and as such of potential interest for a range of applications including fuel cell membranes, nanofiltration membranes, smart surfaces with switchable wettability, and

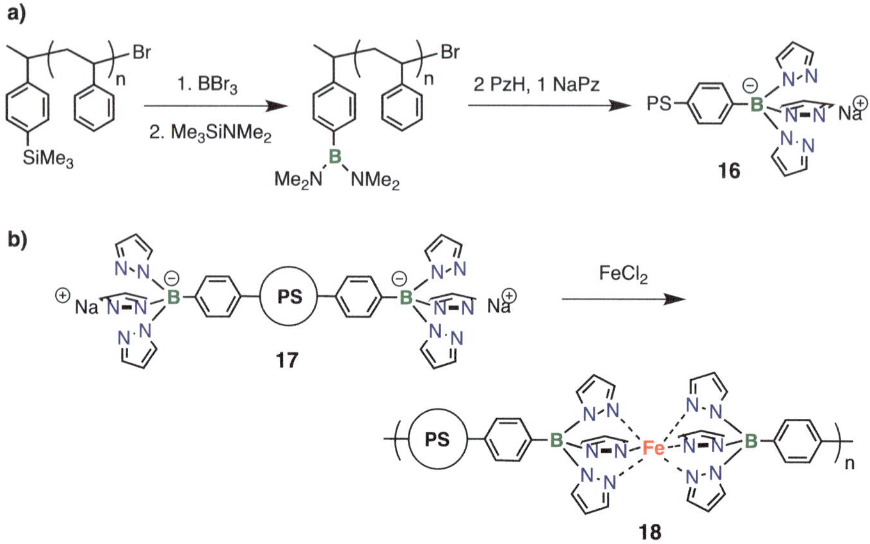

Fig. 4.9 (**a**) Synthesis of tris(pyrazolyl)borate (Tp) end-functionalized polystyrene; (**b**) ditelechelic polymer **17** and its complexation with $FeCl_2$ to give **18**

antimicrobial surfaces. Materials based on boronium moieties are also investigated as novel ionic liquids and multistep redox-active materials [77, 78].

When introducing Lewis basic sites on the pendent organoborate moieties, another interesting class of functional polymers that act as "poly-ligands" arises. We have developed routes to polymers containing "scorpionate"-type tridentate tris (pyrazolyl)borate (Tp) or tris(2-pyridyl)borate (Tpyb) chelate ligands that enable strong metal ion complexation [16, 79–82]. For instance, post-polymerization modification of silyl-terminated polystyrene by reaction with BBr_3 and subsequent treatment with an excess of Me_3SiNMe_2 gave rise to a bis(dimethylamino)boryl-modified polymer (Fig. 4.9a) [79]. The latter was readily converted to the monotelechelic Tp-functionalized polymer **16** by reaction with 2 equivalents of pyrazole and 1 equivalent of sodium pyrazolide. The ditelechelic polymer **17** (Fig. 4.9b) was prepared through a similar procedure. Complexation of monotelechelic **16** with $FeCl_2$ led to a metal-bridged "dimer," while complexation of the ditelechelic polymer **17** resulted in a metal-bridged "polymer of polymers" (**18**). The metal complexation and formation of an extended polymeric system were verified by GPC analysis with a photodiode array detector to detect the presence of metal-to-ligand charge transfer (MLCT) bands.

Using a similar approach, we also prepared side-chain Tp-functionalized polymer **19** with varying degrees of functional group loading (Fig. 4.10a) [16]. Metal complexation with $[CpRu(CH_3CN)_3]PF_6$ resulted in decoration of the polymers with "TpRuCp" metal complexes (**20**). A corresponding molecular model compound was

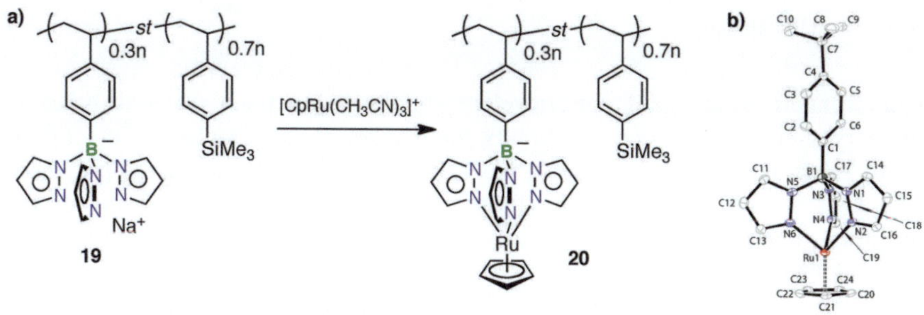

Fig. 4.10 (**a**) Synthesis of side-chain Tp-functionalized polymer **19** and its complexation with CpRu moieties to give metallopolymer **20**; (**b**) X-ray crystal structure of monomeric model complex. (Adapted with permission from Ref. [16]. Copyright © 2008 American Chemical Society)

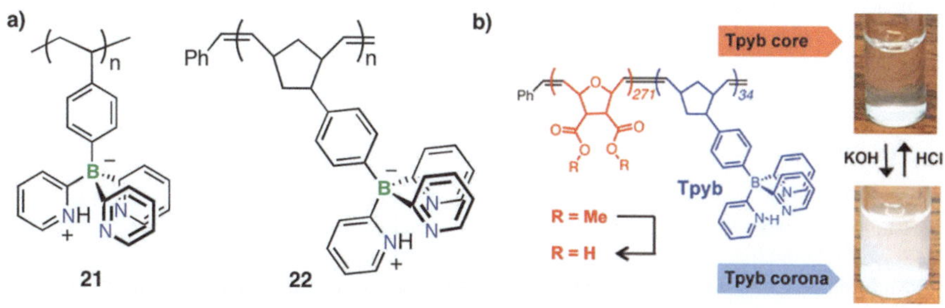

Fig. 4.11 (**a**) Structures of Tpyb-functionalized polystyrene (**21**) and polynorbornene (**22**); (**b**) illustration of the pH-dependent micellization of a Tpyb-functionalized amphiphilic block copolymer. (Adapted with permission from Ref. [81]. Copyright © 2015 American Chemical Society)

also synthesized, and its X-ray crystal structure is displayed in Fig. 4.10b. Analysis of the Ru polymer and the model compound by cyclic voltammetry revealed similar redox properties, indicating reversible oxidation of the metal complexes in the side chain at a similar potential to that of the monomer.

In 2012 we introduced a new class of powerful and highly robust "scorpionate"-- type ligands, the tris(2-pyridylborate)s [82–84]. Attachment of vinyl and norbornyl moieties gave rise to polymerizable monomers. Utilizing these monomers, polymer **21** [80] was obtained by nitroxide-mediated free radical polymerization (NMP) at high temperatures, whereas polymer **22** [81] could be generated at room temperature by ring-opening metathesis polymerization (ROMP) with Grubbs' third-generation catalyst (Fig. 4.11a). A related amphiphilic block copolymer was obtained by sequential polymerization of an oxanorbornene and the Tpyb-norbornene monomer (Fig. 4.11b). Deprotection of the ester functionalities in the oxanorbornene gave rise to a double pH-responsive block copolymer where the carboxylic acid groups are deprotonated at high pH, whereas the Tpyb moieties are protonated at low pH. Reversible metal ion complexation of polymer **22** was also demonstrated.

4.7 Conclusions

Our recent efforts in regard to the development of new functional materials based on the borylation of polystyrene are briefly summarized in this chapter. Efficient post-polymerization modification approaches and new methods for the direct polymerization of borane monomers are illustrated. Different application fields of the resulting polymers are also discussed. We hope that this chapter will provide a sense of the breadth of new functional materials that can be accessed using relatively simple boron-based element blocks while also inspiring new research into boron-functionalized polymeric materials.

Acknowledgments This material is based upon work supported by the National Science Foundation under Grants CHE-1362460 and CHE-1609043. F. J. thanks all his current and former group members who contributed to our research on boron-containing polymers discussed in here.

References

1. Chujo Y, Tanaka K (2015) New polymeric materials based on element-blocks. Bull Chem Soc Jap 88:633–643
2. Matsumi N, Chujo Y (2008) π-conjugated organoboron polymers via the vacant p-orbital of the boron atom. Polym J 40:77–89
3. Tanaka K, Chujo Y (2012) Advanced luminescent materials based on organoboron polymers. Macromol Rapid Commun 33:1235–1255
4. Jäkle F (2006) Lewis acidic organoboron polymers. Coord Chem Rev 250:1107–1121
5. Jäkle F (2010) Advances in the synthesis of organoborane polymers for optical, electronic and sensory applications. Chem Rev 110:3985–4022
6. Abd-El-Aziz AS, Carraher CE Jr, Pittman CU Jr, Zeldin M (eds) (2007) Macromolecules containing metal and metal-like elements, Boron-Containing Polymers, vol 8. Wiley, Hoboken
7. Dash BP, Satapathy R, Maguire JA, Hosmane NS (2011) Polyhedral boron clusters in materials science. New J Chem 35:1955–1972
8. Brooks WLA, Sumerlin BS (2016) Synthesis and applications of boronic acid-containing polymers: from materials to medicine. Chem Rev 116:1375–1397
9. Helten H (2016) B=N units as part of extended π-conjugated oligomers and polymers. Chem Eur J 22:12972–12982
10. Jäkle F (2005) Borylated polyolefins and their application. J Inorg Organomet Polym Mater 15:293–307
11. Cheng F, Jäkle F (2011) Boron-containing polymers as versatile building blocks for functional nanostructured materials. Polym Chem 2:2122–2132
12. Qin Y, Cheng G, Sundararaman A, Jäkle F (2002) Well-defined boron-containing polymeric Lewis acids. J Am Chem Soc 124:12672–12673
13. Qin Y, Cheng G, Achara O, Parab K, Jäkle F (2004a) A new route to organoboron polymers via highly selective polymer modification reactions. Macromolecules 37:7123–7131
14. Qin Y, Sukul V, Pagakos D, Cui C, Jäkle F (2005) Preparation of organoboron block copolymers via ATRP of silicon and boron-functionalized monomers. Macromolecules 38:8987–8990
15. Qin Y, Cui C, Jäkle F (2007) Silylated initiators for the efficient preparation of borane-end-functionalized polymers via ATRP. Macromolecules 40:1413–1420

16. Qin Y, Cui C, Jäkle F (2008) Tris(1-pyrazolyl)borate (scorpionate) functionalized polymers as scaffolds for metallopolymers. Macromolecules 41:2972–2974
17. Yamaguchi S, Akiyama S, Tamao K (2001) Colorimetric fluoride sensing by boron-containing π -electron systems. J Am Chem Soc 123:11372–11375
18. Wade CR, Broomsgrove AEJ, Aldridge S, Gabbai FP (2010) Fluoride ion complexation and sensing using organoboron compounds. Chem Rev 110:3958–3984
19. Miyata M, Chujo Y (2002) π-conjugated organoboron polymer as an anion sensor. Polym J 34:967–969
20. Chen P, Jäkle F (2011) Highly luminescent, electron-deficient bora-cyclophanes. J Am Chem Soc 133:20142–20145
21. Chen PK, Lalancette RA, Jäkle F (2012) π-expanded borazine: an ambipolar conjugated B-π -N macrocycle. Angew Chem Int Ed 51:7994–7998
22. Parab K, Venkatasubbaiah K, Jäkle F (2006) Luminescent triarylborane-functionalized polystyrene: synthesis, photophysical characterization, and anion-binding studies. J Am Chem Soc 128:12879–12885
23. Parab K, Doshi A, Cheng F, Jäkle F (2011) Synthesis and characterization of luminescent polystyrene derivatives with sterically protected fluorenyl- and carbazolylborane moieties. Macromolecules 44:5961–5967
24. Parab K, Jäkle F (2009) Synthesis, characterization, and anion binding of redox-active triarylborane polymers. Macromolecules 42:4002–4007
25. Park MH, Kim T, Huh JO, Do Y, Lee MH (2011) Luminescent polyethylene with side-chain triarylboranes: synthesis and fluoride sensing properties. Polymer 52:1510–1514
26. Sung WY, Park MH, Park JH, Eo M, Yu M-S, Do Y, Lee MH (2012) Triarylborane-functionalized polynorbornenes: direct polymerization and signal amplification in fluoride sensing. Polymer 53:1857–1863
27. Cheng F, Bonder EM, Jäkle F (2013a) Electron-deficient triarylborane block copolymers: synthesis by controlled free radical polymerization and application in the detection of fluoride ions. J Am Chem Soc 135:17286–17289
28. Rao YL, Amarne H, Wang SN (2012) Photochromic four-coordinate N,C-chelate boron compounds. Coord Chem Rev 256:759–770
29. Ulrich G, Ziessel R, Harriman A (2008) The chemistry of fluorescent bodipy dyes: versatility unsurpassed. Angew Chem Int Ed 47:1184–1201
30. Qin Y, Pagba C, Piotrowiak P, Jäkle F (2004b) Luminescent organoboron quinolate polymers. J Am Chem Soc 126:7015–7018
31. Qin Y, Kiburu I, Shah S, Jäkle F (2006) Synthesis and characterization of organoboron quinolate polymers with tunable luminescence properties. Macromolecules 39:9041–9048
32. Cheng F, Jäkle F (2010) RAFT polymerization of luminescent boron quinolate monomers. Chem Commun 46:3717–3719
33. Cheng F, Bonder EM, Doshi A, Jäkle F (2012a) Organoboron star polymers via arm-first RAFT polymerization: synthesis, luminescent behavior, and aqueous self-assembly. Polym Chem 3:596–600
34. Cheng F, Bonder EM, Jäkle F (2012b) Luminescent boron quinolate block copolymers via RAFT polymerization. Macromolecules 45:3078–3085
35. Cheng F, Bonder EM, Salem S, Jäkle F (2013b) Pyridine-functionalized luminescent organoboron quinolate block copolymers as versatile building blocks for assembled nanostructures. Macromolecules 46:2905–2915
36. Amat-Guerri F, Liras M, Carrascoso ML, Sastre R (2003) Methacrylate-tethered analogs of the laser dye PM567 - synthesis, copolymerization with methyl methacrylate and photostability of the copolymers. Photochem Photobiol 77:577–584
37. López Arbeloa F, Bañuelos Prieto J, López Arbeloa I, Costela A, García-Moreno I, Gómez C, Amat-Guerri F, Liras M, Sastre R (2003) Photophysical and lasing properties of new analogs of the boron-dipyrromethene laser dye pyrromethene 567 incorporated into or covalently bounded to solid matrices of poly(methyl methacrylate). Photochem Photobiol 78:30–36

38. Costela A, García-Moreno I, Gómez C, Amat-Guerri F, Liras M, Sastre R (2003) Efficient and highly photostable solid-state dye lasers based on modified dipyrromethene•BF_2 complexes incorporated into solid matrices of poly(methyl methacrylate). Appl Phys B Lasers Opt 76:365–369
39. Nagai A, Miyake J, Kokado K, Nagata Y, Chujo Y (2008) Highly luminescent bodipy-based organoboron polymer exhibiting supramolecular self-assemble structure. J Am Chem Soc 130:15276–15278
40. Nagai A, Kokado K, Miyake J, Chujo Y (2009) Highly luminescent nanoparticles: self-assembly of well-defined block copolymers by π –π stacked bodipy dyes as only a driving force. Macromolecules 42:5446–5452
41. Nagai A, Kokado K, Miyake J, Chujo Y (2010) Thermoresponsive fluorescent water-soluble copolymers containing bodipy dye: inhibition of Haggregation of the bodipy units in their copolymers by LCST. J Polym Sci A Polym Chem 48:627–634
42. Paris R, Quijada-Garrido I, Garcia O, Liras M (2011) Bodipy-conjugated thermo-sensitive fluorescent polymers based on 2-(2-methoxyethoxy)ethyl methacrylate. Macromolecules 44:80–86
43. Novoa S, Paquette JA, Barbon SM, Maar RR, Gilroy JB (2016) Side-chain boron difluoride formazanate polymers via ring-opening metathesis polymerization. J Mater Chem C 4:3987–3994
44. Liu ZQ, Marder TB (2008) B-N versus C-C: how similar are they? Angew Chem Int Ed 47:242–244
45. Bosdet MJD, Piers WE (2009) B-N as a C-C substitute in aromatic systems. Can J Chem 87:8–29
46. Campbell PG, Marwitz AJV, Liu SY (2012) Recent advances in azaborine chemistry. Angew Chem Int Ed 51:6074–6092
47. Abbey ER, Lamm AN, Baggett AW, Zakharov LN, Liu SY (2013) Protecting group-free synthesis of 1,2-azaborines: a simple approach to the construction of BN-benzenoids. J Am Chem Soc 135:12908–12913
48. Xu SM, Zakharov LN, Liu SY (2011) A 1,3-dihydro-1,3-azaborine debuts. J Am Chem Soc 133:20152–20155
49. Schafer M, Beattie NA, Geetharani K, Schafer J, Ewing WC, Krahfuss M, Horl C, Dewhurst RD, Macgregor SA, Lambert C, Braunschweig H (2016) Synthesis of functionalized 1,4-azaborinines by the cyclization of di-tert-butyliminoborane and alkynes. J Am Chem Soc 138:8212–8220
50. Bosdet MJD, Piers WE, Sorensen TS, Parvez M (2007) 10a-aza-10b-Borapyrenes: heterocyclic analogues of pyrene with internalized BN moieties. Angew Chem Int Ed 46:4940–4943
51. Wang XY, Wang JY, Pei J (2015a) BN heterosuperbenzenes: synthesis and properties. Chem Eur J 21:3528–3539
52. Krieg M, Reicherter F, Haiss P, Strobele M, Eichele K, Treanor MJ, Schaub R, Bettinger HF (2015) Construction of an internally B N –doped nanographene molecule. Angew Chem Int Ed 54:8284–8286
53. Wang XY, Zhang F, Schellharnmer KS, Machata P, Ortmann F, Cuniberti G, Fu YB, Hunger J, Tang RZ, Popov AA, Berger R, Mullen K, Feng XL (2016) Synthesis of NBN-type zigzag-edged polycyclic aromatic hydrocarbons: 1,9-diaza-9a-boraphenalene as a structural motif. J Am Chem Soc 138:11606–11615
54. Baggett AW, Guo F, Li B, Liu S-Y, Jäkle F (2015) Regioregular synthesis of azaborine oligomers and a polymer with a syn conformation stabilized by N-H···π interactions. Angew Chem Int Ed 54:11191–11195
55. Wang X-Y, Zhuang F-D, Wang J-Y, Pei J (2015b) Incorporation of polycyclic azaborine compounds into polythiophene-type conjugated polymers for organic field-effect transistors. Chem Commun 51:17532–17535

56. Su K, Remsen EE, Thompson HM, Sneddon LG (1991) Syntheses and properties of poly(B-vinylborazine) and poly(styrene-co-Bvinylborazine) copolymers. Macromolecules 24:3760–3766
57. Jackson LA, Allen CW (1992) Organoborazines. iii. Homo- and copolymerization of p-vinylphenylcyclotriborazines. J Polym Sci A Polym Chem 30:577–581
58. Wan WM, Baggett AW, Cheng F, Lin H, Lamm AN, Liu SY, Jäkle F (2016) Synthesis by free radical polymerization and properties of BNpolystyrene and BN-poly(vinylbiphenyl). Chem Commun 52:13616–13619
59. Thiedemann B, Gliese PJ, Hoffmann J, Lawrence PG, Sönnichsen FD, Staubitz A (2017) High molecular weight poly(N-methyl-Bvinylazaborine) – a semi-inorganic B–N polystyrene analogue. Chem Commun 53:7258–7261
60. van de Wouw HL, Lee JY, Klausen RS (2017) Gram-scale free radical polymerization of an azaborine vinyl monomer. Chem Commun 53:7262–7265
61. Hoffmann AK, Thomas WB (1959) The synthesis of p-vinylphenylboronic acid and some of its derivatives. J Am Chem Soc 81:580–582
62. Gilman H, Moore LO (1958) Some studies on the preparation of arylboronic acids. J Am Chem Soc 80:3609–3611
63. Cui CZ, Bonder EM, Qin Y, Jäkle F (2010a) Synthesis and solvent-dependent micellization of the amphiphilic block copolymer poly(styreneboronic acid)-block-polystyrene. J Polym Sci A Polym Chem 48:2438–2445
64. Kim KT, Cornelissen J, Nolte RJM, van Hest JCM (2009) A polymersome nanoreactor with controllable permeability induced by stimuli-responsive block copolymers. Adv Mater 21:2787–2791
65. Vancoillie G, Hoogenboom R (2016a) Synthesis and polymerization of boronic acid containing monomers. Polym Chem 7:5484–5495
66. Vancoillie G, Hoogenboom R (2016b) Responsive boronic acid-decorated (co)polymers: from glucose sensors to autonomous drug delivery. Sensors-Basel 16:1736
67. Chudzinski MG, Chi YC, Taylor MS (2011) Borinic acids: a neglected class of organoboron compounds for recognition of diols in aqueous solution. Aust J Chem 64:1466–1469
68. Wan W-M, Cheng F, Jäkle F (2014) A borinic acid polymer with fluoride ion- and thermo-responsive properties that are tunable over a wide temperature range. Angew Chem Int Ed 53:8934–8938
69. Wan WM, Zhou P, Cheng F, Sun XL, Lv XH, Li KK, Xu H, Sun M, Jäkle F (2015) Thermo-responsive behavior of borinic acid polymers: experimental and molecular dynamics studies. Soft Matter 11:7159–7164
70. Cheng F, Wan WM, Zhou Y, Sun XL, Bonder EM, Jäkle F (2015) Borinic acid block copolymers: new building blocks for supramolecular assembly and sensory applications. Polym Chem 6:4650–4656
71. Sablong R, van der Vlugt JI, Thomann R, Mecking S, Vogt D (2005) Disperse amphiphilic submicron particles as non-covalent supports for cationic homogeneous catalysts. Adv Synth Catal 347:633–636
72. Matsumi N, Sugai K, Miyake M, Ohno H (2006) Polymerized ionic liquids via hydroboration polymerization as single ion conductive polymer electrolytes. Macromolecules 39:6924–6927
73. Berven BM, Oviasuyi RO, Klassen RJ, Idacavage M, Gillies ER, Ragogna PJ (2013) Self-crosslinking borate anions for the production of tough uv-cured polyelectrolyte surfaces. J Polym Sci A Polym Chem 51:499–508
74. Cui C, Bonder EM, Jäkle F (2010b) Weakly coordinating amphiphilic organoborate block copolymers. J Am Chem Soc 132:1810–1812
75. Cui C, Bonder EM, Jäkle F (2009) Organoboronium amphiphilic block copolymers. J Polym Sci Part A: Polym Chem 47:6612–6618
76. Cui C, Jäkle F (2009) Organoboronium-functionalized polystyrenes as a new class of polycations. Chem Commun 0:2744–2746

77. Fox PA, Griffin ST, Reichert WM, Salter EA, Smith AB, Tickell MD, Wicker BF, Cioffi EA, Davis JH, Rogers RD, Wierzbicki A (2005) Exploiting isolobal relationships to create new ionic liquids: novel room-temperature ionic liquids based upon (n-alkylimidazole)(amine) BH "boronium" ions. Chem Commun 0:3679–3681
78. Cui C, Heilmann-Brohl J, Perucha AS, Thomson MD, Roskos HG, Wagner M, Jäkle F (2010c) Redox-active ferrocenylboronium polyelectrolytes with main chain charge-transfer structure. Macromolecules 43:5256–5261
79. Qin Y, Shipman P, Jäkle F (2012) Self-assembly of borane end-functionalized polystyrene through tris(1-pyrazolyl)borate (Tp) iron(ii) linkages. Macromol Rapid Commun 33:562–567
80. Shipman P, Cui C, Lupinska P, Lalancette RA, Sheridan JB, Jäkle F (2013) Nitroxide-mediated controlled free radical polymerization of thechelate monomer tris(2-pyridyl)(4-styryl)borate (StTpyb) and supramolecular assembly via metal complexation. ACS Macro Lett 2:1056–1060
81. Pawar GM, Lalancette RA, Bonder EM, Sheridan JB, Jäkle F (2015) ROMP-derived pyridylborate block copolymers: self-assembly, pHresponsive properties, and metal-containing nanostructures. Macromolecules 48:6508–6515
82. Pawar GM, Sheridan JB, Jäkle F (2016) Pyridylborates as a new type of robust scorpionate ligand: from metal complexes to polymeric materials. Eur J Inorg Chem:2227–2235
83. Cui C, Lalancette RA, Jäkle F (2012) The elusive tripodal tris(2-pyridyl)borate ligand: a strongly coordinating tetraarylborate. Chem Commun 48:6930–6932
84. Cui C, Shipman P, Lalancette RA, Jäkle F (2013) Tris(2-pyridylborate) (Tpyb) metal complexes: synthesis, characterization and formation of extrinsically porous structures with large cylindrical channels. Inorg Chem 52:9440–9447

Chapter 5
Element-Block Polymeric Materials Based on Cage Silsesquioxane Frameworks

Kensuke Naka

Abstract This chapter focuses on recent efforts to prepare single-component element-block materials based on cage silsesquioxane frameworks. Polyhedral octasilsesquioxanes (POSSs), denoted as $(RSiO_{1.5})_8$ or labeled T_8 cages, are used here as the cage silsesquioxane frameworks. Thermoplastic optically transparent silsesquioxane materials derived from a single cage compound can be achieved by dumbbell- and star-shaped cage structures, allowing precise design of their structures for tuning properties. Incompletely condensed POSS exhibited lower crystallinity without loss of thermal stability in comparison with a completely condensed POSS. Difunctional POSS monomers, which were prepared by a selective corner-opening reaction and a subsequent corner-capping reaction, significantly reduce their crystallinity in comparison with those of monofunctionalized T_8 cages. Several examples for polymerization of the difunctional POSS monomers are described.

Keywords Cage silsesquioxane · POSS · Incompletely condensed POSS · Difunctional POSS monomer

5.1 Introduction

Silsesquioxanes are a general term of organosilicon compounds with the chemical formula $[RSiO_{1.5}]_n$, where R is hydrogen, any alkyl, alkylene, aryl, or organo-functional derivatives [1]. The most common silsesquioxanes are "random" structures. Random silsesquioxanes provide optically transparent hybrid polymeric materials consisting of organic substituents and inorganic backbones, which have attracted widespread interest for industrial applications because of their excellent thermal, mechanical, optical, and electrical properties [2, 3]. The molecular level

K. Naka (✉)
Faculty of Molecular Chemistry and Engineering, Graduate School of Science and Technology, Kyoto Institute of Technology, Kyoto, Japan
e-mail: kenaka@kit.ac.jp

Y. Chujo (ed.), *New Polymeric Materials Based on Element-Blocks*,
https://doi.org/10.1007/978-981-13-2889-3_5

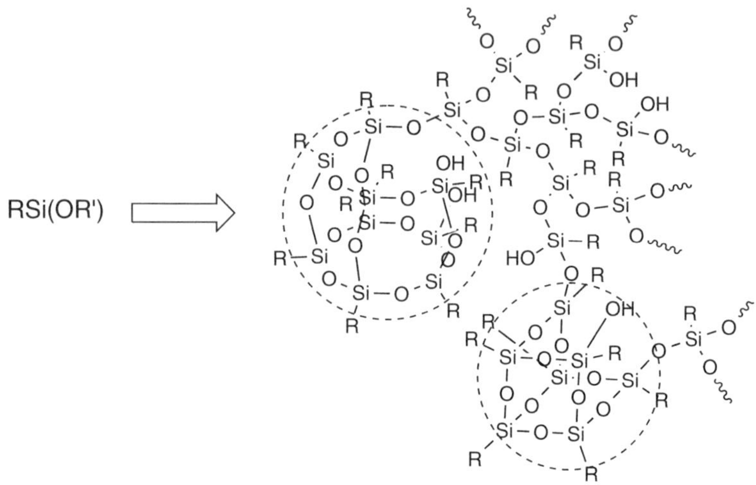

Fig. 5.1 Structure example of random silsesquioxane materials by sol-gel reaction

structures of the random silsesquioxanes significantly affect their physical properties (Fig. 5.1). Lee and co-workers demonstrated that both their refractive indices and dielectric constants decreased with increasing cage/random ratios because the free volume of a cage is larger than that of a random structure [4]. However, the molecular level structures of the random silsesquioxanes are difficult to control by usual sol-gel processing. Another disadvantage is that a high-temperature curing process for preparation of these materials often inhibits to apply them as coating on thermally unstable polymers.

Polyhedral oligosilsesquioxanes (POSSs), especially, polyhedral octasilsesquioxanes denoted as $(RSiO_{1.5})_8$ or labeled T_8 cages, have received increasing interest over the past decade as important well-defined nano-building blocks. They include a 0.53 nm, rigid, inorganic core that can be linked to eight functional groups to produce molecular level organic-inorganic hybrid structures [2, 5, 6]. Octadimethylsiloxy-Q_8-cages denoted as $(RSiMe_2OSiO_{1.5})_8$ are alternate siloxane-based cage frameworks and have been also used as nano-building blocks for various organic-inorganic hybrid materials [7, 8, 9]. These siloxane-based cage compounds have been used as fillers in polymer matrices to form organic-inorganic hybrid materials [10–13, 15, 16]. Their use as molecular fillers is an attractive strategy to control functional properties of the polymer matrices such as refractive index, dielectric constant, mechanical properties, and thermal stability. Another method of using cage silsesquioxane frameworks involves direct cross-linking with small organic molecules to form three-dimensional networks [14–19].

Recently, "polymeric materials based on element-blocks" have been proposed as a new concept in hybrid materials and can be expected to promote new research and to present new ideas for materials design involving all elements in the periodic table [20]. Among them, the use of cage silsesquioxane frameworks as element-blocks has been demonstrated to be an efficient method for designing polymeric materials based on element-blocks, which possess three-dimensional well-defined molecular level structures. This chapter shows several examples from our recent work related to this direction.

5.2 Single-Component Cage Silsesquioxane Materials

5.2.1 Dumbbell-Shaped Cage Silsesquioxane Derivatives

To develop low-temperature processable silsesquioxane materials precisely designed on the molecular level, structurally well-defined cage silsesquioxanes are considered attractive candidates. Due to their high symmetry and crystallinity, however, optically transparent films of a single cage silsesquioxane rarely form without cross-linking reagents. A simple approach to obtaining optically transparent single-component hybrid films is to lower cage symmetries thereby decreasing crystallinity and providing optically transparent film-forming properties. Such materials are regarded as thermoplastic silsesquioxane materials, in contrast to usual random silsesquioxanes, which are regarded as thermosetting materials.

A simple concept to decrease cage symmetries is a dumbbell-shaped structure, in which two POSS units are connected with an appropriate linker segment. Dumbbell-shaped trifluoropropyl-substituted C2-linked-, C3-linked-, and C6-linked-T_8 cage silsesquioxane (**F-DE**, **F-DP**, and **F-DH**, respectively) were prepared by corner-capping of heptatrifluoropropyl-substituted incompletely condensed cage silsesquioxane (**F7-Na**) with 1,2-bis(trichlorosilyl)ethane, 1,3-bis(trichlorosilyl)propane, and 1,6-bis(trichlorosilyl)hexane, respectively (Fig. 5.1a) [21]. Film-forming properties of the dumbbell-shaped POSS derivatives were studied by spin coating of hexafluorobenzene solutions on soda-lime glasses and subsequent heating at 100 °C for 3 min. **F-DP** and **F-DH** form transparent films. However, an opaque film forms from **F-DE** (Fig. 5.2b). These POSS derivatives were the first examples of optically transparent T_8 cage silsesquioxane films that show thermoplastic properties.

Refractive indices of optical transparent films of **F-DP** and **F-DH** were compared with those of random silsesquioxanes with the same components as those of **F-DP** and **F-DH**. The values of the films of the dumbbell-shaped POSS derivatives are in the range of 1.38–1.39, which are the same as those of the corresponding random silsesquioxanes. As described in the introduction, it was reported that refractive indices of random silsesquioxanes decreased with increasing their cage/random ratios, because the free volume of a cage is larger than that of a random structure [4]. However, the present data suggests that the refractive indices of the silsesquioxane films composed of only the cage structures were the same as those of the corresponding random silsesquioxanes. Although the random silsesquioxanes are required to be cured at the high temperature, **F-DP** and **F-DH** showed excellent film-forming properties under the lower-temperature processing, which would open a way to apply coating on various thermally unstable materials.

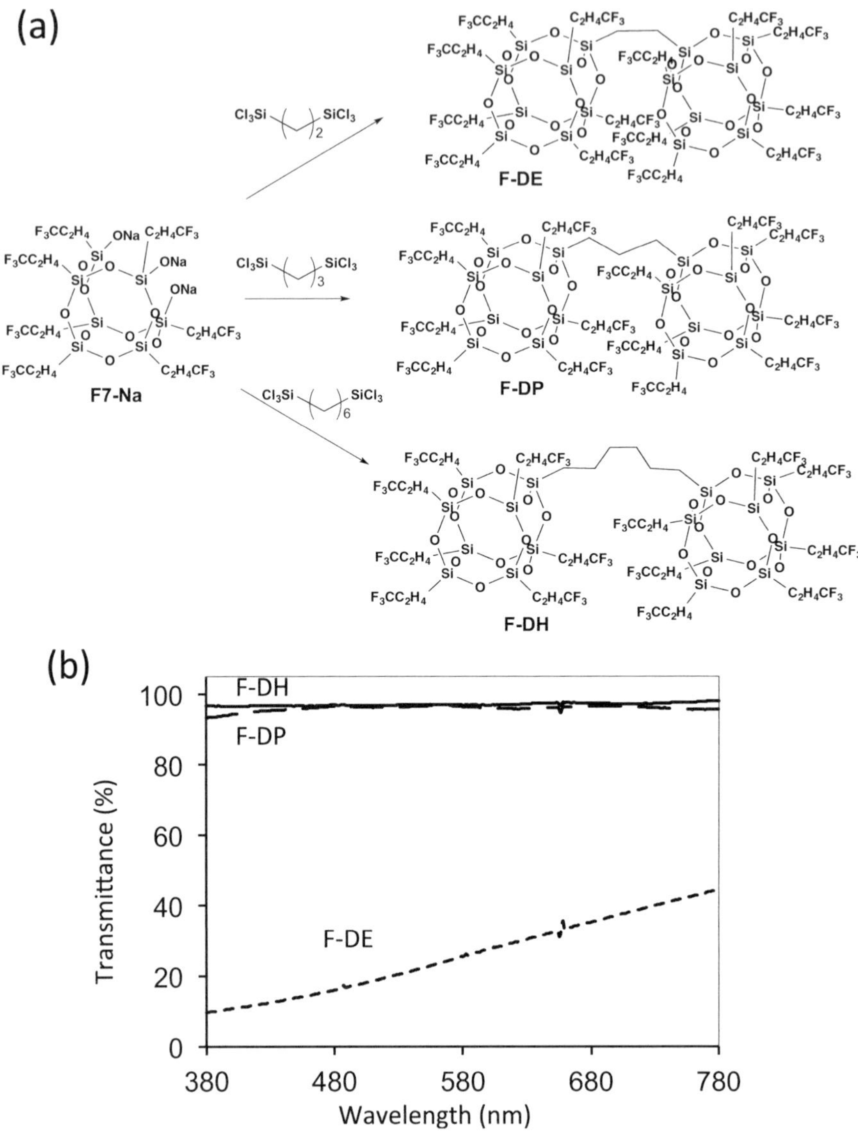

Fig. 5.2 (**a**) Synthesis of **F-DE**, **F-DP**, and **F-DH** by corner-capping of **F7-Na** and (**b**) UV-vis spectra of the films obtained from **F-DE**, **F-DP**, and **F-DH** on soda-lime glass

Crystallinity and thermal behavior of cage silsesquioxanes are dependent on their organic substituents. Dumbbell-shaped isobutyl-substituted C2-linked-, C3-linked-, and C6-linked cage silsesquioxanes (**IB-DE**, **IB-DP**, and **IB-DH**, respectively) were prepared from heptaisobutyl-substituted incompletely condensed cage

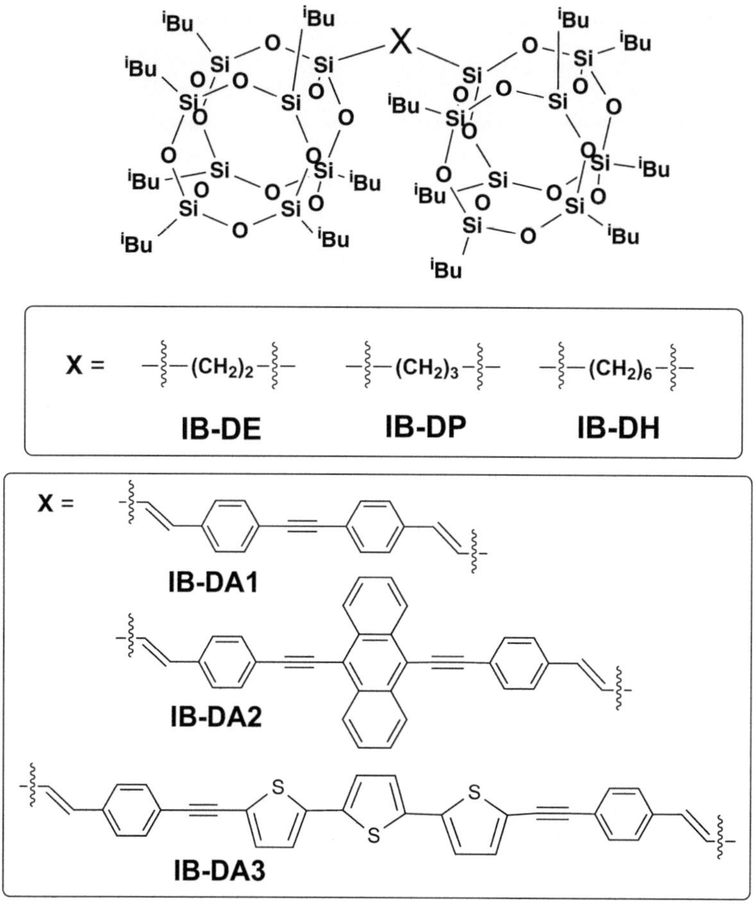

Fig. 5.3 Structures of dumbbell-shaped isobutyl-substituted cage silsesquioxane derivatives with simple aliphatic and rigid π-conjugated linkers

silsesquioxane (Fig. 5.3) [22]. Substitution of trifluoropropyl with isobutyl groups in the dumbbell structure reduces the optical transparency in films producing opaque whitish films formed in all cases. The use of rigid π-conjugated linkers (**IB-DA1**, **IB-DA2**, and **IB-DA3**) provided good optically transparent films due to reduction of their symmetries and molecular mobilities [23]. These results suggest that rigid linkers decrease the crystallinity of isobutyl-substituted cages compared with the flexible linkers. These obtained films were emitted under UV irradiation. The dumbbell structures minimize longer wavelength shifts and improve emission efficiency of the luminescent π-conjugated linker units in their solid states. The dumbbell-shaped isobutyl-substituted POSSs linked by the π-conjugated linkers are an effective approach to achieve amorphous luminescent molecules.

5.2.2 Star-Shaped Cage Silsesquioxane Derivatives

An alternative and effective way to obtain optically transparent films of a single cage silsesquioxane derivative is star-shaped cage silsesquioxanes linked by flexible dimethylsiloxy spacers, which provide optically transparent films depending on the length of the linking aliphatic chains [22]. The C2-, C3-, and C8-linked star-shaped cage silsesquioxane derivatives (**Star-C2**, **Star-C3**, and **Star-C8**, respectively) were prepared by hydrosilylation of heptaisobutyl-vinyl-, allyl-, and octenyl-T_8-silsesquioxanes and octadimethylsiloxy-Q_8-silsesquioxane ($Q_8M_8^H$) using Karstedt's catalyst, respectively (Fig. 5.4a) [24]. Compounds **Star-C2** and **Star-C3** form transparent films. However, an opaque film forms for **Star-C8**, due to the crystallization of the T_8-silsesquioxane units (Fig. 5.4b). The star-shaped structures significantly decrease melting points compared with the dumbbell-shaped derivatives (Fig. 5.4c). This is likely due to the strong influence of the flexible dimethylsiloxy spacer groups [25, 26].

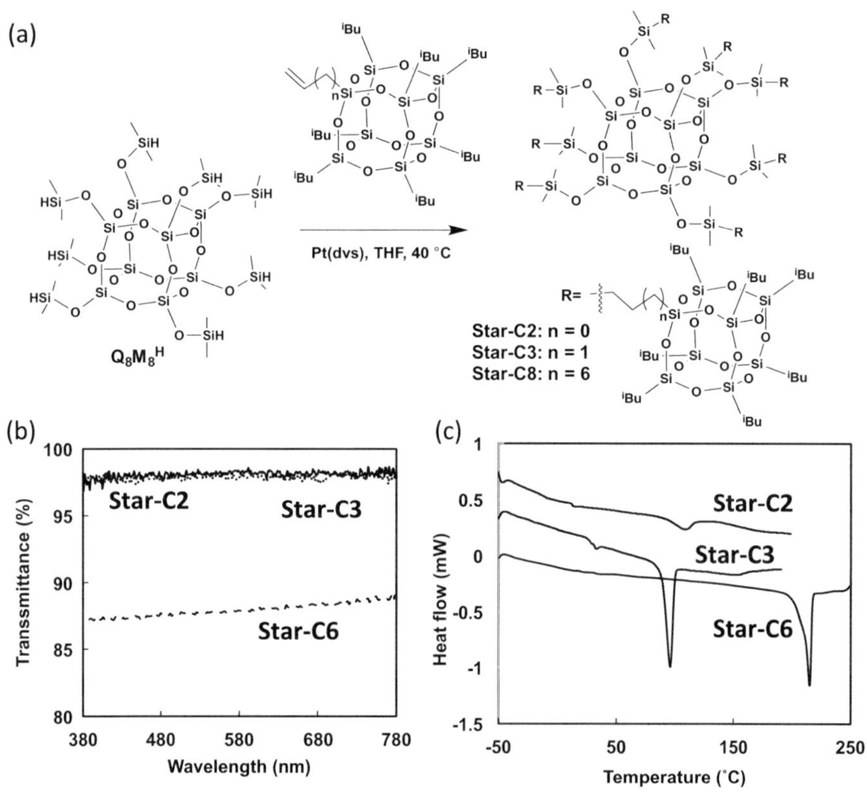

Fig. 5.4 (**a**) Synthesis, (**b**) UV-vis spectra of the films on soda-lime glass, and (**c**) DSC curves of the star-shaped cage silsesquioxane derivatives

The refractive indices of the cast films from solutions of **Star-C2** and **Star-C3** were 1.4529 ± 0.0005 and 1.4556 ± 0.0007, respectively. Decreasing the length of the methylene spacer decreases the refractive index. The values of the coefficient of molecular packing (K_p) for **Star-C2** and **Star-C3**, which were estimated using the experimental data and the calculated van der Waals volume of the molecule, are 0.821 and 0.951, respectively, suggesting that **Star-C2** has more free volume than **Star-C3b** does. The present star-shaped cage silsesquioxane derivatives are regarded as thermoplastic optically transparent silsesquioxane materials derived from a single cage silsesquioxane compound.

5.3 Silsesquioxane Element-Blocks Based on Incompletely Condensed POSSs

Incompletely condensed POSS (IC-POSS) derivatives are easily prepared from appropriate trialkoxysilanes, such as isobutyl-, cyclopentyl-, and trifluoropropyl-alkoxysilanes, which contain three silanol groups on the cage silsesquioxane skeletons. IC-POSSs have generally been used as precursors for monofunctionalized POSSs [27, 28]. They can also serve as three-dimensional scaffolds of tripodal molecules such as initiators [29], ligands for transition metals [30], and cross-linking agents [31, 32]. However, the thermal properties of IC-POSSs have never been investigated because they have not been previously recognized as candidates for building blocks of organic-inorganic hybrid materials. The three silanol groups of heptaisobutyl incompletely condensed POSS (**IC-3OH**) were reacted with chlorodimethylsilane to obtain tris(dimethylsilyl) heptaisobutyl POSS (**IC-3M^{H}**) (Scheme 5.1) [33]. The degradation temperatures for 5% and 10% weight loss (T_{d5} and T_{d10}, respectively) of **IC-3M^{H}** measured by thermogravimetric analysis (TGA) were 203 °C and 214 °C, respectively (Fig. 5.5a). Those of the corresponding completely condensed POSS, heptaisobutyl POSS (**CC-H**), were 207 °C and 217 °C, respectively (Fig. 5.5a). These data confirm that an incompletely condensed POSS exhibits the thermal stability equivalent to that of a completely condensed POSS. Differential scanning calorimetry (DSC) analysis for **IC-3M^{H}** and **CC-H** showed the detected melting points of **IC-3M^{H}** and **CC-H** were −18 °C and 130 °C, respectively, indicating that the completely condensed POSS formed much more stable crystals than the incompletely condensed POSS (Fig. 5.5b). The incompletely condensed POSSs can effectively restrict crystallinity by lowering the rigidity to gain solubility, without losing thermal stability, in comparison to completely condensed POSSs.

The monofunctionalized completely condensed cubic POSSs are often used to synthesize amphiphilic organic-inorganic element-block molecules [34–38]. In comparison with completely condensed POSSs (CC-POSSs), IC-POSSs drastically reduce symmetry and crystallinity, thus leading to the control of various self-assembly structures. Tripodal amphiphilic POSS derivatives based on isobutyl-

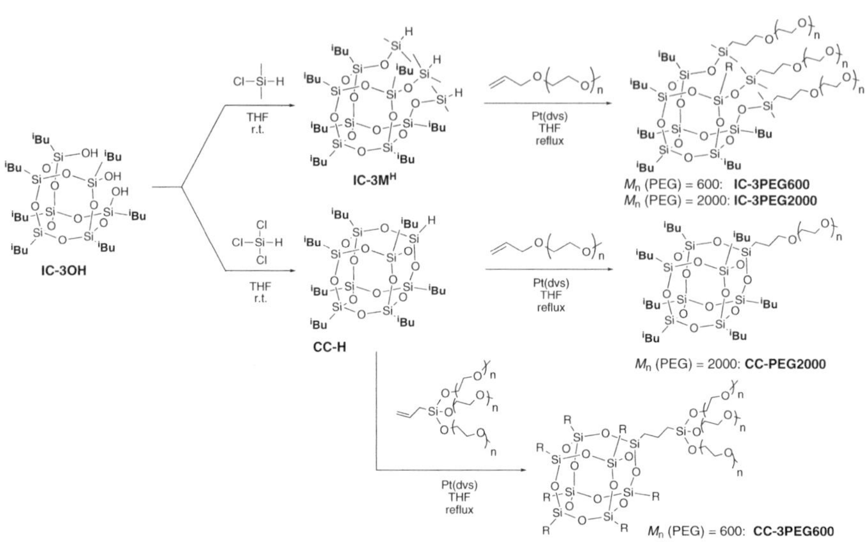

Scheme 5.1 Synthesis of tris(dimethylsilyl) heptaisobutyl POSS (**IC-3M^{H}**), heptaisobutyl POSS (**CC-H**), and amphiphilic cage silsesquioxane derivatives

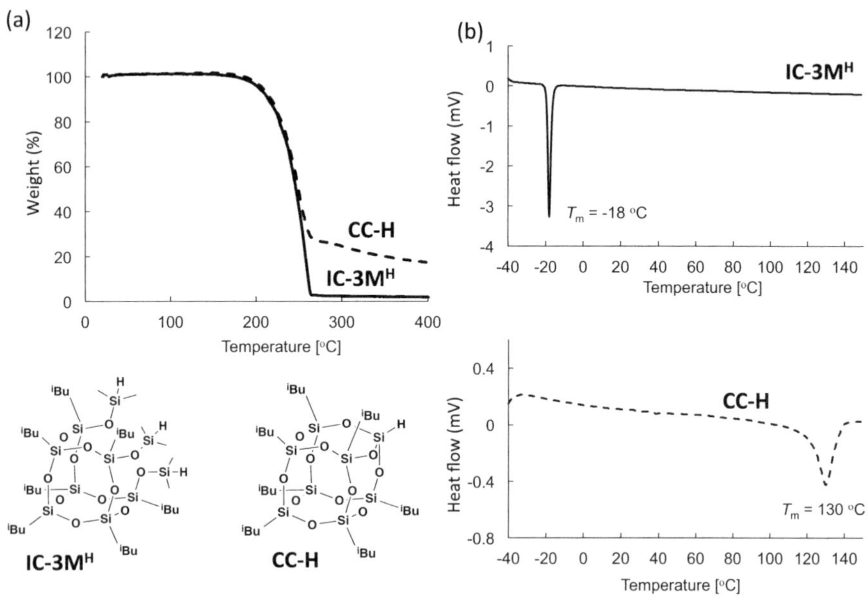

Fig. 5.5 (**a**) TGA thermograms and (**b**) DSC curves of **IC-3M^{H}** and **CC-H**

substituted IC-POSS with hydrophilic poly(ethylene glycol) (PEG) tails possessed good solubility in water and effectively stabilized oil-in-water emulsions, while the monosubstituted amphiphilic CC-POSS did not work as an emulsifier because of its

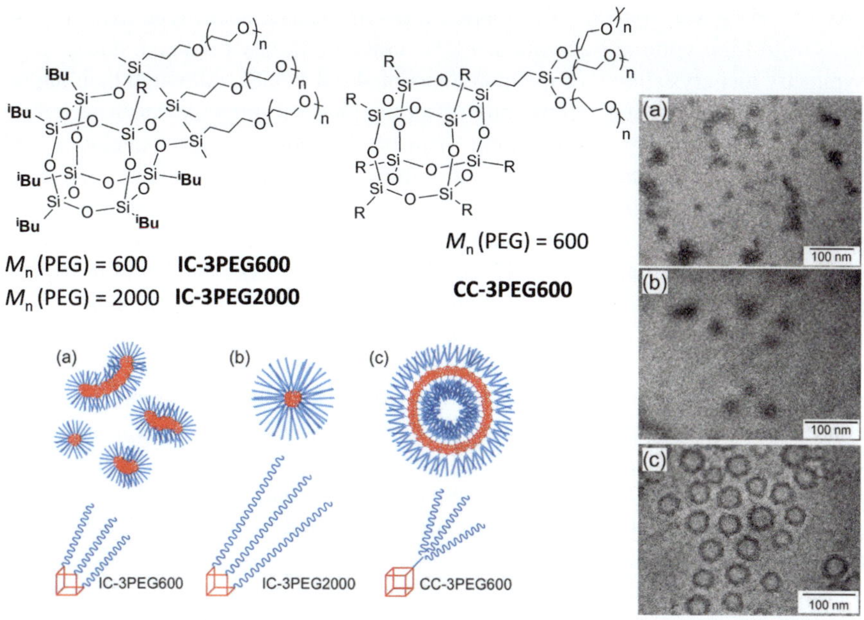

Fig. 5.6 Schematic and TEM images of the aggregates formed from IC-3POSS600 (**a**), IC-3PEG2000 (**b**), and CC-3PEG600 (**c**) in pure water

lack of water solubility [33]. The prepared organic-inorganic hybrid emulsions were stable against coalescence, and no demulsification occurred over 1 month.

The tripodal amphiphilic POSS derivatives based on isobutyl-substituted IC-POSS form spherical micelles consisting of a hydrophobic IC-POSS core and hydrophilic PEG chain shell (Fig. 5.6) [39]. On the other hand, tris(poly(ethylene glycol)) CC-POSS (CC-3PEG600) was prepared by platinum-catalyzed hydrosilylation[28] of tris(poly(ethylene glycol)) allylsilane, and heptaisobutyl POSS forms a vesicle structure in spite of the same PEG number and length. These results strongly indicate that the length of the PEG chain and the shape of the POSS head group play a crucial role in determining the self-assembly structures.

5.4 Element-Block Polymers with the Cage Silsesquioxanes in the Main Chain

Numerous hybrid polymers featuring cage silsesquioxane units in their side chain are well studied, because various monosubstituted POSSs are easily prepared by corner-capping of the incompletely condensed cage silsesquioxanes. They lead to an enhancement in their physical properties, i.e., improved thermal and mechanical stability, flammability, oxidative resistance, and a low dielectric constant

[40–46]. However, optically transparent films of the side chain-type hybrid polymers with high volume fractions of POSS units are hardly obtained, due to high symmetry and crystallinity of the monofunctionalized POSSs. Although their aggregation and crystallization significantly affect the physical properties of these hybrid materials, the difficulty in controlling them on the molecular level inhibits a real understanding of structure-property relationships in these hybrid polymers. To solve these drawbacks, polymers incorporating the cage silsesquioxanes in the main chain are expected to obtain optically transparent films even if the volume fraction of the cage silsesquioxane units is high [47–52]. In contrast to a large number of reports on polymers containing the cage silsesquioxanes in the side chain, few papers have reported the synthesis of polymers with cage silsesquioxanes in the main chain, which are prepared by a step polymerization system of well-defined cage silsesquioxane monomers having two polymerizable functional groups. Partially condensed disilanol silsesquioxanes are obtained by hydrolytic condensation of cyclohexyltrichlorosilane or partially hydrolyzed T_8 cages and used as monomers for polymerization [37, 38, 53, 54]. Di- and trifunctional T_{10} and T_{12} cages are prepared by rearrangement of T_8 cages caused by F^- and employed to obtain soluble and processable polymers incorporating the cage silsesquioxanes in the main chain [55–58]. A soluble polymer containing cage silsesquioxanes in its main chain is prepared in one step by hydrolytic condensation of amino group-containing organotrialkoxysilanes [59]. The use of double-decker-shaped phenyl-substituted silsesquioxanes (DDSQs) possessing precisely two reactive hydrosilane groups is the most successful approach to making linear hybrid polymers [47–52]. Introduction of DDSQ in the main chain of polymer backbones showed outstanding thermal stability and significantly increased in glass transition temperatures.

A *para*-substituted bis(3-aminopropyl)hexaisobutyl-substituted T_8 cage (**CC-2NH$_2$**) was successfully synthesized via a selective corner-opening reaction of 3-aminopropylheptaisobutyl-substituted T_8 cage (**CC-NH$_2$**) and a subsequent corner-capping reaction (Scheme 5.2) [60]. The key stage of this route is the corner-opening of **CC-NH$_2$** with aqueous tetraethylammonium hydroxide (TEAOH), affording trisilanol aminopropyl hexaisobutyl-substituted T_8 cage (**IC-NH$_2$**) as the predominant product [61, 62]. The selective corner-opening of vinylheptaisobutyl-substituted T_8 cage (**CC-vinyl**) also occurs under an appropriate condition. Subsequent corner-capping of resulting trisilanol vinylhexaisobutyl-

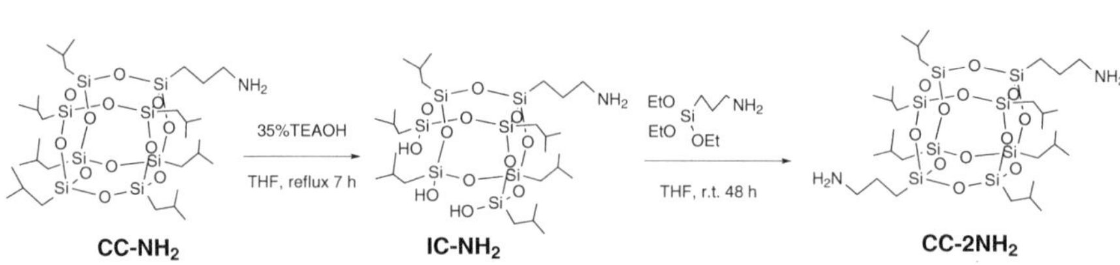

Scheme 5.2 Synthesis of *para*-substituted bis(3-aminopropyl)hexaisobutyl-substituted T_8 cage (**CC-2NH$_2$**)

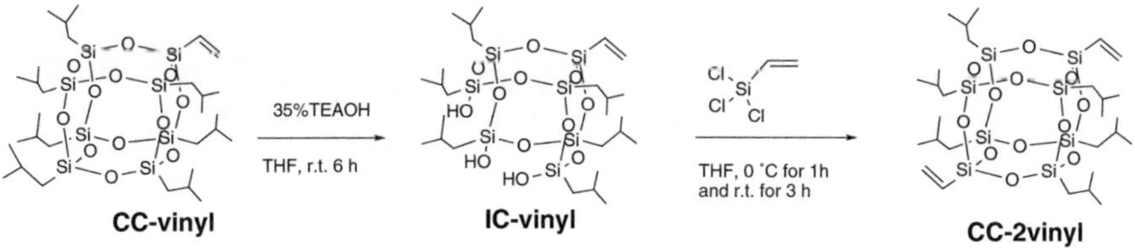

Scheme 5.3 Synthesis of *para*-substituted bisvinylhexaisobutyl-T_8 cage (**CC-2vinyl**)

substituted T_8 cage (**IC-vinyl**) afforded *para*-substituted bisvinylhexaisobutyl-T_8 cage (**CC-2vinyl**) (Scheme 5.3) [63].

Because the hydrolysis by TEAOH should be initiated by a nucleophilic attack, the patterns of the LUMO and its neighboring unoccupied MOs of the optimized molecular structure of 3-aminopropyl- and vinylheptaisobutyl-substituted T_8 cage (**CC-NH$_2$** and **CC-vinyl**, respectively) were examined by DFT calculations [60, 63]. The energy levels of the LUMO+1 and LUMO+2 in **CC-NH$_2$** are almost degenerate, and a detailed analysis shows that the largest contribution to these MOs comes from the atomic orbitals of silicon at the *para*-position to the Si atom substituted with the 3-aminopropyl group (Fig. 5.7a). The energy levels of the LUMO+2 and LUMO+3 in **CC-vinyl** are almost degenerate and shows that the largest contribution to the MO in LUMO+3 comes from the atomic orbitals of silicon (5Si) at the *para*-position to the Si atom substituted with the vinyl group in **CC-vinyl** (Fig. 5.7b). This arrangement favors the nucleophilic attack to **CC-vinyl** that leads to intermediate **IC-vinyl**.

Polymerization of **CC-2NH$_2$** with pyromellitic dianhydride (PMDA) and 4,4′-(hexafluoroisopropylidene)diphthalic anhydride (6FDA) resulted in corresponding polyimides containing T_8 cages in the main chains (T_8/PMDA and T_8/6FDA, respectively) (Scheme 5.4) [60, 64]. The optical transmittance of the yellow-colored freestanding film of T_8/PMDA was over 90% in the visible region between 780 and 475 nm with a film thickness of 0.1 mm, and the absorption edge was observed at approximately 590 nm. On the other hand, the film of T_8/6FDA showed excellent transparency in the visible region with excellent optical transparency of over 90% even at 360 nm. DSC analysis of both the polyimides showed no glass transition and melt behavior between room temperature and 400°C. Contact angles of water for the films of T_8/PMDA and T_8/6FDA were $97 \pm 2°$ and $96 \pm 5°$, respectively, and both are significantly higher than that of poly(pyromellitic dianhydride-oxydianiline) (PMDA-ODA) polyimide and comparable to that of high-density polyethylene (HDPE) (Table 5.1). Martens' hardness and coefficient of thermal expansion (CTE) of T_8/PMDA were significantly lower and higher than that of PMDA-ODA, respectively, and these values are comparable to those of HDPE. Replacement of PMDA to 6FDA significantly improved Martens' hardness and CTE. Density data were obtained by gas displacement pycnometry. Densities (*d*) of T_8/PMDA and T_8/6FDA are lower than that of PMDA-ODA. Density of octaisobutyl-substituted T_8 cage is 1.248 g/cm^3 according to crystal packing [65]. High

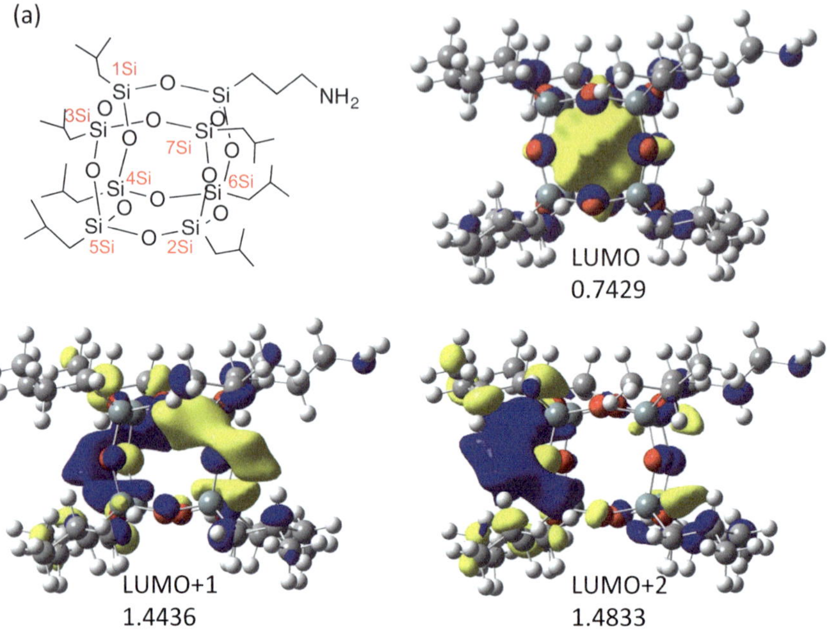

Fig. 5.7 Chemical structures and patterns of the LUMO, LUMO+1, and LUMO+2 of **CC-NH$_2$** (**a**) and the LUMO, LUMO+1, LUMO+2, and LUMO+3 of **CC-vinyl** (**b**) along with the orbital energies in eV

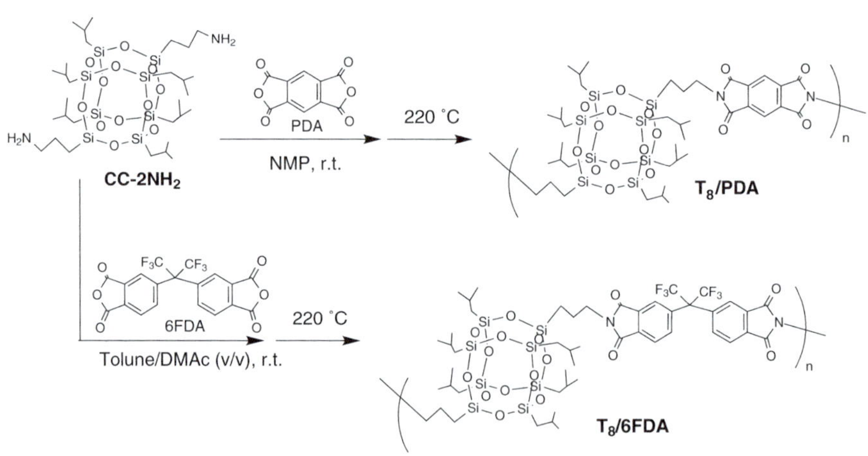

Scheme 5.4 Synthesis of T$_8$/PMDA and T$_8$/6FDA

Table 5.1 Hardness, CTE, density, and contact angle of the polyimides

	Martens' hardness	CTE	Density (d)	Contact angle
	N/mm^2	/K	g/cm^3	°
T_8/PMDA	37.8 ± 0.2	18.2×10^{-5}	1.24 ± 0.00	97 ± 2
T_8/6FDA	140.5 ± 19.1	8.13×10^{-5}	1.12 ± 0.00	96 ± 5
PMDA-ODA	219.7 ± 0.5	5.0×10^{-5}	1.48 ± 0.01	76 ± 2
HDPE	42.1 ± 6.2	22.2×10^{-5}	0.946 ± 0.002	–

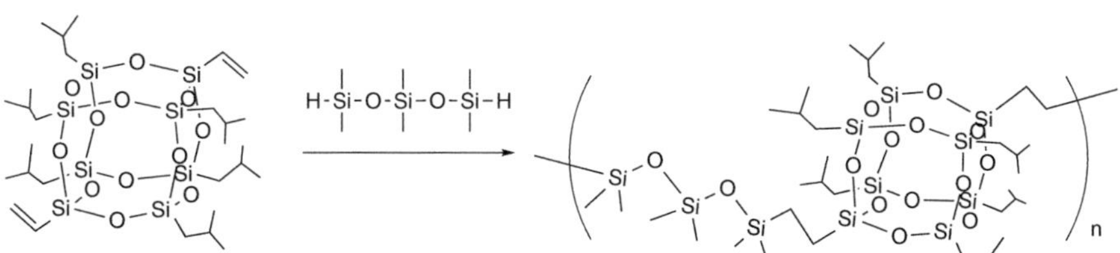

CC-2vinyl

Scheme 5.5 Polymerization of **CC-2vinyl** with 1,1,3,3,5,5-hexamethyltrisiloxane

volume fraction of the T_8 cages in the present polyimides decreased their densities in comparison with common polyimides. Generally, introducing fluorinated substituents such as 6FDA has been recognized as one of the most promising methods to synthesize colorless and hydrophobic polyimides. However, these types of polyimides often reduce T_g [66, 67]. These observations suggest that introduction of the T_8-unit in the main chain significantly improved hydrophobicity as well as increased the T_g, and physical properties of polyimides containing the hexaisobutyl-substituted T_8 cage in their main chains can be tuned by changing their linkage structures without decreasing their T_g.

CC-2vinyl was polymerized with a trisiloxane monomer, 1,1,3,3,5,5-hexamethyltrisiloxane, in the presence of Karstedt's catalyst (Scheme 5.5). The obtained polymer was soluble in organic solvents, including $CHCl_3$ and THF. GPC analysis ($CHCl_3$, PSt standards) showed that number average molecular weight (M_n) and molecular weight distribution (M_w/M_n) are 1.9×10^4 and 2.3, respectively. The polymer showed 1% and 5% weight losses at 352 and 474 °C, respectively, under N_2. DSC analysis of the polymer showed an unclear baseline shift at 30 °C, suggesting glass transition temperature (T_g). However, no obvious softening was observed even heating the film at 100 °C. Introduction of the T_8-unit in the main chain of polysiloxane significantly increases in T_g. The film showed excellent transparency in the visible region as well as UV region. The optical transmittance of the film was over 98% in the visible region between 780 and 330 nm with a film thickness of 35 μm. The polymer exhibits excellent UV transparency over 80% even at 250 nm with low cutoff wavelength (219 nm).

The XRD pattern of the polymer shows a broad peak centered at a 2θ value of 18°, indicating an amorphous polymer (Fig. 5.3b). The polymer also shows a broad diffraction peak at a 2θ value of 7.4°. The spacing evaluated from the 2θ value was

1.2 nm, corresponding to the isobutyl-substituted T_8 cage unit. This value is comparable with the size of the isobutyl-T_8 cage unit [65]. This suggests highly packing of the isobutyl-substituted T_8 cage moieties in the polymer without proceeding crystal packing of them. The difunctional T_8 cages significantly reduce their crystallinity in comparison with those of the monofunctionalized T_8 cages.

5.5 Conclusions

The concept of "polymeric materials based on element-blocks" is a promising approach to create new polymeric materials with well-defined and controlled structures. Cage silsesquioxane frameworks are attractive element-blocks to produce optically transparent silsesquioxane materials with predominantly volume fraction of cage silsesquioxane frameworks having well-defined molecular level structures. The most cage silsesquioxanes decorated with simple alkyl, alkylene, and aryl substituents possess high symmetry and crystallinity. Thus, optically transparent films of a single cage silsesquioxane rarely form. This can be achieved by reducing crystallinity of the cage silsesquioxane frameworks by several approaches. The dumbbell- and star-shaped cage silsesquioxane derivatives are regarded as thermoplastic optically transparent silsesquioxane materials derived from a single cage silsesquioxane compound, which can precisely design their structures for tuning their properties. This approach enables us to understand structure-property relationship of silsesquioxane-based materials at molecular level and design their properties more precisely than conventional sol-gel method. Single-component element-block materials based on cage silsesquioxane frameworks are a promising approach to control the sequence of organic and inorganic segments in organic-inorganic hybrids at molecular level.

The incompletely condensed cage silsesquioxane frameworks are attractive candidate to reduce crystallinity without loss of thermal stability in comparison to the completely condensed cage silsesquioxane frameworks. The incompletely condensed cage silsesquioxane frameworks are potential candidates for easy-to-handle building blocks of organic-inorganic hybrid materials. The tripodal amphiphilic POSS derivatives worked as effective emulsifiers. The present molecular design will open a new pathway toward the construction of a novel class of emulsifiers to achieve unique organic-inorganic architectures.

The difunctional T_8 cages significantly reduce their crystallinity in comparison with those of monofunctionalized T_8 cages. Thus, polymers incorporating the cage silsesquioxanes in the main chain are expected to obtain optically transparent films even if the volume fraction of the cage silsesquioxane units is high.

References

1. Baney RH, Itoh M, Sakakibara A, Suzuki T (1995) Synthetic 6FDA–ODA copolyimide membranes for gas separation and pervaporation: functional groups and separation properties. Chem Rev 95:1409–1430
2. Tanaka K, Chujo Y (2012) Advanced functional materials based on polyhedral oligomeric silsesquioxane (POSS). J Mater Chem 22:1733–1746
3. Mikoshiba S, Hayase S (1999) Preparation of low density poly (methylsilsesquioxane)s for LSI interlayer dielectrics with low dielectric constant. Fabrication of angstrom size pores prepared by baking trifluoropropylsilyl copolymers. J Mater Chem 9:591–598
4. Lee JH, Kim WC, Min SK, Ree HW, Yoon DY (2003) Synthesis of poly(methyl-co-trifluoropropyl) silsesquioxanes and their thin films for low dielectric application. Macromol Mater Eng 288:455–461
5. Cordes DB, Lickiss PD, Rataboul F (2010) Recent developments in the chemistry of cubic polyhedral oligosilsesquioxanes. Chem Rev 110:2081–2173
6. Laine RM (2005) Nanobuilding blocks based on the $[OSiO_{1.5}]$ (x = 6, 8, 10) octasilsesquioxanes. J Mater Chem 15:3725–3744
7. Hasegawa I, Ino K, Ohnishi H (2003) An improved procedure for syntheses of silyl derivatives of the cubeoctameric silicate anion. Appl Organomet Chem 17:287–290
8. Choi J, Yee AF, Laine RM (2003) Organic/inorganic hybrid composites from cubic silsesquioxanes. Epoxy resins of octa(dimethylsiloxyethylcyclohexylepoxide) silsesquioxane. Macromolecules 36:5666–5682
9. Sasi kumar R, Ariraman M, Alagar M (2014) Design of lamellar structured POSS/BPZ polybenzoxazine nanocomposites as a novel class of ultra low–k dielectric materials. RSC Adv 4:19127–19136
10. Kim KM, Chujo Y (2001) Liquid–crystalline organic-inorganic hybrid polymers with functionalized silsesquioxanes. J Polym Sci A Polym Chem 39:4035–4043
11. Mitsuishi M, Zhao F, Kim Y, Watanabe A, Miyashita T (2008) Preparation of ultrathin silsesquioxane nanofilms via polymer langmuir–Blodgett films. Chem Mater 20:4310–4316
12. Wahab MA, Mya KY, He CO (2008) Synthesis, morphology, and properties of hydroxyl terminated-POSS/polyimide low–k nanocomposite films. J Polym Sci A Polym Chem 46:5887–5896
13. Tanaka K, Adachi S, Chujo Y (2009) Structure–property relationship of octa–substituted POSS in thermal and mechanical reinforcements of conventional polymers. J Polym Sci A Polym Chem 47:5690–5697
14. Choi J, Yee AF, Laine RM (2004) Toughening of cubic silsesquioxane epoxy nanocomposites using core–shell rubber particles: a three–component hybrid system. Macromolecules 37:3267–3276
15. Zhang C, Babonneau F, Bonhomme C, Laine RM, Soles CL, Hristov HA, Yee AL (1998) Highly porous polyhedral silsesquioxane polymers. Synthesis and characterization. J Am Chem Soc 120:8380–8391
16. Lin H, Qu J, Zhang Z, Dong J, Zou H (2013) Ring-opening polymerization reaction of polyhedral oligomeric silsesquioxanes (POSSs) for preparation of well–controlled 3D skeletal hybrid monoliths. Chem Commun 49:231–233
17. Jeon JH, Tanaka K, Chujo Y (2014) Light-driven artificial enzymes for selective oxidation of guanosine triphosphate using water–soluble POSS network polymers. Org Biomol Chem 12:6500–6506
18. Jeon JH, Kakuta T, Tanaka K, Chujo Y (2015) Facile design of organic–inorganic hybrid gels for molecular recognition of nucleoside triphosphates. Bioorg Med Chem Lett 25:2050–2055
19. Kakuta T, Tanaka K, Chujo Y (2015) Synthesis of emissive water–soluble network polymers based on polyhedral oligomeric silsesquioxane and their application as optical sensors for discriminating the particle size. J Mater Chem C 3:12539–12545
20. Chujo Y, Tanaka K (2015) New polymeric materials based on element–blocks. Bull Chem Soc Jpn 86:633–643

21. Araki H, Naka K (2011) Syntheses of dumbbell–shaped trifluoropropyl-substituted POSS derivatives linked by simple aliphatic chains and their optical transparent thermoplastic films. Macromolecules 44:6039–6045
22. Araki H, Naka K (2012) Syntheses and properties of star– and dumbbell–shaped POSS derivatives containing isobutyl groups. Polym J 44:340–346
23. Araki H, Naka K (2012) Syntheses and properties of dumbbell–shaped POSS derivatives linked by luminescent π –conjugated units. J Polym Sci A Polym Chem 50:4170–4181
24. Yasumoto Y, Yamanaka T, Sakurai S, Imoto H, Naka K (2016) Design of low–crystalline and – density isobutyl–substituted caged silsesquioxane derivatives by star–shaped architectures linked with short aliphatic chains. Polym J 48:281–287
25. Perrin FX, Viet Nguyen TB, Margaillan A (2011) Linear and branched alkyl substituted octakis (dimethylsiloxy)octasilsesquioxanes: WAXS and thermal properties. Eur Polym J 47:1370–1382
26. Perrin FX, Panaitescu DM, Frone AN, Radovici C, Nicolae C (2013) The influence of alkyl substituents of POSS in polyethylene nanocomposites. Polymer 54:2347–2354
27. Di Iulio C, Jones MD, Mahon MF, Apperley DC (2010) Zinc(II) silsesquioxane complexes and their application for the ring–opening polymerization of rac–Lactide. Inorg Chem 49:10232–10234
28. Zhou J, Zhao Y, Yu K, Zhou X, Xie X (2011) Synthesis, thermal stability and photoresponsive behaviors of azobenzene–tethered polyhedral oligomeric silsesquioxanes. New J Chem 35:2781–2792
29. Yamahiro M, Oikawa H, Yoshida K, Ito K, Yamamoto Y, Tanaka M, Ootake N, Watanabe K, Ohno K, Tsujii Y, Fukuda T (2004) PCT Int. Appl. WO 2004026883 A1 20040401
30. Ionescu G, van der Vlugt JI, Abbenhuis HCL, Vogt D (2005) Synthesis and applications of chiral phosphite ligands derived from incompletely condensed silsesquioxane backbones. Tetrahedron Asymmetry 16:3970–3975
31. Bian Y, Mijović J (2009) Effect of side chain architecture on dielectric relaxation in polyhedral oligomeric silsesquioxane/polypropylene oxide nanocomposites. Polymer 50:1541–1547
32. Miyasaka M, Fujiwara Y, Kudo H, Nishikubo T (2010) Synthesis of hyperbranched fluorinated polymers with controllable refractive indices. Polym J 42:799–803
33. Imoto H, Nakao Y, Nishizawa N, Fujii S, Nakamura Y, Naka K (2015) Tripodal polyhedral oligomeric silsesquioxanes as novel class of three–dimensional emulsifiers. Polym J 47:609–615
34. Brown JF, Vogt LH (1965) The polycondensation of cyclohexylsilanetriol. J Am Chem Soc 87:4313–4317
35. Brown JF (1965) The polycondensation of phenylsilanetriol. J Am Chem Soc 87:4317–4324
36. Feher FJ, Newman DA, Walzer JF (1989) Silsesquioxanes as models for silica surfaces. J Am Chem Soc 111:1741–1748
37. Feher FJ, Budzichowski TA, Blanski RL, Weller KJ, Ziller JW (1991) Facile syntheses of new incompletely condensed polyhedral oligosilsesquioxanes: $[(c\text{-}C_5H_9)_7Si_7O_9(OH)_3]$, $[(c\text{-}C_7H_{13})_7Si_7O_9(OH)_3]$, and $[(c\text{-}C_7H_{13})_6Si_6O_7(OH)_4]$. Organometallics 10:2526–2528
38. Feher FJ, Terroba R, Ziller JW (1999) A new route to incompletely–condensed silsesquioxanes: base-mediated cleavage of polyhedral oligosilsesquioxanes. Chem Commun 69:2309–2310
39. Yusa S, Ohno S, Honda T, Imoto H, Nakao Y, Naka K, Nakamura Y, Fujii S (2016) Synthesis of silsesquioxane–based element–block amphiphiles and their self–assembly in water. RSC Adv 6:73006–73012
40. Lichtenhan JD, Otonari YA, Carr MJ (1995) Linear hybrid polymer building blocks: methacrylate–functionalized polyhedral oligomeric silsesquioxane monomers and polymers. Macromolecules 28:8435–8437
41. Zheng L, Hong S, Cardoen G, Burgaz E, Gido SP, Coughlin EB (2004) Polymer nanocomposites through controlled self–assembly of cubic silsesquioxane scaffolds. Macromolecules 37:8606–8611
42. Ahn B, Hirai T, Jin S, Rho Y, Kim KW, Kakimoto M, Gopalan P, Hayakawa T, Ree M (2010) Hierarchical structure in nanoscale thin films of a poly(styrene–*b*– methacrylate grafted with POSS) (PS_{214}–*b*–$PMAPOSS_{27}$). Macromolecules 43:10568–10581

43. Wu J, Ge Q, Mather PT (2010) PEG–POSS multiblock polyurethanes: synthesis, characterization, and hydrogel formation. Macromolecules 43:7637–7649
44. Lee J, Cho HJ, Jung BJ, Cho NS, Shim HK (2004) Stabilized blue luminescent polyfluorenes: introducing polyhedral oligomeric silsesquioxane. Macromolecules 37:8523–8529
45. Pyun J, Matyjaszewski K (2000) The synthesis of hybrid polymers using atom transfer radical polymerization: homopolymers and block copolymers from polyhedral oligomeric silsesquioxane monomers. Macromolecules 33:217–220
46. Escudé NC, Chen EYX (2009) Stereoregular methacrylate–POSS hybrid polymers: syntheses and nanostructured assemblies. Chem Mater 21:5743–5753
47. Wright ME, Schorzman DA, Feher FJ, Jin RZ (2003) Synthesis and thermal curing of aryl–ethynyl–terminated coPOSS imide oligomers: new inorganic/organic hybrid resins. Chem Mater 15:264–268
48. Wu S, Hayakawa T, Kikuchi R, Grunzinger SJ, Kakimoto M, Oikawa H (2007) Synthesis and characterization of semiaromatic polyimides containing POSS in main chain derived from double–decker–shaped silsesquioxane. Macromolecules 40:5698–5705
49. Wu S, Hayakawa T, Kakimoto M, Oikawa H (2008) Synthesis and characterization of organosoluble aromatic polyimides containing POSS in main chain derived from double–decker–shaped silsesquioxane. Macromolecules 41:3481–3487
50. Hoque MA, Kakihana Y, Shinke S, Kawakami Y (2009) Polysiloxanes with periodically distributed isomeric double–decker silsesquioxane in the main chain. Macromolecules 42:3309–3315
51. Wang L, Zhang C, Zheng S (2011) Organic–inorganic poly(hydroxyether of bisphenol A) copolymers with double–decker silsesquioxane in the main chains. J Mater Chem 21:19344–19352
52. Yoshimatsu M, Komori K, Ohnagamitsu Y, Sueyoshi N, Kawashima N, Chinen S, Murakami Y, Izumi J, Inoki D, Sakai K, Matsuo T, Watanabe K, Kunitake M (2012) Necklace-shaped dimethylsiloxane polymers bearing a polyhedral oligomeric silsesquioxane cage prepared by polycondensation and ring-opening polymerization. Chem Lett 41:622–624
53. Lichtenhan JD, Vu NQ, Carter JA, Gilman JW, Feher FJ (1993) Silsesquioxane–siloxane copolymers from polyhedral silsesquioxanes. Macromolecules 26:2141–2142
54. Raftopoulos KN, Jancia M, Aravopoulou D, Hebda E, Pielichowski K, Pissis P (2013) POSS along the hard segments of polyurethane. Phase separation and molecular dynamics. Macromolecules 46:7378–7386
55. Asuncion MZ, Laine RM (2010) Fluoride rearrangement reactions of polyphenyl– and polyvinylsilsesquioxanes as a facile route to mixed functional phenyl, vinyl T_{10} and T_{12} silsesquioxanes. J Am Chem Soc 132:3723–3736
56. Jung JH, Laine RM (2011) Polymers formed from the reaction of [NH2PhSiO1.5]x[PhSiO1.5]10–x and [NH2PhSiO1.5]x[PhSiO1.5]12–x mixtures (x = 2–4) with the Diglycidyl ether of Bisphenol A. Macromolecules 44:7263–7272
57. Jung JH, Furgal JC, Clark S, Schwartz M, Chou K, Laine RM (2013) Beads on a Chain (BoC) polymers with model dendronized beads. Copolymerization of [(4-$NH_2C_6H_4SiO_{1.5})_6(IPhSiO_{1.5})_2$] and [(4-$CH_3OC_6H_4SiO_{1.5})_6(IPhSiO_{1.5})_2$] with 1,4-Diethynylbenzene (DEB) gives through–chain, extended 3–D conjugation in the excited state that is an average of the corresponding homopolymers. Macromolecules 46:7580–7590
58. Furgal JC, Jung JH, Clark S, Richard M (2013) Beads on a Chain (BoC) phenylsilsesquioxane (SQ) polymers via F–catalyzed rearrangements and ADMET or reverse heck cross–coupling reactions: through chain, extended conjugation in 3-D with potential for dendronization. Macromolecules 46:7591–7604
59. Tokunaga T, Koga S, Mizumo T, Ohshita J, Kaneko Y (2015) Facile preparation of a soluble polymer containing polyhedral oligomeric silsesquioxane units in its main chain. Polym Chem 6:3039–3045
60. Maegawa T, Irie Y, Fueno H, Tanaka K, Naka K (2014) Synthesis and polymerization of a Para–disubstituted T_8–caged hexaisobutyl–POSS monomer. Chem Lett 43:1532–1534

61. Carniato F, Boccaleri E, Marchese L (2008) A versatile route to bifunctionalized silsesquioxane (POSS): synthesis and characterisation of Ticontaining aminopropylisobutyl–POSS. Dalton Trans 1:36–39
62. Olivero F, Renò F, Carniato F, Rizzi M, Cannas M, Marchese L (2012) A novel luminescent bifunctional POSS as a molecular platform for biomedical applications. Dalton Trans 41:7467–7473
63. Maegawa T, Irie Y, Imoto H, Fueno H, Naka K (2015) Para–bisvinylhexaisobutyl–substituted T_8 caged monomer: synthesis and hydrosilylation polymerization. Polym Chem 6:7487–7632
64. Maegawa T, Miyashita O, Irie Y, Imoto H, Naka K (2016) Synthesis and properties of polyimides containing hexaisobutyl–substituted T_8 cages in their main chains. RSC Adv 6:31751–31757
65. Bassindale AR, Liu Z, MacKinnon IA, Taylor PG, Yang Y, Light ME, Horton PN, Hursthouse MB (2003) A higher yielding route for T_8 silsesquioxane cages and X–ray crystal structures of some novel spherosilicates. Dalton Trans 14:2945–2949
66. Xiao X, Kong D, Qui X, Zhang W, Zhang F, Liu L, Liu Y, Zhang S, Hu Y, Leng J (2015) Shape–memory polymers with adjustable high glass transition temperatures. Macromolecules 48:3582–3589
67. Xiao S, Huang RYM, Feng X (2007) Synthetic 6FDA–ODA copolyimide membranes for gas separation and pervaporation: functional groups and separation properties. Polymer 48:5355–5368

Chapter 6
Silsesquioxane-Based Hierarchical and Hybrid Materials

Fuping Dong and Chang-Sik Ha

Abstract Much attention in scientific community in silsesquioxanes (SSQs) during the recent years has been attracted by the design of hierarchical structure and hybrid composite. The wide variety of SSQ materials with hierarchical morphology including core/shell, hollow, yolk-shell, nanorod and so on have been developed. The present chapter demonstrates the most recent progress made in the design of this morphology by utilizing different preparation methods such as sol-gel, in situ polymerization, electrospinning, etc. In this chapter, special attention is paid to the fabrication and application of the silsesquioxane hybrid composite. The interest in the hybrid nanocomposite is illustrated by a large variety of hybrid materials including polymers, silica, carbon materials, noble metals, and quantum dots.

Keywords Silsesquioxane · Hybrid · Hierarchical morphology · Nanocomposite

6.1 Introduction

Silsesquioxanes, with a general formula $(RSiO_{1.5})_n$, are one kind of molecular-level organic/inorganic hybrid silica-based materials. They combine many unique physical properties (thermal, mechanical, and structural stability) from inorganic part and chemical properties (possibility for functionalization and high flexibility) from organic part that traditional composite materials do not exhibit. Silsesquioxanes conventionally are synthesized from the hydrolysis and condensation of trialkoxysilane $[RSi(OR')_3]$ or trichlorosilane $(RSiCl_3)$ monomers with active or inactive organic groups. By utilizing various silane precursors, various functional

F. Dong · C.-S. Ha (✉)
Department of Polymer Materials and Engineering, College of Materials and Metallurgy, Guizhou University, Guiyang, China

Department of Polymer Science and Engineering, Pusan National University, Busan, South Korea
e-mail: fpdong@gzu.edu.cn; csha@pnu.edu

Y. Chujo (ed.), *New Polymeric Materials Based on Element-Blocks*,
https://doi.org/10.1007/978-981-13-2889-3_6

groups can be introduced to control the structures including ladderlike polysilsesquioxane (LPSQ) and cage-like polyhedral oligomeric silsesquioxane (POSS) in which a cubic core is connected with eight armlike organic substituents [1, 2].

The construction of hierarchical structured SSQs with different length scales usually is necessary to transfer the properties of molecule or atom to macroscopic materials. There are some reports on SSQ materials, focusing on the structure-composition-performance relationship of this kind of materials with well-defined morphology and nanostructure [3, 4]. Recently, design, fabrication, and applications of hierarchically structured porous SSQ materials have become a rapidly developing field due to their promising applications in separation, sensor, catalysis, and drug delivery [5]. A large series of synthesis methods have been developed involving templating method, conventional technique, sol-gel method, and self-formation method [6].

In recent decades, considerable research effort has been focused on the preparation of inorganic silicon-based materials modified with organic functional groups [7]. The integration or hybrid of different building block with silsesquioxanes has drawn much research attention, as this approach offers novel functionalities to the hybrid composite. In fact, the synthesis of hybrid silsesquioxane nanocomposites for potential application has becoming one of the attractive fields. SSQ-containing hybrid composites, mostly hybrid with polymers, have been designed due to their excellent thermomechanical properties [8]. SSQs could offer inorganic materials with biocompatible or physicochemical property which make them potential in biomedical applications. In fact, SSQs also could encapsulate and protect guest molecules from destruction or degradation.

In this chapter, we review some significant progresses made recently in the field silsesquioxane-based materials with well-controlled hierarchical structures. The fabrication process, structure-forming mechanism, the enhanced properties, and the final applications have been emphasized. We also review the main synthetic processes and applications of silsesquioxane hybrid materials. We will focus on the interaction between the SSQs and the hybrid materials, in order to reveal the hybrid mechanism.

6.2 Hierarchical Silsesquioxanes

Considerable attention has been paid to the design and fabrication of materials containing hierarchical silsesquioxane with improved physical and chemical properties. Hierarchical silsesquioxanes refer to the structure with morphology on different length scales. In detail, the SSQs involved hierarchical structures including core/shell, hollow, bowl-shaped, golf ball-like, etc.

6.2.1 SSQs with Core/Shell Structure

Core/shell structured materials based on silsesquioxanes have been extensively explored through the process such as sol-gel process, emulsion polymerization, template methods, electrospinning process, etc. The silsesquioxanes could not only offer the final materials with enhanced physical or chemical properties but also could protect the molecules in the core from leaching or degradation. Conventionally, the core/shell structured SSQs could be prepared from emulsion polymerization process. For example, the polysilsesquioxane/polyacrylate/polydimethylsiloxane core/shell particles have been successfully prepared via seeded emulsion polymerization of acrylate monomers and octamethylcyclotetrasiloxane with emulsifier as seeds [9]. Similarly, polystyrene/poly(γ-methacryloxypropyltrimethoxysilane) core/shell latex particles also were obtained through emulsion polymerization by adding methacryloxypropylene functionalized SSQ sol into the emulsion system of styrene monomer [10].

As Ha et al. reported, core/shell structured microspheres with raspberrylike to flowerlike morphology were fabricated with polysilsesquioxane (PSQ) shell grown stepwise on polystyrene template (Fig. 6.1). Time-dependent study on the forming of the core/shell structures demonstrates that the diameter and the topography of the microspheres could be well tailored by adjusting the silane precursor content. The

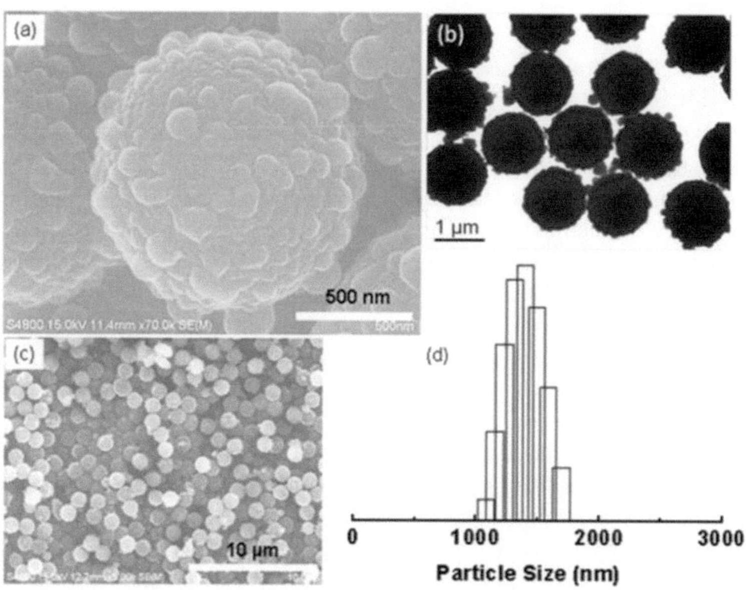

Fig. 6.1 The core/shell structured polysilsesquioxane/polystyrene microspheres. SEM (**a**, **c**) and TEM (**b**) images of the hierarchical microspheres with (**d**) the corresponding size distribution measured by DLS technique. (Reprinted with permission from Ref. [11]. Copyright 2017 Royal Society of Chemistry)

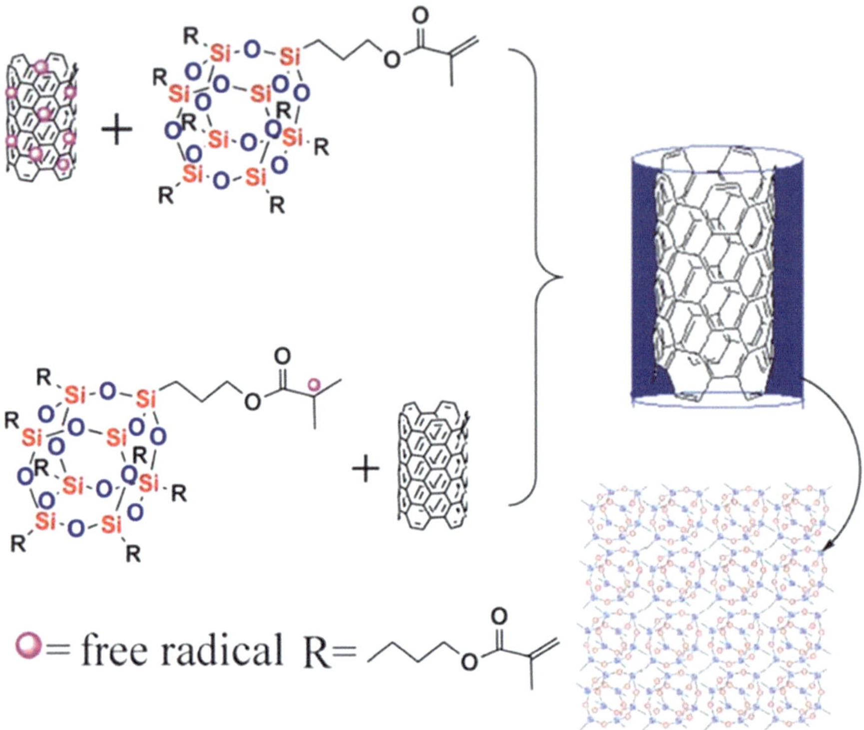

Fig. 6.2 Cross-linked silsesquioxane shell coated on carbon nanotube. (Reprinted with permission from Ref. [13]. Copyright 2014, American Chemical Society)

core/shell structure and the low surface energy from PSQs enable the assembled particulate film to exhibit superhydrophobic property [11]. By adding the mercaptopropyl trimethoxysilane precursor into PS aqueous suspension, Deng et al. also prepared PS@mercaptopropyl-PSQs core/shell structured materials [12].

In addition, core/shell structured hybrids with cross-linked octa-methacrylate-POSS (MA-POSS) as shell and multi-walled carbon nanotubes (MWCNTs) as center have been fabricated through an in situ free radical polymerization process (Fig. 6.2). Due to the coating of PSQs, the obtained materials could disperse well in organic solvent and show a controlled electrical performance [13, 14].

In the core/shell structured materials, silsesquioxanes could be used as a layer to protect the guest molecules in the core. For example, Tolbert et al. encapsulated hydrophobic sunscreens in the polysilsesquioxane-silica shell to reduce the leaking and photodegradation of the materials [15]. 3-Glycidoxypropyl-silsesquioxane materials were also coated on ferrite nanoparticles to help the particles' dispersion well in the epoxy and prevented sedimentation of the nanoparticles [16]. Core/shell structured polymer/POSS composites also have been fabricated through electrospinning process and been utilized to protect bioactive molecules in the core from thermal and chemical destruction [17, 18].

6.2.2 SSQs with Hollow Structures

As a specific member of SSQ-based materials, hollow SSQs have attracted much attention due to their unique micro- or nano-sized free volume in the structure, which can act as large reservoirs for guest molecule including catalysts or drugs [19]. For example, hybrid hollow mesoporous silsesquioxane nanoparticles have shown improved therapeutic performance and enhanced biomedical property [20].

Hollow-structured silsesquioxanes can be fabricated through the procedures like chemical etching, emulsion templating, hard templating, etc. Usually, hard template method for hollow polysilsesquioxane involves the formation of PSQ shell onto a template core which could be removed to leave behind a hollow shell. Ha et al. have offered a green method to fabricate a series of shape uniform and monodisperse hollow spheres based on organoalkoxysilanes and polystyrene template [21, 22]. With silica as a template, Koike et al. also have fabricated one kind of hollow silsesquioxane nanoparticles by stirring a biphasic mixture of organoalkoxysilane precursors and a silica aqueous suspension (Fig. 6.3). The tunable particle size, shell thickness, high surface areas, and large pore volumes of the hollow particles make them potential for practical application in many fields [23]. Similarly, with silica nanospheres as hard template, Zou et al. prepared organic groups, modified periodic mesoporous organosilica (PMO) hollow spheres, which

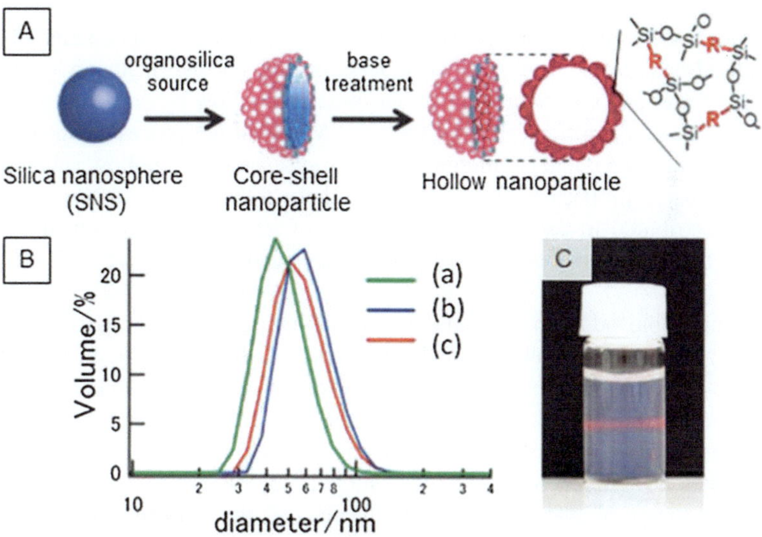

Fig. 6.3 (**a**) Schematic of the formation of hollow organosilica nanoparticles using silica nanoparticles as templates. (**b**) Dynamic light scattering curves of the dispersions of (*a*) silica particles sized ca. 40 nm, (*b*) silica core-organosilica shell nanoparticles, and (*c*) hollow organosilica nanoparticles. (**c**) Photograph of the dispersion of hollow organosilica nanoparticles (ca. 1.2 wt.%). (Reprinted with permission from Ref. [23]. Copyright 2013, Royal Society of Chemistry)

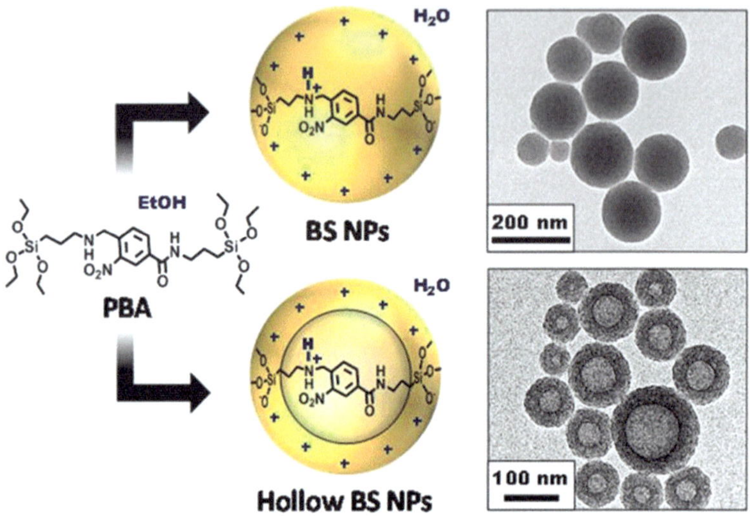

Fig. 6.4 Design of bridged silsesquioxane (BS) and hollow BS NPs via the sol-gel reactions of the photoresponsive bridged alkoxysilane (PBA) precursor, as shown by TEM micrographs. (Reprinted with permission from Ref. [26]. Copyright 2015, American Chemical Society)

are potential in the fields of drug delivery, bio-imaging, sensing, and heterogeneous catalysis [24].

Through an in situ polymerization method, Xing et al. prepared one kind of well-defined silsesquioxane hollow nanospheres based on the methacryloxypropyl silsesquioxanes and styrene monomer. The obtained hollow spheres show excellent methylene blue adsorption performance [25]. Via a sol-gel process, Fatieiev et al. prepared o-nitrophenylene-ammonium bridged silsesquioxane hollow spheres with photoresponsive properties (Fig. 6.4) [26]. About 50% organic content homogeneously distributed in the hybrid composition of silsesquioxanes, which make it potential for on-demand delivery of plasmid DNA in HeLa cancer cells via light actuation.

Via reversible addition-fragmentation chain transfer (RAFT) polymerization, Zhang et al. prepared hollow polymeric capsules by self-assembly of an amphiphilic POSS-based block copolymer (Fig. 6.5). The obtained hollow polymeric capsules are responsive to pH and redox potential, so the capsules could be further utilized in the responsive drug release and photodynamic therapy [27]. In addition, Jiang et al. fabricated hollow mono adamantane-functionalized POSS/β-cyclodextrin spheres by the assistance of interface of H_2O/toluene, and the obtained materials show excellent behavior of adhesion and proliferation of cells [28].

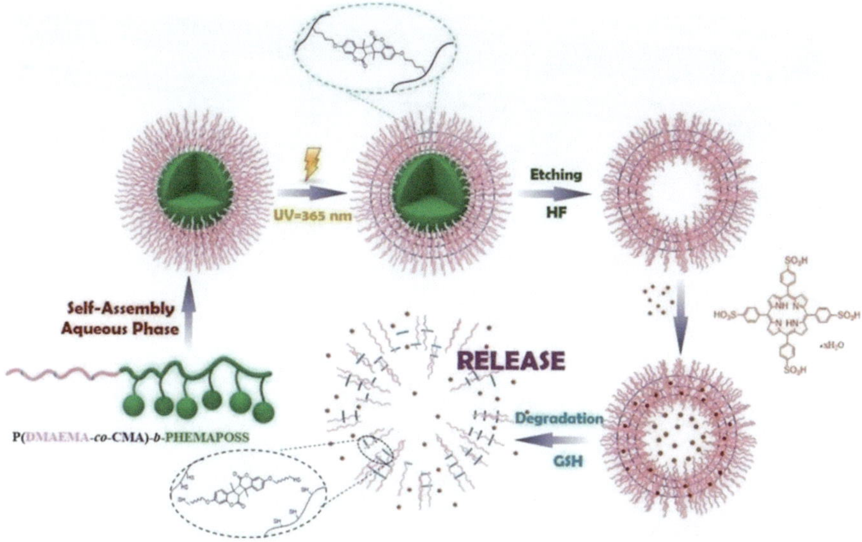

Fig. 6.5 Illustration of polymeric capsule preparation and the process of loading and release of drugs. (Reprinted with permission from Ref. [27]. Copyright 2016, American Chemical Society)

6.2.3 SSQs with Other Hierarchical Structures

Besides core/shell and hollow-structured silsesquioxane-based materials, there are still some other specially structured materials such as bowl-shaped, golf-like, grape-like materials, nanofibrillar micelles, etc. [29]. Bowl-shaped polysilsesquioxane particles have been also fabricated from two kinds of organoalkoxysilane precursors through a facile and controllable sol-gel polymerization procedure. The key factors for the formation of a bowl-shaped structure are the flexibility of the shell, osmotic pressure, as well as the centrifugal force. The concentrations of the two silane precursors, methyltriethoxylsilane (MTES) and phenyltriethoxysilane (PTES) monomers, as well as the ammonia concentration show a significant effect on the morphology of final particles [30].

Golf ball-like polymethylsilsesquioxane microspheres were fabricated through the hydrolysis and co-condensation of methyltrimethoxysilane (MTMS) and tetraethoxysilane (TEOS) (Fig. 6.6). The golf ball-like wrinkled surface was proved to be obtained from the self-assembly of polymethylsilsesquioxane and silica under Ostwald ripening process [31].

With polystyrene colloidal crystals as templates, grapelike silica-based hierarchical porous interlocked aminopropyl polysilsesquioxanes have been fabricated, whose structure is the integration of open-mouthed structure, hierarchical porous nanostructure, and interlocked architecture (Fig. 6.7) [32].

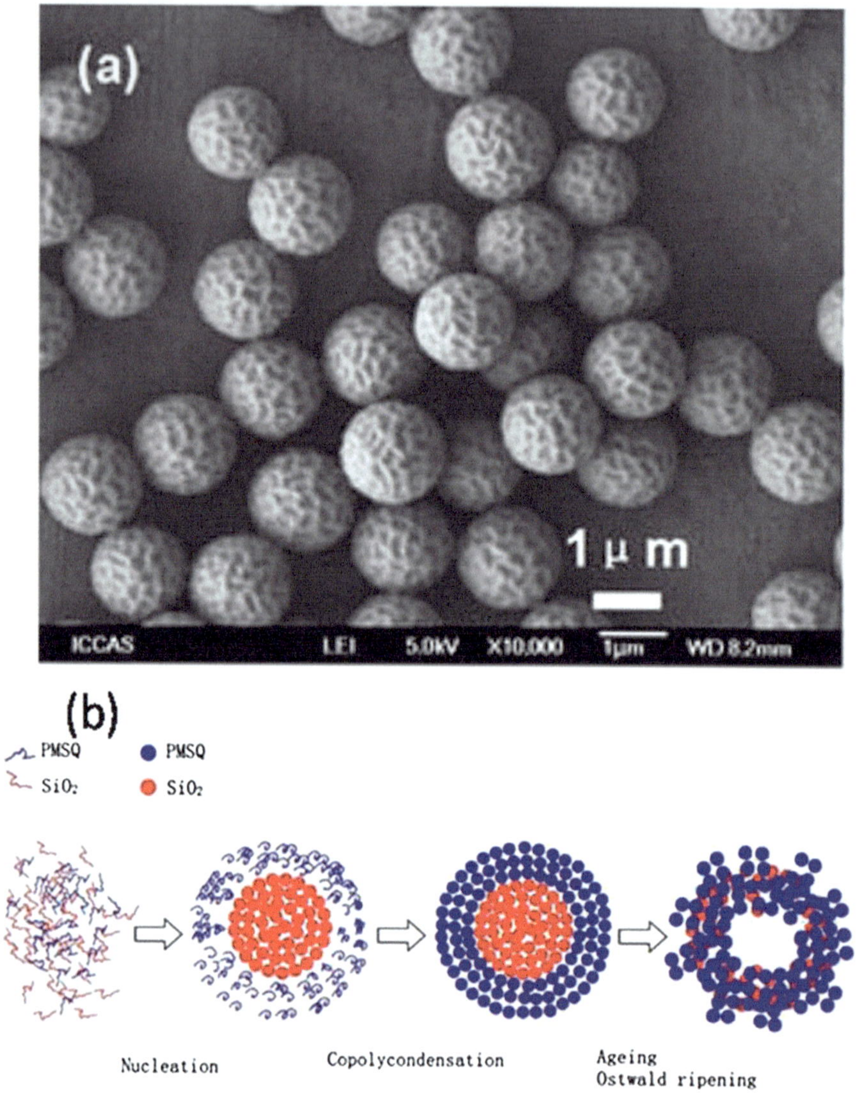

Fig. 6.6 (**a**) SEM image of the polysilsesquioxane/silica spheres and (**b**) the formation mechanism of the golf-ball like spheres. (Reprinted with permission from Ref. [31]. Copyright 2011, Royal Society of Chemistry)

Yolk-shell structured mesoporous inorganic-organic hybrid spheres with well-controlled size have been fabricated by Teng et al. with TEOS and 1,2-bis (triethoxysilyl) ethane (BTSE) as precursors. The monodisperse hybrid spheres with tunable diameter, shell thickness, and core size show excellent hemocompatibility and have a great promise for various applications [33].

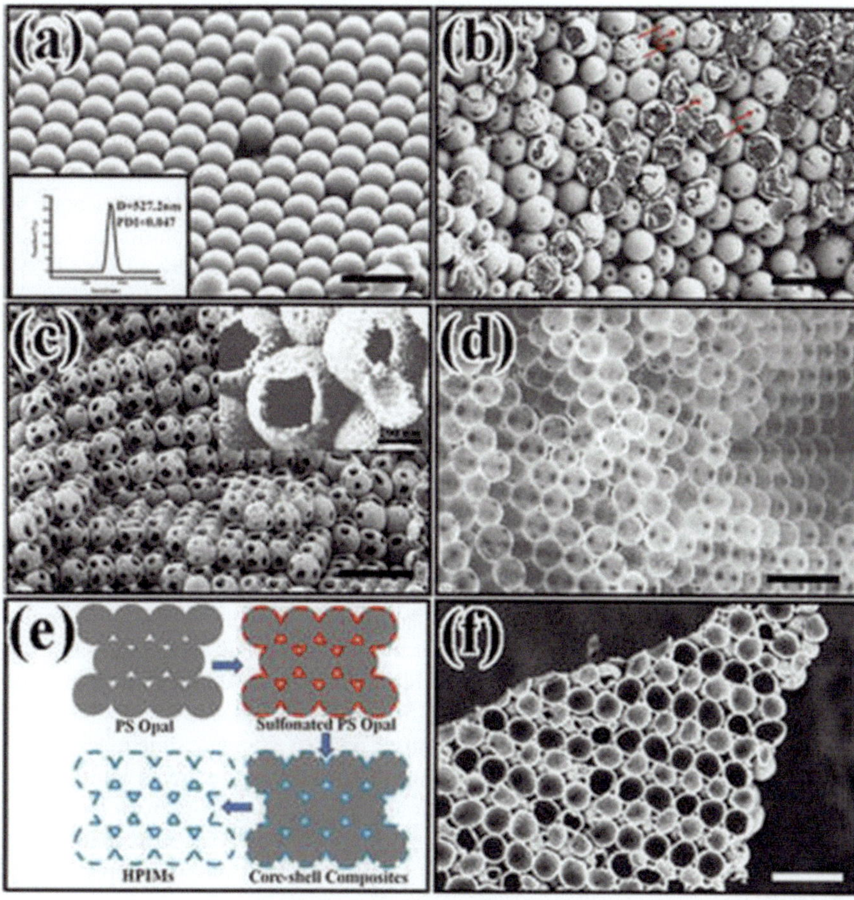

Fig. 6.7 (**a**) SEM image of polystyrenes and average size distribution of the microspheres (inset); (**b**) SEM image of polystyrene@silica composites; (**c**) SEM (1.00 kV) image and (**d**) STEM (5.00 kV) image of grapelike microspheres; (**e**) schematic illustration of the preparation process of the grapelike microspheres; (**f**) cross-sectional SEM image of grapelike microspheres. All the scale bars are 1 μm. (Reprinted with permission from Ref. [32]. Copyright 2015, American Chemical Society)

Rodlike polysilsesquioxane also has been successfully prepared by oxidation and hydrolytic polycondensation of 3-mercaptopropyltrimethoxysilane (MPTMS) in a mixed aqueous solution of NaOH and H_2O_2. The obtained rodlike sulfo-group containing PSQs have a hexagonally stacked structure and show high proton conductivity [34].

Besides one-dimensional shape, two dimension layers also have been developed. Kataoka et al. fabricated microporous layered perovskites from metal halides and cage-like silsesquioxane. In the forming process, POSS could form micropores between the metal halide perovskite layers, which may offer new property for perovskites [35]. Similarly, hierarchical porous structured films on PET matrix were also achieved with cross-linked POSS as the silica source and a copolymer surfactant as the porogen. The final obtained films, with high surface areas and well-controlled pore sizes, could be used in many fields such as supercapacitors, sensors, filtration, and catalysis [36]. Much attention also has been drawn to three-dimensional porous scaffold. Through sol-gel polymerization from bridged silane precursors, microscale aerogels containing nanoparticle-constructed networks have been fabricated by vacuum drying process [37]. Through a facile cross-linking and solution extraction process, ionic gels and scaffolds with interconnected mesopores derived from POSS have been fabricated. The obtained POSS-based porous materials show excellent catalytic performance and superior lithium ion battery performance [38].

6.3 Silsesquioxane Hybrid Composites

6.3.1 SSQ Hybrid with Polymer

Recently, much attention has been attracted by POSS-based molecules and polymers, especially the design, preparation, and applications of the POSS hybrid polymers [39–45]. Due to the organic/inorganic hybrid structure in the molecular level, silsesquioxane has become an ideal building for polymer/SSQ nanocomposite through sol-gel or melt mixing process [46–51]. POSS consists of a stable silica core and eight active/inactive functional arms. The active organic functional groups on the POSS could work as a nucleus for covalent bonding to create multi-armed polymers with enhanced mechanical and biological properties [52]. In fact, polymer incorporated with suitable organic functionalized POSS derivatives as side chains could be obtained for targeted applications [53]. For example, the star-shaped POSS-polycaprolactone- polyurethane (POSS-PCL-PU) film with high porosity has been fabricated. The nanocomposites show unique surface nanotopography and excellent biocompatibility which make it a great candidate as a tissue engineering scaffold biomaterial [54]. Through atom transfer radical polymerization (ATRP), Qiang et al. synthesized two different eight-arm star-shaped POSS fluorinated acrylates, which show a great potential in filtration, cell culture, tissue engineering, and marine antifouling applications [55]. Zhang et al. prepared the star-shaped organic/inorganic hybrid poly (L-lactide) (PLLA) based on octa(3-hydroxypropyl) polyhedral oligomeric silsesquioxane via ring-opening polymerization (ROP) of L-lactide (LLA) for biological and medical applications [56]. In electrochromic applications, PANI is the most widely used due to their good environmental stability and electrical property, which could be improved by hybrid with POSS. Jia et al. prepared polyaniline (PANI)-tethered cubic POSS via oxidative copolymerization of octa(aminophenyl) silsesquioxane and aniline in the presence of HCl. The as-prepared POSS-PANI copolymer in emeraldine salt (ES) was filtered, washed, and treated with triethylamine to achieve POSS-PANI copolymer in emeraldine base (EB) form. The films fabricated via layer-by-layer (LBL) assembly show enhanced electrical

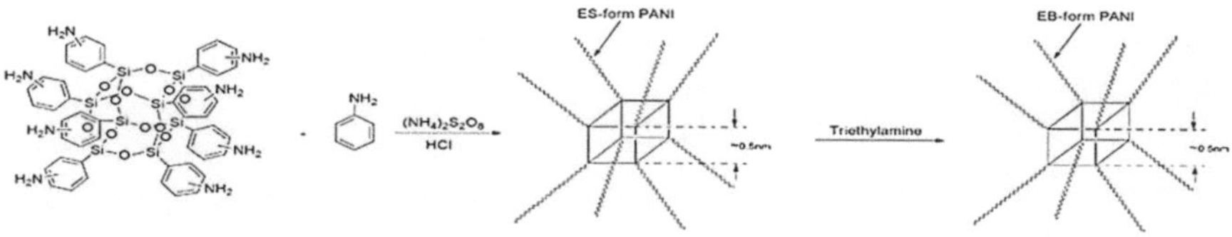

Fig. 6.8 Oxidative copolymerization of multi-armed polyaniline (PANI)-tethered cubic POSS. POSS-PANI-ES was formed from POSS and aniline, followed by conversion to POSS-PANI-EB. (Reprinted with permission from Ref. [57]. Copyright 2009, American Chemical Society)

conductivity (Fig. 6.8) [57]. Lin et al. also fabricated multi-armed polyaniline-octa-aminophenylsilsesquioxane conjugates (PANI-SSQ) with hierarchical porous structure, exhibiting excellent specific capacitance and stability, which make it potential for electrode material applications [58].

Lin et al. fabricated one kind of carbon nanofiber-silsesquioxane-polyaniline nanohybrids with hierarchical structure and used as flexible supercapacitor electrodes [59]. Firstly, through amide linkage, octa-aminophenylsilsesquioxane was first attached onto the carboxylated nanocarbon (CNF-COOH). And then phenylamino-modified CNF surface was copolymerized with aniline via chemical oxidative polymerization to create a unique CNF-conjugated polymer hybrid (CNFS-PANI). The obtained flexible CNFS-PANI nanohybrid shows excellent electric conductivity and specific capacitance, suggesting its potential as electrochemical electrode material (Fig. 6.9).

Based on poly(styrene-b-butadiene-b-styrene) (SBS) cross-linked by POSS, Bai et al. have fabricated nanostructured thermoplastic elastomeric composites [60]. Through ATRP and the copper-catalyzed azide-alkyne "click" reaction process, Zhang et al. have prepared dumbbell-shaped POSS/poly(tert-butyl acrylate) (PtBA) from alkyne-functionalized POSS and azido-terminated PtBA [61]. Through electrospinning process, Pisuchpen et al. have fabricated polystyrene polyhedral oligomeric silsesquioxane-derived methacrylate (PS-co-PMAPOSS) copolymer fibers. The obtained fibers with highly porous structure are potential for further novel applications [62]. This polymerization process is a combination of reversible addition-fragmentation chain transfer (RAFT) and activator regenerated by electron transfer for ATRP (ARGET ATRP).

6.3.2 SSQ Hybrid with Silica

Silsesquioxane mesoporous frameworks are a kind of synergistic combination of inorganic silica, mesopores, and organics with some physicochemical and biocompatible properties which could be used as bio-imaging agent and drug delivery system [63]. Via the hydrolysis and condensation polymerization from the mixture of TEOS and MTMS, Hayashi et al. have prepared mesoporous and hydrophobic silicate silsesquioxane hybrid copolymers under a strong basic condition [64].

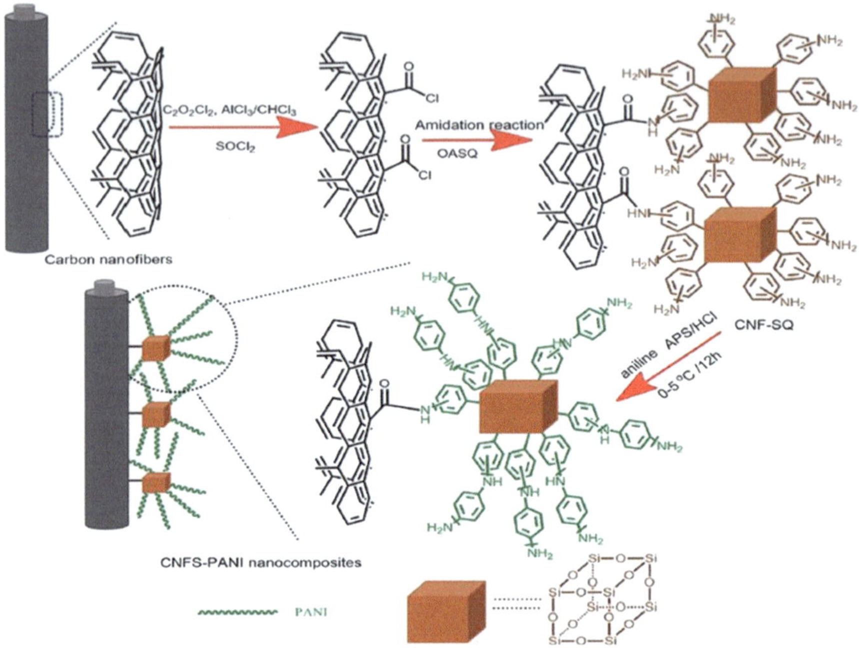

Fig. 6.9 Schematic design for the preparation of carbon nanofiber-silsesquioxane-polyaniline (CNFS-PANI) nanohybrids by functionalizing CNF with octa-aminophenylsilsesquioxane (OASQ) through amidation followed by a reaction with aniline to form a hyperbranched PANI-grafted CNF. (Reprinted with permission from Ref. [59]. Copyright 2015, Royal Society of Chemistry)

Azo-functionalized silsesquioxanes have been used as gatekeeper for drug-loaded mesoporous silica, and the obtained drug release system shows enzyme-responsive drug release behavior [65]. Dopierala et al. have reported hydrophobic POSS materials containing silica nanoparticles which could be used as self-cleaning coating. In this system, a hybrid of silica nanoparticles make the POSS monolayer more condensed and rigid; meanwhile, POSS molecules prevent silica nanoparticles from aggregations (Fig. 6.10) [66].

6.3.3 SSQ Hybrid with Fe_3O_4

Magnetic nanoparticles could offer SSQ materials with interesting properties for applications such as medical diagnostic, wastewater treatment, catalyst carrier, and drug delivery [67–69]. Via a one-pot coprecipitation and surface grafting approach, Ha et al. have fabricated magnetite-polysilsesquioxane hybrid nanoparticles with ferrous, ferric chloride and various silane monomers as raw materials (Fig. 6.11) [70]. The magnetic PSSQ hybrid nanoparticle composites have excellent adsorption

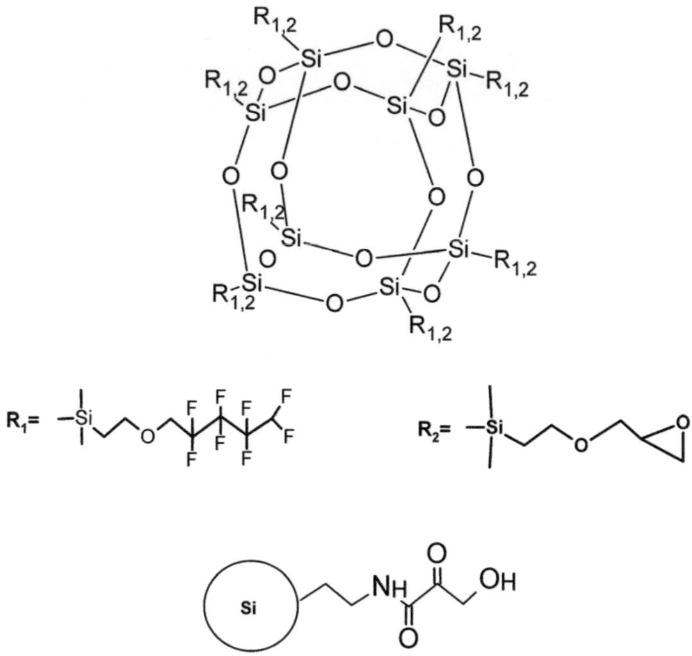

Fig. 6.10 The chemical structures of the silsesquioxane cage with substituents (on the top) and 3-aminopropyl- (3-oxobutanoic acid) functionalized silica (on the bottom). (Reprinted with permission from Ref. [66]. Copyright 2016, Royal Society of Chemistry)

and selectivity behavior for iron (Fe^{3+}). The magnetic and functional magnetic nanoparticles can be separated easily from the adsorbed solution by a bar magnet and reused for the repeated cycles of adsorption.

In another report, octavinyl POSS was first constructed on the Fe_3O_4 nanoparticles by surface polymerization process. After modification with dithiol via thiol-ene addition reaction on the as-prepared Fe_3O_4@POSS hybrid composite, the ultimate material (Fe_3O_4@POSS-SH) could also be used to remove heavy metal ions and organic dyes from wastewater (Fig. 6.12) [71].

6.3.4 SSQ Hybrid with Noble Metals

Hierarchically porous SSQ with well-defined macropores or mesopores could be embedded with noble metal nanoparticles, which are promising as heterogeneous catalysts or antimicrobial materials [72].

Silver nanoparticles could be hybrid with POSS and used for antimicrobial applications [73]. To avoid the aggregation of silver nanoparticles, hydrophobic POSS have been used to protect the particles [74]. Schneid et al. synthesized

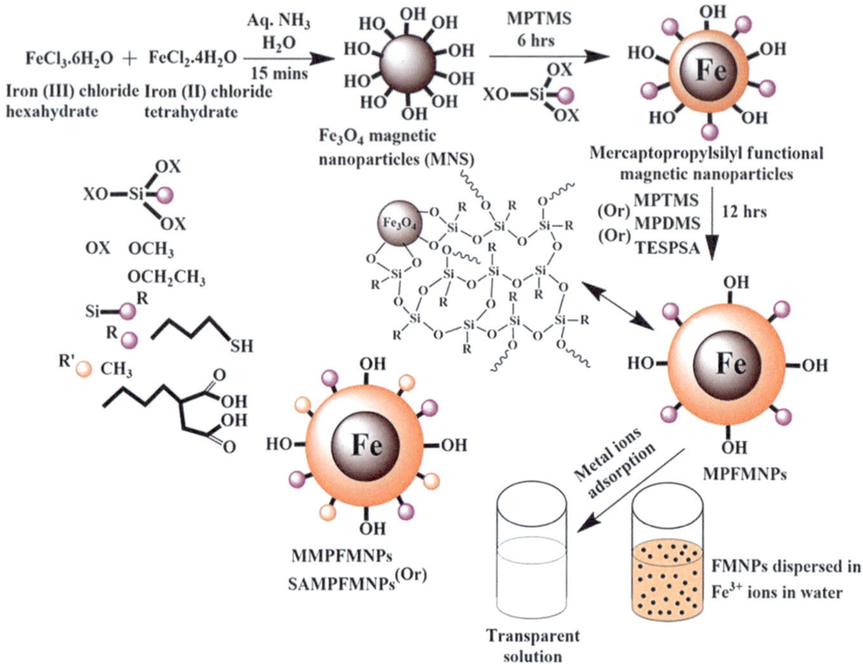

Fig. 6.11 Schematic diagram of the formation of magnetite-polysilsesquioxane hybrid nanoparticles. (Reprinted with permission from Ref. [70]. Copyright 2017, Royal Society of Chemistry)

spherical silver nanoparticles hybrid with charged silsesquioxane containing a quaternary ammonium group and applied as an antibacterial agent. The cytotoxicity assay showed that the system also is safe for mammalian cells at the studied concentrations [75].

The SSQ hybrid silver particles could also be utilized as a smart colorimetric probe for rapid and accurate detection of hydrogen sulfide. With a poly-POSS formaldehyde polymer (PPF) cage as a ligand and reductant, Zhang et al. developed a novel strategy for the fabrication of a positively charged silver nanoparticle probe. POSS cage works as a capping ligand and reducing agent for the fabrication of well-dispersed silver nanoparticles; the final system has excellent performance on the rapid and accurate detection of hydrogen sulfide (Fig. 6.13) [76].

Functional polysilsesquioxanes containing different organic groups could be utilized as carriers to load and protect gold nanoparticles for applications such as catalyst, sensors, DNA assays, etc. [77, 78]. Ha's group has loaded gold nanoparticles on the shell of cyanopropyl polysilsesquioxane (CPSQ) hollow spheres which initially combined cyanopropyl groups and pores on the shell (Fig. 6.14). The pores make it possible for guest molecules to diffuse through the shell, and the cyanopropyl groups make the noble metals stable due to high affinity between cyanopropyl group and gold. PSQ hollow spheres with well-dispersed gold

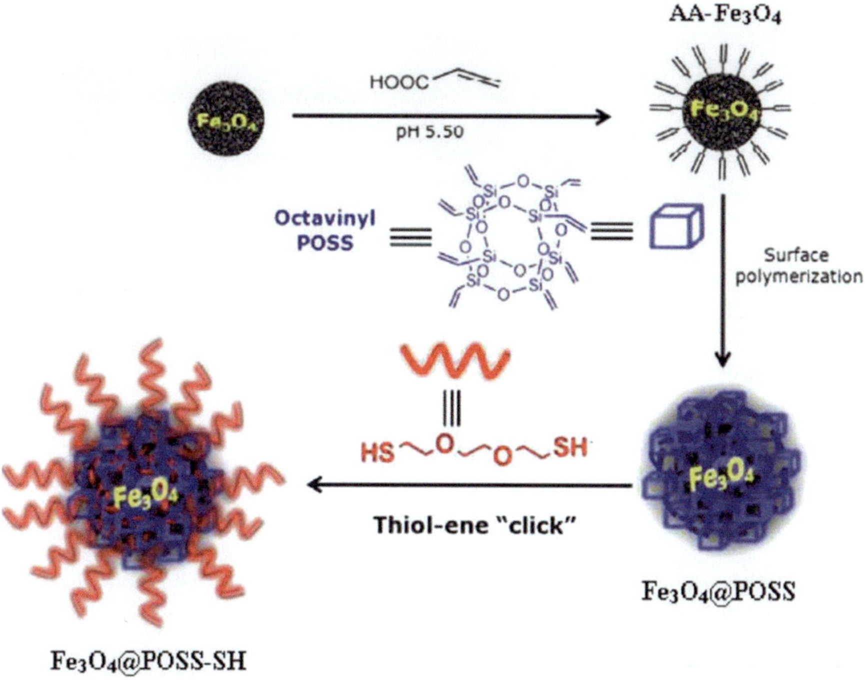

Fig. 6.12 Schematic representation of the route for synthesis of Fe_3O_4@POSS-SH. (Reprinted with permission from Ref. [71]. Copyright 2013, American Chemical Society)

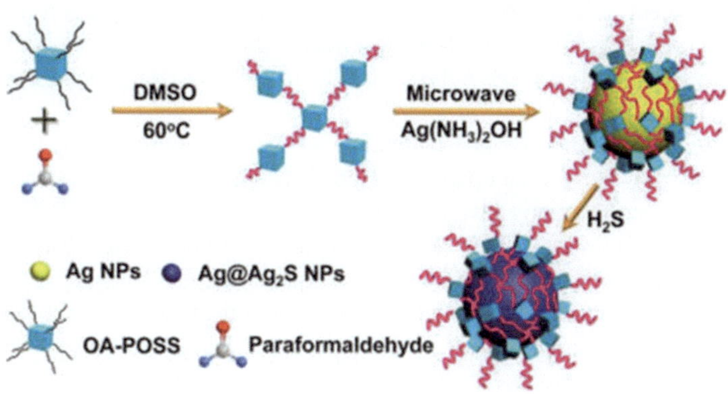

Fig. 6.13 Schematic illustration of the preparation of poly-POSS formaldehyde polymer-AgNPs and its colorimetric detection of H_2S. (Reprinted with permission from Ref. [76]. Copyright 2017, American Chemical Society)

nanoparticles demonstrate excellent catalytic performances [79]. Scholder et al. have fabricated gold nanoparticles on hydrophilic dithiol-modified POSS scaffolds, showing highly efficient catalytic performance [80].

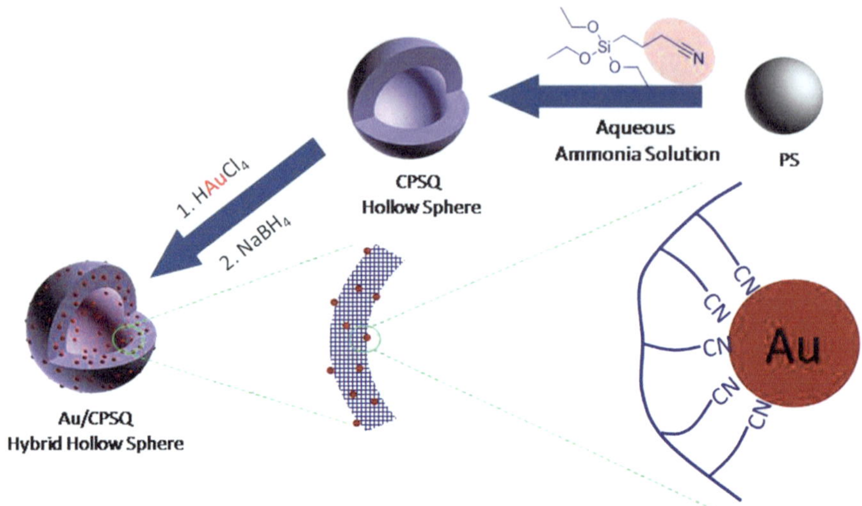

Fig. 6.14 Fabrication of CPSQ hollow spheres and Au/CPSQ hybrid nanocomposites loaded with highly dispersed Au nanoparticles. (Reprinted with permission from Ref. [79]. Copyright 2012, Royal Society of Chemistry)

Silva et al. have designed nanocomposites by water-soluble silsesquioxane polymer hybrid with gold nanoparticles for detection of pollutant (nitrophenol isomers) [81]. Zapp et al. loaded gold nanoparticles on silsesquioxane based on liquid crystal and applied them as nanostructured immunosensor to detect a protein troponin T [82]. Brigo et al. deposited an aryl-bridged polysilsesquioxane film on Au nanoparticle to form an aryl-bridged polysilsesquioxane system for xylene gas optical sensors [83].

Silsesquioxane hybrid with palladium (Pd) was utilized to protect the nanoparticles [84–86]. Pd nanoparticles capped with SSQ possessing stable reactivity of Pd-catalyzed reactions have been quickly and easily synthesized [87]. Tanabe et al. have produced a palladium complex with an O,O-chelating silsesquioxanate ligand from an incompletely condensed silsesquioxane which reacts with a palladium precursor [88]. Through on-site reduction-based methodology, Moitra et al. have loaded Pd nanoparticles onto the hierarchically porous hydrogen silsesquioxane (HSQ) monolithic (Fig. 6.15). The obtained Pd@HSQ catalyst shows high catalytic activity, reusability, and easy handling [89].

6.3.5 SSQ Hybrid with Carbon Nanomaterials

Silsesquioxane nanocomposites hybrid with carbon through covalent or uncovalent linkage will improve the physical or chemical property of the carbon materials

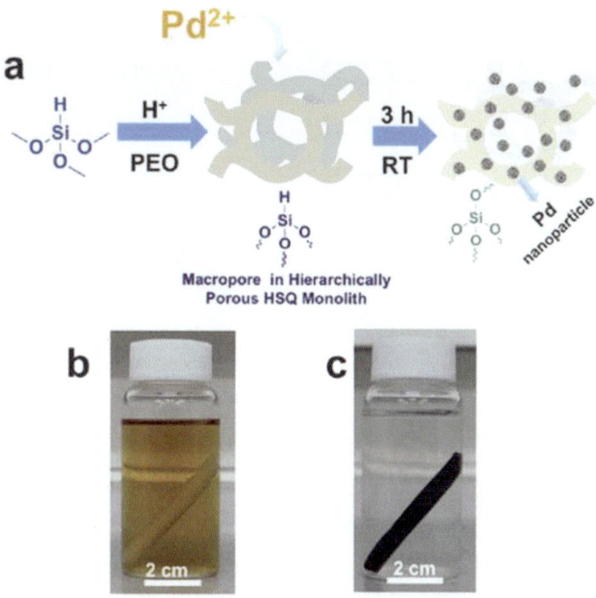

Fig. 6.15 (**a**) Schematic representation of on-site reduction of $PdCl_2$ on a hydrogen silsesquioxane (HSQ) monolith; simultaneous transformation of Pd2+ to Pd(0) nanoparticles and Si-H to Si-OB. (**b**, **c**) Digital camera images of the HSQ monolith right after immersing in $PdCl_2$ solution and after reacting for 3 h, respectively. (Reprinted with permission from Ref. [89]. Copyright 2014, Royal Society of Chemistry)

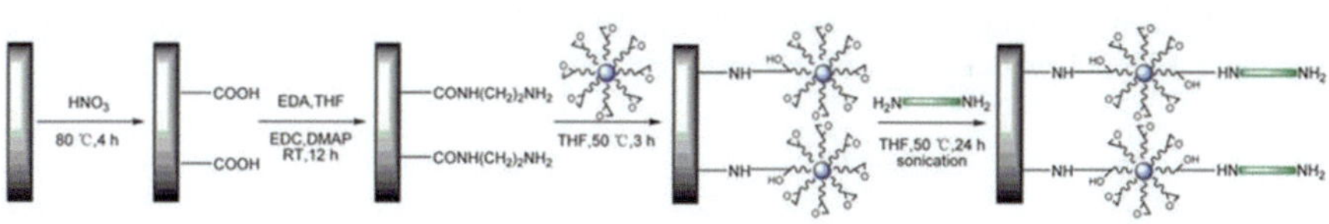

Fig. 6.16 The schematic of binary grafting procedure including grafting carbon nanotubes onto carbon fibers using polyhedral oligomeric silsesquioxanes as coupling agents. (Reprinted with permission from Ref. [90]. Copyright 2011, Royal Society of Chemistry)

[69]. Octaglycidyldimethylsilyl POSS worked as coupling agents to graft carbon nanotubes onto carbon fibers (Fig. 6.16) [90]. Due to the introduction of POSS, the properties of the resulting composites such as toughness, service temperature, oxidation, and chemical resistance have been enhanced.

Chemical functionalization of carbon nanotube with POSS, as a kind of innovative 0D–1D nanohybrid structure, is an important strategy to improve the thermal or electrical property of the composite. Zhang et al. prepared a core/shell structured composite with a MWCNT at the center and cross-linked octa-acrylate POSS as shell, which could improve the dielectric permittivity and low dielectric loss of

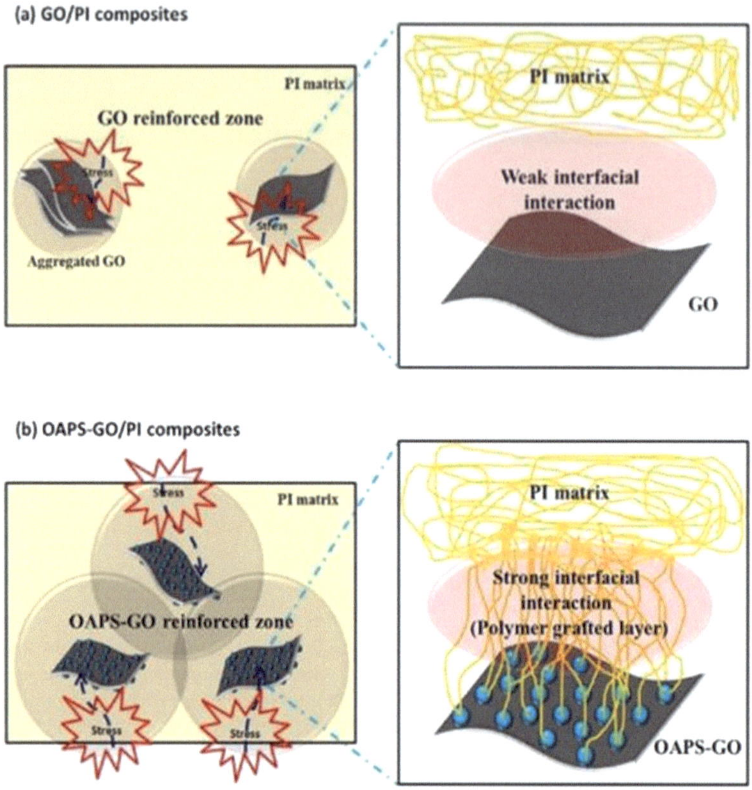

Fig. 6.17 Proposed model of the reinforcing (**a**) GO and (**b**) octa(aminopropyl) POSS-GO in polyimide (PI) polymeric matrix. (Reprinted with permission from Ref. [97]. Copyright 2014, American Chemical Society)

polymers [91]. Through the formation of amide bonds, Sabet et al. fabricated POSS covalent-bonded MWCNT, and the obtained material has the high thermal stability which could effectively reinforce polymer materials [92]. Through Friedel-Crafts acylation and amidation chemistry, cube-like octa-organsilsesquioxane hybrid with one-dimensional single-walled carbon nanotubes (SWNTs) has been reported by Xu et al. [93].

Polyoligomeric silsesquioxane (POSS)-decorated graphene or graphene oxide (GO) nanoplatelets have been obtained by various methods [94, 95]. Typically, POSS-GO could be successfully synthesized via amide formation between the octa (aminopropyl) POSS (OAPOSS) and GO [96]. Utilizing similar amide formation process, POSS-functionalized graphene nanosheets have been fabricated from amine-functionalized POSS and oxygen-containing groups, graphene oxide (GO). The obtained nanocomposites are highly soluble in various organic solvents and show potential in multifunctional applications (Fig. 6.17) [97, 98]. Through a

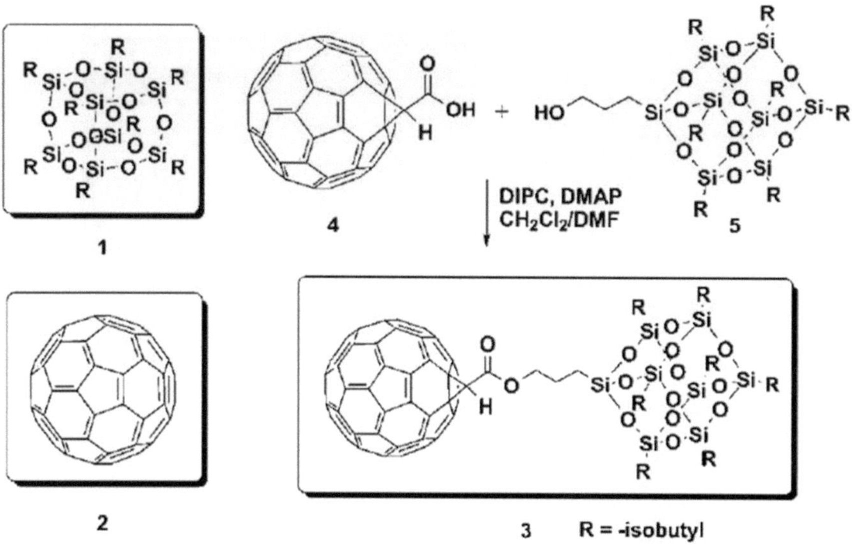

Fig. 6.18 Chemical structures of octaisobutyl-POSS (1), C60 (2), and POSS-C_{60} (3) and the synthetic route to POSS-C_{60} (3). (Reprinted with permission from Ref. [100]. Copyright 2011, Royal Society of Chemistry)

one-step hydrothermal process, Bai et al. have prepared POSS/reduced graphene oxide nanocomposite (POSS/rGO) as a hydrophobic electrochemical sensor for nitrite detection [99].

Cheng's group designed and fabricated a sphere-cubic-shaped amphiphile based on 3-hydroxypropylheptaisobutyl-POSS and modified fullerene (Fig. 6.18). To connect the two building blocks and prevent macroscopic phase separation, a short covalent linkage was thus selected. The hybrid nanocomposite of layered structure with an alternating conductive fullerene and insulating POSS layer structure is of great interest for various potential applications such as nano-capacitors [100]. In the same group, POSS-[60]fullerene (POSS-C_{60}) dyad was designed and used as a novel electron acceptor for polymer solar cells with an inverted device configuration [101].

6.3.6 *Hybrid with Quantum Dot*

Utilizing organic substituted POSS as stabilizer, hybrid CdSe quantum dots (QDs) could be fabricated [102, 103]. By embedding water-soluble N, S-co-doped carbon dots into a POSS matrix, highly efficient solid-state luminophores with strong deep blue emission could be obtained [104]. Due to the bulkiness of the siloxane core, mercapto-substituted POSS(SH-POSS) was utilized as a ligand and a steric stabilizer

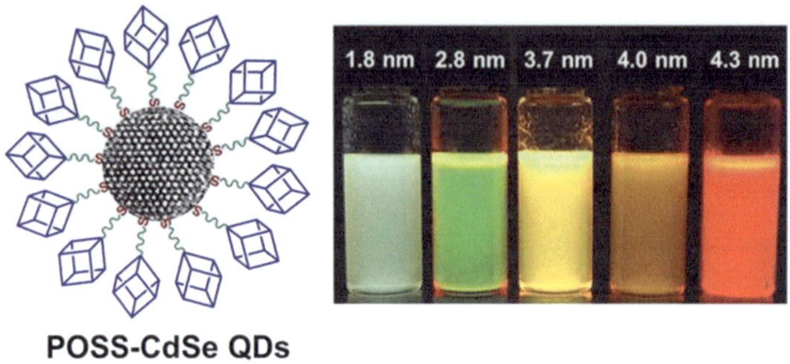

Fig. 6.19 Schematic drawing of a CdSe QD capped with mercaptopropylisobutyl-POSS and a photograph showing the photoluminescence color change of the POSS-CdSe QDs under a UV lamp with different sizes of CdSe QDs. (Reprinted with permission from Ref. [105]. Copyright 2013, American Chemical Society)

to produce wurtzite phase CdSe QDs. The obtained SH-POSS capped QDs show excellent optical properties including photoluminescence quantum efficiencies and fluorescence lifetimes (Fig. 6.19) [105].

6.4 Conclusions

In summary, recent advances and applications in various aspects of the silsesquioxane-based hierarchical and hybrid materials have been highlighted. From the recent trend, it is clear that more and more hierarchical silsesquioxanes have been designed and fabricated for unique chemical and physical properties. At the same time, a wide category of SSQ hybrid materials has been developed to meet the requirement for a versatile application. It is believed that, in the near future, more sophisticated silsesquioxane composite with various morphologies will be created for more and better high performance.

References

1. Schramm C, Rinderer B, Tessadri R (2017) Ladder-like aromatic imide-functionalized poly (silsesquioxane): preparation and characterization via the sol-gel route. Adv Polym Technol 36(1):77–85
2. Dong F, Ha C-S (2012) Multifunctional materials based on polysilsesquioxanes. Macromol Res 20(4):335–343
3. Kuo S-W, Chang F-C (2011) POSS related polymer nanocomposites. Prog Polym Sci 36(12):1649–1696
4. Chen Y, Shi J (2016) Chemistry of mesoporous organosilica in nanotechnology: molecularly organic-inorganic hybridization into frameworks. Adv Mater 28(17):3235–3272

5. Kannan RY, Salacinski HJ, Butler PE, Seifalian AM (2005) Polyhedral oligomeric silsesquioxane nanocomposites: the next generation material for biomedical applications. Acc Chem Res 38(11):879–884
6. Cordes DB, Lickiss PD, Rataboul F (2010) Recent developments in the chemistry of cubic polyhedral oligosilsesquioxanes. Chem Rev 110(4):2081–2173
7. Hu L-C, Shea KJ (2011) Organo-silica hybrid functional nanomaterials: how do organic bridging groups and silsesquioxane moieties work hand-in-hand? Chem Soc Rev 40(2):688–695
8. Ayandele E, Sarkar B, Alexandridis P (2012) Polyhedral oligomeric silsesquioxane (POSS)-containing polymer nanocomposites. Nanomaterials 2(4):445–475
9. Bai RQ, Qiu T, Han F, He LF, Li XY (2012) Preparation and characterization of inorganic-organic trilayer core-shell polysilsesquioxane/polyacrylate/polydimethylsiloxane hybrid latex particles. Appl Surf Sci 258(19):7683–7688
10. Yang SL, Song CF, Qiu T, Guo LH, Li XY (2013) Synthesis of polystyrene/polysilsesquioxane core/shell composite particles via emulsion polymerization in the existence of poly(gamma-methacryloxypropyl trimethoxysilane) sol. Langmuir 29(1):92–101
11. Dong F, Xie H, Zheng Q, Ha C-S (2017) Superhydrophobic polysilsesquioxane/polystyrene microspheres with controllable morphology: from raspberry-like to flower-like structure. RSC Adv 7(11):6685–6690
12. Deng TS, Bongard HJ, Marlow F (2015) A one-step method to coat polystyrene particles with an organo-silica shell and their functionalization. Mater Chem Phys 162:548–554
13. Sun D, Zhou Z, Chen G-X, Li Q (2014) Regulated dielectric loss of polymer composites from coating carbon nanotubes with a cross-linked silsesquioxane shell through free-radical polymerization. ACS Appl Mater Interfaces 6(21):18635–18643
14. Sun D, Li Q, Chen G-X (2014) Preparation of core-shell structured carbon nanotube-silsesquioxane hybrids by a direct free-radical reaction. Mater Lett 120:90–93
15. Tolbert SH, McFadden PD, Loy DA (2016) New hybrid organic/inorganic polysilsesquioxane-silica particles as sunscreens. ACS Appl Mater Interfaces 8(5):3160–3174
16. Olsson RT, Hedenqvist MS, Strom V, Deng J, Savage SJ, Gedde UW (2011) Core-Shell structured ferrite-silsesquioxane-epoxy nanocomposites: composite homogeneity and mechanical and magnetic properties. Polym Eng Sci 51(5):862–874
17. Bauer AJP, Wu Y, Li B (2016) Electrospun poly(epsilon-caprolactone)/polyhedral oligomeric silsesquioxane-based copolymer blends: evolution of fiber internal structures. Macromol Biosci 16(5):705–716
18. Bauer AJP, Zeng T, Liu J, Uthaisar C, Li B (2014) The enhanced encapsulation capacity of polyhedral oligomeric silsesquioxane-based copolymers for the fabrication of electrospun core/shell fibers. Macromol Rapid Commun 35(7):715–720
19. Wei K, Wang L, Li L, Zheng S (2015) Synthesis and characterization of bead-like poly (N-isopropylacrylamide) copolymers with double decker silsesquioxane in the main chains. Polym Chem 6(2):256–269
20. Chen Y, Meng Q, Wu M, Wang S, Xu P, Chen H, Li Y, Zhang L, Wang L, Shi J (2014) Hollow mesoporous organosilica nanoparticles: a generic intelligent framework-hybridization approach for biomedicine. J Am Chem Soc 136(46):16326–16334
21. Dong F, Guo W, Chu S-W, Ha C-S (2010) Novel fluorinated polysilsesquioxane hollow spheres: synthesis and application in drug release. Chem Commun 46(40):7498–7500
22. Dong F, Guo W, Park SS, Ha CS (2011) Uniform and monodisperse polysilsesquioxane hollow spheres: synthesis from aqueous solution and use in pollutant removal. J Mater Chem 21(29):10744–10749
23. Koike N, Ikuno T, Okubo T, Shimojima A (2013) Synthesis of monodisperse organosilica nanoparticles with hollow interiors and porous shells using silica nanospheres as templates. Chem Commun 49(44):4998–5000
24. Zou H, Wang R, Li X, Wang X, Zeng S, Ding S, Li L, Zhang Z, Qiu S (2014) An organosilane-directed growth-induced etching strategy for preparing hollow/yolk-shell mesoporous

organosilica nanospheres with perpendicular mesochannels and amphiphilic frameworks. J Mater Chem A 2(31):12403–12412

25. Xing Y, Peng J, Xu K, Lin W, Gao S, Ren Y, Gui X, Liang S, Chen M (2016) Polymerizable molecular silsesquioxane cage armored hybrid microcapsules with in situ shell functionalization. Chem A Eur J 22(6):2114–2126
26. Fatieiev Y, Croissant JG, Alsaiari S, Moosa BA, Anjum DH, Khashab NM (2015) Photoresponsive bridged silsesquioxane nanoparticles with tunable morphology for light-triggered plasmid DNA delivery. ACS Appl Mater Interfaces 7(45):24993–24997
27. Zhang Z, Xue Y, Zhang P, Muller AHE, Zhang W (2016) Hollow polymeric capsules from POSS-based block copolymer for photodynamic therapy. Macromolecules 49(22):8440–8448
28. Jiang B, Tao W, Lu X, Liu Y, Jin H, Pang Y, Sun X, Yan D, Zhou Y (2012) A POSS-based supramolecular amphiphile and its hierarchical self-assembly behaviors. Macromol Rapid Commun 33(9):767–772
29. Simionescu B, Bordianu I-E, Aflori M, Doroftei F, Mares M, Patras X, Nicolescu A, Olaru M (2012) Hierarchically structured polymer blends based on silsesquioxane hybrid nanocomposites with quaternary ammonium units for antimicrobial coatings. Mater Chem Phys 134(1):190–199
30. Cao JJ, Zhu QZ, Dou JT, Li CX, Chen WK, Li ZQ (2013) Controlling sol-gel polymerization to create bowl-shaped polysilsesquioxane particles with a kippah structure. Polymer 54(10):2493–2497
31. Zhou Q, Xiang H, Fan H, Yang X, Zhao N, Xu J (2011) Facile fabrication of golf ball-like hollow microspheres of organic-inorganic silica. J Mater Chem 21(34):13056–13061
32. Gu JY, Wang XM, Tian L, Feng L, Qu JY, Liu PG, Zhang X (2015) Construction of grape-like silica-based hierarchical porous interlocked microcapsules by colloidal crystals templates. Langmuir 31(45):12530–12536
33. Teng Z, Wang S, Su X, Chen G, Liu Y, Luo Z, Luo W, Tang Y, Ju H, Zhao D, Lu G (2014) Facile synthesis of yolk-shell structured inorganic-organic hybrid spheres with ordered radial mesochannels. Adv Mater 26(22):3741–3747
34. Kaneko Y, Toyodome H, Mizumo T, Shikinaka K, Iyi N (2014) Preparation of a sulfo-group-containing rod-like polysilsesquioxane with a hexagonally stacked structure and its proton conductivity. Chem A Eur J 20(30):9394–9399
35. Kataoka S, Banerjee S, Kawai A, Kamimura Y, Choi J-C, Kodaira T, Sato K, Endo A (2015) Layered hybrid perovskites with micropores created by alkylammonium functional silsesquioxane interlayers. J Am Chem Soc 137(12):4158–4163
36. Song D-P, Naik A, Li S, Ribbe A, Watkins JJ (2016) Rapid, large-area synthesis of hierarchical nanoporous silica hybrid films on flexible substrates. J Am Chem Soc 138(41):13473–13476
37. Wang Z, Dai Z, Wu J, Zhao N, Xu J (2013) Vacuum-dried robust bridged silsesquioxane aerogels. Adv Mater 25(32):4494–4497
38. Lee JH, Lee AS, Lee J-C, Hong SM, Hwang SS, Koo CM (2017) Multifunctional mesoporous ionic gels and scaffolds derived from polyhedral oligomeric silsesquioxanes. ACS Appl Mater Interfaces 9(4):3616–3623
39. Kim D-G, Kang H, Han S, Lee J-C (2012) Dual effective organic/inorganic hybrid star-shaped polymer coatings on ultrafiltration membrane for bio- and oil-fouling resistance. ACS Appl Mater Interfaces 4(11):5898–5906
40. Zheng Y, Wang L, Zheng S (2012) Synthesis and characterization of heptaphenyl polyhedral oligomeric silsesquioxane-capped poly(N-isopropylacrylamide)s. Eur Polym J 48(5):945–955
41. Ma Y, He L, Pan A, Zhao C (2015) Poly(glycidyl methacrylate-POSS)-co-poly(methyl methacrylate) latex by epoxide opening reaction and emulsion polymerization. J Mater Sci 50(5):2158–2166
42. Cosgrove T, Swier S, Schmidt RG, Muangpil S, Espidel Y, Griffiths PC, Prescott SW (2015) Impact of end-tethered polyhedral nanoparticles on the mobility of poly(dimethylsiloxane). Langmuir 31(30):8469–8477

43. Tong C, Tian Z, Chen C, Li Z, Modzelewski T, Allcock HR (2016) Synthesis and characterization of trifluoroethoxy polyphosphazenes containing polyhedral oligomeric silsesquioxane (POSS) side groups. Macromolecules 49(4):1313–1320
44. Wang D, Feng S, Liu H (2016) Fluorescence-tuned polyhedral oligomeric silsesquioxane-based porous polymers. Chem A Eur J 22(40):14319–14327
45. Crowley C, KlanritB P, Butler CR, Varanou A, Plate M, Hynds RE, Chambers RC, Seifalian AM, Birchall MA, Janes SM (2016) Surface modification of a POSS-nanocomposite material to enhance cellular integration of a synthetic bioscaffold. Biomaterials 83:283–293
46. Wang F, Lu X, He C (2011) Some recent developments of polyhedral oligomeric silsesquioxane (POSS)-based polymeric materials. J Mater Chem 21(9):2775–2782
47. Cheng B, Li X, Hao J, Yang R (2016) Rheological behavior of polycarbonate/ultrafine Octaphenyl Silsesquioxane (OPS) composites. J Appl Polym Sci 33(27):1–7
48. Ye Y, Tian M, Zhang C, Du Z, Mi J (2016) Understanding controls on wetting at fluorinated polyhedral oligomeric silsesquioxane/polymer surfaces. Langmuir 32(1):230–238
49. Yang B, Li M, Hung Z, Wu Y (2011) Preparation and properties of polymethyl mathacrylate-b-polyhedral oligomeric silsesquioxane nanocomposites. Asian J Chem 23(5):2243–2246
50. Yi Y, Li L, Zheng S (2014) Poly(epsilon-caprolactone)-block-poly(N-vinyl pyrrolidone) diblock copolymers grafted from macrocyclic oligomeric silsesquioxane. Polymer 55(16):3925–3935
51. Mehdi A, Cerclier CV, Le Bideau J, Guyomard D, Dalmas F, Chenal J-M, Chazeau L, Fontaine O, Vioux A (2017) PEO-silsesquioxane flexible membranes: organic-inorganic solid electrolytes with controlled homogeneity and nanostructure. Chemistryselect 2(6):2088–2093
52. Kuo S-W, Tsai H-T (2010) Control of peptide secondary structure on star shape polypeptides tethered to polyhedral oligomeric silsesquioxane nanoparticle through click chemistry. Polymer 51(24):5695–5704
53. Ye Q, Zhou H, Xu J (2016) Cubic polyhedral oligomeric silsesquioxane based functional materials: synthesis, assembly, and applications. Chem Asian J 11(9):1322–1337
54. Teng CP, Mya KY, Win KY, Yeo CC, Low M, He C, Han M-Y (2014) Star-shaped polyhedral oligomeric silsesquioxane-polycaprolactone-polyurethane as biomaterials for tissue engineering application. NPG Asia Mater 6:1–10
55. Qiang X, Ma X, Li Z, Hou X (2014) Synthesis of star-shaped polyhedral oligomeric silsesquioxane (POSS) fluorinated acrylates for hydrophobic honeycomb porous film application. Colloid Polym Sci 292(7):1531–1544
56. Zhang W, Wang S, Li X, Yuan J, Wang S (2012) Organic/inorganic hybrid star-shaped block copolymers of poly(L-lactide) and poly(N-isopropylacrylamide) with a polyhedral oligomeric silsesquioxane core: synthesis and self-assembly. Eur Polym J 48(4):720–729
57. Jia P, Argun AA, Xu J, Xiong S, Ma J, Hammond PT, Lu X (2009) Enhanced electrochromic switching in multilayer thin films of polyaniline-tethered silsesquioxane nanocage. Chem Mater 21(19):4434–4441
58. Lin W, Xu K, Xin M, Peng J, Xing Y, Chen M (2014) Hierarchical porous polyaniline-silsesquioxane conjugated hybrids with enhanced electrochemical capacitance. RSC Adv 4(74):39508–39518
59. Lin W, Xu K, Peng J, Xing Y, Gao S, Ren Y, Chen M (2015) Hierarchically structured carbon nanofiber-silsesquioxane-polyaniline nanohybrids for flexible supercapacitor electrodes. J Mater Chem A 3(16):8438–8449
60. Bai J, Shi Z, Yin J, Tian M (2014) A simple approach to preparation of polyhedral oligomeric silsesquioxane crosslinked poly(styrene-b-butadiene-b-styrene) elastomers with a unique micro-morphology via UV-induced thiol-ene reaction. Polym Chem 5(23):6761–6769
61. Zhang W, Yuan J, Weiss S, Ye X, Li C, Mueller AHE (2011) Telechelic hybrid poly(acrylic acid)s containing polyhedral oligomeric silsesquioxane (POSS) and their self-assembly in water. Macromolecules 44(17):6891–6898

62. Pisuchpen T, Intasanta V, Hoven VP (2014) Highly porous organic-inorganic hybrid fiber from copolymers of styrene and polyhedral oligomeric silsesquioxane-derived methacrylate: syntheses, fiber formation and potential modification. Eur Polym J 60:38–48
63. Du X, Li X, Xiong L, Zhang X, Kleitz F, Qiao SZ (2016) Mesoporous silica nanoparticles with organo-bridged silsesquioxane framework as innovative platforms for bioimaging and therapeutic agent delivery. Biomaterials 91:90–127
64. Hayashi E, Shimokawa T (2016) Synthesis and characterization of hydrophobic and mesoporous silicate-silsesquioxane hybrid copolymers. Microporous Mesoporous Mater 219:178–185
65. Li X, Tang T, Zhou Y, Zhang Y, Sun Y (2014) Applicability of enzyme-responsive mesoporous silica supports capped with bridged silsesquioxane for colon-specific drug delivery. Microporous Mesoporous Mater 184:83–89
66. Dopierala K, Bojakowska K, Karasiewicz J, Maciejewski H, Prochaska K (2016) Interfacial behaviour of cubic silsesquioxane and silica nanoparticles in Langmuir and Langmuir-Blodgett films. RSC Adv 6(97):94934–94941
67. Nagappan S, Ha C-S (2015) Emerging trends in superhydrophobic surface based magnetic materials: fabrications and their potential applications. J Mater Chem A 3(7):3224–3251
68. Mi Y, Li K, Liu Y, Pu K-Y, Liu B, Feng S-S (2011) Herceptin functionalized polyhedral oligomeric silsesquioxane – conjugated oligomers – silica/iron oxide nanoparticles for tumor cell sorting and detection. Biomaterials 32(32):8226–8233
69. Osorio AG, Machado GB, Pereira MB, Benvenutti EV, Pereira LG, Bergmann CP, de Oliveira AH, Haas Costa TM (2016) Synthesis and characterization of magnetic carbon nanotubes/silsesquioxane nanocomposite thin films. Appl Surf Sci 371:9–15
70. Nagappan S, Ha HM, Park SS, Jo N-J, Ha C-S (2017) One-pot synthesis of multi-functional magnetite-polysilsesquioxane hybrid nanoparticles for the selective Fe3+ and some heavy metal ions adsorption. RSC Adv 7(31):19106–19116
71. He H-B, Li B, Dong J-P, Lei Y-Y, Wang T-L, Yu Q-W, Feng Y-Q, Sun Y-B (2013) Mesostructured nanomagnetic polyhedral oligomeric silsesquioxanes (POSS) incorporated with dithiol organic anchors for multiple pollutants capturing in wastewater. ACS Appl Mater Interfaces 5(16):8058–8066
72. Moitra N, Kanamori K, Shimada T, Takeda K, Ikuhara YH, Gao X, Nakanishi K (2013) Synthesis of hierarchically porous hydrogen silsesquioxane monoliths and embedding of metal nanoparticles by on-site reduction. Adv Funct Mater 23(21):2714–2722
73. Zhang H, Oyanedel-Craver V (2013) Comparison of the bacterial removal performance of silver nanoparticles and a polymer based quaternary amine functionalized silsesquioxane coated point-of-use ceramic water filters. J Hazard Mater 260:272–277
74. Bai W, Sheng Q, Ma X, Zheng J (2015) Synthesis of silver nanoparticles based on hydrophobic interface regulation and its application of electrochemical catalysis. ACS Sustain Chem Eng 3(7):1600–1609
75. Schneid AC, Roesch EW, Sperb F, Matte U, da Silveira NP, Costa TMH, Benvenutti EV, de Menezes EW (2014) Silver nanoparticle-ionic silsesquioxane: a new system proposed as an antibacterial agent. J Mater Chem B 2(8):1079–1086
76. Zhang Y, Shen H-Y, Hai X, Chen X-W, Wang J-H (2017) Polyhedral oligomeric silsesquioxane polymer-caged silver nanoparticle as a smart colorimetric probe for the detection of hydrogen sulfide. Anal Chem 89(2):1346–1352
77. Silambarasan K, Kumar AVN, Sivakumar C, Joseph J (2014) Formation of nanogap Au-polysilsesquioxane 1D chains for SERS application. RSC Adv 4(75):40003–40007
78. Jung JA, Kim YB, Kim YA, Ryu SB, Kim V (2011) Preparation of functional spherical polysilsesquioxane/gold nanoparticle composites and their applications in DNA assay. J Nanopart Res 13(6):2361–2374
79. Dong F, Guo W, Park SK, Ha CS (2012) Controlled synthesis of novel cyanopropyl polysilsesquioxane hollow spheres loaded with highly dispersed Au nanoparticles for catalytic applications. Chem Commun 48(8):1108–1110

80. Scholder P, Hafner M, Hassel AW, Nischang I (2016) Gold nanoparticle@polyhedral oligomeric silsesquioxane hybrid scaffolds in microfluidic format – highly efficient and green catalytic platforms. Eur J Inorg Chem 7:951–955
81. Silva PS d, Gasparini BC, Magosso HA, Spinelli A (2014) Gold nanoparticles hosted in a water-soluble silsesquioxane polymer applied as a catalytic material onto an electrochemical sensor for detection of nitrophenol isomers. J Hazard Mater 273:70–77
82. Zapp E, da Silva PS, Westphal E, Gallardo H, Spinelli A, Vieira IC (2014) Troponin T immunosensor based on liquid crystal and silsesquioxane-supported gold nanoparticles. Bioconjug Chem 25(9):1638–1643
83. Brigo L, Cittadini M, Artiglia L, Rizzi GA, Granozzi G, Guglielmi M, Martucci A, Brusatin G (2013) Xylene sensing properties of aryl-bridged polysilsesquioxane thin films coupled to gold nanoparticles. J Mater Chem C 1(27):4252–4260
84. Hwang IS, Kim K-Y, Lim J-H, Kim K-M (2016) Creation of spherical aggregates of Pd nanoparticles on the surface of POSS-modified graphene oxide. Polym Bull (Heidelb Ger) 73(9):2557–2565
85. Park DS, Ha TS, Kim KY, Lim JH, Kim KM (2014) New composites of spherical bridged polysilsesquioxanes and aggregates of Pd nanoparticles with POSS via ionic interactions. Polym Bull 71(4):819–828
86. Arsalani N, Akbari A, Amini M, Jabbari E, Gautam S, Chae KH (2017) POSS-based covalent networks: supporting and stabilizing Pd for heck reaction in aqueous media. Catal Lett 147(4):1086–1094
87. Lu C-H, Chang F-C (2011) Polyhedral oligomeric silsesquioxane-encapsulating amorphous palladium nanoclusters as catalysts for heck reactions. ACS Catal 1(5):481–488
88. Tanabe M, Mutou K, Mintcheva N, Osakada K (2011) Preparation and reactivity of an O, O-chelating silsesquioxane-palladium complex. J Organomet Chem 696(6):1211–1215
89. Moitra N, Matsushima A, Kamei T, Kanamori K, Ikuhara YH, Gao X, Takeda K, Zhu Y, Nakanishi K, Shimada T (2014) A new hierarchically porous Pd@HSQ monolithic catalyst for Mizoroki-Heck cross-coupling reactions. New J Chem 38(3):1144–1149
90. Zhao F, Huang Y (2011) Preparation and properties of polyhedral oligomeric silsesquioxane and carbon nanotube grafted carbon fiber hierarchical reinforcing structure. J Mater Chem 21(9):2867–2870
91. Zhang W, Zhou Z, Li Q, Chen G-X (2014) Controlled dielectric properties of polymer composites from coating multiwalled carbon nanotubes with octa-acrylate silsesquioxane through diels-alder cycloaddition and atom transfer radical polymerization. Ind Eng Chem Res 53(16):6699–6707
92. Sabet SM, Mahfuz H, Terentis AC, Hashemi J, Boesl B (2016) A facile approach to the synthesis of multi-walled carbon nanotube-polyhedral oligomeric silsesquioxane (POSS) nanohybrids. Mater Lett 168:9–12
93. Xu K, Lin W, Wu J, Peng J, Xing Y, Gao S, Ren Y, Chen M (2015) Construction and electronic properties of carbon nanotube hybrids with conjugated cubic silsesquioxane. New J Chem 39(11):8405–8415
94. Estevam RB, Ferreira RT, Bischof ABH, dos Santos FS, Santos CS, Fujiwara ST, Wohnrath K, Lazaro SR, Garcia JR, Pessoa CA (2015) Preparation and characterization of LbL films based on graphene oxide nanoparticles interacting with 3-n-propylpyridinium silsesquioxane chloride. Surf Coatings Technol 275:2–8
95. Mondal T, Bhowmick AK, Krishnamoorti R (2014) Butyl lithium assisted direct grafting of polyoligomeric silsesquioxane onto graphene. RSC Adv 4(17):8649–8656
96. Hu L, Jiang P, Bian G, Huang M, Haryono A, Zhang P, Bao Y, Xia J (2017) Effect of octa (aminopropyl) polyhedral oligomeric silsesquioxane (OapPOSS) functionalized graphene oxide on the mechanical, thermal, and hydrophobic properties of waterborne polyurethane composites. J Appl Polym Sci 134(6):1–11
97. Liao W-H, Yang S-Y, Hsiao S-T, Wang Y-S, Li S-M, Ma C-CM, Tien H-W, Zeng S-J (2014) Effect of octa(aminophenyl) polyhedral oligomeric silsesquioxane functionalized graphene

oxide on the mechanical and dielectric properties of polyimide composites. ACS Appl Mater Interfaces 6(18):15802–15812

98. Xue Y, Liu Y, Lu F, Qu J, Chen H, Dai L (2012) Functionalization of graphene oxide with polyhedral oligomeric silsesquioxane (POSS) for multifunctional applications. J Phys Chem Lett 3(12):1607–1612
99. Bai W, Sheng Q, Zheng J (2016) Hydrophobic interface controlled electrochemical sensing of nitrite based on one step synthesis of polyhedral oligomeric silsesquioxane/reduced graphene oxide nanocomposite. Talanta 150:302–309
100. Sun H-J, Tu Y, Wang C-L, Van Horn RM, Tsai C-C, Graham MJ, Sun B, Lotz B, Zhang W-B, Cheng SZD (2011) Hierarchical structure and polymorphism of a sphere-cubic shape amphiphile based on a polyhedral oligomeric silsesquioxane- 60 fullerene conjugate. J Mater Chem 21(37):14240–14247
101. Zhang W-B, Tu Y, Sun H-J, Yue K, Gong X, Cheng SZD (2012) Polymer solar cells with an inverted device configuration using polyhedral oligomeric silsesquioxane- 60 fullerene dyad as a novel electron acceptor. Sci Chin Chem 55(5):749–754
102. Semenov SG, Bedrina ME (2013) A quantum chemical study of silsesquioxanes: H8Si8O12, Me8Si8O12, H@Me8Si8O12, He@Me8Si8O, and He@Me8Si8O12. J Struct Chem 54(1):159–163
103. Zeng Y, Kuo C-I, Hsu C, Najmzadeh M, Sachid A, Kapadia R, Yeung C, Chang E, Hu C, Javey A (2015) Quantum well InAs/AlSb/GaSb vertical tunnel FET with HSQ mechanical support. IEEE Trans Nanotechnol 14(3):580–584
104. Wang Y, Kalytchuk S, Wang L, Zhovtiuk O, Cepe K, Zboril R, Rogach AL (2015) Carbon dot hybrids with oligomeric silsesquioxane: solid-state luminophores with high photoluminescence quantum yield and applicability in white light emitting devices. Chem Commun 51(14):2950–2953
105. Wang Y, Vaneski A, Yang H, Gupta S, Hetsch F, Kershaw SV, Teoh WY, Li H, Rogach AL (2013) Polyhedral oligomeric silsesquioxane as a ligand for CdSe quantum dots. J Phys Chem C 117(4):1857–1862

Chapter 7
Nanocomposite Materials Properties of Aminophenylsilsesquioxanes

R. M. Laine, K. Takahasi, R. Tamaki, J. Choi, S. G. Kim, C. Brick, M. Z. Asuncion, E. Chetioui, S. Sulaiman, and R. Basheer

Abstract A series of aminophenylsilsesquioxanes were synthesized and cross-linked with a set of epoxy resins to form 3-D epoxy resins with completely defined interfaces. The objectives of this work were to make materials with very low coefficients of thermal expansions (CTEs) such that these materials could be used for flip-chip underfill in the manufacture of chips on printed circuit boards. Related polyimides were made as oxygen barrier materials. We were able to make epoxy resin hybrid composites with CTEs of $\approx$30 ppm/°C. Likewise with polyimides we were able to make composite resins with oxygen transmission rates of 3–-5 $\pm$ 0.5 cc·20 μm/m^2·day·atm in films that were only 0.5 mm thick.

Keywords Hybrid materials · Flip-chip underfill · Low-viscosity epoxy resins · Composite oxygen barrier materials

7.1 Introduction

Cubic silsesquioxanes of the types $[RSiO_{1.5}]_8$ (T_8 systems) and $[RSiMe_2OSiO_{1.5}]_8$ (Q_8 systems) exhibit perfect symmetry placing each R group in a different octant in Cartesian space [1–6]. These compounds can be viewed as being spherical with eight functional groups placed as far apart from each other as possible (orthogonally) with diameters of 1.2–1.4 nm. The combination of high symmetry with nanometer size suggests that they can be used as nanobuilding blocks for the assembly of macro-scale materials but with control of global properties extending to the nanometer

R. M. Laine (✉) · K. Takahasi · R. Tamaki · J. Choi · S. G. Kim · C. Brick · M. Z. Asuncion
E. Chetioui · S. Sulaiman
Departments of Materials Science and Engineering, Chemistry and The Macromolecular Science and Engineering Center, University of Michigan, Ann Arbor, MI, USA
e-mail: talsdad@umich.edu

R. Basheer
Delphi Research Laboratories, Shelby Township, MI, USA

Y. Chujo (ed.), *New Polymeric Materials Based on Element-Blocks*,
https://doi.org/10.1007/978-981-13-2889-3_7

length scale. In addition, the presence of a core that can be considered to be the smallest single crystal of silica offers the rigidity of an inorganic component coupled with heat capacity similar to silica rather than to typical organic frameworks. Thus, the assembly of materials based on such nanoscale building blocks offers the potential to incur very novel properties. Indeed, these materials are prime examples of "element block" components as discussed in detail recently [7]. As such, cage-type silsesquioxanes and materials made from them have been the subject of diverse studies ranging from models of catalyst support surfaces to catalysts, microporous materials, NMR standards, novel encapsulants, dendrimer cores, luminescent materials, and even coatings on spacecraft [8–35].

Our own efforts in this area have focused on developing chemistries (construction tools) that allow the assembly of these nanobuilding blocks, both the Q_8 systems [36–42] and more recently the T_8 systems where R = phenyl [43–49], with complete control of nanometer periodicity at macroscopic length scales. In particular, the goal has been to identify functionality that permits the connection of individual building blocks through organic tethers that link individual vertices on each cube, in principle preserving the cubic symmetry in the macroscopic nanocomposite. Figure 7.1 represents a schematic of the anticipated structures.

The rationale behind creating organic/inorganic hybrid nanocomposites as suggested in Fig. 7.1 is that the silica core represents a perfect hard particle completely separated (dispersed) from any other hard particle. Thus, the potential influence of interparticle contacts on the global properties of the nanocomposite should be nil. Second, if all of the organic "tethers" are independent of each other and the nanoparticles are a constant, then the global properties of this material should reflect only the structure of the individual tethers. As such, it should then be possible to tailor global properties through changes in the structure of the organic component only. In the following background section, we briefly describe earlier efforts to define criteria and chemistries that could be used to demonstrate the feasibility of this approach to entirely new materials.

7.2 Background

Two attractive chemistries that could provide nanocomposites with tailored tether architectures in turn providing tailored global properties are those based on epoxy and imide derivatives of the compounds, $[NH_2C_6H_5SiO_{1.5}]_x$ where x = 8 (OAPS) or 12 (DAPS) with substitution patterns, 60% meta, 20% ortho, and 10% para, or $[(NH_2)_2C_6H_5SiO_{1.5}]_x$ HDAPS (hexadecaminooctphenylsilsesquioxane) with primarily a 2,5 substitution pattern [50].

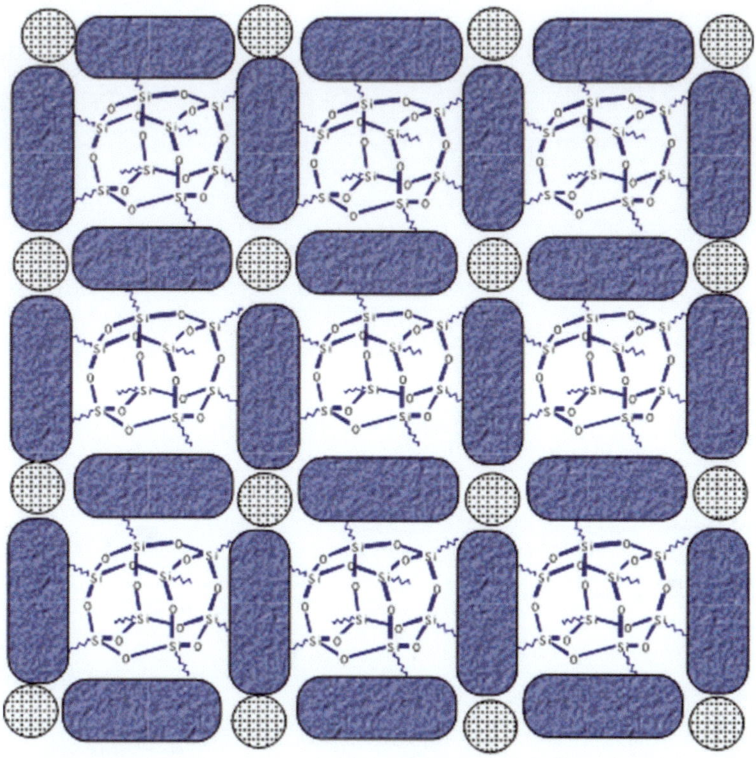

Fig. 7.1 Nanocomposite structure with organic components separating inorganic components, with all phases completely discontinuous. Gray circles represent nanopores

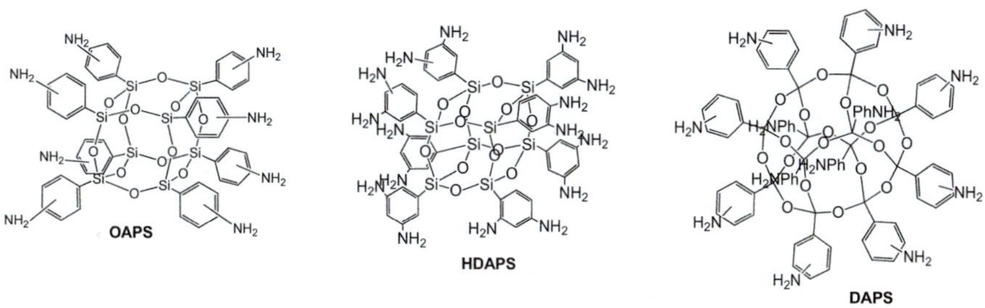

There are three basic reactions can occur as illustrated below.

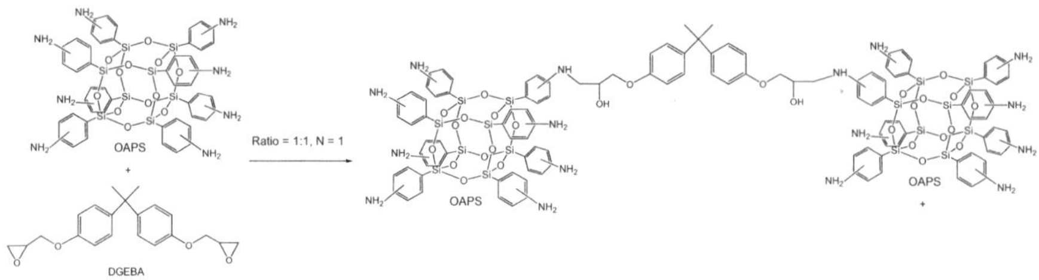

(7.1)

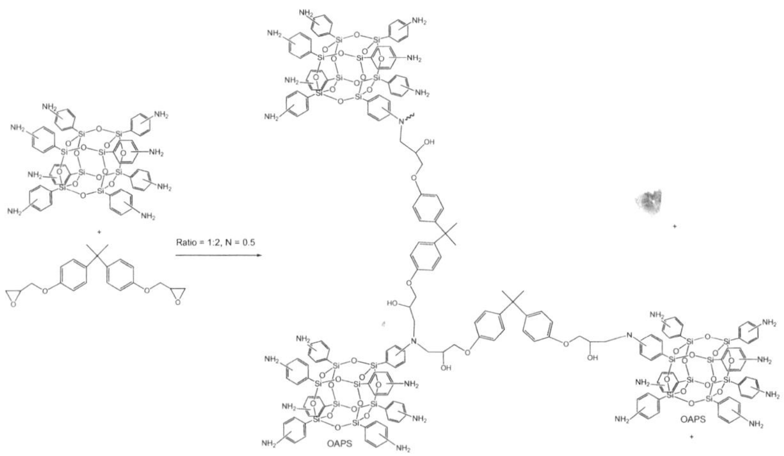

(7.2)

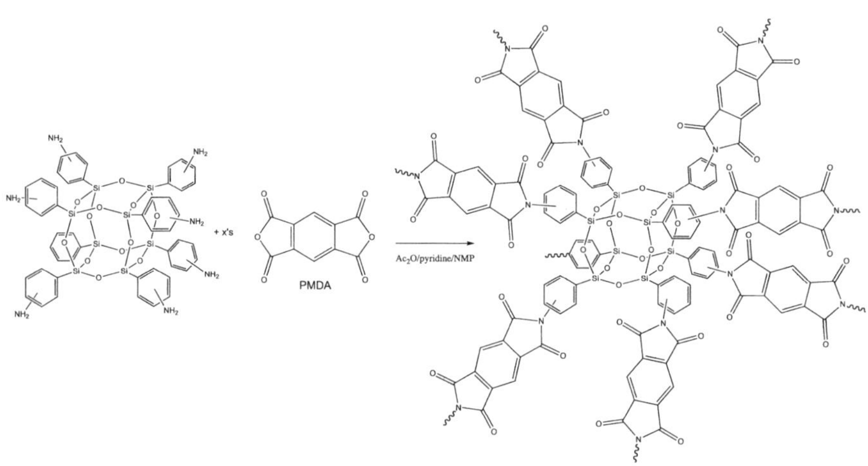

(7.3)

Reaction (7.1) results in linear tethers that in principle should lead to completely discontinuous nanocomposites per Fig. 7.1. However, it is difficult to imagine that one could indeed form bonds between each and every vertex. One might presume that this would be improbable given the changes in viscosities that must occur during curing as well as the likelihood of misaligning individual cubes in the final matrix that would lead to incomplete curing with defects. Defects are defined as dangling reactive groups or cubes missing in the periodic array generated during assembly.

Despite this apparent impossibility, we have shown that in reaction (7.4) [49], when we use a stoichiometry of $N = 1$, one epoxide to one NH_2 group, we can dissolve away the silica cage using HF and extract the silylated tethers. GPC studies suggest that ≈90% of the 4,4′-DDM has reacted with just one epoxy group. The number of unreacted NH_2 groups is less than 10%. The number of doubly reacted NH_2- groups is even smaller [49]. Hence, it appears that under some conditions, one can obtain nearly complete reaction and furthermore limit the number of defects in the nanocomposite formed. This in turn points to the ability to obtain highly reproducible properties, predict properties, and on that basis tailor properties.

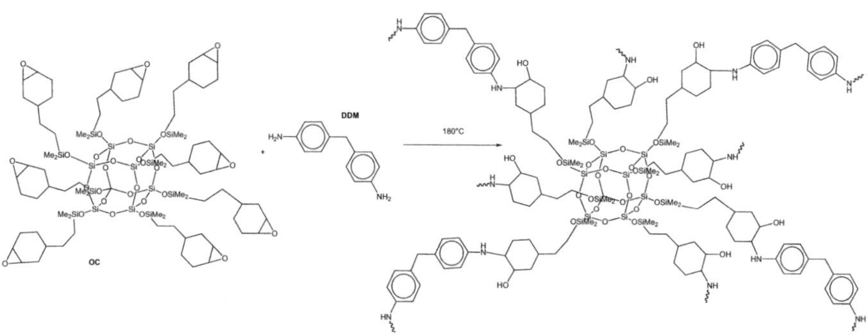

(7.4)

One of the more interesting discoveries is that by changing the stoichiometry from the traditional $N = 0.5$ to $N = 1$, it is possible to greatly improve the mechanical properties of epoxy resins. Thus in recent studies, we compared the elastic moduli of OC/DDM of [4] with OG/DDM and DGEBA/DDM, a commercial aircraft epoxy resin (see below) shown in Fig. 7.2.

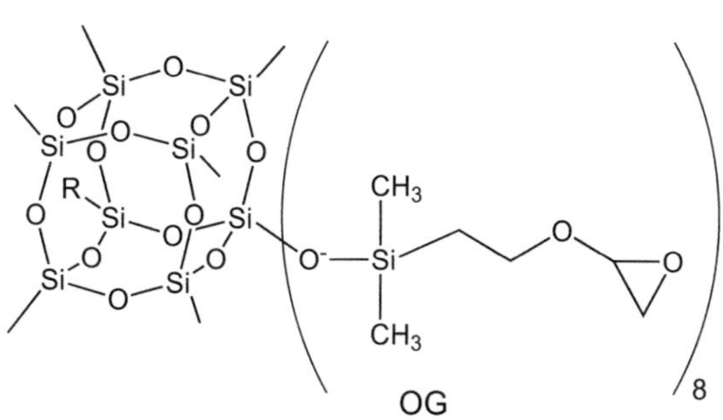

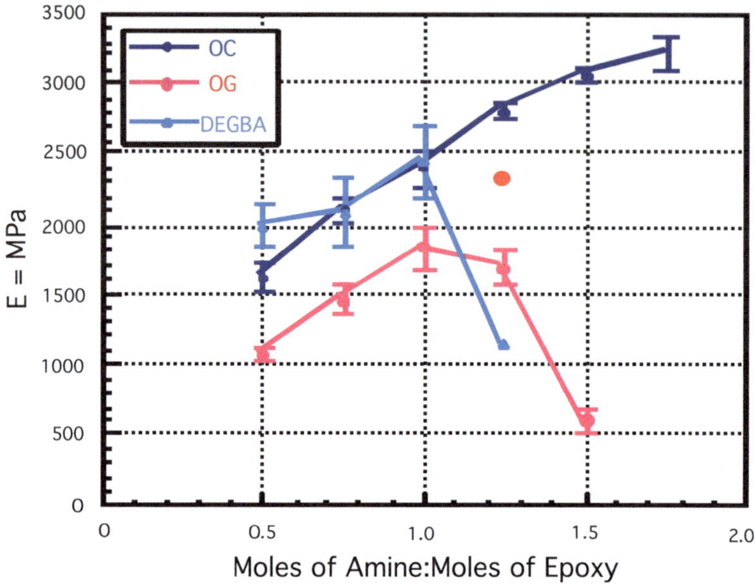

Fig. 7.2 Tensile moduli of OG, OC, and DGEBA/DDM composites at various compositions cured at 150 °C/10 h/N_2. *DGEBA* = diglycidyl ether of bisphenol A. Red dot = 75:25 OG/OC

The choice of OG and OC was made based on the fact that each epoxy functionality has essentially the same number of atoms between the $SiMe_2$ unit and the epoxy ring but differ by the presence of an extra carbon that forms the ring structure in OC. Thus, OG is anticipated to offer more flexible functionality than OC. This results in OG being able to react with secondary amines, whereas OC does not. Consequently, OC cannot form bifurcated structures as shown in Reaction (7.2).

Three observations can be made from Fig. 7.2 results. First, the mechanical properties of the traditional aircraft epoxy resin made with a standard $N = 0.5$ composition are superior to the properties of both the OG and OC equivalents made with the same stoichiometries. One might infer that our results are not very interesting. However, given that OC cannot form bifurcated structures, this composition for OC actually has 50% defects because roughly half the epoxy groups present cannot form bonds with secondary NH species. In addition, because OG can form bifurcated structures, it most likely has one of the highest cross-link densities of any known epoxy resin and therefore should be brittle at $N = 0.5$. This is attested to by the fact that at this composition, it has no Tg.

The second important point is that our efforts to make completely discontinuous nanocomposites, $N = 1$, led to the discovery that all of these epoxy resins show superior properties at this composition, even the aircraft epoxy. So, why use the other composition? We are not entirely sure but can suggest that other issues including Tg, typically lower for $N = 1$, or resistance to moisture uptake or high hardness are all desirable and likely greater at $N = 0.5$ compositions. However, as we show below other factors could drive the switch to $N = 1$ compositions, perhaps though only with silsesquioxane resins. The fracture toughness behavior [50] of these materials also follows suit. That is, the properties for OG and DGEBA also improve at $N = 1$. For OC, these materials are so rigid that there is no change in fracture toughness over *any* composition tested.

The third point is that it is possible to predict the behavior of these materials through modeling studies that indicate that global properties follow directly on the flexibility of the two tethers [48]. Thus, OG shows reasonable tensile strength at $N = 1$, slightly less than DGEBA, but has fracture toughness behavior equivalent to DGEBA. In contrast, the OC materials have very poor fracture toughness but exhibit tensile strengths that increase by 50% above $N = 1$. This behavior was determined to arise because the OC tether actually is not linear in nanocomposites with $N \leq 1$ but coiled. Consequently, on adding more amine (defects), the coils unfurl, and the tensile strengths reflect the strengths of chemical bonds rather than single-coiled tethers. We were able to demonstrate this through these modeling studies [48].

Thereafter, we determined that we could tailor properties at nanometer length scales. By taking the best properties of both OG and OC, we developed a material that offers superior properties. Thus, a 75:25 mixture of OG:OC gives the tensile strength shown in Fig. 7.2 at $N = 1.25$ (red dot). This value of N was chosen because both the tensile strength and the fracture toughness of all of the materials, especially DGEBA/DDM, are quite poor at this highly defective composition. Thus the tensile strength is essentially equivalent to the best found for the $N = 1$ materials. Although not shown, the fracture toughness exceeds all materials by 50%, but the Tg drops to near 70 °C.

Thus, these studies demonstrated that it is indeed possible to make organic/inorganic hybrid nanocomposites with excellent control of periodicity at nanometer length scales, while maintaining this periodicity over centimeter length scales. Furthermore, it is possible to develop materials that offer properties akin to normally epoxy resins. Indeed, most of the cube epoxy resins are 40–70 °C more stable than their purely organic counterparts. Finally, it is possible to tailor their properties as discussed in the following sections.

7.3 Recent Results on Nanotailoring

The above background provides the basis for discussing further efforts to employ what we have learned to tailor global properties in epoxy resin systems targeting other possible applications. We present two examples here: one concerned with developing novel flip-chip underfills and the other oxygen barrier membranes.

7.3.1 Flip-Chip Underfills [51–62]

Silicon integrated circuits (ICs) are assembled by mounting, via solder bumps, to substrates. One of the key problems with this process is that the Si ICs generate considerable amounts of heat that must be removed rapidly during operation. Thus, an air layer between the Si IC and the substrate does not provide good thermal management. Consequently, the current processing steps involve those shown in Fig. 7.3 wherein an epoxy resin is introduced between the layers, after the solder reflow process (to establish electrical connections) to provide a mechanism for heat removal.

The key problem with this process arise because the coefficient of thermal expansion (CTE) of Si is 2–4 ppm/°C; however, the substrate has typical CTEs of 20–40 ppm/°C. To avoid fatigue cracking of the solder bumps because of thermal cycling, the CTE of the epoxy underfill is manipulated to be 20–25 ppm/°C by filling with silica. Unfortunately, sufficient silica must be added that the viscosities of these materials are frequently new 50,000 MPa-sec making them like toothpaste and making the filling process difficult. Furthermore, to successfully underfill as shown in Fig. 7.3, the individual chips must be placed sufficiently far apart to allow side filling.

Side filling defeats efforts to make the circuitry more compact. Therefore, a new approach has been developed called "no-flow underfills." In this process, illustrated in Fig. 7.4, the underfill is applied first, then the Si IC is passed into the epoxy resin to contact the solder bumps, and then curing and reflow occur together. This obviates the need to space the Si ICs to allow for side filling. The problem now becomes one of controlling CTE with a silica filler. One could imagine that silica particles in the underfill might interfere with the solder bumps making complete contact with the

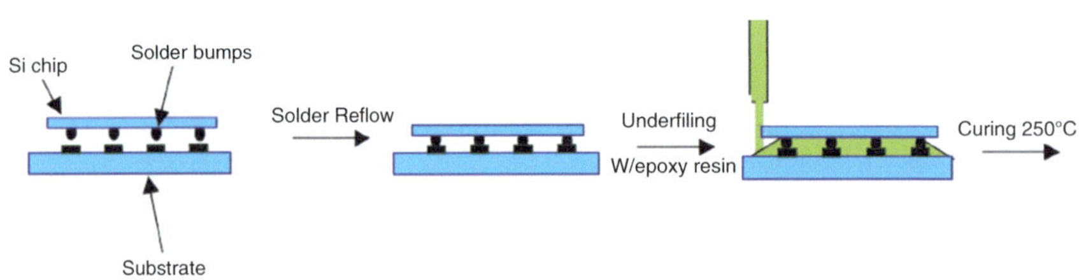

Fig. 7.3 Schematic of standard underfill processing

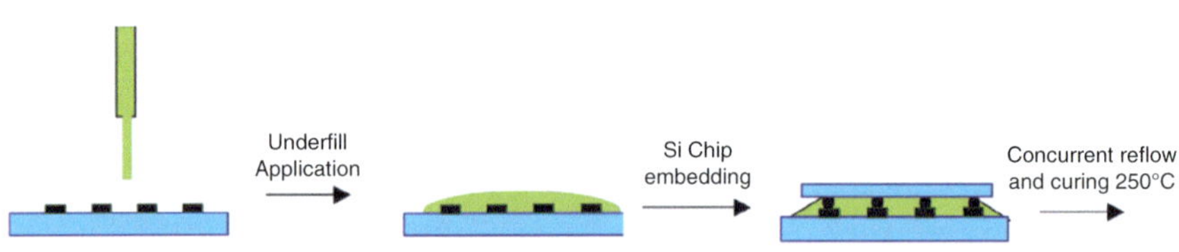

Fig. 7.4 Schematic of no-flow underfill processing

Table 7.1 Published CTEs of selected epoxy resins where N = NH_2s/epoxy group = 0.5

Epoxy	Curing agent	Time/ max. cure (°C)	CTE (ppm/°C)	Ref.
4,4′-Bis (2,3-epoxypropoxy)-α-methylstilbene	Sulfanilamide	4 h/200	53	[30]
DGEBA	Polyamide	1 h/100	78	[31]
DGEBA	4,4-diaminodiphenylmethane	4 h/150	82	[32]
DGEBA	Diaminodiphenylsulfone	2 h/250	68	[33]
4,4′-Di(2,3-epoxypropyloxy)-biphenyl	4-aminoacetophenone azine	160	96	[34]
4,4′-Di(2,3-epoxypropyloxy)-biphenyl	Diaminodiphenylsulfone	4 h/220	64	[35]
3,4-epoxycyclohexylmethyl-3,4-epoxy cyclohexyl carboxylate	Methylhexahydrophthalic anhydride	15 min/ 250	87	[36]
DGEBA	Triethylenetetramine	100 °C	107	[37]
DGEBA	Phenol formaldehyde	2 h/165	75	[38]
DGEBA	Cresol formaldehyde	2 h/165	87	[38]
Triglycidyl p-aminophenol	Phenol formaldehyde	2 h/165	63	[38]
Naphthalene diepoxide	Phenol formaldehyde	2 h/165	63	[38]

substrate leading to loss of electrical connections. When mounting multiple chips, the chances for failure must be sufficiently high that this becomes a serious problem.

Recognizing this problem, it occurred to us that it might be possible to use our nanostructured resins to overcome this serious problem. However, it is important to recognize that the CTEs of most common epoxy resins are much higher than the 20–25 ppm/°C values sought and those that might have good values have very high viscosities are near room temperatures. Table 7.1 illustrates this problem.

Thus, the challenge was to identify a system that both provides low viscosities at near room temperature, but that then cures at temperatures compatible with IC circuitry processing ($\leq$ 400 °C) while remaining stable and also providing CTEs 20–25 ppm/°C. Recognizing the high cross-link densities accessible with cubic silsesquioxanes, coupled with their high rigidity and heat capacity, we sought to explore their development for use as no-flow underfills. Here the silica filler is part of the molecule.

Table 7.2 provides CTE data for the series of epoxies shown in Fig. 7.5. The data indicate that it is indeed possible to control CTEs for cube resins over an order of magnitude. Furthermore, we believe the values for the TGMX system of 25 ppm/°C are better than any previously reported "unfilled" liquid epoxy resin system cured under these conditions. We again note that the nontraditional $N = 1$ composition gives better CTEs than the traditional $N = 0.5$ values used for commercial resins, especially for resins that offer RT viscosities of $\approx$1000 MPa-s.

Table 7.2 CTEs of selected epoxy resins where N = number of NH_2s/epoxy group

Aminophenyl silsesquioxane	Epoxy	Molar ratio N	CTE (ppm/°C) below	Inflection point[a]	CTE (ppm/°C) above
OAPS	TGMX	0.5	55	120	103
OAPS	TGMX	1	25 ± 2	110	46
OAPS	TGMX	1/10 wt % Al_2O_3	22 ± 2	125	33
DAPS	TGMX	1	27	–	–
HADPS	TGMX	1	36	–	–
OAPS	ECHX	0.5	91	110	191
OAPS	ECHX	1	55	110	92
DAPS	ECHX	1	54	–	–
HDAPS	ECHX	1	46	–	–
OAPS	DGEBA	0.5	80	95	176
OAPS	DGEBA	1	72	115	145
OAPS	OG	0.5	196	100	179
OAPS	OG	1	141	100	162
OAPS	OC	0.5	185	120	240
OAPS	OC	1	220 ± 10	100	329

All samples cured 20 h/200 °C/N_2

[a]No true T_gs were observed for any of the above resins by TMA or DSC. Inflection points were observed and measured as the intersection of slopes on the extreme ends of TMA plots

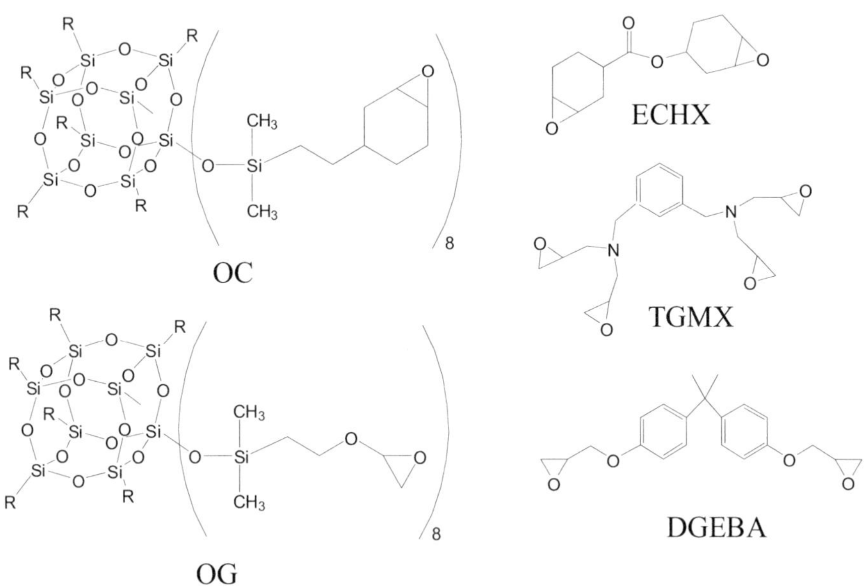

Fig. 7.5 Sets of epoxies tested in OAPS resins for tailoring CTEs

Table 7.3 Experimental parameters and observations on the solid imides resins

Resin	Curing schedule	Solvent (15 mL)	Solution color	Resin characteristics	CTE (ppm/°C)
DDM/ PMDA	150 °C/24 h; 250 °C/24 h	NMP/ DMF	NMP: brown DMF: yellow	Segregated	40
DDM/ ODPA	NMP: 150 °C/4 h, DMF: 100 °C/2 h, 120 °C/2 h	NMP/ DMF	NMP: brown DMF: yellow	Very brittle	36
OAPS/ PMDA	RT/12 h, 60 °C/48 h	NMP	Dark yellow	Very tough, transparent	66
OAPS/ ODPA	RT/12 h, 60 °C/48 h	NMP	Dark brown	Very tough, transparent	65

Reasons for the observed results are believe to be twofold. First, cage steric effects are such that fewer amine hydrogens are likely to be available to react at the $N = 0.5$ stoichiometry. In contrast in the $N = 1$ system, many more amine hydrogens are available, leading to higher cross-link densities and/or more complete curing. Second, at $N = 1$, the mole fraction of silsesquioxane cage is twice as high as in the $N = 0.5$ system, hence the lower CTE. Also, it is likely that the $N = 1$ stoichiometry also provides lower viscosities that in turn lead to higher degrees of curing.

Finally, the use of the dodecamer DAPS and/or the hexadecaaminoOPS, HDAPS, in place of OAPS provides little if any changes to the observed CTE values, although thermal stabilities with DAPS might be higher as the cage has more silica in it [63]. This remains untested.

For comparative purposes, we also conducted CTE studies on polyimides, Reaction (7.3), using pyromellitic dianhydride (PMDA) or oxydiphthalic anhydride (ODPA) shown below (see Table 7.3). From previous studies, we know that these imides cure at ≥250 °C. Thus the model studies down using DDM rather than OAPS were done to demonstrate the final possible CTEs with these model compounds.

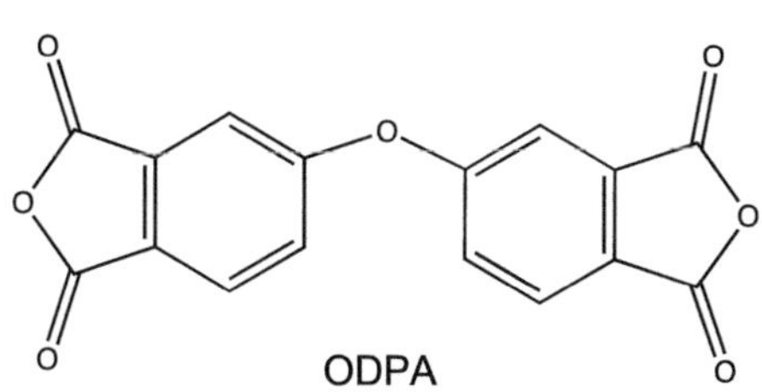
ODPA

However, we are also interested in examining the properties of the cage materials on low-temperature curing as some substrates cannot be heated to ≥250 °C. As can be seen, the CTEs of the PMDA/OAPS and ODPA/OAPS systems, even though cured to only 60 °C, are reasonably low given that it is likely that these systems are really only at the amic acid level of conversion. It is likely that still lower CTEs can be achieved by curing to ≈250 °C.

7.3.2 Barrier Materials

In related studies, we examined the same epoxy resins and polyimides as used above in the CTE experiments as oxygen barrier materials finding that they can be processed to give materials that offer properties at least as good as commercial materials [64–68]. Some of the data for this work is shown in Tables 7.4 and 7.5. For comparison purposes we also include data for EVAl F, a commercial product produced by Kuraray [64].

The rationale for these studies comes from the fact that superior barrier properties are obtained from polymeric materials modified using an assortment of approaches but typically through the creation or introduction of dense components in the polymer [66–68]. For example, good barrier properties in polymers arise from strong intrachain forces and high chain packing densities that hinder gas diffusion. Such densities are achieved through strong H-bonding, chain alignment by extrusion, high degrees of crystallinity, or liquid crystallinity. An alternative is to add easily dispersed

Table 7.4 Oxygen transmission rates (OTR) of cast silsesquioxane films (at 20 °C and 65% RH; cc·20 μm/m^2·day·atm-O_2, i.e., standardized to 20 μm thickness)

Sample	N	Curing	OTR (±0.5) (cc·20 μm/m^2 ·day·atm)
50% OAPS/PMDA	–	120 °C/4 h, 205 °C/4 h, 215 °C/4 h	31
50% OAPS/PMDA	–	120 °C/4 h, 205 °C/4 h, 215 °C/8 h, 240 °C/8	27
50% OAPS/ODPA		120 °C/4 h, 205 °C/4 h	35
50% OAPS/ODPA	–	120 °C/4 h, 205 °C/4 h, 215 °C/8 h, 225 °C/8 h	25
50% OAPS/ODPA	–	120 °C/4 h, 205 °C/4 h, 215 °C/8 h, 240 °C/8 h	27
50% OAPS/ODPA /ODA	–	120 °C/4 h, 205 °C/4 h, 215 °C/8 h, 225 °C/8 h	29
OAPS/DGEBA	0.5	130 °C/5 h	110
OAPS/DGEBA	0.5	130 °C/5 h, 150 °C/5 h	21
OAPS/DGEBA	1.0	130 °C /5 h, 150 °C /5 h	24
OAPS/TGMX	0.5	90 °C/2 h	14
OAPS/ECHX	0.5	100 °C/1 h, 130 °C/4 h	24
OAPS/ECHX	1.0	100 °C/1 h, 130 °C/4 h	8
OAPS/ECHX	0.5	100 °C/1 h, 130 °C/4 h, 180 °C/4 h	6
OAPS/RDGE	0.5	95 °C/4 h, 115 °C/4 h	105
OAPS/RDGE	1.0	95 °C/4 h, 115 °C/4 h	115
EVAL F grade†	–	–	<1.0

[a]Eval F grade is 32% ethylene-vinyl alcohol copolymer biaxially orientated (3 × 3) at 140 °C [64]

Table 7.5 Oxygen transmission rates (OTR) of warm-pressed silsesquioxane films (at 20 °C and 65% RH; cc·20 μm/m^2·day·atm-O_2, i.e., standardized to 20 μm thickness)

Sample	N	Initial curing	Curing w/ pressure	OTR (±0.5) (cc·20 μm/m^2 ·day·atm)
50% OAPS/PMDA	–	120 °C/4 h	240 °C/8 h at 150 psi	17
50% OAPS/ODPA	–	120 °C/4 h	240 °C/8 h at 150 psi	12
50% OAPS/ODPA/ ODA	–	120 °C/4 h	240 °C/8 h at 150 psi	13
OAPS/DGEBA	0.5	120 °C/4 h	200 °C / 10 h at 100 psi	7
OAPS/DGEBA	0.5	120 °C/4 h	200 °C / 10 h at 125 psi	5
OAPS/DGEBA	0.5	120 °C/4 h	200 °C / 10 h at 150 psi	3.9
OAPS/TGMX	0.5	100 °C/1 h, 130 °C/4 h	200 °C/4 h at 150 psi	3.2
OAPS/ECHX	0.5	100 °C/1 h, 130 °C/4 h	200 °C/4 h at 150 psi	5.2
OAPS/TGMX and OAPS/ECHX bilayer	0.5 0.5	100 °C/1 h, 130 °C/4 h (100 °C/1 h, 130 °C/4 h) × 2	200 °C/4 h at 150 psi	1.2 ± 0.4

second phases that can be organic or inorganic inclusions such as exfoliated inorganics such as clay particles or simply silica. An extension of this approach is to make bilayer films where transport across the interface is impossible such as in aluminum coated polymers. Clearly, the need to disperse or coat a second phase adds to the difficulty and expense of processing as does extrusion or other processing methods that align polymer chains. This is the basic motivation to use silsesquioxane molecules to create perfectly dispersed nanosilica nanocomposites leading to the data found in Tables 7.4 and 7.5.

From Table 7.4, we can observe that the simple epoxy systems are sufficiently poor barriers in as cast films that the O_2 transport rates are one to two orders of magnitude greater than the commercial Kuraray EVAL material. However, it is important to recognize that the EVAL films are optimized and have been biaxially stretched during heating.

Thus, in a second set of studies resulting in Table 7.5 data, we warm-pressed these films and were pleased to find that the O_2 permeabilities dropped to the same order of magnitude without biaxially orienting these materials. Furthermore, efforts to make bilayer films provided at least one system that offers properties comparable to the commercial EVAL system. Of further importance is the fact that these materials were not biaxially oriented. Finally, the epoxy and imide silsesquioxane barrier films are more stable by approximately 100 °C than the EVAl films which are a copolymer of ethylene and vinyl alcohol.

7.4 Summary

Octafunctional cubic silsesquioxanes because of their cubic symmetry place single functional groups in each octant in Cartesian space. As such they offer the potential to serve as nanobuilding blocks for the assembly of nanocomposite materials with exceptional control of properties. By sensible selection of the chemistries of the functional groups used to assemble these nanocomposites, it appears possible to tailor global properties using subtle changes in organic components to tailor periodicity at nanometer length scales.

In earlier work, we demonstrated that epoxy resins provided good examples of the assembly of nanocomposites with control of periodicity at nm length scales over centimeter lengths to tailor properties. Furthermore, we also discovered that traditional epoxy resin stoichiometries of two epoxy groups per one NH_2 provided poorer mechanical properties that using a stoichiometry of 1:1. In the current paper, we find that shown that by careful choice of epoxy and using the 1:1 stoichiometry, it is possible to control the CTEs of cured resins over a range of almost one order of magnitude, 25–220 ppm/°C. These same epoxy resins systems also provide good-to-excellent barriers to O_2 transport, but in this case there is no clear-cut evidence for an effect of stoichiometry on barrier efficiencies.

Related CTE studies were done using dianhydrides in place of epoxy components. In these studies, with curing at only 60 °C, relatively low CTEs were also observed. The reason for these low curing temperatures was to see if these materials might be coated on aluminum alloys for use as corrosion resistant coatings for among other things, aircraft fuselages.

In general, the resulting materials are robust and extend the utility of silsesquioxane nanocomposites as materials that are highly tailorable permitting extensive changes in global material properties by manipulation of nanoscale architectures.

Acknowledgments The authors would like to thank Delphi Inc., Kuraray Ltd., Matsushita Electric, and AFRL Wright Patterson Air Force Base through subcontract from Mayaterials on SBIR Contract Number F33615 03M 5018 for support of this work.

References

1. Voronkov MG, Lavrent'yev VI (1982) Polyhedral oligosilsesquioxanes and their Homo derivatives. Top Curr Chem 102:199–236
2. Baney RH, Itoh M, Sakakibara A, Suzuki T (1995) Silsesquioxanes. Chem Rev 95:1409–1430

3. Provatas A, Matisons JG (1997) Synthesis and applications of silsesquioxanes. Trends Polym Sci 5:327–333
4. Loy DA, Shea KJ (1995) Bridged Polysilsesquioxanes. Highly porous hybrid organic-inorganic materials. Chem Rev 95:1431–1442
5. Lichtenhan J (1996) Silsesquioxane-based polymers. In: Salmone JC (ed) Polymeric materials encyc, vol 10. CRC Press, N.Y, pp 7768–7777
6. Laine RM (2005) Nano-building blocks based on the $[OSiO_{1.5}]_8$ silsesquioxanes. J Mater Chem 15:3725–3744
7. Chujo Y, Tanaka K (2015) New polymeric materials based on element-blocks. Bull Chem Soc Jpn 88:633–643
8. Feher FJ, Newman DA, Walzer JF (1989) Silsesquioxanes as models for silica surfaces. J Am Chem Soc 111:1741
9. Feher FJ, Budzichowski TA, Blanski RL, Weller KJ, Ziller JW (1991) Facile syntheses of new incompletely condensed polyhedral oligosilsesquioxanes: $[(c\text{-}C_5H_9)_7Si_7O_9(OH)_3]$, $[(c\text{-}C_7H_{13})_7Si_7O_9(OH)_3]$, $[(c\text{-}C_7H_{13})_6Si_6O_7(OH)_4]$. Organometallics 10:2526–2528
10. Maschmeyer T, Klunduk MC, Martin CM, Shephard DS, Thomas JM, Johnson BFG (1997) Modelling the active sites of heterogeneous titanium-centred epoxidation catalysts with soluble silsesquioxane analogues. Chem Commun:1847–1848
11. Feher FJ, Blanski RL (1992) Olefin polymerization by vanadium-containing silsesquioxanes: synthesis of a dialkyl-oxo-vanadium(V) complex that initiates ethylene polymerization. J Am Chem Soc 114:5886–5887
12. Feher FJ, Budzichowski TA (1995) Silsesquioxanes as ligands in inorganic and organometallic chemistry. Polyhedron 14:3239–3253
13. Severn JR, Duchateau R, van Santen RA, Ellis DD, Spek AL (2002) Homogeneous models for chemically tethered silica-supported olefin polymerization catalysts. Organometallics 21:4–6
14. Duchateau R, Abbenhuis HCL, van Santen RA, Meetsma A, Thiele SK-H, van Tol MFH (1998) Half-Sandwich titanium complexes stabilized by a novel Silsesquioxane ligand: soluble model systems for silica-grafted olefin polymerization catalysts. Organometallics 17:5222–5224
15. Maxim N, Magusin PCMM, Kooyman PJ, van Wolput JHMC, van Santen RA, Abbenhuis HCL (2002) Synthesis and characterization of microporous Fe-Si-O materials with tailored iron content from silsesquioxane precursors. J Phys Chem B 106:2203–2206
16. Bonhomme C, Toledano P, Livage MJ, Bonhomme-Coury L. Studies of octameric vinylsilasilsesquioxane by carbon-13 and silicon-29 cross polarization magic angle spinning and inversion recovery cross polarization nuclear magnetic resonance spectroscopy. J Chem Soc Dalton Trans. 1997, 1617–26
17. Bassindale AR, Pourny M, Taylor PG, Hursthouse MB, Light ME (2003) Angew Chem Int Ed 42:3488
18. Bassindale AR, Parker DJ, Pourny M, Taylor PG, Horton PN, Hursthouse MB (2004) Fluoride ion entrapment in octasilsesquioxane cages as models for ion entrapment in zeolites. Further examples, x-ray crystal structure studies, and investigations into how and why they may be formed. Organometallics 23:4400–4405
19. Tsuchida A, Bolln C, Sernetz FG, Frey H, Mülhaupt R (1997) Ethene and propene copolymers containing Silsesquioxane side groups. Macromolecules 30:2818–2824
20. Lichtenhan JD, Vu HQ, Carter JA, Gilman JW, Feher FJ (1993) Silsesquioxane-siloxane copolymers from polyhedral silsesquioxanes. Macromolecules 26:2141–2142
21. Gilman JW, Schlitzer DS, Lichtenhan JD (1996) Low earth orbit resistant siloxane copolymers. J Appl Polym Sci 60:591–596
22. Brunsvold AL, Minton TK, Gouzman I, Grossman E, Gonzalez RI (2004) An investigation of the resistance of POSS polyimide to atomic oxygen attack. High Perform Polym 16:303–318
23. Tomczak SJ, Marchant D, Svejda S, Minton TK, Brunsvold AL, Gouzman I, Grossman E, Schatz GC, Troya D, Sun L, Gonzalez RI, Properties and improved space survivability of POSS (Polyhedral Oligomeric Silsesquioxane) polyimides. Private communication from S. Tomczak

24. Weidner R, Zeller N, Deubzer B, Frey V. Organooligosilsesquioxanes. U.S. Patent 5,047,492, 1991
25. Dathe S, Popowski E, Sonnek G, Feiher T, Jancke H, Schelm U, Euro. Patent 0,348,705 A1 1989
26. Freyer C, Wolferseder J, Peetz U, Euro. Patent 0,624, 691 A1 1993
27. Calzaferri G (1996) Silsesquioxanes. In: Corriu R, Jutzi P (eds) Tailor-made silicon-oxygen compounds, from molecules to materials. Publ. Friedr. Vieweg & Sohn mbH, Braunshweig, pp 149–169
28. Hong B, Thoms TPS, Murfee HJ, Lebrun HJ (1997) Highly branched dendritic macromolecules with core polyhedral silsesquioxane functionalities. Inorg Chem 36:6146–6147
29. Feher FJ, Wyndham KD (1998) Amine and ester-substituted silsesquioxanes; synthesis, characterization and use as a core for starburst dendrimers. Chem Commun:323–324
30. Dvornic PR, Hartmann-Thompson C, Keinath SE, Hill EJ (2004) Organic-inorganic polyamidoamine (PAMAM) dendrimer-polyhedral oligosilsesquoixane (POSS) nanohybrids. Macromolecules 37:7818–7831
31. Sellinger A, Tamaki R, Laine RM, Ueno K, Tanabe H, Williams E, Jabbour GE (2005) Solution processable nanocomposites based on silsesquioxane cores for use in organic light emitting diodes (OLEDS). Mater Res Soc Symp Proc 847:399–403
32. Lo MY, Ueno K, Tanabe H, Sellinger A (2006) Silsesquioxane-based nanocomposite dendrimers with photoluminescent and charge transport properties. Chem Rec 6:157–168
33. Waddon AJ, Coughlin EB (2003) Crystal structure of polyhedral oligomeric silsequioxane (POSS) nano-materials: a study by x-ray diffraction and electron microscopy. Chem Mater 15:4555–4561
34. Deng J, Farmer-Creely CE, Viers BD, Esker AR (2004) Unique rodlike surface morphologies in trisilanolcyclohexyl polyhedral oligomeric silsesquioxane films. Langmuir 20:2527–2530
35. Deng J, Polidan JT, Hottle JR, Farmer-Creely CE, Viers BD, Esker AR (2002) Polyhedral oligomeric silsesquioxanes: a new class of amphiphiles at the air/water interface. J Am Chem Soc 124:15194–15195
36. Sellinger A, Laine RM, Chu V, Viney C (1994) Palladium and platinum catalyzed coupling reactions of allyloxy aromatics with hydridosilanes and hydridosiloxanes: novel liquid crystalline/organosilane materials. J Polym Sci A Polym Chem 32:3069–3089
37. Sellinger A, Laine RM (1996) Silsesquioxanes as synthetic platforms. Thermally and photo curable inorganic/organic hybrids. Macromolecules 29:2327–2330
38. Sellinger A, Laine RM (1996) Silsesquioxanes as synthetic platforms. III. Photocurable, liquid epoxides as inorganic/organic hybrid precursors. Chem Mater 8:1592–1593
39. Zhang C, Laine RM (1996) Silsesquioxanes as synthetic platforms. II. Epoxy-functionalized inorganic/organic hybrid species. J Organomet Chem 521:199–201
40. Zhang C, Babonneau F, Bonhomme C, Laine RM, Soles CL, Hristov HA, Yee AF (1998) Highly porous polyhedral silsesquioxane polymers. Synthesis and characterization. J Am Chem Soc 120:8380–8391
41. Zhang C, Laine RM (2000) Hydrosilylation of allyl alcohol with $[HSiMe_2OSiO1.5]_8$. Octa (3-hydroxypropyldimethylsiloxy)octasilsesquioxane and its octamethacrylate derivative as potential precursors to hybrid nanocomposites. J Am Chem Soc 122:6979–6988
42. Tamaki R, Tanaka Y, Asuncion MZ, Choi J, Laine RM (2001) Octa(aminophenyl)silsesquioxane as a nanoconstruction site. J Am Chem Soc 123:12416–12417
43. Laine RM, Tamaki R, Choi J. Well-defined nanosized building blocks for organic/inorganic nanocomposites. WO 02/100867 A1 2002
44. Tamaki R, Choi J, Laine RM (2003) A polyimide nanocomposite from Octa(aminophenyl)-silsesquioxane. Chem Mater 15:793–797
45. Choi J, Tamaki R, Kim SG, Laine RM (2003) Organic/inorganic imide nanocomposites from Aminophenylsilsesquioxanes. Chem Mater 15:3365–3375
46. Choi J, Kim SG, Laine RM (2004) Organic/inorganic hybrid epoxy nanocomposites from Aminophenylsilsesquioxanes. Macromolecules 37:99–109

47. Choi J, Yee AF, Laine RM (2004) Toughening of cubic silsesquioxane epoxy nanocomposites using core shell rubber particles; a three component hybrid system. Macromolecules 37:3267–3276
48. Laine RM, Choi J, Lee I (2001) Organic-inorganic nanocomposites with completely defined interfacial interactions. Adv Mater 13:800–803
49. Choi J, Yee AF, Laine RM (2003) Organic/inorganic hybrid composites from cubic silsesquioxanes. Epoxy resins of octa-(dimethylsiloxyethylcyclohexylepoxide) silsesquioxane. Macromolecules 15:5666–5682
50. Takahashi K, Sulaiman S, Katzenstein JM, Snoblen S, Laine RM (2006) New aminophenylsilsesquioxanes, synthesis, properties and epoxy nanocomposites. Aust J Chem 59:564–570
51. Lin EK, Zhang CX, Wu WL, Laine RM. Materials characterization of model epoxy-functionalized silsesquioxanes as potential underfill encapsulants. Proc Int Symp Adv Packag Mater: Process Prop Interfaces 1999, 63–66
52. Sulaiman S, Brick CM, De Sana CM, Katzenstein JM, Laine RM, Basheer RA. Tailoring the global properties of nanocomposites. Epoxy resins with very low coefficients of thermal expansion. Macromol. web published 15-Jul-2006
53. Benicewicz BC, Smith ME, Earls JD, Duran RS, Setz SM, Douglas P (1998) Magnetic field orientation of liquid crystalline epoxy thermosets. Macromolecules 31(15):4730–4738
54. Tsuchida K, Bell JP (2000) A new epoxy/episulfide resin system for coating applications: curing mechanism and properties. Int J Adhes Adhes 20(6):449–456
55. Farren C, Akatsuka M, Takezawa Y, Itoh Y (2001) Thermal and mechanical properties of liquid crystalline epoxy resins as a function of mesogen concentration. Polymer 42(4):1507–1514
56. Su W-FA (1993) Thermoplastic and thermoset main chain liquid crystal polymers prepared from biphenyl mesogens. J Polym Sci Polym Chem 31(13):3251–3256
57. Carfagna C, Amendola E, Giamberini M, D'Amore A, Priola A, Malucelli G (1999) The effect of prepolymer composition of amino-hardened liquid crystalline epoxy resins on physical properties of cured thermoset. Macromol Symp 148:197
58. Lee JY, Jang J, Hwang SS, Hong SM, Kim KU (1998) Synthesis and curing of liquid crystalline epoxy resins based on 4,4′-biphenol. Polymer 39(24):6121–6126
59. Wong CP, Vincent MB, Shi S (1998) Fast-flow underfill encapsulant: flow rate and coefficient of thermal expansion. IEEE Trans Comp Pack Manuf Tech A 21(2):360
60. Katz HS (1987) Handbook of fillers for plastics. Van Nostrand Reinhold, New York
61. Miyagawa H, Rich MJ, Drzal LT (2004) Amine-cured epoxy/clay nanocomposites. II. The effect of the nanoclay aspect ratio. J Polym Sci Polym Phys 42(23):4391–4400
62. Zhang Z, Fan L (2002) Development of environmentally friendly nonanhydride no-flow underfills. IEEE Trans Comp Pack Manuf Tech A 25(1):140–147
63. Takahashi K, Sulaiman S, Katzenstein JM, Snoblen S, Laine RM (2006) New Aminophenylsilsesquioxanes, synthesis, properties and epoxy nanocomposites. Aust J Chem 59:564–570
64. Asuncion MZ, Laine RM (2007) Silsesquioxane barrier materials. Macromolecules 40:555–562
65. Eval Americas Technical Bulletin No. 110 Barrier properties of eval™ resins (2000)
66. Leterrier Y (2003) Durability of nanosized oxygen-barrier coatings on polymers. Prog Mater Sci 48:1–55
67. Hu YS, Prattipati V, Mehta S, Schiraldi DS, Hiltner A, Baer E (2005) Improving gas barrier of PET by blending with aromatic polyamides. Polymer 46(2685)
68. Hiltner A, Liu RYF, Hu YS, Baer E (2005) Oxygen transport as a solid state structure probe for polymeric materials: a review. J Polym Sci B Phys 43:1047

Chapter 8
Necklace-Shaped Dimethylsiloxane Polymers Bearing Polyhedral Oligomeric Silsesquioxane Cages as a New Type of Organic–Inorganic Hybrid

Masashi Kunitake

Abstract A series of necklace-shaped alternating siloxane copolymers that consisted of a bifunctional polyhedral oligomeric silsesquioxane (POSS) cage and flexible linear dimethyloligosiloxane (DMS) chain segment have been synthesized from bifunctional POSS molecules by two different synthetic methodologies, polycon-densation and ring-opening polymerization. Necklace-shaped POSS-DMS polymers with three different chain arrangements, "constant chain," "random chain," and "alternating modulated chain," have been synthesized. The necklace-shaped POSS-DMS polymers gave transparent thermoplastic and thermosetting plastics with high heat resistance upon cross-linking at the end of the polymers. The controllable structural diversity of these polymers allows control of their physical properties such as flexibility and glass transition temperature. These necklace-shaped polymers consisting of bulky functional inorganic units and alternately connected short soft chains will pave the way to inorganic soft materials, which are novel inorganic materials that can be handled like an organic polymer possessing solubility, plasticity, or entropic elasticity.

Keywords Necklace-shaped alternating siloxane copolymers · Polyhedral oligomeric silsesquioxane (POSS) · Heat-resistant polymers

8.1 Introduction

Organosilicon polymer compounds with siloxane bonds as the main skeleton have structural diversity comparable to that of organic compounds. The large degree of structural freedom of siloxane-based molecules allows the possibility of structural

M. Kunitake (✉)
Faculty of Advanced Science and Technology, Kumamoto University, Kumamoto, Japan
e-mail: kunitake@kumamoto-u.ac.jp

Y. Chujo (ed.), *New Polymeric Materials Based on Element-Blocks*,
https://doi.org/10.1007/978-981-13-2889-3_8

diversity similar to that of carbon materials, but it tends to produce disordered structures, making rational synthesis of precisely controlled structures difficult. Discrete nanostructures such as polyhedral oligomeric silsesquioxane (POSS) molecules [1–6], which possess discrete nanocage structures, have been studied extensively for both basic research and industrial applications as the basic structure of silsesquioxane nanomaterials. POSS cage units are composed of polymers with dendritic cores, pendant side chains with terminating groups, and multifunctional cross-linkers. Organosilicon polymers are unique compounds that simultaneously possess an inorganic-like "hard" structure based on three-dimensional connection and an organic polymer-like "soft" structure with flexibility or plasticity, respectively. Inorganic polymers consisting of POSS units and dimethyloligosiloxane (DMS) chains are of interest because of their structural specificity and useful physical properties including high transparency, flexibility, and heat resistance [7, 8]. In this chapter, synthetic tactics and basic physical properties of necklace-shaped POSS-DMS alternating polymers are introduced as a model polymer series with tunable structure diversity from the viewpoint of hierarchical control of the primary structures of the necklace-type polymers.

8.2 Polymers with Silsesquioxane Cages

Organic–inorganic hybrid materials designed to improve polymer properties, such as thermal, mechanical, optical, and electric properties, have been investigated extensively. Many organic–inorganic hybrid materials consist of a polymer matrix with incorporated inorganic fillers. POSS molecules are frequently used as nanofillers in polymer matrices because of their useful characteristics including high thermal stability, low dielectric properties [9], and good transparency [10–12].

Increasing the content of filler added to a polymer matrix leads to improvement of the desired properties from the general aspect of the ideal dispersion state of the fillers. As seen in Fig. 8.1, there is naturally a limit to the amount of filler that can be introduced, and in most cases, it is inevitable that characteristic polymer deterioration occurs. The deterioration from a certain threshold of filler loading would be related to formation of a heterogeneous structure based on the agglomeration of the filler in a polymer matrix. Moreover, the threshold would be attributed to the balancing point between entropic dispersion and enthalpic agglomeration. The dispersion state of the fillers in a polymer matrix is essentially ruled by the compatibility between the filler and polymer matrix from the viewpoint of thermodynamics.

Kinetic and/or thermodynamic approaches to overcome the above threshold problem are always conducted in the development of functional hybrid materials. The chemical modification of fillers to enhance their compatibility with the polymer matrix is a typical thermodynamic approach. In addition, smaller filler particles are advantageous to increase their compatibility with the polymer matrix. The miniaturization of the filler has been studied for micro-, meso-, and nanocomposite

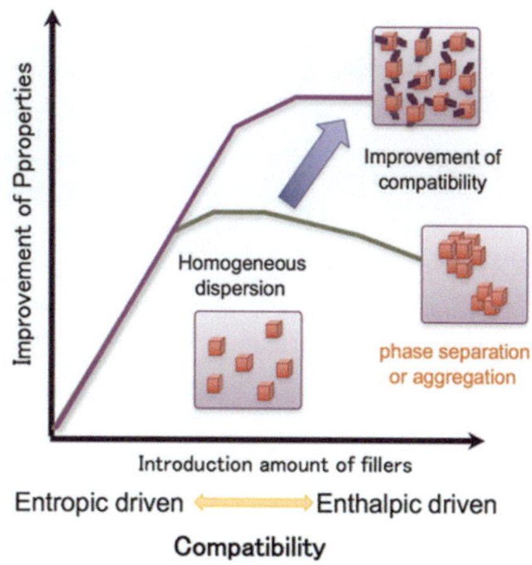

Fig. 8.1 Compatibility of nanofillers in a polymer matrix

fillers. As a kinetic approach, direct covalent bonding, especially cross-linking, between fillers and matrix and the containment of the fillers in a polymer network has often been studied.

Even for polymeric hybrid materials containing POSS as the nanofiller, the performance degradation caused by the aggregation of POSS in the polymer matrix has been universally reported. A design involving the alternating direct connection rigid and bulky POSS cages and soft polymer chains in a necklace-shaped polymer might be a fundamental answer to solve the problem of agglomeration of nanofillers in organic–inorganic hybrids (Fig. 8.2).

Most common POSS molecules possess a symmetrical cube (general hexahedron) shape with eight corners, which are relatively easy to chemically modify. However, regioselective introduction of different substituents on the eight corners of POSS is extremely difficult from the viewpoint of synthetic chemistry, although POSS intermediates bearing only one substituent or all eight substituents are often used as a component of POSS polymers.

POSS molecules with many functional groups at their corners are often utilized as multipoint cross-linking agents to produce thermosetting resins [12–14] and network polymers. Some studies employed dendrimers [9, 15] as core molecules. Monosubstituted POSS may also be used as a terminating group [16–18]. Many pendant-type polymers [16, 19–21] with a POSS unit in the side chain as a pendant unit have also been synthesized using POSS monomers. Monovinyl-substituted POSSs [16–18] are often used in copolymers through copolymerization with other vinyl monomers. At first glance, the dispersibility of the pendant POSS unit in the polymer seems to be high, because the POSS units in such a side chain are covalently linked. However, aggregation of POSS side chains tends to occur when the concentration of POSS units is high.

Polymers bearing POSS cages in their main chains [20, 22–28] have recently been reported and are referred to as necklace-shaped linear polymers or thermoplastic polymers containing POSS. To design the polymers, a POSS molecule with two

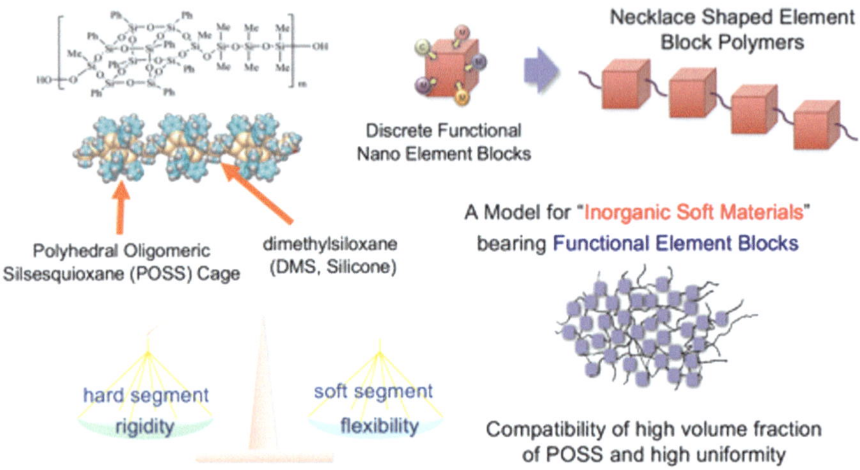

Fig. 8.2 Nanofusion of "soft" segments and "hard and bulky" inorganic functional units in necklace-shaped element block polymers

reactive groups at symmetrical positions is needed. Connecting this bifunctional POSS unit with a short chain or linker leads to a necklace-shaped polymer.

As a pioneering work with a bifunctional silsesquioxane cage, Lichtenhan and colleagues synthesized copolymers with a backbone bearing a defective octahedral cube containing an open section as a discrete bifunctional POSS cage [22]. Yoshida et al. [29] achieved the industrial production of a tetrafunctional double-decker-type silsesquioxane bearing eight phenyl groups. The sealing reaction of the silsesquioxane with trichloro alkylsilane produces ring-closed bifunctional cage-type silsesquioxane molecules (Fig. 8.3). Several necklace-shaped polymers consisting of the phenyl-substituted POSS units have been synthesized [3, 8, 24, 30–33]. Recently, Naka and coworkers [34] reported the synthesis of a bifunctional POSS cage from octahedron-shaped POSS molecules and the subsequent synthesis of siloxane polymers bearing the bifunctional POSS units.

Among the polymer designs in which POSS units are incorporated into a linear polymer main chain, the combination between a rigid and bulky POSS cage and soft dimetylsiloxane chain [35, 36] is interesting as a nanofusion of hard and soft segments.

8.3 Hierarchical Structural Design and Synthesis of Necklace-Shaped POSS-DMS Polymers

Necklace-shaped inorganic polymers consisting of alternating POSS units and DMS chains have been investigated by several groups [8, 23, 34]. Kawakami reported polysiloxanes with periodically distributed isomeric double-decker silsesquioxane

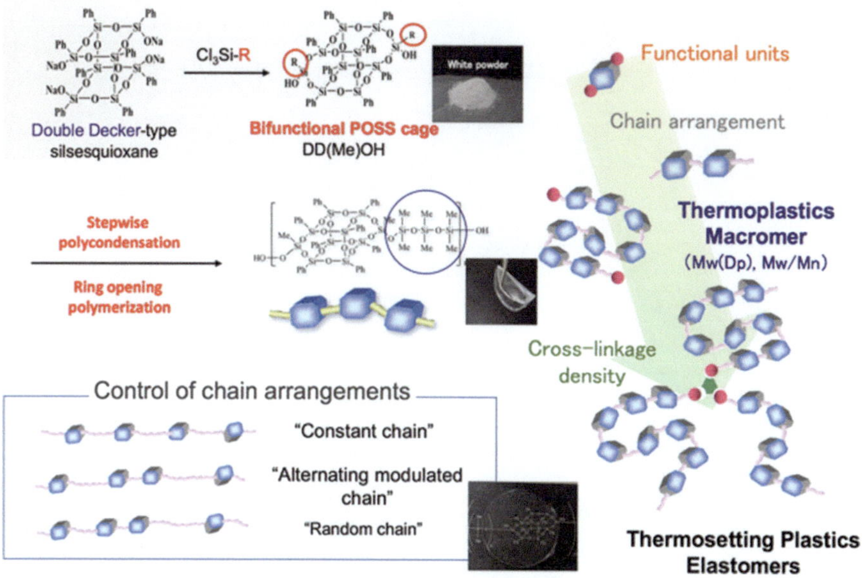

Fig. 8.3 Hierarchical design of necklace-shaped POSS-DMS polymer systems with systematic structural diversity

[23, 37] in the main chain. We reported two synthetic methodologies [8, 30] based on polycondensation and ring-opening polymerization to obtain necklace-shaped POSS-DMS polymers. By selectively using two polymerization methods, hierarchical control of primary polymer structures in a necklace-shaped polymer was achieved over not only the degree of polymerization but also the length and arrangement of DMS chains between POSS, providing polymers with "constant chain," "random chain," and "alternating modulated chain" arrangements [8].

8.3.1 Synthesis of "Constant Chain" POSS-DMS Polymers by Polycondensation

Figure 8.4 outlines the synthetic pathway to necklace-shaped POSS siloxane polymers starting from a bifunctional POSS. The most straightforward synthetic route to produce necklace-shaped polymers that possess an alternating structure is polycondensation between bulky POSS units and DMS units.

Polycondensation between two silanol groups of bifunctional POSS and dichloro-oligo-dimethylsilane (DCMS) produces a necklace-shaped POSS-DMS alternating polymer with constant chain length between POSS cages. In the case of polycondensation, the DMS chain length between POSS cages can be controlled by selecting a DCMS with a constant chain length. Polymers with chain lengths of

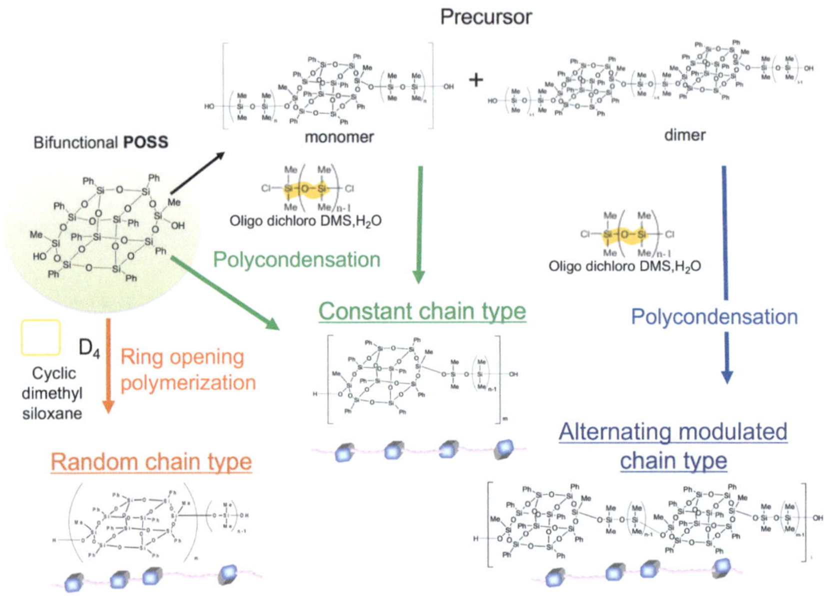

Fig. 8.4 Synthetic routes to necklace-shaped POSS-DMS polymers with three types of chain arrangements

up to four have been synthesized by direct one-step polycondensation of POSS with DCMS. Alternating POSS polymers with longer chain length can also be synthesized with POSS precursors prepared by discrete reaction with DCMS. Polycondensation using DCMS requires a very strict dehydration operation because water contamination leads to a marked decrease in the average molecular weight. However, this approach has succeeded in producing polymers with a molecular weight of more than one million.

8.3.2 Synthesis of "Alternating Modulated Chain" POSS-DMS Polymers by Multi-step Polycondensation

Dimers with two POSS molecules are also generated when POSS molecules with siloxane chains at both ends are synthesized by condensation reaction of bifunctional POSS and DCMS. Depending on the reaction conditions, the dimers may be synthesized as the main product and purified by recrystallization, because most of the dimers have higher crystallinity than the monomers. Polycondensation of POSS dimer molecules with DCMS can produce polymers with a modulated chain length in which the lengths of siloxane chains between POSS units alternate. The combination of the length of the siloxane chain at the time of dimer synthesis and the chain length of DCMS used at the time of condensation makes it possible to freely design the lengths of long and short chains.

8.3.3 Synthesis of "Random Chain" POSS-DMS Polymers by Ring-Opening Equilibrium Polymerization

Ring-opening polymerization of cyclic siloxanes such as octamethylcyclotetrasiloxane (D4) with an acid or alkali catalyst is commonly used as an industrial synthesis method of siloxane polymers. In this ring-opening polymerization reaction with a bifunctional POSS, POSS cages are introduced into the siloxane chain to generate a POSS-DMS alternating polymer. In fact, this ring-opening polymerization is an equilibrium reaction that forms POSS-DMS polymers with random chain length. During the reaction, coupling and cleavage of siloxane chains proceed in both directions competitively as a reversible reaction. Because the cleavage point in the chain is not controlled, the inter-POSS cage chain length is not a simple multiple of four opened by D4, but a random structure with a distribution of chain lengths.

This ring-opening equilibrium polymerization does not require strict dehydration conditions as those of the polycondensation using DCMS, and it is relatively easy to synthesize a polymer with an average molecular weight of more than 100,000 in one step. It must be emphasized that control of the average chain length between POSS units is easily achieved by means of the composition ratio of POSS and DMS in the reactor.

Interestingly, insoluble gel products resulting from cross-linking hardly form during the ring-opening polymerization with POSS, indicating the chemical robustness of the T3-type silsesquioxane bond in the POSS cage compared with the bond in DMS chains. The attack by the acid catalyst also may extend to the cage structure, but the selective recombination returning to the original cage structure must occur simultaneously.

8.3.4 Thermosetting Resins Based on POSS-DMS Polymers

In either the polycondensation method or equilibrium polymerization method, the terminal functional group of the synthesized POSS-DMS necklace polymers is a reactive silanol unit. Through the cross-linking reaction of these end groups with a polyfunctional agent, a POSS-DMS thermoplastic useful as a thermosetting network polymer can be obtained. By limiting the cross-linking reaction to only the ends of polymer chains, the polymer is transformed into a thermosetting resin that microscopically retains the mesostructure and thermal properties of the POSS-DMS linear polymer structure.

As shown in Fig. 8.4, a series of necklace polymers with hierarchical diversity of their primary structures have been synthesized systematically from the following points: (1) the sealing group, (2) the average chain length and arrangement between POSS units, (3) the average molecular weight (degree of polymerization) of the macromer (linear polymer), and (4) the cross-linking density in thermosetting polymers. Such precise stepwise design and control of the polymer structure would be attractive to tune characteristics of the polymer.

8.4 Correlation Between the Primary Structures and Thermal Properties of Necklace-Shaped POSS-DMS Polymers

Necklace-shaped POSS-DMS polymers show very good solubility in common organic solvents because of their unique alternating structure with soft siloxane chains placed between bulky POSS cage units. Therefore, the polymer films are easily prepared by simple casting. The films generally possess an amorphous structure and are obtained as a flexible and transparent polymer sheet. The agglomeration of the POSS cage unit is not observed, although the expected occupied volume of the POSS is extremely high (about 80%), suggesting an amorphous structure with homogeneous dispersion of POSS. Despite the poor compatibility between the POSS units and DMS chains, their alternating arrangement in necklace-shaped polymers allows us to achieve a homogeneous POSS dispersion in the overall polymer matrix. The series of POSS-DMS polymers were intentionally designed as amorphous polymers. With the exception of polymers possessing DMS chains shorter than two units, the necklace-shaped POSS-DMS polymers were all amorphous. These amorphous structures with controlled structural diversity allow us to elucidate the correlation between the polymer primary structures and various physical properties without needing to consider the influence of heterogeneous mesostructures.

When thermosetting resins are prepared by cross-linking at the ends of the polymers, insoluble polymer sheets are obtained. The original structural features and thermal properties of the necklace-shaped thermoplastic polymers are almost maintained even after cross-linking to produce the corresponding thermosetting resins. This is because of the retained homogeneous dispersion and high liquidity of POSS units. Regardless of whether these polymers are thermoplastic or thermosetting, one of the major features of necklace-shaped POSS-DMS polymers is extremely high heat resistance. The 5% decomposition temperature (Td5%), which is a measure of heat resistance, of polydimethylsiloxane (PDMS) is about 350 °C. In contrast, Td5% of many POSS-DMS polymers exceeds over 400 °C. The Td5% values of POSS-DMS polymers vary according to their primary structure, particularly the average chain length between POSS units. POSS-DMS polymers with average chain lengths of three to four showed extremely high Td5% of over 470 °C in air. The highest Td5% of 478 °C was obtained for the POSS-DMS polymer with an average chain length of 2.9. For the polymers bearing longer DMS chain lengths of over three, Td5% decreased and approached that of PDMS with increasing chain length.

In accelerated deterioration tests, no yellowing, which is typically a problem for siloxane polymers, was observed at all. This remarkably high heat resistance performance is considered to result from the extremely high volume and high homogeneity of POSS cages in the polymer matrix, which are allowed by the necklace-shaped polymer structure. A space-filling model indicated that POSS units in POSS cages are close and partially contact each other in the case of the polymer with a constant siloxane chain length of two. In addition, the polymer model for the siloxane chain length of four showed adequate space between POSS units. These results indicate that the balance of the trade-off relation between the composition ratio of POSS and uniformity in the matrix is crucial to design high-performance polymers.

The glass transition temperature (Tg) is one of the most important standard parameters of polymeric materials. In the case of ordinary vinyl polymers, it is well-known that the increase in Tg with respect to molecular weight gradually decreases with rising molecular weight, becoming almost constant over a molecular weight of about 10,000. In contrast, POSS-DMS polymers bearing large POSS units in which the average POSS number (degree of polymerization) was around ten showed an obvious increase in Tg with molecular weight even over 10,000.

Synthesis of high-molecular-weight polymers consisting of necklace-shaped polymers bearing bulky block units, such as the POSS-DMS alternating polymers, is very important for their use as thermoplastics. Therefore, the correlation between Tg and primary polymer structure was determined for polymers with a molecular weight of approximately 100,000 in which the average number of POSS units (degree of polymerization) was roughly 100. As expected, Tg of the POSS-DMS polymers decreased, and the flexibility of the polymer films rose with increasing average chain length of DMS between POSS units. The relationship between the average chain length and Tg was roughly consistent within a series of polymers with "constant chain" and "random chain" arrangements. The Tg values of the polymers with an "alternating modulated chain" arrangement did not follow the same trend.

The relationship between the average chain length and Tg for "random chain"--type polymers was comparable to that of the "constant chain"-type polymers, even though the "random chain"-type polymers contained individual longer chains in their distribution of chain length. A gradual increase in temperature induces the stepwise local melting of DMS chain units in the main chains of the polymers. The glass transition to a rubber phase requires collaborative micro-Brownian movement over the local relaxation in long-chain moieties. In the case of polymers with "alternating modulated chain" structures, half of the chain moieties (the connecting chains) are longer than the rest of the parts. Therefore, cooperative melting, which leads to global micro-Brownian motion, is possible at the melting temperature of the longer-chain moieties. The POSS dimer moieties connected by short chains act as one segment just above Tg. Certainly, two stepwise softening features are observed in the curves of the polymers with “alternating modulated chain” arrangements.

8.5 Conclusions and Outlook

Necklace-shaped POSS-DMS polymers, in which POSS cages are alternately connected to short soft chains to form linear polymers, were introduced. These series of polymers enable hierarchical structure control to achieve systematic diversity of primary polymer structure. The compatibility between an extremely high introduction ratio of element block nanounits and high homogeneous dispersibility is made possible by this necklace structure. Despite the extremely high content of POSS units, these polymers have achieved high transparency, designable flexibility, and extremely high heat resistance as a result of their amorphous structure and homogeneous dispersion of POSS. In addition, necklace-shaped polymers have the potential for free design from thermoplastics to thermosetting plastics to elastomers by means of the reactivity of the terminal groups.

Modern advanced materials are required to simultaneously satisfy multiple demands. Among the properties required, many combinations might have an essentially contradictory relation, for instance, flexibility versus low linear expansion and barrier properties and selectivity versus permeability. To obtain favorable characteristics with such a trade-off relation, nanofusion of materials with conflicting properties is a straightforward tactic, as was originally the case with organic–inorganic hybrids. The hierarchical control of polymer structure in a necklace-shaped polymer architecture will lead to the creation of a new material systems, “element-block polymers,” beyond the boundary between organic and inorganic materials.

The development of necklace-shaped polymers consisting of bulky functional inorganic units and alternately connected short soft chains is a promising direction for organic–inorganic hybrid materials. The research of such polymers will pave the way to develop novel inorganic materials that can be handled like organic polymers with solubility, plasticity, or entropic elasticity, representing the concept of inorganic soft materials.

Acknowledgments I am deeply grateful to the students who engaged in this research and JNC Corporation and JNC Petrochemical Corporation for continued collaboration. This work was supported by a Grant-in-Aid for Scientific Research on Innovative Areas “New Polymeric Materials Based on Element-Blocks” (24102006) from the Ministry of Education, Culture, Sports, Science, and Technology, Japan.

References

1. Chujo Y, Tanaka K (2015) New polymeric materials based on element-blocks. Bull Chem Soc Jpn 88:633–643. https://doi.org/10.1246/bcsj.20150081
2. Kuo SW, Chang FC (2011) POSS related polymer nanocomposites. Prog Polym Sci 36:1649–1696. https://doi.org/10.1016/j.progpolymsci.2011.05.002
3. Han S-Y, Wang X-M, Shao Y, Guo Q-Y, Li Y, Zhang W-B (2016) Janus POSS based on mixed [2:6] octakis-adduct regioisomers. Chem A Eur J 22:6397–6403. https://doi.org/10.1002/chem.201600579
4. Li G, Charles UP (2005) Polyhedral oligomeric silsesquioxane (POSS) polymers, copolymers, and resin nanocomposites. Macromol Contain Met Met Elem Gr IVA Polym 4:79–131. https://doi.org/10.1002/0471712566.ch5
5. Laine RM, Roll MF (2011) Polyhedral phenylsilsesquioxanes. Macromolecules 44:1073–1109. https://doi.org/10.1021/ma102360t
6. Araki H, Naka K (2012) Syntheses and properties of star- and dumbbell-shaped POSS derivatives containing isobutyl groups. J Polym Sci Part A 50:4170–4181
7. Zhang W, Camino G, Yang R (2016) Polymer/polyhedral oligomeric silsesquioxane (POSS) nanocomposites: an overview of fire retardance. Prog Polym Sci 67:77–125. https://doi.org/10.1016/j.progpolymsci.2016.09.011
8. Katsuta N, Yoshimatsu M, Komori K, Natsuaki T, Suwa K, Sakai K, Matsuo T, Ohba T, Uemura S, Watanabe S, Kunitake M (2017) Necklace-shaped dimethylsiloxane polymers bearing polyhedral oligomeric silsesquioxane cages with alternating length chains. Polym (United Kingdom) 127:8–14. https://doi.org/10.1016/j.polymer.2017.08.049
9. Wang J, Sun J, Zhou J, Jin K, Fang Q (2017) Fluorinated and thermo-cross-linked polyhedral oligomeric silsesquioxanes: new organic–inorganic hybrid materials for high-performance dielectric application. ACS Appl Mater Interfaces 9:12782–12790. https://doi.org/10.1021/acsami.7b01415
10. Ghanbari H, Cousins BG, Seifalian AM (2011) A nanocage for nanomedicine: polyhedral oligomeric silsesquioxane (POSS). Macromol Rapid Commun 32:1032–1046. https://doi.org/10.1002/marc.201100126
11. Jeon JH, Tanaka K, Chujo Y (2013) Rational design of polyhedral oligomeric silsesquioxane fillers for simultaneous improvements of thermomechanical properties and lowering refractive indices of polymer films. J Polym Sci A Polym Chem 51:3583–3589. https://doi.org/10.1002/pola.26757
12. Tsukada S, Sekiguchi Y, Takai S, Abe Y, Gunji T (2015) Preparation of POSS derivatives by the dehydrogenative condensation of T 8 H with alcohols. J Ceram Soc Jpn 123:739–743. https://doi.org/10.2109/jcersj2.123.739
13. Huang C, Chang F (2009) Using click chemistry to fabricate ultrathin thermoresponsive microcapsules through direct covalent layer-by-layer assembly. Macromolecules 42:5155–5166. https://doi.org/10.1021/ma900478n
14. Tokunaga T, Koge S, Mizumo T, Ohshita J, Kaneko Y (2015) Facile preparation of a soluble polymer containing polyhedral oligomeric silsesquioxane units in its main chain. Polym C 6:3039–3045. https://doi.org/10.1039/C5PY00192G
15. Naka K, Masuoka S, Shinke R, Yamada M (2012) Synthesis of first- and second-generation imidazole- terminated POSS-core dendrimers and their pH responsive and coordination properties. Polym J 44:353–359. https://doi.org/10.1038/pj.2011.145
16. Yu C-B, Ren L-J, Wang W (2017) Synthesis and self-assembly of a series of n POSS- b -PEO block copolymers with varying shape anisotropy. Molecules 22:622–630. https://doi.org/10.3390/molecules22040622
17. Zhang W, Fang B, Walther A, Mu AHE (2009) Synthesis via RAFT polymerization of tadpole-shaped organic/inorganic hybrid poly (acrylic acid) containing polyhedral oligomeric silsesquioxane (POSS) and their self-assembly in water. Macromolecules 42:2563–2569

18. Shockey EG, Bolf AG, Jones PF, Schwab JJ, Chaffee KP, Haddad TS, Lichtenhan JD (1999) Functionalized polyhedral oligosilsesquioxane (POSS) macromers: new graftable POSS hydride, POSS α-olefin, POSS epoxy, and POSS chlorosilane macromers and POSS-siloxane triblocks. Appl Organomet Chem 13:311–327. https://doi.org/10.1002/(SICI)1099-0739
19. Lichtenhan JD, Otonari YA, Cam MJ, Carr MJ (1995) Linear hybrid polymer building blocks: methacrylate-functionalized polyhedral oligomeric silsesquioxane monomers and polymers. Macromolecules 28:8435–8437. https://doi.org/10.1021/ma00128a067
20. Hirai T, Leolukman M, Hayakawa T, Kakimoto MA, Gopalan P (2008) Hierarchical nanostructures of organosilicate nanosheets within self-organized block copolymer films. Macromolecules 41:4558–4560. https://doi.org/10.1021/ma800872v
21. Pramudya I, Rico CG, Lee C, Chung H (2016) POSS-containing bioinspired adhesives with enhanced mechanical and optical properties for biomedical applications. Biomacromolecules 17:3853–3861. https://doi.org/10.1021/acs.biomac.6b00805
22. Mantz RA, Jones PF, Chaffee KP, Lichtenhan JD, Gilman JW, Ismail IMK, Burmeister MJ (1996) Thermolysis of polyhedral oligomeric silsesquioxane (POSS) macromers and POSS-siloxane copolymers. Chem Mater 8:1250–1259. https://doi.org/10.1021/cm950536x
23. Hoque MA, Kakihana Y, Shinke S, Kawakami Y (2009) Polysiloxanes with periodically distributed isomeric double-decker silsesquioxane in the main chain. Macromolecules 42:3309–3315. https://doi.org/10.1021/ma900124x
24. Liu N, Li L, Wang L, Zheng S (2017) Organic-inorganic polybenzoxazine copolymers with double decker silsesquioxanes in the main chains: synthesis and thermally activated ring-opening polymerization behavior. Polym (Guildf) 109:254–265. https://doi.org/10.1016/j.polymer.2016.12.049
25. Sodkhomkhum R, Ervithayasuporn V (2016) Synthesis of poly(siloxane/double-decker silsesquioxane) via dehydrocarbonative condensation reaction and its functionalization. Polymer (United Kingdom) 86:113–119. https://doi.org/10.1016/j.polymer.2016.01.044
26. Furgal JC, Jung JH, Clark S, Goodson T, Laine RM (2013) Beads on a chain (BoC) phenylsilsesquioxane (SQ) polymers via F− catalyzed rearrangements and ADMET or reverse heck cross- coupling reactions: through chain, extended conjugation in 3-D with potential for dendronization. Macromolecules 46:7591–7604. https://doi.org/10.1021/ma401423f
27. Jung JH, Furgal JC, Clark S, Schwartz M, Chou K, Laine RM (2013) Beads on a chain (BoC) polymers with model dendronized beads. Copolymerization of [(4-NH2C6H4SiO1.5)6 (IPhSiO1.5)2] and [(4-CH3OC6H4SiO1.5)6(IPhSiO1.5)2] with 1,4-Diethynylbenzene (DEB) gives through-chain, extended 3-D conjugation in the excited state tha. Macromolecules 6:7580–7590. https://doi.org/10.1021/ma401422tl
28. Jung JH, Laine RM (2011) Beads on a chain (BOC) polymers formed from the reaction of [NH 2 PhSiO 1.5] x [PhSiO 1.5] 10– x and [NH 2 PhSiO 1.5] x [PhSiO 1.5] 12– x mixtures (x = 2–4) with the Diglycidyl ether of bisphenol A. Macromolecules 44:7263–7272. https://doi.org/10.1021/ma201032r
29. Morimoto Y, Watanabe K, Ootake N, Inagaki J, Yoshida K, Ohguma K (2004) POSS patent: US 20040249103
30. Yoshimatsu M, Komori K, Ohnagamitsu Y, Sueyoshi N, Kawashima N, Chinen S, Murakami Y, Izumi J, Watanabe K, Matuo T, Sakai K, Kunitake M (n.d.) Necklace shaped dimethylsiloxane polymers bearing a polyhedral oligomeric silsesquioxane cage prepared by polycondensation and ring-opening polymerization. pp 2–5
31. Kunitake M, Sakai K, Hirabayashi C, Morimoto Y, Kunitake M, Sakai K, Hirabayashi C, Morimoto Y (2006) JP 2006022207, n.d
32. Seino M, Hayakawa T, Ishida Y, Kakimoto MA, Watanabe K, Oikawa H (2006) Hydrosilylation polymerization of double-decker-shaped silsesquioxane having hydrosilane with diynes. Macromolecules 39:3773–3775. https://doi.org/10.1021/ma052631p
33. Maegawa T, Irie Y, Fueno H, Tanaka K, Naka K (2014) Synthesis and polymerization of a para -disubstituted T8 caged hexaisobutyl-POSS monomer. Chem Lett 43:1532–1534. https://doi.org/10.1246/cl.140515

34. Maegawa T, Irie Y, Imoto H, Fueno H, Tanaka K, Naka K (2015) Para-bisvinylhexaisobutyl-substituted T 8 caged monomer: synthesis and hydrosilylation polymerization. Polym Chem 6:7500–7504. https://doi.org/10.1039/C5PY01262G
35. Kunitake M, Nishi T, Yamamoto H, Nasu K, Manabe O, Nakashima N (1994) Preparation and characterization of two-dimensional cross-linked monolayers and Langmuir-Blodgett films of oligo(dimethylsiloxane) copolymer. Langmuir 10:3207–3212. https://doi.org/10.1021/la00021a051
36. Kato D, Masaike M, Majima T, Hirata Y, Mizutani F, Sakata M, Hirayama C, Kunitake M (2002) Permselective monolayer membrane based on two-dimensional cross-linked polysiloxane LB films for hydrogen peroxide detecting glucose sensors. Chem Commun (Camb):2616–2617. https://doi.org/10.1039/B207957G
37. Ervithayasuporn V, Sodkhomkhum R, Teerawatananond T, Phurat C, Phinyocheep P, Somsook E, Osotchan T (2013) Unprecedented formation of cis- and trans-di[(3-chloropropyl) isopropoxysilyl]-bridged double-decker octaphenylsilsesquioxanes. Eur J Inorg Chem 2013:3292–3296. https://doi.org/10.1002/ejic.201300283

Chapter 9
Alternative Aspects of Polythiophenes

Takashi Nishino, Takuya Matsumoto, and Atsunori Mori

Abstract Syntheses of high-molecular-weight (Mw up to 880 k) and high regioregular (head-to-tail >99%) poly(3-hexylthiophene) (P3HT) were first described. Self-standing film could be only obtained using high *M*w P3HT. Molecular orientation in the P3HT thin film was generally known as edge-on structure; however, thin film with high *M*w P3HT spin-coated on hydrophilic substrate at low temperature was found to show flat-on rich structure. By substituting side chains from hexyl groups to disiloxane moieties (P3SiT), fragile issue of P3HT (elongation at break <15%) could be overcome. This is because high *M*w P3SiT gave low Tg (−10 °C), which resulted in high elongation at break (more than 200%). These alternative novel futures of polythiophenes as element-block polymers were described.

Keywords Polythiophene · Self-standing film · Siloxane moiety · High molecular weight · Face-on

9.1 Introduction

In general, polymers are electric insulating materials, which have been widely useful for tapes, coatings, and substrates. On the contrary, the possibility of high-temperature superconducting was proposed by Little for polymers in 1964 [1]. Shirakawa et al. discovered polyacetylene as high electroconducting polymer, which was awarded with the Nobel Prize in Chemistry in 2000 [2]. Then, many types of electroconducting polymers were proposed theoretically and experimentally in order to utilize advantages of polymers such as low density, high processability, and high resistance against corrosion. Nowadays, their significant improvements in efficiency and performance gather the interests of industries and opened the way to practical

T. Nishino (✉) · T. Matsumoto · A. Mori
Department of Chemical Science and Engineering, Graduate School of Engineering, Kobe University, Kobe, Japan
e-mail: tnishino@kobe-u.ac.jp

Y. Chujo (ed.), *New Polymeric Materials Based on Element-Blocks*,
https://doi.org/10.1007/978-981-13-2889-3_9

applications for organic semiconductors as organic field-effect thin film transistors [3], organic light-emitting diodes [4], and solar cells [5] to replace inorganic material and metals.

The polymers with group 16 heteroaromatic compounds, kinds of element-block polymers, can be expected as good candidates because of their remarkable physical properties as electronic materials. However, owing to high rigidity of their conjugated skeleton, unsubstituted polymers are generally insoluble to any organic solvent and infusible. Thus, in order to give wet/melt processability, flexible side chains were added to the backbone of the conjugated system, and then substituted polymer, typically poly(3-hexylthiophene) (P3HT), has been investigated widely and extensively [6]. In these cases, the formation of head-to-tail (HT) type regioregular arrangement of hexyl groups is particularly important to maintain highly planar and 3-conjugation-extended characteristics [7]. In addition, molecular weight of P3HT was found to play an important role for structure formation and mechanical properties.

In this chapter, syntheses of new series of high-molecular-weight and high head-to-tail-type polythiophenes and the effect of molecular weight on their structure and mechanical properties were described.

9.2 High-Molecular-Weight, High Head-to-Tail P3HT

Table 9.1 shows typical three examples for the syntheses of P3HT. First example (a) is oxidative polymerizations with electrochemically or using $FeCl_3$ (as an oxidant) [8–10]. In this method, HT regularity was not well controlled (~70%), and weight average molecular weight (*M*w) was limited to 27 k with wide molecular weight distribution. These characteristics resulted in relatively low electron conductivity around 0.1–1 S/cm^2. It was shown in (b) that high regioregular P3HT was prepared by nickel-catalyzed polycondensation of the corresponding dibromothiophene with a Grignard reagent (the GRIM method) [11]. However, it was difficult to obtain P3HT with a molecular weight as high as 10 k, which results in failure of preparing self-standing film [12]. In contrast with the extensive spectroscopic and electronic characterizations of P3HT, the mechanical performance of P3HT as a self-standing film has rarely been studied owing to the preparative difficulties involved in the synthesis of high-molecular-weight P3HT. Recently, the synthesis of higher-molecular-weight P3HT could be achieved with [(o-tolyl) NiCl(dppp)] as an initiator, affording polythiophene with a molecular weight up to ca. 200 k [13]. This prompted us to report our separate findings on the development of a new class of nickel catalysts bearing a cyclopentadienyl (Cp) ligand for the polymerization of thiophene derivatives to yield high-molecular-weight P3HT [14]. We have reported recently that nickel-catalyzed dehydrobrominative or dehydrochlorinative polycondensation is also a powerful tool (in addition to the GRIM method) for the synthesis of high regioregular polythiophenes using the

Table 9.1 Three different synthesis routes of P3HT and their properties

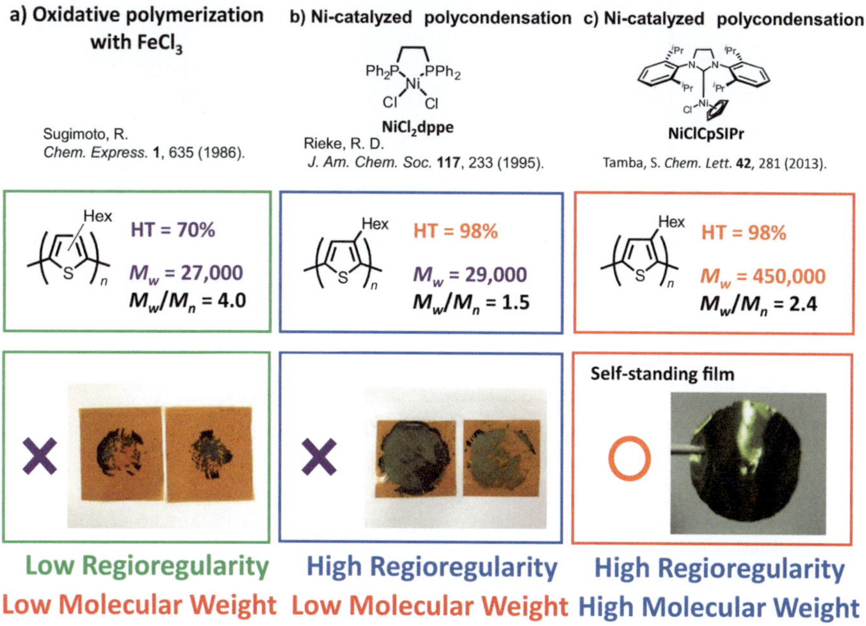

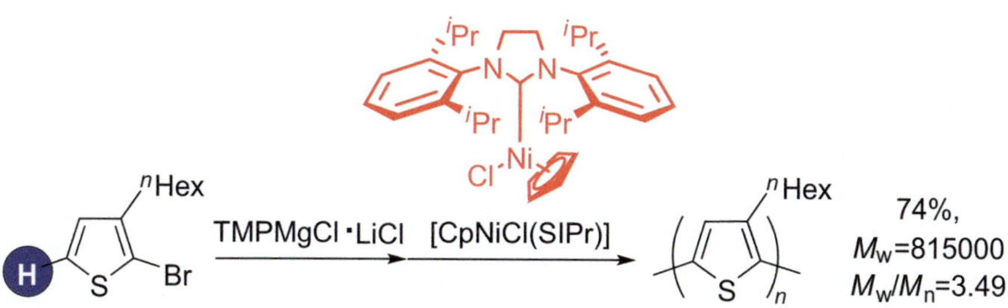

Scheme 9.1 Synthesis of high-molecular-weight and high regioregular polythiophenes

Knochel–Hauser base, chloromagnesium 2,2,6,6-tetramethylpiperidide lithium chloride salt (TMPMgCl·LiCl), in the presence of a nickel(II) catalyst [15] as shown in Scheme 9.1.

The method allows polymerization at a lower reaction temperature within a shorter polymerization time with high atom efficiency. In our studies on the dehydrobrominative synthesis as well as the aforementioned GRIM method, it is remarkable that the molecular weight of P3HT is controllable by the ratio of monomer feed and catalyst loading, because of the characteristics of living polymerization. In the course of our studies on the synthesis of P3HT, we found

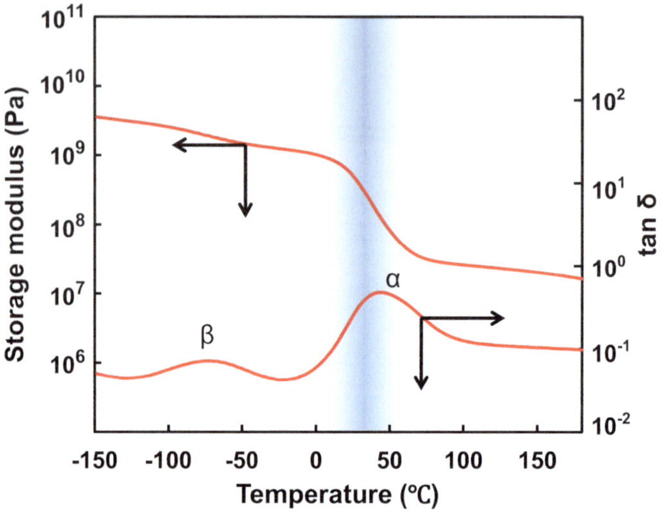

Fig. 9.1 Temperature dependence of the storage modulus and mechanical tanδ of P3HT self-standing film

unexpectedly that P3HT of an extremely high molecular weight (> 800 k) was obtained irrespective of the catalyst loading when a nickel catalyst bearing a cyclopentadienyl (Cp) and an *N*-heterocyclic carbene ligand (NHC: SIPr) [CpNiCl (SIPr)] was employed. This finding opened extensive characterizations of the mechanical properties of P3HT as a self-standing film without using supporting substrate, which, to the best of our knowledge, have rarely been studied owing to the lack of a methodology to afford high-molecular-weight polymers.

Figure 9.1 shows the temperature dependence of the storage modulus and mechanical tan δ of high *M*w and high H-T P3HT self-standing film. The melt-pressed, followed by quenching film with the thickness of 200 μm, was employed for the measurement. The tan δ peaks clearly showed two mechanical dispersions indexed as α and β. The higher one (α-dispersion) indicates that glass transition temperature of P3HT is 45 °C. From the stress–strain curve at room temperature, Young's modulus, tensile strength, and elongation at break were 3.3 GPa, 165 MPa, and 14%, respectively. These show P3HT can be comparable to conventional glassy polymers mechanically.

Additional effect of high molecular weight also appeared in thermal behavior of P3HT.

Figure 9.2 shows the differential scanning thermograms of P3HT with different molecular weights. With increasing Mw, melting endotherm shifted to higher temperature (255 °C), which results in the increase of the melting point *Tm* around 50°. *Tm* is well known to correlate to the crystallite size and/or defects for semi-crystalline polymers. The increase of *Tm* reveals that the crystalline perfection

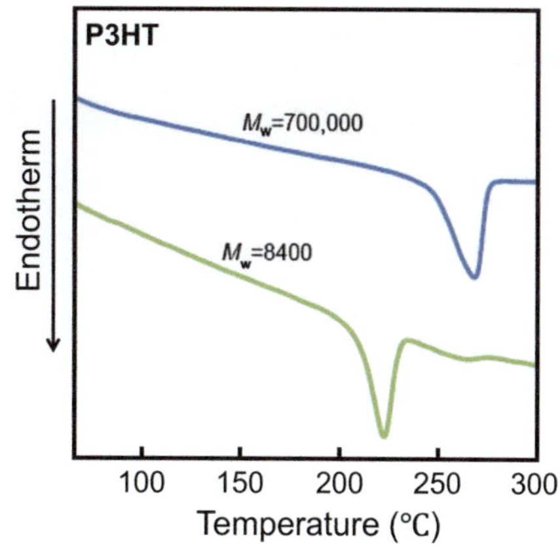

Fig. 9.2 Differential scanning thermograms of P3HT with different *M*w

increased for P3HT with higher Mw. One reason of this is that the number of chain ends (act as defects in the crystalline regions) reduced, and then its effect is getting negligible with increasing molecular weight.

9.3 Molecular Weight Effect on the Structure of P3HT

The carrier mobility of P3HT is drastically influenced by the morphology and crystallite orientation in solid states. In particular, P3HT crystal shows anisotropic carrier mobility. The highest mobility appears along the main chain's direction in the crystalline regions. On the other hand, the mobility in the direction to π–π stacking of P3HT is higher relative to that parallel to the side chains as shown in Fig. 9.3. Thus, the control of the crystallite orientation is a key factor for the electronic performance of devices [16, 17]. The crystallinity and the crystallite orientation have been considered to be variable according to regioregularity, molecular weight, and end groups of P3HT. Temperature and procedures for device fabrication and surface futures of the substrates for spin-coating were also investigated. However, P3HTs with molecular weight less than 10 k have been employed, and then the structure at the surface is assumed to be the same as that at bulk so far. We prepared thin films of P3HT with a variety of average molecular weight ($M_w = 13$ k, 128 k, 214 k, 346 k, 828 k) on silicon wafer. Their structures were investigated by the small-angle incidence/grazing incidence X-ray diffraction.

Figure 9.4 shows the schematic of the small-angle incidence/grazing incidence X-ray diffraction with two geometries. The X-ray beam was irradiated on the sample surface with very small incident angle α, and then diffractions were detected with two different geometries; (a) the detector was scanned in the direction perpendicular to the sample surface and (b) in the parallel direction [18]. The former was called "out-of-plane" measurement and the latter was "in-plane" measurement. In the "out-of-plane" geometry, the lattice planes nearly parallel to the surface of thin films were

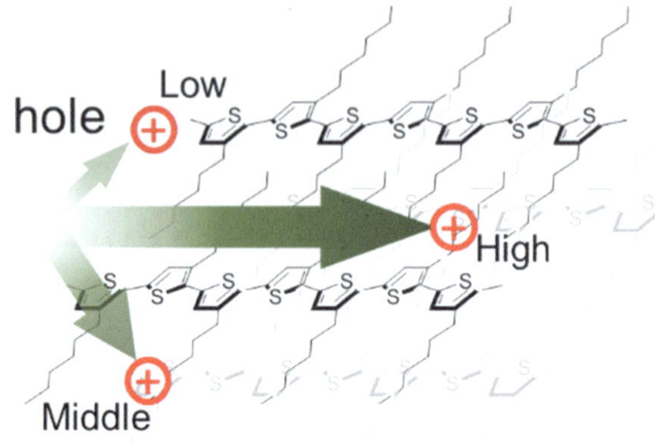

Fig. 9.3 Anisotropy of hole mobilities of P3HT

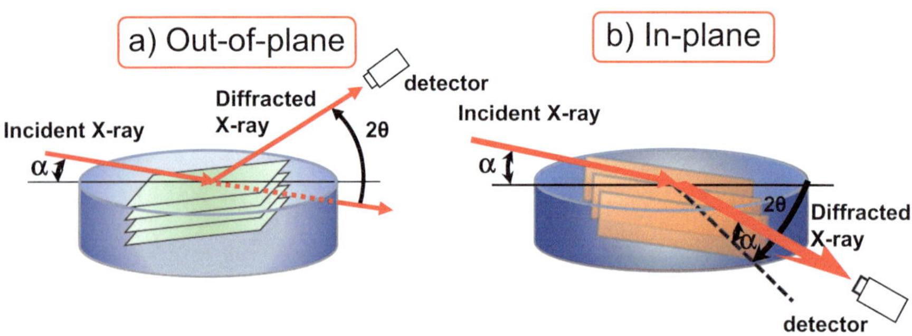

Fig. 9.4 Schematic illustration of the small-angle incidence/grazing incidence X-ray diffraction with two geometries. (**a**) Out-of-plane, (**b**) in-plane

detected. In contrast, in the "in-plane" geometry, the lattice planes nearly perpendicular to the surface were observed. The penetration depth of X-ray beam from the surface was around 1.7 μm at $\alpha = 0.20°$, which is called as small-angle incidence X-ray diffraction.

Figure 9.5 shows X-ray diffraction profiles of P3HT films with various molecular weights through (a) "out-of-plane" and (b) "in-plane" geometries [19]. In all the profiles of P3HT thin films, the peaks originated from P3HT crystal were observed, and they could be well indexed with the reported crystal structure of P3HT [20] as shown in the figure. The 100 and 020 reflections were observed at $2\theta = 5°$, $23°$ respectively. From the powder diffraction of P3HT annealed at 100 °C for 1 h, the intensity ratio from 100 to 020 reflection (I_{100}/I_{020}) of 3.0 was represented with the dotted line for the non-oriented one. The intensity ratio I_{100}/I_{020} in each profile is shown in Fig. 9.5c. With increasing the molecular weight, the ratios of I_{100}/I_{020} (filled circles) in the "out-of-plane" geometry decreased obviously. In contrast, the ratios of I_{100}/I_{020} (open circles) in the "in-plane" increased. In the low-molecular-weight P3HT ($M_w \leq 346$ k), the I_{100}/I_{020} were larger in the "out-of-plane" geometry and smaller in

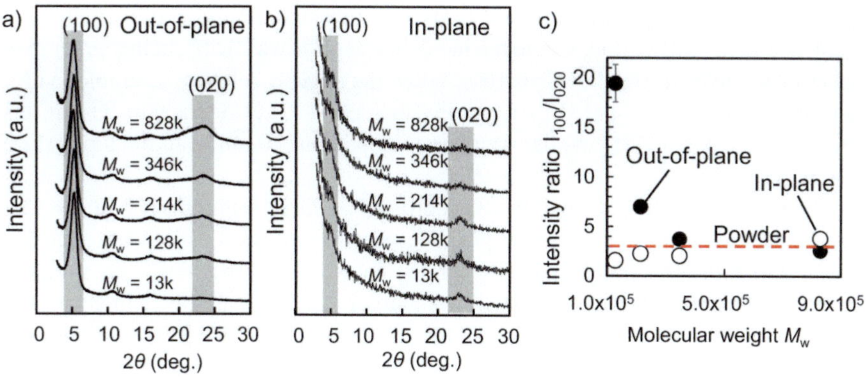

Fig. 9.5 X-ray diffraction profiles of P3HT films with various *M*w through (**a**) "out-of-plane" and (**b**) "in-plane" geometries. (**c**) Indicates the effect of *M*w on the intensity ratio I_{100}/I_{020}

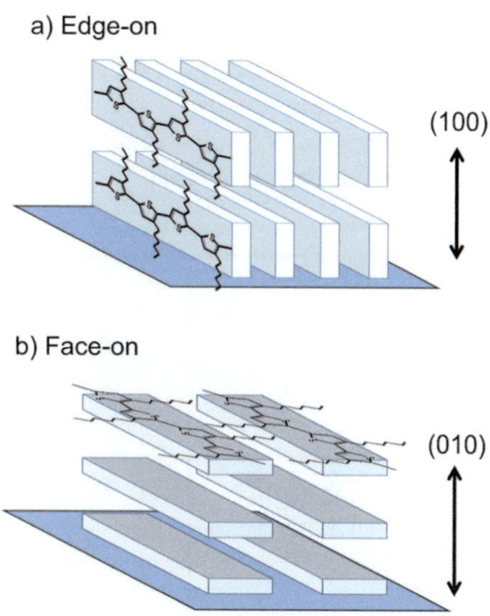

Fig. 9.6 Schematic representation of two different orientations, so-called (**a**) edge-on and (**b**) face-on of P3HT

the "in-plane" than that of the powder sample. This means that the crystallite orientation in the thin film changed depending on the molecular weight of P3HT.

Figure 9.6 shows schematic representation of two different orientations, so-called (a) edge-on and (b) face-on of P3HT. Optimum performance of transistor is achieved when the π-stacking direction in a film aligns with the direction of charge transport. For example, hole mobility of P3HT is 100 times higher for the edge-on structure. On the contrary, flat-on structure is suitable for solar cells since they have vertically arranged electrodes. The P3HT with lower molecular weight had "edge-on"-rich structures, as previously reported. As the molecular weights increase, the structures in thin films transformed from "edge-on" to "face-on" gradually. In the thin film of P3HT ($M_w = 828$ k), the ratio of "edge-on" and "flat-on" structures was inverted.

So far, the bulk structure of thin films was investigated using the small-angle incidence X-ray diffraction, where the incidence angle α was 0.20°, being larger than the critical angles α_c (for P3HT; 0.16°). When the angle α below α_c is employed for the measurement, it is called as a grazing incidence X-ray diffraction. Using this geometry, we can investigate the structure within about 10 nm from the top surface of thin films.

Figure 9.7 shows the X-ray diffraction profiles of P3HT thin films [(a) $M_w = 13$ k and (b) $M_w = 828$ k] in the "out-of-plane" geometry with various incidence angles. P3HT thin film ($M_w = 13$ k) at the surface showed the "edge-on"-rich structure similar to the bulk. The I_{100}/I_{020} ratios were larger than 20 at all the incident angles. In contrast, the I_{100}/I_{020} ratio of P3HT thin film ($M_w = 828$ k) decreased with the lower incidence angles of X-ray beams as shown in Fig. 9.7c. These results mean that the thin film possessed the more "edge-on"-rich structure near the topmost surface. Accordingly, P3HT thin film ($M_w = 828$ k) possesses heterogeneous structure in the thickness direction of the same film, that is, "face-on"-rich was observed in the bulk and "edge-on"-rich was observed at the surface.

Table 9.2 summarizes the various factors that affect the thin film morphology of P3HT. From the structure of thin films prepared under various conditions, the main attribution is the speeds of sedimentation and inhabitation of the crystallization. These fundamental results would lead the suggestion in fabrication and performance of organic electronic devices.

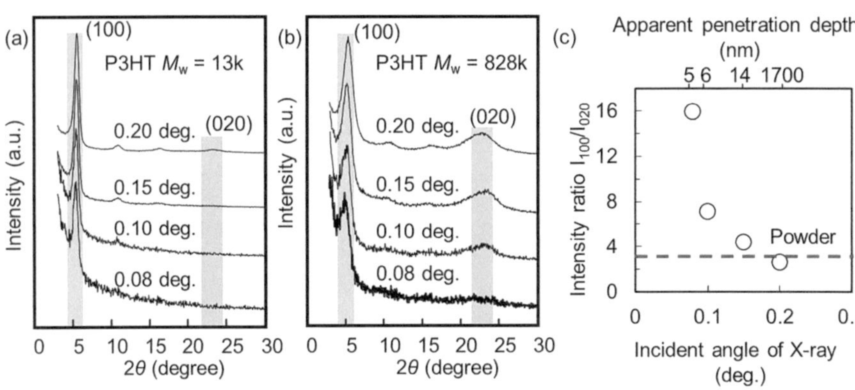

Fig. 9.7 X-ray diffraction profiles of P3HT thin films. (**a**) $M_w = 13$ k and (**b**) $M_w = 828$ k in the "out-of-plane" geometry with various incidence angles. (**c**) The relationship between the I_{100}/I_{020} ratio of P3HT thin film ($M_w = 828$ k) and the incidence angles of X-ray beams

Table 9.2 Various factors affect the spin-coated thin film morphology of P3HT

	Edge-on	Face-on
Molecular weight *Mw*	Low	High
Thin film portion	Surface	Bulk
Substrate surface	Hydrophobic	Hydrophilic
Spin-coating temperature	High	Low

9.4 Flexible Polythiophene with Disiloxane Side Chains

For various applications in industrial fields, the mechanical properties and processability of a material are equally important to their electrical properties for the fabrication of semiconductive polymer-based devices. However, the flexibility and bendability of P3HT are not sufficient to tolerate sequential times of multiple bending processes. Moreover, the electrical properties of P3HT decrease drastically after bending of their devices. Accordingly, mechanical durability of P3HT remains a challenge for material applications and processing of stretchable devices [21]. Herein, we described high-molecular-weight and high regioregular poly (3-substituted thiophene) with disiloxane moieties in the side chains (P3SiT) [22]. The Si–O bond distance in disiloxane in the side chains (1.66 Å) is longer than the C–C bonding distance (1.54 Å). Moreover, the low rotation barrier of Si–O–Si bonding (3.3 kJ/mol) is comparable to the free rotation energy. These features of the disiloxane groups can be expected to contribute great flexibility from the view of mechanical performance. Additionally, it is well known that polymers with higher molecular weights have better flexibility and achieve increased elongation at break.

High-molecular-weight and high regioregular P3SiT was polymerized by Kumada coupling based on a nickel(II) catalyst as shown in Scheme 9.2.

The *M*w of P3SiT was 300 k, and a head-to-tail structure of over 99% was confirmed by ^{1}H NMR. After casting from hexane solution, the mechanical properties of P3SiT film were examined and compared with that of high-molecular-weight P3HT film. During casting, cracks were observed for P3SiT with lower molecular weight ($M_w = 60$ k or 130 k). This again recalls the significance of high molecular weight for getting self-standing film. Additional advantage of P3SiT is that the dissolubility and processability of polymer are particularly improved and easily dissolved in organic solvent like hexane or tetrahydrofuran.

Figure 9.8 shows the temperature dependence of storage modulus and mechanical tanδ of P3SiT. In the tanδ curve, four distinct dispersion peaks were observed at different temperatures. At the lowest temperature (−110 °C), the dispersion involves a small change in modulus which was observed, which can be probably attributed to motions of the side chain groups, namely, the disiloxane groups in P3SiT. The main peak at higher temperature (−10 °C) is a primary dispersion involving a large change in modulus, corresponded to the glass transition temperature (T_g) of P3SiT. A huge difference was found between T_g of P3SiT and P3HT ($T_g = 45$ °C, see Fig. 9.1), and this low T_g of P3SiT, below ambient temperature, means that P3SiT is in a rubbery state at room temperature. The weaker intermolecular interactions and flexible disiloxane side chains of P3SiT induced more amount of free volume, which allows this polymer to maintain the rubbery state, even at room temperature.

Figure 9.9 shows the stress–strain curves of P3SiT together with that of P3HT cast film at room temperature. As mentioned above, the P3HT film exhibited high tensile modulus and high tensile strength. On the other hand, the tensile fracture occurred catastrophically, which gave elongation at break as low as 14%. In contrast, for P3SiT, the tensile strength decreased, but the elongation at break became 226%,

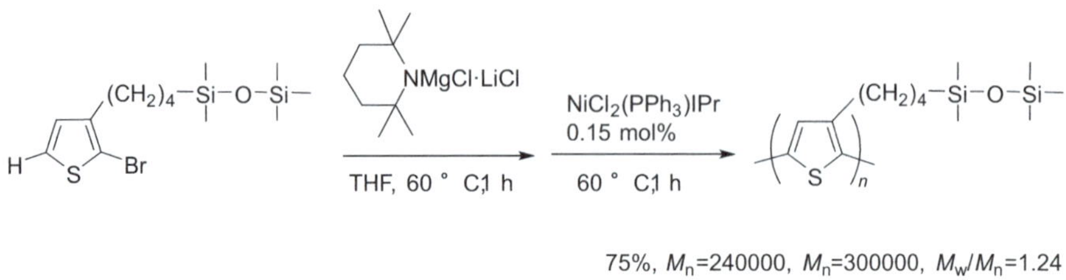

Scheme 9.2 Synthesis of high-molecular-weight and high regioregular P3SiT

Fig. 9.8 Temperature dependence of the storage modulus and mechanical tanδ of P3SiT self-standing film

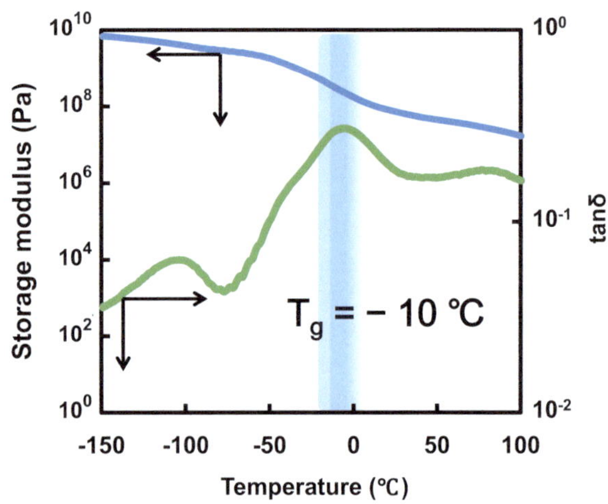

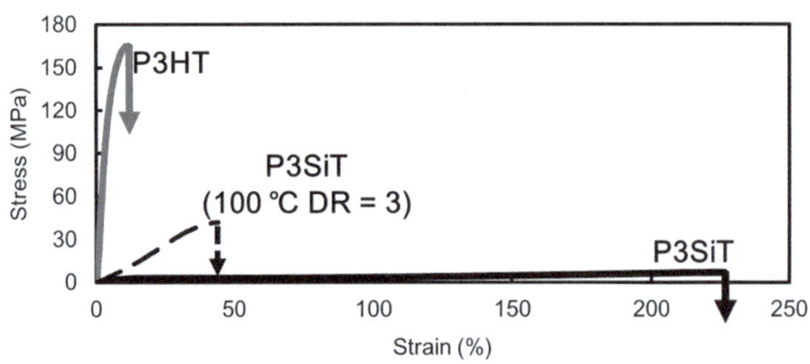

Fig. 9.9 Stress–strain curves of P3HT, P3SiT, and three-times-drawn P3SiT

which was more than 15 times higher than that of P3HT. As mentioned above, P3SiT is in a rubbery state at ambient temperature, which contributes a great increase in flexibility relative to P3HT. For investigation of the elastomeric performance of P3SiT, we also carried out repeated loading–unloading tensile tests. Below 10% of the tensile strain, the P3SiT film demonstrated high elastic recovery. As the strain

Fig. 9.10 X-ray diffraction photographs of as-cast and drawn (draw ratio of three times at 100 °C) P3SiT

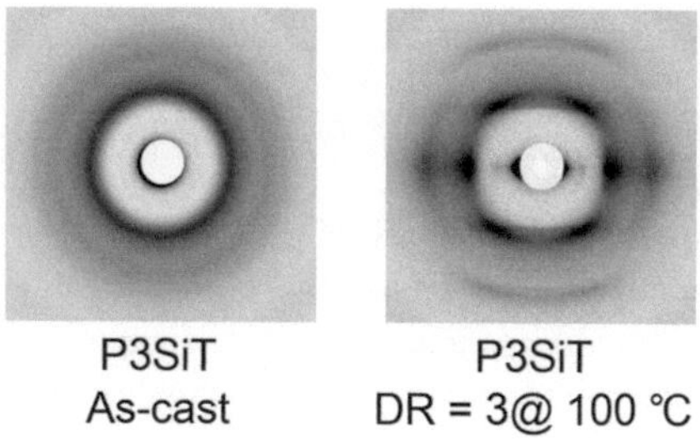

increased above 10%, the permanent residual deformation occurred in the P3SiT film, and it finally lost the elastic recovery when the tensile strain was up to 50%. For the achievement of the high elasticity and recovery in polythiophene derivatives, one of the best ways should be the introduction of the physical or chemical cross-linking to this P3SiT.

We also found P3SiT can be drawn to high draw ratio.

Figure 9.10 shows X-ray diffraction photographs of as-cast, then uniaxial drawn (three times at 100 °C), P3SiT. The Debye–Scherrer rings were observed for as-cast P3SiT. After drawing, several streak-like diffractions were clearly observed. This finding indicates that the crystallites of P3SiT were oriented by drawing. The stress–strain curve of three-times-drawn P3SiT is also superimposed in Fig. 9.9. In contrast with as-cast P3SiT, three-times-drawn P3SiT showed Young's modulus of 129 MPa and tensile strength of 42 MPa, which were two and seven times higher than those of undrawn P3SiT, respectively. These are due to the increases of the crystallite orientation and crystallinity during the drawing processes. Accordingly, the mechanical performance of P3SiT could be controllable by hot drawing. Owing to the larger and longer side chains of P3SiT, the sheet resistivity (1.8×10^6 Ω/square) was larger than that (1.3×10^6 Ω/square) of P3HT. However, after drawing, the absolute value in the parallel direction (1.1×10^6 Ω/square) was comparable to that of P3HT. These results indicate that the electrical resistivity of polythiophene derivatives is also controllable by hot drawing.

9.5 Conclusions

Electron properties as semiconducting polymer have been extensively investigated for P3HT. In these investigations, relatively low-molecular-weight P3HT (Mw<50 k) were used, so the applications have been limited as thin films. In this chapter, we first described the syntheses of high-molecular-weight (Mw up to 880 k) and high regioregular (head-to-tail >99%) P3HT. Self-standing film could be only obtained using high *M*w P3HT, which enable to investigate mechanical performance of P3HT. Molecular orientation in the P3HT thin film was generally known as edge-on structure, where π-stacking direction aligns with the direction parallel to the substrate. On the other hand, thin film fabricated using high *M*w P3HT, low spin-coating temperature, and hydrophilic substrate were found to give flat-on rich structure, being suitable for solar cells. In addition, fragile issue of P3HT (elongation at break <15%) could be overcome by substituting side chains from hexyl groups to disiloxane moieties. High *M*w P3SiT gave low Tg (−10 °C), which resulted in high elongation at break (more than 200%) at room temperature. In addition to these alternative structures and mechanical properties of new types of polythiophenes, we also found very high water repellency/high hydrophobicity of P3HT. This is also an alternative noble feature of polythiophenes as element-block polymers [23].

References

1. Little WA (1964) Possibility of synthesizing an organic superconductor. Phys Rev 134(6A): A1416–A1424
2. Shirakawa H, Ikeda S (1971) Infrared spectra of poly (acetylene). Polym J 2:231–244
3. Dimitrakopoulos CD, Malenfant PRL (2002) Organic thin film transistors for large area electronics. Adv Mater 14:99–117
4. Friend RH, Gymer RW, Holmes AB, Burroughes JH, Marks RN, Taliani C, Bradley DDC, Dos Santos DA, Brédas JL, Lögdlund M, Salaneck WR (1999) Electroluminescence in conjugated polymers. Nature 397:121–128
5. Park SH, Roy A, Beaupré S, Cho S, Coates N, Moon JS, Moses D, Leclerc M, Lee K, Heeger AJ (2009) Bulk heterojunction solar cells with internal quantum efficiency approaching 100%. Nat Photonics 3:297–302
6. Roncali J (1992) Conjugated poly (thiophenes): synthesis, functionalization, and applications. Chem Rev 92:711–738
7. Kim JY, Lee K, Coates NE, Moses D, Nguyen T-Q, Dante M, Heeger AJ (2007) Efficient tandem polymer solar cells fabricated by all-solution processing. Science 317:222–225
8. Sato M, Tanaka S, Kaeriyama K (1986) Soluble conducting polythiophenes. J Chem Soc Chem Commun:873–874
9. Jen KY, Miller GG, Elsenbaumer RL (1986) Highly conducting, soluble, and environmentally-stable poly(3-alkylthiophenes). J Chem Soc Chem Commun:1346–1347
10. Sugimoto R (1986) Preparation of soluble polythiophene derivatives utilizing transition metal halides as catalysts and their property. Chem Express 1:635–638
11. Chen TA, Wu X, Rieke RD (1995) Regiocontrolled synthesis of poly(3-alkylthiophenes) mediated by Rieke zinc: their characterization and solid-state properties. J Am Chem Soc 117:233–244
12. Lamps J-P, Catala J-M (2011) Kinetic study, by UV–vis spectroscopy, on the strong effect of LiCl on the controlled polymerization of 2-Bromo-3-hexyl-5-iodothiophene and 2-Iodo-3-hexyl-5-bromothiophene: determination of the propagation rate constants, application to the synthesis of high molecular weight Polydodecylthiophene. Macromolecules 44:7962–7968
13. Osaka I, McCullough RD (2008) Advances in molecular design and synthesis of regioregular polythiophenes. Acc Chem Res 41:1202–1214
14. Tamba S, Shono K, Sugie A, Mori A (2011) C–H functionalization polycondensation of chlorothiophenes in the presence of nickel catalyst with stoichiometric or catalytically generated magnesium amide. J Am Chem Soc 133:9700–9703
15. Tamba S, Fuji K, Meguro H, Okamoto S, Tendo T, Komobuchi R, Sugie A, Nishino T, Mori A (2013) Synthesis of high-molecular-weight head-to-tail-type poly(3-substituted-thiophene)s by cross-coupling Polycondensation with [CpNiCl(NHC)] as a catalyst. Chem Lett 42:281–283

16. Kline RJ, McGehee MD, Toney MF (2006) Highly oriented crystals at the buried interface in polythiophene thin-film transistors. Nat Mater 5:222–228
17. DeLongchamp DM, Vogel BM, Jung Y, Gurau MC, Richter CA, Kirillov OA, Obrzut J, Fischer DA, Sambasivan S, Richter LJ, Lin EK (2005) Variations in semiconducting polymer microstructure and hole mobility with spin-coating speed. Chem Mater 17:5610–5612
18. Nishino T, Matsumoto T, Nakamae K (2000) Surface structure of isotactic polypropylene by X-ray diffraction. Polym Eng Sci 40:336–343
19. Matsumoto T, Nishi K, Tamba S, Kotera M, Hongo C, Mori A, Nishino T (2017) Molecular weight effect on. Surface and bulk structure of poly(3-hexylthiophene) thin films. Polymer 119:76–82
20. Sirringhaus H, Brown PJ, Friend RH, Nielsen MM, Bechgaard K, Langeveld-Voss BMW, Spiering AJH, Janssen RAJ, Meijer EW, Herwig P, de Leeuw DM (1999) Two-dimensional charge transport in self-organized, high-mobility conjugated polymers. Nature 401:685–688
21. Kim JS, Kim JH, Lee W, Yu H, Kim HJ, Song I, Shin M, Oh JH, Jeong U, Kim TS, Kim BJ (2015) Tuning mechanical and optoelectrical properties of poly(3-hexylthiophene) through systematic regioregularity control. Macromolecules 48:4339–4346
22. Shen J, Fujita K, Matsumoto T, Hongo C, Misaki M, Ishida K, Mori A, Nishino T (2017) Mechanical, thermal, and electrical properties of flexible polythiophene with disiloxane side chains. Macromol Chem Phys 218:1700197
23. Chujo Y, Tanaka K (2015) New polymeric materials based on element-blocks. Bull Chem Soc Jpn 88:633–643

Chapter 10
Polyphosphazenes as an Example of the Element-Blocks Approach to New Materials

Harry R. Allcock

Abstract Polyphosphazenes are inorganic-organic high polymers with a backbone of alternating phosphorus and nitrogen atoms and two organic or organometallic side groups attached to each phosphorus. Most of these polymers are synthesized by macromolecular substitution reactions carried out on poly(dichlorophosphazene), $(NPCl_2)_n$. The chlorine substitution reactions involve alkoxides, aryloxides, primary or secondary amines, or a range of organometallic reagents. Structural variations are accomplished via the use of one, two, or more different nucleophiles and substituents along the polymer chain and by the employment of reagent size and reactivity to control polymer properties and emphasize specific uses. Applications have been developed for these polymers as elastomers, thermoplastics, biostable or bioerodible medical materials, fire-resistant lithium battery electrolytes, films, or foams, and gas and liquid separation membranes.

Keywords Polymers · Polyphosphazenes · Macromolecular substitution · Properties · Uses

10.1 Introduction

Numerous advantages exist for the incorporation of inorganic elements into polymers. The inorganic elements broaden the range of structures and properties that are accessible. They also allow the use of nonclassical synthesis reactions and utilize starting materials other than oil or natural products. Moreover, the presence of inorganic elements in polymers provides access to materials that are resistant to extreme conditions. Examples of these advantages are provided by the polyphosphazene platform – a group of several hundred different polymers with a wide range of properties that in many cases cannot be generated from classic

H. R. Allcock (✉)
Department of Chemistry, The Pennsylvania State University, University Park, PA, USA
e-mail: hra1@psu.edu

Y. Chujo (ed.), *New Polymeric Materials Based on Element-Blocks*,
https://doi.org/10.1007/978-981-13-2889-3_10

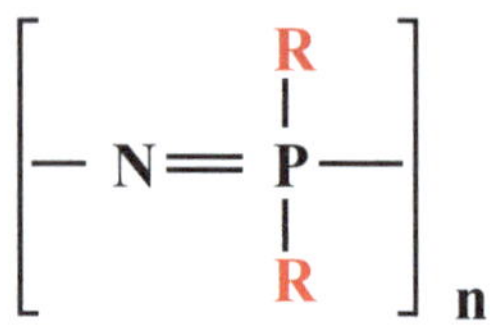

Fig. 10.1 The basic polyphosphazene structure

hydrocarbon-based macromolecules. These advantages have led to advances in elastomer technology, biomedical materials, non-flammable lithium battery electrolytes, gas transmission membranes, coatings, and textiles [1].

Polyphosphazenes are high molecular weight polymers with a backbone of alternating phosphorus and nitrogen atoms with two organic inorganic or organometallic side groups attached to every phosphorus atom (Fig. 10.1). The synthesis of most of these polymers is almost unique since it involves the use of macromolecular substitution, a process in which a broad range of different side groups and combinations of different groups are linked to the polyphosphazene chain *after the polymerization process has been completed.* Thus, one of the main features of this system is the use of the polyphosphazene backbone as a *platform* for the introduction of a wide variety of useful side group structures into a polymer system.

10.2 History

The origins of this field can be traced to the early 1800s when Liebig and Rose [2, 3] reported that phosphorus pentachloride and ammonia or ammonium chloride react to yield crystalline solids with the composition $NPCl_2$. Sporadic studies in Europe were followed by the work of H. N. Stokes in the USA in the 1890s [4], who proposed that these compounds are ring systems of formula $(NPCl_2)_{3,\ 4,\ 5}\ \ldots$ with alternating single and double phosphorus-nitrogen bonds. Stokes also found that if any of these ring systems was heated to ~250 °C, the molten material gelled to a rubbery elastomer. However, the elastomer was insoluble in all solvents and slowly decomposed on exposure to atmospheric moisture – a characteristic of phosphorus-chlorine bonds. Moreover, with modern insight, it is clear that its insolubility was a strong indication of a cross-linked structure.

We began our studies by showing that the chlorine atoms in the cyclic trimer, $(NPCl_2)_3$, could be replaced by organic groups to produce water-stable derivatives, and this raised the question of whether Stokes' high polymeric elastomer would react in a similar manner. Unfortunately, although some chlorine replacement occurred, the substitution was incomplete, and the insoluble cross-linked structure remained. As a result, it became clear that the preparation of stable, high polymeric polyphosphazenes from poly(dichlorophosphazene) depends on avoiding the cross-linking process. Thus, the breakthrough experiment was the demonstration that careful purification of the cyclic trimer and protection of it from the atmosphere, followed by heating at 200–250 °C, resulted in the formation of a polymer that was soluble in benzene, toluene, tetrahydrofuran (THF), and other solvents. Once this had been accomplished, the path was open to replacement of all the approximately 2000–30,000 chlorine atoms in each high polymer molecule by organic nucleophiles. This remains today the main method for the synthesis of several hundred different poly(organophosphazenes) [5–7].

10.3 Polyphosphazene Synthesis

The conventional synthesis of classical organic polymers is accomplished by addition, condensation, or ring-opening polymerization processes [8, 9] in which the side groups destined for the final polymer are already covalently linked to the monomer. A number of poly(organophosphazenes) have been synthesized by similar methods, especially by condensation reactions [10–15] or by the ring-opening polymerization of partially organic-substituted small-molecule phosphazene rings. The ring-opening method is effective only if three or four of the six chlorine atoms remain linked to the cyclic trimer. However, the molecular weights tend to be much lower than those obtained via the polymerization of $(NPCl_2)_3$ followed by halogen replacement reactions carried out on poly(dichlorophosphazene). Thus, this last method has evolved as the preferred route to exploit the broad scope of this system. Condensation routes to poly(dichlorophosphazene) are an alternative to the $(NPCl_2)_3$ ring-opening polymerization process but have a more specialized utility [10–17]. For example, the living cationic polymerization route to $(NPCl_2)_n$ is a vehicle for the synthesis of block copolymers [16, 18].

The macromolecular substitution route (Fig. 10.2) [5–7] places few limits on the types of side groups or the combinations of different side groups that can be incorporated into the polymers, and this allows a broader range of properties and potential uses than in almost any other existing synthetic macromolecular system. Moreover, the main chain lengths are very high – typically from 1000 to 15,000 repeating units depending on polymerization conditions. Each different side group arrangement gives a polymer with different properties and potential uses.

Thus, to summarize, this synthesis protocol is a two-step process in which a small-molecule cyclic phosphazene with chlorine side groups is first thermally melt-polymerized to a reactive polymer, and the chlorine atoms in this high molecular weight macromolecule are then replaced by a wide variety of different nucleophiles such as alkoxides, amines, or organometallic reagents either to give single-substituent or mixed-substituent polymers. Many of these side groups would not survive the polymerization process if they were present initially on the cyclic trimer, or they would inhibit the polymerization process. For some phosphazene polymers, a third step is also possible in which deprotection chemistry is subsequently carried out on the organic side groups to introduce specific functional units such as hydroxyl, carboxylic acid, amino, or organometallic units. An idea of the scope of this approach can be gauged by the abbreviated list of side groups that have been linked to a polyphosphazene chain shown in Table 10.1.

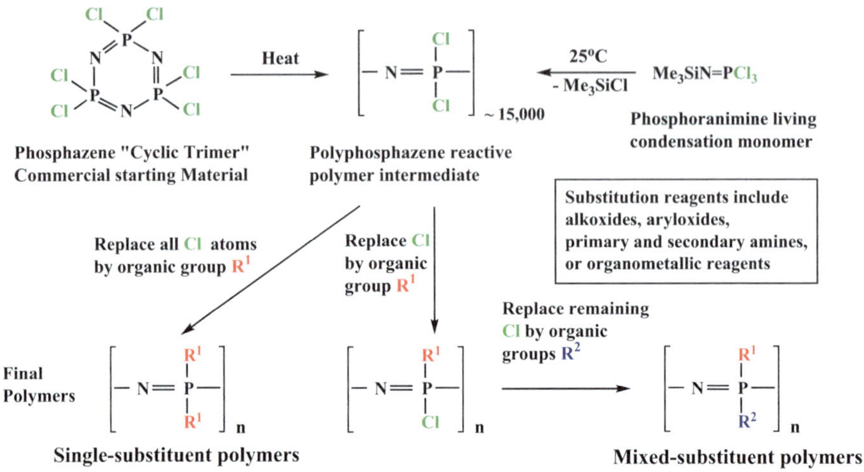

Fig. 10.2 The polyphosphazene macromolecular substitution approach

Table 10.1 Polyphosphazene molecular and materials design by side group variations[a, b]

For films and fibers	*For elastomer formation*	*For hydrophobicity*
OCH_2CF_3	OCH_2CF_3/ $OCH_2(CF_2)_xCF_2H$	OCH_2CF_3
$OCH_2(CF_2)_xCF_2H$	$OC_6H_5/OC_6H_4CH_3$	$OCH_2(CF_2)_xCF_2H$
OC_6H_5	$OCH_2CF_3/OCH_2Si(CH_3)_3$	OC_6H_5
OC_5H_9 (cyclo)	OCH_2CF_3/ cyclophosphazenes	OC_6F_5, $OC_6H_xF_y$
OC_6H_{11} (cyclo)	OCH_2CF_3/oligophenylenes	OC_4H_7 (cyclo)
	$OCH_2CF_3/OCH_2)_3Si$ $(CH_3)_3$	OC_5H_9 (cyclo)
		OC_6H_{11} (cyclo)
		$OC_6H_xF_{5-x}$
		$OC_6H_4C_6H_5$
For surface hydrophilicity	*Bulky side groups with special properties (1)*	*Bulky side groups with special properties (2)*
$OCH_2CH_2OCH_2CH_2OCH_3$ and branched analogs	Adamantane	Organic dyes
OC_6H_4COOH (PCPP) and Na salt	Cyclodextrin	Phthalocyanines
$OC_6H_4SO_3Na$	Cyclotriphosphazene	Phthalocyanine-metal complexes
$OC_6H_4SO_3H$	Cyclotetraphosphazene	Tetraphenylporphyrin
$OC_6H_4NH_2$	Oligophenylenes	Ferrocene
Glucosyl,	POSS	Phenanthrene
Glyceryl	Phosphazo	Carboranes
$NHCH_3$	Azo dyes	
For ionic coordination (Li ion batteries and fuel cell membranes)	*For high Tg's (above 37 °C)*	*For solubility in water*

(continued)

Table 10.1 (continued)

$OCH_2CH_2OCH_2CH_2OCH_3$ and branched analogs	NHC_6H_5	$NHCH_3$
Crown ethers	$OC_6H_4C_6H_5$	OCH_3
$OC_6H_5/OC_6H_4SO_3H$	$OC_6H4COOH$	$OCH_2CH_2OCH_2CH_2OCH_3$ and branched analogs
$OC_6H_5/OC_6H_4P(O)$-$(OR)_2OH$	OC_6H_4COONa	OC_6H_4COONa
$OC_6H_5/OC_6H_4P(O)NHS$-O_2CF_3	$OC_6H_4N{=}CHR$	$OC_6H_4SO_3H$
	Glucosyl	Glucosyl
		Glyceryl
High refractive index glasses	***For pyrolysis to ceramics***	***For electronic conductivity***
$OC_6H_4C_6H_5$	Borazine side groups	TCNQ derivatives
O-naphthalene	Amino side groups	Pyrrole, thiophene, furan groups
O-anthracene		Ferrocene
Aryloxy-Cl, Br, I		
S-containing side groups		
Metal coordination groups	***For micelle formation***	***For fire retardance***
Phenanthrolene-metal	Block copolymers with hydrophobic and hydrophilic blocks	Most low- or medium-carbon side groups such as:
Bipyridine-metal		OCH_2CCl_3
Terpyridine-metal		OCH_2CF_3
Phthalocyanine-metal		OC_6H_5
Carboranyl		$OCH_2CH_2OCH_2CH_2OCH_3$
Metallocenes		
Imidazolyl		
Pyridinyl		
$OCH_2CH_2SCH_3$		
$OC_6H_4PPh_2$		
Photonic/NLO/liquid crystalline polymers	***Organoboron and organosilicon side groups***	***Organometallic side groups***
$OC_6H_4C_6H_5$	Carboranes	Ferrocene
$OC_6H_4N{=}NC_6H_4OCH_3$	Borazines	Metal carbonyls such as
$O(CH_2CH_2O)_x$- $C_6H_4N{=}N$-aryl	$OCH_2Si(CH_3)_3$	$Co_2(CO)_8$
$OCH_2CH_2N(Et)$ $C_6H_4N{=}NC_6H_4N\ O_2$	$O(CH_2)_3Si(CH_3)_3$	$FeCp(CO)_2$
$O(CH_2CH_2O)_x$-$C_6H_4N{=}NC_6H_5C_4H_9$	$OC_6H_4Si(CH_3)_3$	
Spiropyrans	$NH(CH_2)_3Si(CH_3)_2Si(CH_3)_3$	
O-naphthalene	$NH(CH_2)_3POSS$	
O-anthracene		
Halogenoaryloxy		
S-containing groups		
Tissue engineering applications (1)	***Tissue engineering applications (2)***	***For biostability***

(continued)

Table 10.1 (continued)

Ferulic acid	Serine	OCH_2CF_3
Aldehyded dextran	Threonine	$OCH_2(CF_2)_xCF_2H$
Citronellol	Dipeptide	OC_6H_5R
OCH_2CH_3	Vitamins	$OCH_2Si(CH_3)_3$
Ethyl glycinate	Phosphoesters	$OCH_2CH_2OCH_2CH_2OCH_3$
Ethyl alanate	Choline	
Ethyl phenylalanate	Glucosyl	
Bioerodible polymers (1)	***Bioerodible polymers (2)***	***Bio-imaging and drug delivery***
Ethyl glycinate	Ferulic acid	Gold nanospheres with PCPP coatings
Ethyl alanate	Aldehyded dextran	Iodine-containing amino acids
Ethyl phenylalanate	Citronellol	
Serine	OCH_2CH_3	
Threonine	Vitamins	
Dipeptides	Phosphoesters	
Imidazolyl	Choline	
Glucosyl	Glyceryl	
For linkage to bioactive agents	***Polymeric drugs***	***Microsphere or micelle formation***
$OC_6H_4NH_2$	Ciprofloxacin	OC_6H_4COOH
OC_6H_4COOH	Norfloxacin	$OC_6H_4SO_3H$
$OC_6H_4N^+R_3Br^-$	Cis-platinum	Hydrophobic polyphosphazene plus hydrophilic organic block copolymer
$OCH_2CH_2OCH_2CH_2NH_2$	Catecholamines	Hydrophilic polyphosphazene plus hydrophobic block copolymer
Proteins		

[a]Details regarding these polymers can be found in Ref. [1] and in the other sources listed here
[b]Note that all of these side groups can also be accompanied along the polymer chain by other groups to allow different properties to be combined. Block copolymers are also utilized to combine two sets of properties

10.4 Challenges in the Macromolecular Substitution Process

The macromolecular substitution route to poly(organophosphazenes) is feasible mainly because of the high reactivity of the phosphorus-chlorine bonds in poly (dichlorophosphazene). However, this advantage must be balanced by several restrictions. First, by the end of the macromolecular substitution process, all the chlorine atoms must have been replaced by organic groups. Failure to accomplish this would leave the polymer liable to at least partial hydrolytic decomposition as residual P-Cl groups were hydrolyzed to P-OH units, which could then react to form cross-links or trigger hydrolytic cleavage at various sites along the chain. Perhaps the most remarkable thing about this synthesis is the fact that thousands of substitution reactions can be carried out efficiently on each macromolecule. Full substitution usually requires that a small excess of reagent is needed to ensure complete chlorine replacement.

Second, di- or multifunctional reagents cannot be employed directly for the macromolecular substitution reactions. The presence of more than one functional group in each nucleophile would allow it to react with two or more polyphosphazene chains to cross-link the system before halogen replacement is complete. This limitation can be circumvented by using protection techniques for some of the reactive units on a multifunctional reagent, followed by deprotection once the organic group has been linked to the main chain [19]. For example, the alkyl ester of p-hydroxy sodium benzoate reacts cleanly with poly(dichlorophosphazene) to yield a polymer with aryloxy ester side groups. Subsequent hydrolysis of the ester groups to carboxylic acid units generates the functional polymer $[NP(OC_6H_4COOH)_2]_n$. This could not have been accomplished by the direct reaction of p-hydroxybenzoic acid with poly(dichlorophosphazene). Protection-deprotection techniques have been used extensively in our laboratory for the linkage of biologically useful organic groups to the polymer chain [19].

Third, macromolecular S_N2 substitution reactions are sensitive to steric inhibition if bulky nucleophiles such as steroids [20], iron porphyrins [21], or carboranes [22] are employed. However, in such systems typically ~10 to 50% chlorine substitution by bulky nucleophiles can be achieved, and this can then be followed by replacement of all the remaining chlorine atoms by less-hindered reagents. Some of the most interesting polymer produced in our laboratory in recent years have utilized this technique to yield mixed-substituent polymers with unexpected and useful properties [23–28]. Inherent in this technique is the question of the *pattern* of chlorine replacement (Fig. 10.3) and the effect of the substitution pattern on properties.

For example, following the replacement of the first chlorine atom somewhere along the polyphosphazene chain, the possibility exists that the second substitution will occur either at that same phosphorus atom or at a distant location along the chain. Steric hindrance will play a role in this process, which should (and does) become evident with the bulkiest nucleophiles. Alternatively, strong electron withdrawal by that first side group could favor the second substitution at the same phosphorus atom (geminal) or at a nearby phosphorus atom. Identifying these substitution patterns with a high polymer is a major challenge that can sometimes be approached through the study of ^{31}P NMR spectra as the reaction proceeds.

In spite of these limitations, more than 250 different organic side groups have been linked to the polyphosphazene chain, to give one of the largest known synthetic polymer systems.

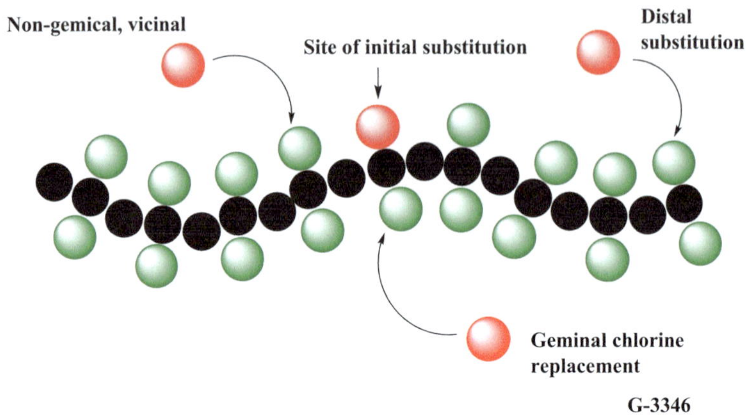

Fig. 10.3 Alternative substitution sites along the polyphosphazene chain following the first chlorine replacement reaction

10.5 Broader Perspective

The differences between this polymer chemistry and traditional macromolecular science will be clear from the preceding discussion. However, it is necessary to view the polyphosphazene field further from two related but different points of view. First, there are inherent properties that arise because of the presence of inorganic elements in the backbone. Second, there are properties and design opportunities that result from the backbone being used as a *carrier* for specific side groups. The properties uniquely depend on both the backbone and the side groups, and only in a few cases, where the backbone is buried beneath massive side group structures, is the influence of the backbone relatively unimportant. The following discussion is organized into these two aspects.

10.6 Properties Generated by the Inorganic Backbone

The following special properties are a consequence of the phosphorus-nitrogen backbone.

1. *High skeletal flexibility*. Unlike organic polymers such as polyacetylene that have conjugated electronic unsaturation along the main chain, the polyphosphazene skeleton has a high flexibility based on an unusually low barrier to torsion of the P-N bonds. This becomes manifest in low glass transition temperatures in the range of −100 to −60 °C, provided the side groups are themselves flexible or have small dimensions. For example, with flexible side groups such as ethoxy, propoxy, butoxy, etc., T_g's in the −90 to −100 °C are accessible [29]. Even with fluoroalkoxy side groups, the T_g's are generally below −40 °C [30]. Only when

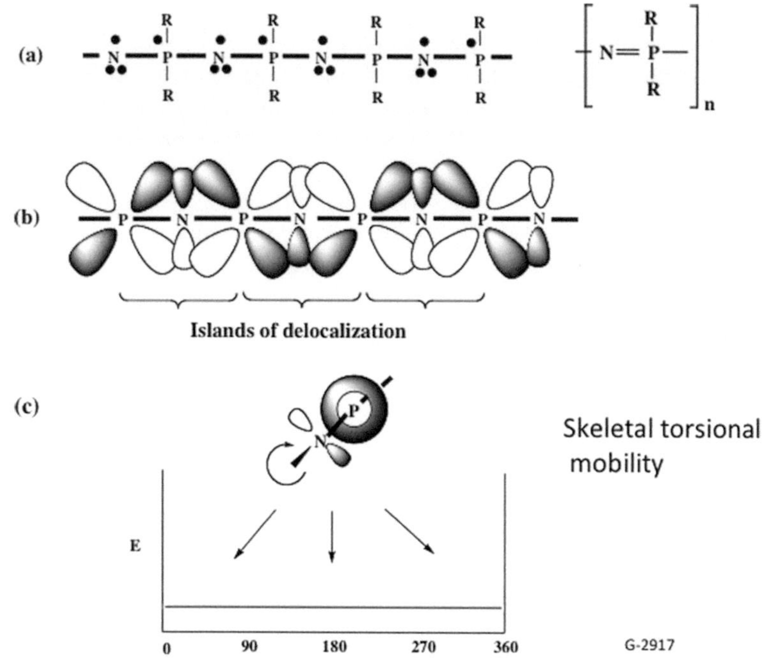

Fig. 10.4 Highly simplified bonding scheme for polyphosphazenes in which a mismatch of orbital signs at the 3d orbitals of phosphorus prevents long-range electron delocalization. (**a**) Electrons beyond those required for the sigma-bonded skeleton, which may interact in highly unusual ways to stabilize the P-N bonds. (**b**) The pi-bond "island" model proposed to explain why long-range delocalization of pi-electrons is apparently absent. (**c**) Illustration of why the barrier to torsion of the backbone bonds is so low, because the nitrogen lone pair orbitals can overlap different d-orbitals as the bond undergoes torsion

rigid, bulky side groups are linked to the chain do the T_g values start to climb. Poly(diphenoxyphosphazene) has a T_g near 0 °C [31], and the related polymer with p-phenylphenoxy side groups has a T_g near 90 °C [32]. 2-Naphthoxy side groups raise the T_g to ~40 °C. In the simple valence bond model, the skeletal flexibility is attributed to the presence of P-N-P three-center "islands" of d_π-p_π character (Fig. 10.4) rather than the stiff, broadly delocalized p_π-p_π chains as found in organic semiconductors. In the polyphosphazene chain, there is a mismatch of orbital signs at every phosphorus atom, which restricts electron delocalization to just three skeletal atoms.

2. *Radiation stability*. The polyphosphazene backbone is transparent throughout the visible region of the spectrum and into the 220 nm region of the ultraviolet, again a consequence of limited electron delocalization and a large band gap. This transparency provides resistance to photolytic decomposition in sunlight. Moreover, the free radical skeletal decomposition pathways that are common in classical polymers appear to be absent in this system.

3. *Nonmetallic appearance.* Polymers with broadly delocalized electrons in the main chain, such as polyacetylene or polythiazyl, have a metallic appearance. Polyacetylene resembles either silver or gold in appearance and polythiazyl is gold. No polyphosphazenes synthesized to date have this characteristic, which is a strong indicator of a high band gap and poor electron delocalization along the backbone.
4. *Fire resistance.* One of the most serious defects associated with many classical organic polymers is their combustibility, due to the thermo-oxidative free radical sensitivity of C-C and C-H bonds. Attempts to confer fire resistance on classical polymers are often made by the addition of organophosphorus small molecules. These small molecules can migrate in the solid or volatilize from the material over time. However, many phosphorus-nitrogen-based polymers have been shown to be inherently fire resistant throughout their lifetime [33–35].

10.7 Properties Generated by the Organic Side Groups

The sensitivity of P-Cl bonds to nucleophilic substitution is well known for small-molecule phosphorus compounds such as PCl_3 and PCl_5, and this is also the key to the macromolecular substitution pathways on which most of the chemistry of polyphosphazenes is based. Thus, this inorganic backbone is an excellent platform for the linkage of a broad range of different organic substituents that confer special properties on the polymers. In this way the characteristics of the final polymers can be controlled by the broad range of available organic nucleophiles, including properties that result from the permutations of two or more different side groups along the same chain. The following observations summarize some of the main trends. Specific examples are shown in Table 10.1.

1. *Alkoxy side groups.* Short, non-fluorinated alkoxy side groups have a high torsional mobility, and this, combined with the flexibility of the backbone, leads to the formation of polymers with low glass transition temperatures (T_g). For example, the following T_g values are associated with polyphosphazenes that bear side chains such as ethoxy (−84 °C), propoxy (−99 °C), n-pentoxy (−102 °C), n-hexoxy (−104 °C), and n-octoxy (−104 °C) [36]. Ethoxy side groups impart hydrolytic sensitivity [37], a property that suggests biomedical uses since the hydrolysis products are an innocuous mixture of phosphate, ammonia, and ethanol.
2. *Cycloalkanoxy side groups.* This series of polymers [38] (Fig. 10.5) provides an excellent example of how an increase in the dimensions of side groups linked to a polyphosphazene chain affects the ease of chlorine replacement and the glass transition temperatures. The glass transition temperatures rise steadily from −60 °C when all the side groups are cyclo-butanoxy to +40 °C for cyclo-octanoxy groups. Cyclobutanoxide through cyclohexanoxide anions react with poly(dichlorophosphazene) to give transparent, hydrophobic, film-forming

polymers. However, the increased steric hindrance along this series requires increasingly more forceful conditions to replace all the chlorine atoms. The reaction of poly(dichlorophosphazene) with the cyclohexanoxy anion is one of the slowest reactions encountered in all the macromolecular substitution reactions studied, and forcing conditions (high temperatures, a large excess of nucleophile, and prolonged reaction times) are needed to ensure complete chlorine replacement. However, only the cyclo-octanoxy anion failed to replace all the chlorine atoms along the chain even under the most forcing reaction conditions. Molecular modeling of this system [38] illustrates clearly how the remaining P-Cl bonds are effectively buried within the electron density of the organic side groups already present. Nevertheless, separation of these cyclic structures from the reactive site by the introduction of methylene spacer groups allows even the most hindered of these rings to be linked to all of the reactive sites along the polymer chain. Partial substitution with cycloalkanoxy groups leaves P-Cl units that can be replaced by other less-hindered groups, such as trifluoroethoxy groups.

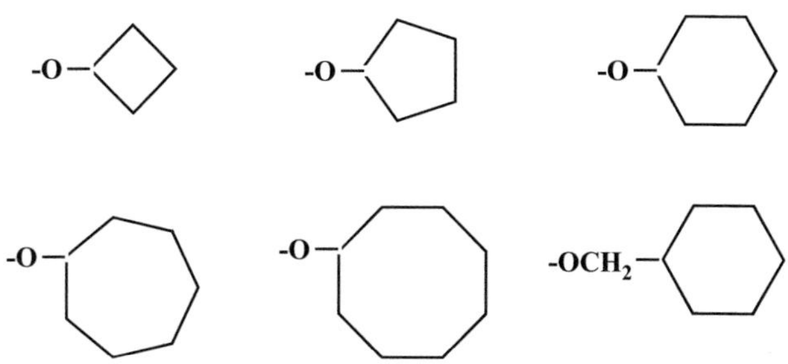

G-3391

3. *Fluoroalkoxy side groups.* Fluoroalkoxy side groups such as trifluoroethoxy, CF_3CH_2O-, or octafluoropentoxy ($HCF_2(CF_2)_3CH_2O$- have proved to be some of the most useful side units in the polyphosphazene series [1, 5, 6, 39, 40]. The polymer with trifluoroethoxy side groups is a hydrophobic, opalescent, microcrystalline, and film and fiber former ($T_g = -66$ °C, $T_m = 242$ °C) with numerous intermediate phase transitions, whereas the counterpart with octafluoropentoxy side groups is a soft, transparent film former (Tg = −64 °C). Combined as a 50-50 ratio in one polymer, these side groups yield amorphous, gum-like materials, but when lightly cross-linked, they are converted to elastomers [40]. These mixed-substituent, cross-linked polymers have been commercialized under the trade names such as PN-F® or Eypel-F® and have been utilized both for severe environment aerospace applications and as dental materials. Research is currently ongoing to optimize them as cardiovascular elastomers.
4. *Aryloxy groups.* Numerous poly(organophosphazenes) with aryloxy side groups have been reported. As mentioned above, non-fluorinated phenoxy side groups, with their moderately bulky profile, have the effect of raising the glass transition temperatures to −8 °C [15, 41, 42]. They yield flexible fiber- and film-forming

materials rather than soft elastomers. Polyphosphazenes with substituted phenoxy side groups have been studied in detail [41]. Larger aryloxy units such as biphenyloxy (T_g = 93 °C), triphenyleneoxy, tetra-phenyleneoxy, etc. progressively reduce the chain mobility and raise the glass transition temperatures. With linear tri- and tetra-phenyloxy groups, 100% replacement of the chlorine atoms in $(NPCl_2)_n$ is not possible due to steric hindrance, although partial substitution below 50% is feasible [24–26]. The remaining chlorine atoms can be replaced by smaller side groups such as trifluoroethoxy. Noncrystalline elastomers are formed from normally microcrystalline polymers when ~5 to 8% of the side groups are bulky co-substituents.

*Fluoro*aryloxy groups have two influences on polymer properties. First, they withdraw electrons from the skeleton, which is a stabilizing effect. 4-Fluorophenoxy side groups increase hydrophobicity but have little effect on the T_g (−6.6 °C). However, as the level of side group fluorination increases beyond this point, the T_g values rise and the solubility in organic solvents falls. Pentafluorophenoxy side groups can replace all the chlorine atoms in $(NPCl_2)_n$ but only under forcing reaction condition. Moreover, the molecular rigidity of this system is such that the detection of T_g and T_m transitions becomes difficult. These side groups confer special properties on the polymers including resistance to non-polar solvents. Trifluoromethyl side groups linked to the phenoxy units have only a marginal effect on the chlorine replacement reactions but have a marked effect through an increase in hydrophobicity and general chemical stability.

5. *Alkyl ether side groups.* Oligo-ethyleneoxy-type side groups can be readily linked to a polyphosphazene chain via the sodium salts of the appropriate alcohols [43–45]. For example, complete chlorine replacement occurs with nucleophiles such as $Na0CH_2CH_2OCH_2CH_2OCH_3$ and its branched isomers. The resultant polymers are water-soluble and water-stable and form hydrogels when lightly cross-linked by exposure to gamma irradiation. In the anhydrous state, these polymers are excellent solid ionic conductors for lithium ions, with conductivities of ~10^{-5} S/cm in the absence of etheric solvents, but have values higher than 10^{-3} S/cm when small amounts (~10%) of etheric solvents are present. Experimental non-flammable lithium ion batteries with either graphite or metallic lithium electrodes and these electrolytes have been fabricated and tested in our program.
6. *Alkylamino and arylamino groups.* As shown in Fig. 10.2, chlorine replacement in $(NPCl_2)_n$ is also possible using primary or secondary amines as nucleophiles [7]. In most reactions a tertiary amine such as triethylamine is present to capture and insolubilize the hydrogen chloride formed in these processes. These reactions are particularly important for the linkage of biologically interesting side groups to the polyphosphazene skeleton. Typical amines that have been used in these reactions include ethylamine, propylamine, etc.; amino acid esters such as the ethyl esters of glycine, alanine, and phenylalanine [19]; and a range of bioactive molecules with NH functional units. Hydrogen bonding between NH bridging units within the polymer matrix is believed to be responsible for higher T_g values

than in the case of similar side groups linked to the skeleton through alkoxy or aryloxy bonds.

7. *Organometallic side groups.* A variety of different organometallic side groups such as metal carbonyls or metallocenes have been linked to the phosphazene skeleton. These substitutions are somewhat more challenging than the reactions of oxo- or nitrogen nucleophiles with poly(dichlorophosphazene) because some organometallic nucleophiles can coordinate to the polymer backbone nitrogen atoms via their lone pair orbitals. Thus, the following special techniques were developed to circumvent this problem.

First, small-molecule model reactions were explored to optimize reaction conditions [46–48].

Second, an avoidance of organometallic side reactions was accomplished by the use of an alternative strategy, using the following logic. Because coordination of metallo species to the backbone nitrogen lone pair orbitals leads to side reactions, withdrawal of those electrons from nitrogen by the use of *fluoro*phosphazenes should lead to halogen replacement rather than nitrogen coordination. The linkage of ferrocenyl side groups to a polyphosphazene chain was accomplished by an unusual variation of the polyphosphazene substitution route. Thus ferrocene was first linked to a *fluorophosphazene cyclic trimer*, either through a single linkage or through a transannular structure, and that trimer was thermally polymerized [48]. The remaining fluorine atoms in the resultant polymer were then replaced by trifluoroethoxy groups. Ring strain in the transannular trimer provides an extra driving force for the ring-opening polymerization.

Polymerization of $(NPF_2)_3$ to $(NPF_2)_n$, followed by replacement of fluorine, in the manner discussed for chlorophosphazenes, is not possible because the high polymer is insoluble in all except a few rare fluorocarbon solvents. Thus, this pathway has not been developed. However, the cyclic trimer $N_3P_3F_5C_6H_5$ does polymerize to an organic soluble polymer, and this then allows direct replacement of P-F bonds by organic or organometallic anions.

8. *Metal coordination side groups.* Coordination of metals to a polyphosphazene can be accomplished either through the skeletal nitrogen atoms or through nitrogen, sulfur, or phosphine ligands in the side group structure. Thus, pendent macrocyclic rings, thioethers, phosphine, or amino ligands provide a facile means for the immobilization of metal atoms or clusters [49]. Some of these are of interest as immobilized catalysts.
9. *Substituent exchange.* Replacement of all the chlorine atoms along a polyphosphazene chain by organic groups is not necessarily the final process since some organic side groups already present may be displaced by other alkoxides [50, 51]. For example, the aryloxyphosphazene polymer $[NP(OC_6H_5)_2]_n$ in solution can be modified by exposure to nucleophiles such as sodium trifluoroethoxide. Trifluoroethoxy groups have been displaced in solution-state reactions by longer-chain fluoroalkoxy groups to give mixed-substituent species. Of additional interest is the possibility that such side group exchange reactions can take place selectively at the *surface* of a solid polymer

fiber or film leaving the interior polymer intact. Thus, a hydrophobic polymer may be modified by surface exchange to have a hydrophilic interface. Such reactions become important for the modification of bio-implantable polymers in the body, where the deposition of biomolecules on the surface is required for maximum biocompatibility.

An alternative approach for surface modification is to use an environmental oxygen or nitrogen plasma [52] to implant hydrophilic amino, oxo, or hydroxy groups into the surface of a hydrophobic material. The relative stability of the inorganic polymer backbone to plasmas reduces the possibility of general degradation.

10.8 Design Opportunities Connected with the Introduction of Bulky Side Groups

As mentioned earlier, the *dimensions* of the reactants influence the progress of nucleophilic substitutions, even when two small molecules are involved in the process. Steric influences become even more important when a small-molecule nucleophile reacts with a macromolecule. The twisting and coiling of a polymer molecule in solution present a hindered target for the nucleophile especially when many of the reactive sites on the polymer are shielded by substituents already present.

This is certainly the case for the reactions of poly(dichlorophosphazene) with bulky aryloxides or amines. In many cases the steric hindrance will be inconsequential at the start of the chlorine replacement process, but it will become increasingly restrictive as fewer and fewer P-Cl units remain along the chain. Thus, studies with bulky nucleophiles, such as those shown in Fig. 10.5, assume a special significance if these groups are designed to confer special properties on the polymers. Such groups include multi-ring aromatic chromophores, transition metal catalysts, or photonic species. Large amino acid esters or steroids and other biologically interesting units are bulky structures that are attractive side groups for linkage to a polyphosphazene chain.

Recent work in our program has revealed another aspect of this principle – the design and synthesis of new polyphosphazene elastomers based on what appear to be the intermolecular interactions between bulky side groups. This becomes manifest in polyphosphazenes that contain minor percentages of bulky side groups interspersed among a majority of smaller side groups. Some of the bulky side groups shown in the polymers in Fig. 10.5 illustrate this principle.

A striking example of the influence of bulky side groups is where the polyphosphazene with all trifluoroethoxy side groups is a microcrystalline, film-, or fiber-forming polymer. It can be oriented by stretching, but it is not elastomeric. However, the presence of roughly 5–8% of bulkier side groups (such as the two

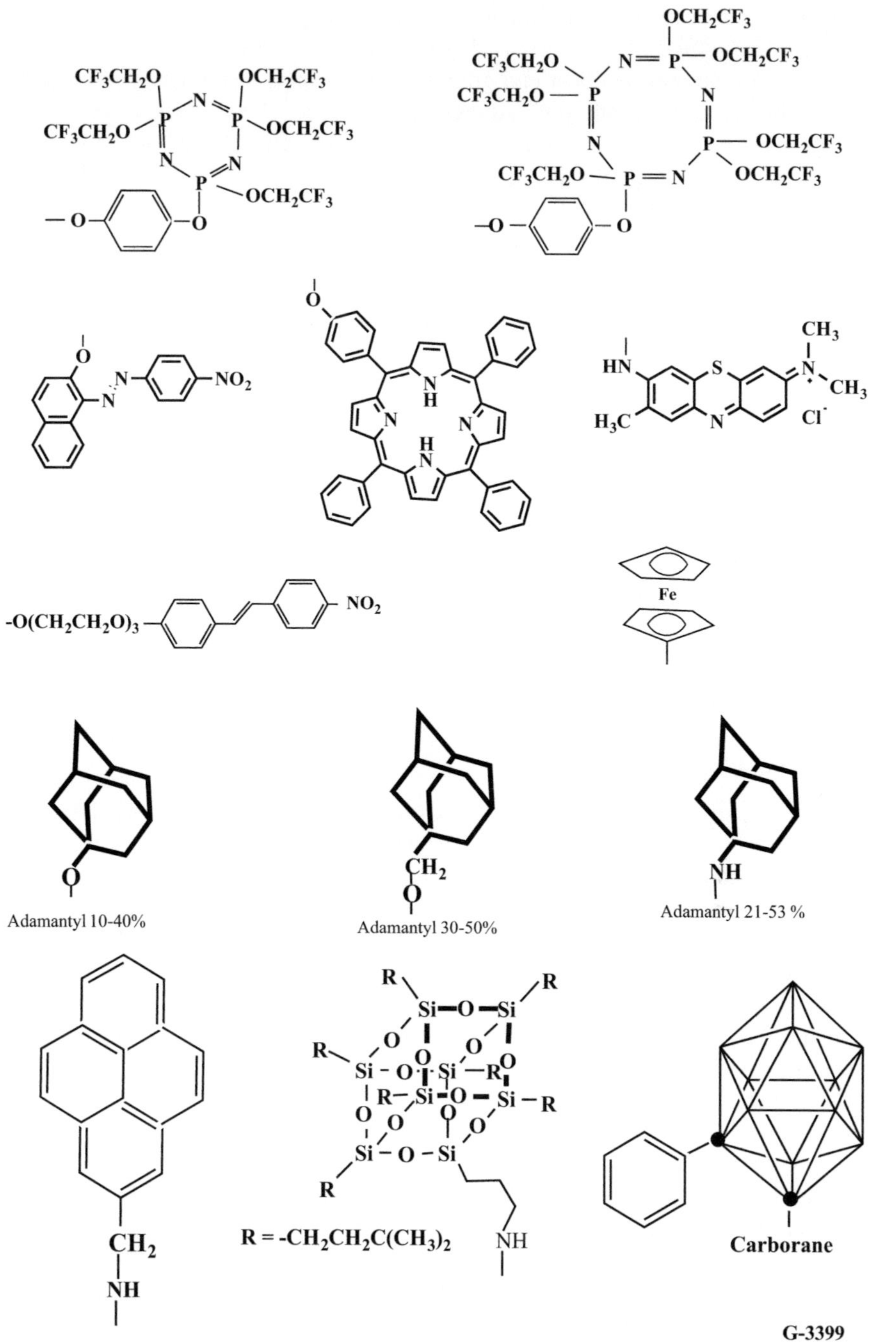

Fig. 10.5 Examples of bulky side groups that have been linked to a polyphosphazene chain. The largest nucleophiles typically replace fewer than 50% of the available reaction sites, but the remaining chlorine atoms can be replaced by less-hindered side groups such as trifluoroethoxy groups

shown at the top of Fig. 10.5) eliminates the crystallinity and converts the material to a rubbery elastomer, *without chemical cross-linking* [23–26].

Polyphosphazenes that bear end groups or side groups that associate with other molecules via a "lock and key" arrangement have also been studied [53]. Thus, polyphosphazenes that bear hydrophobic adamantane units linked to the ends or middle units of a chain form host-guest complexes with beta-cyclodextrin units arrayed along an organic polymer. These complexes form a variety of structures such as "palm tree" or reversible cross-linked arrangements.

10.9 Practical Applications of Polyphosphazenes

The ability to use the polyphosphazene chain as a carrier for different side groups opens numerous avenues for addressing practical problems, and many of these possibilities remain to be exploited. Two summaries that address this topic were published in 2003 [1, 54]. The following are a few examples that illustrate both the earlier and more recent initiatives.

1. *Biomedical Applications.* First, the coupling of biologically useful side groups to the phosphazene polymer chain can be utilized for the controlled delivery of drugs either for subcutaneous implantation or as soluble molecules or nanospheres or microspheres that circulate in the blood. Polymers have been also designed to self-destruct by hydrolysis to release the drug following changes in pH or other stimuli [55–58].

Second, pH- or ion strength-responsive alkyl ether-substituted polyphosphazene membranes have been fabricated for the controlled release of drug molecules, or for the incorporation of enzymes, the permeability of which can be controlled by pH variations in the body [59].

Third, an appropriate choice of amino acid ester side groups can yield a polymer that serves a structural role in the body before bioeroding by hydrolysis to harmless products in what is called a tissue engineering procedure. Such polymers can in principle be used as resorbable surgical sutures, woven surgical mesh, or bone regeneration matrices. Bone regeneration has been a particular target in our collaborative program with the group of C. Laurencin at the University of Connecticut [60–64].

In another medical application, gold nanospheres coated with a polyphosphazene with ionically cross-linkable functional groups (PCPP) have been utilized by our collaborators at the University of Pennsylvania both for the imaging and identification of diseased tissue or for targeted drug delivery [55].

Dentistry is an area of biomedical materials development that has long utilized polyphosphazene elastomers [65]. The elastomers in question are mixed-substituent fluoroalkoxy polymers of the "PN-F" type (see earlier) which are incorporated into an interpenetrating network with a second polymer such as ethylene glycol dimethacrylate or trimethylolpropane trimethacrylate. The composite material has advantages over other materials because of its antifungal properties and its ability to absorb energy without instant rebound. It is also considered to be a favorable material for maxillofacial prosthetics.

Finally, cardiovascular devices and microspheres have been coated with the hydrophobic $[NP(OCH_2CF_3)_2]_n$ or similar fluorophosphazenes to retard detrimental blood/materials interactions. This application is in semi-commercial use for drug delivery applications and for coating implanted anti-thrombogenic treated devices [66].

2. *Battery and Solar Cell Electrolytes.* As mentioned earlier, the MEEP family of polyphosphazene (with alkyl ether side groups) has been investigated in some detail for rechargeable lithium battery electrolyte applications. These polymers are good solid or gel solvents for lithium salts, and the gel forms provide conductivities above 10^{-3} C/cm. The fire-retardant characteristics of these polymers are particularly interesting in view of the known flammability of most organic liquid electrolytes [67, 69].
3. *Hydrophobic and Super-hydrophobic Films, Fibers, and Foams*. The combination of water repellency and limited organic solvent solubility means that fluoroalkoxy and fluoroaryloxy polyphosphazenes have some advantages over classical fluorocarbon polymers. In particular, nanofiber mats made by electrospinning or surface coatings or wire coatings applied from solution have potential uses in aircraft anti-icing surfaces and (as mentioned above) in biomedical devices. The solubility of $[NP(OCH_2CF_3)_2]_n$ in supercritical carbon dioxide has facilitated the fabrication of expanded hydrophobic foams for possible use as fire-resistant flotation constructs [70].
4. *Traditional Photographic Applications*. Patents have been issued by Fujii Photo Film, Eastman Kodak, and Konica for the use of alkyl ether- or phenyl alkyl ether-substituted polyphosphazenes as antistatic agents in silver halide films and as components in processing solutions [71–73]. It appears that these applications are based on the cation coordination properties of the polymers and the amphilicity.
5. *Fire-Resistant Applications.* Small-molecule cyclic phosphazenes with aryloxy side groups have been offered commercially for several years as fire retardants additives for conventional organic polymers. An application for the polymers is as fire retardants for polyurethanes in which the polyphosphazene becomes chemically bonded into the polyurethane network [74].
6. *Optical and Photonic Applications*. The control of refractive index by side group variations is being considered for a number of optical applications [75, 76]. Moreover the linkage of nonlinear optical side groups suggests applications in photonic switches. Dye molecules linked to the polyphosphazene chain have been proposed as non-light-scattering primary color filters for imaging sensors [28]. In earlier work polyphosphazenes with liquid crystalline and photochromic side groups have been studied in some detail [76]. Light-emitting polymers with phosphazene components have also been described [77].

7. *High-Performance Elastomers.* One of the earliest uses for fluoroalkoxy or aryloxy polyphosphazenes was as energy-absorbing, solvent- and fire-resistant elastomers for uses in aircraft, satellites, or fuel lines in automobiles. This interest continues as one of the major applications of polyphosphazenes [1].
8. *Membranes.* Numerous technological membrane applications for both gas and liquid separations have been studied [78, 79]. The most recent interest is in polyphosphazenes for carbon dioxide separations as part of the CO_2 sequestration program for counteracting global warming [79]. As with many other phosphazene applications, the ease of manipulating properties by side group variations is considered to be a major advantage compared to other polymers.

10.10 The Future

The broad span of side groups and skeletal architectures currently known for the polyphosphazene platform already exceeds the scope of most other synthesized macromolecular systems. However, this is just a fraction of the number of different polymers and architectures that will probably become accessible in the future based on an expansion of current knowledge. Moreover, the practical applications of these polymers are only just beginning to be investigated, and this aspect also foretells a promising prospect in the coming years. The diversity of the polyphosphazene platform is also an indicator of what can be expected for other, still undeveloped polymers and materials that combine the attributes of the inorganic and organic elements.

References

1. Allcock HR (2003) Chemistry and applications of polyphosphazenes. John Wiley & Sons, Hoboken
2. Liebig J (1834) A Compound of Phosphorus with Nitrogen (Supplement). Ann Chem Pharm 11:139
3. Rose H (1834) A Compound of Phosphorus with Nitrogen. Ann Chem Pharm 11:129
4. Stokes HN (1897) On the Chloronitrides of Phosphorus (II). Amer Chem J 19:782
5. Allcock HR, Kugel RL (1965) Synthesis of High Polymeric Alkoxy- and Aryloxyphosphonitriles. J Am Chem Soc 87:4216–4217
6. Allcock HR, Kugel RL, Valan KJ (1966) High Molecular Weight Poly[alkoxy- and aryloxyphosphazenes]. Inorg Chem 5:1709–1715
7. Allcock HR, Kugel RL (1966) High Molecular Weight Poly(diamino-phosphazenes). Inorg Chem 5:1716–1718
8. Allcock HR, Lampe FW, Mark JE (2003) Contemporary polymer chemistry, 3rd edn. Prentice-Hall, Englewood Cliffs
9. Odian G (2004) Principles of polymerization, 4th edn. John Wiley & Sons, Hoboken
10. Flindt EP, Rose H (1977) Trivalent-pentavalente Phosphosverbindi-dungen/Phosphazene. IV. Z Anorg Allg Chem 428:204

11. Wisian-Neilson P, Neilson RH (1980) Poly(dimethylphosphazene), $(Me2)PN)_n$. J Am Chem Soc 102:2848
12. Neilson RH, Wisian-Neilson P (1988) Poly(alkyl/aryl phosphazenes). Chem Rev 88:54
13. Wisian-Neilson P, Jung J-H, Potluri SK (2006) Cyclic and Polymeric Alky/Aryl Phosphazenes in *Modern Aspects of Main Group Chemistry*. ACS symp. series 917, Washington, DC, pp 335–346
14. Montague RA, Matyjaszewski K (1990) Synthesis of Poly[bis(trifluoroethoxy)-phosphazene] Under Mild Conditions Using a Fluoride Initiator. J Am Chem Soc 112:6721
15. Matyjaszewski K, Moore MM, White ML (1993) Synthesis of Polyphosphazene Block Copolymers Bearing Alkoxyethoxy and Trifluoroethoxy Groups 26:6741–6748
16. (a) Honeyman CH, Manners I, Morrissey CT, Allcock HR (1995) Ambient Temperature Synthesis of Poly(dichlorophosphazene) with Molecular Weight Control. J Am Chem Soc 117:7035–7036. (b) Allcock HR, Reeves SD, Nelson JM, Crane CA, Manners I (1997) Polyphosphazene Block Copolymers via the Living Cationic, Ambient Temperature Polymerization of Phosphoranimines. Macromolecules 30:2213–2215
17. De Jaeger R, Potin P (2004) Ch.2. In: Gleria M, De Jaeger R (eds) Synthesis and characterization of poly(organophosphazenes). Nova Publishers, New York
18. Allcock HR (2004) Ch. 3. In Gleria M, De Jaeger R (eds) Synthesis and characterization of poly (organophosphazenes). Nova Publishers, New York
19. Allcock HR, Morozowich NL (2012) Bioerodible Polyphosphazenes and their Medical Potential. RSC Polym Chem 3:578–590
20. Allcock HR, Fuller TJ (1980) Phosphazene High Polymers with Steroidal Side Groups. Macromolecules 13:1338–1345
21. Allcock HR, Greigger PP, Gardner JE, Schmutz JL (1979) Water Soluble Polyphosphazenes as Carrier Molecules for Iron(III) and Iron(II) Porphyrins. J Am Chem Soc 101:606–611
22. Allcock HR, Scopelianos AG, O'Brien JP, Bernheim MY (1981) Synthesis and Structure of Carborane-Substituted Cyclic and Polymeric Phosphazenes. J Am Chem Soc 103:350–357
23. Modzelewski T, Allcock HR (2014) An Unusual Polymer Architecture for the Generation of Elastomeric Properties in Fluorinated Polyphosphazenes. Macromolecules 47:6776–6782
24. Modzelewski T, Wilts E, Allcock HR (2015) Elastomeric Polyphosphazenes with Phenoxy-cyclotriphosphazene Side Groups. Macromolecules 48:7543–7549
25. Li Z, Chen C, Tian Z, Modzelewski T, Allcock HR (2016) Polyphosphazenes with Cyclotetraphosphazene Side Groups: Synthesis and Elastomeric Properties. J Inorg Organomet Mater Polym 26:667–674
26. Modzelewski T, Wonderling NM, Allcock HR (2015) Polyphosphazene Elastomers Containing Interdigitated Oligo-p-phenyleneoxy Side Groups: Synthesis, Mechanical Properties and X-ray Scattering Studies. Macromolecules 48:4882–4890
27. Tong C, Tian Z, Chen C, Li Z, Modzelewski T, Allcock HR (2016) Synthesis and Characterization of Trifluoroethoxy Polyphosphazenes Containing POSS Side Groups. Macromolecules 49:1313–1320
28. Li Z, Allcock HR (2015) Polyphosphazenes with Immobilized Dyes as Potential Color Filter Materials. Appl Mater Interface 1:13518–13523
29. Weikel AL, Lee D, Krogman NR, Allcock HR (2010) Phase Changes of Poly (alkoxyphosphazene) and their Behavior in the Presence of Oligoisobutylene. J Polym Eng Sci 92A:114–125
30. Tian Z, Liu X, Manseri A, Ameduri B, Allcock HR (2013) Limits to Expanding the PN-F Series of Polyphosphazene Elastomers. Polym Eng Sci 54:1827–1832
31. Singler RE, Schneider NS, Hagnauer GL (1975) Polyphosphazenes: Synthesis-Properties-Applications. Polym Eng Sci 51:321–338
32. Allcock HR, Mang MN, Dembek AA, Wynne KJ (1989) Poly(aryloxyphosphazenes) with Phenylphenoxy and Related Bulky Side Groups. Synthesis, Thermal Transition Behavior and Optical Properties. Macromolecules 22:4179–4190

33. Reed CA, Taylor GP, Guigley KS, Kully KS, Bernheim KA, Coleman MM, Allcock HR (2000) Polyurethane/Poly[bis(carboxylato-phenoxy)phosphazene] Blends and their Potential as Flame Retardant Materials. J Polym Sci Eng 40:465–472
34. Chen C, Liu X, Tian Z, Allcock HR (2012) 2,2,2-Trichloroethoxy-Substituted Polyphosphazenes: Synthesis, Characterization, and Properties. Macromolecules 45:9085–9091
35. Weikel AL, Owens SG, Fushimi T, Allcock HR (2012) Synthesis and Characterization of Methionine- and Cysteine-Substituted Phosphazenes. Macromolecules 4:5205–5210
36. Weikel AL, Lee D, Krogman NR, Allcock HR (2010) Phase Changes of Poly (alkoxyphosphazenes) and Their Behavior in the Presence of Oligoisobutylene. J Polym Sci Eng 92A:114–125
37. Nichol JL, Hotham IT, Allcock HR (2014) Ethoxyphosphazene Polymers and their Hydrolytic Behavior. Polym Degrad Stab 109:92–96
38. Tian Z, Hess A, Fellin CR, Nulwala H, Allcock HR (2015) Phosphazene High Polymers and Models with Cyclic Aliphatic Side Structure-Property Relationships. Macromolecules 48:4301–4311
39. Kojima M, Magill J (1985) Phase Transitions in Polyphosphazene Films: Poly[bis (trifluoroethoxy)phosphazene]. Macromol Chem Phys 186:649–663
40. Rose SH, Cable J (1969) U.S. Govt Research Report, AD 693,28
41. Schneider NS, Desper CR, Singler RE (1976) The Thermal Transition Behavior of Polyorganophosphazenes. J Appl Polym Sci 20:3087–3103
42. Allcock HR, Kim C (1990) Liquid Crystalline Phosphazenes Bearing Biphenyl Mesogenic Groups. Macromolecules 23:3881–3887
43. Blonsky PM, Shriver DF, Austin PE, Allcock HR (1984) Polyphosphazene Solid Electrolytes. J Am Chem Soc 106:6854–6855
44. Allcock HR, Austin PE, Neenan TX, Sisko JT, Blonsky PM, Shriver DF (1986) Polyphosphazenes with Etheric Side Groups: Prospective Biomedical and Solid Electrolyte Polymers. Macromolecules 19:1508–1512
45. Allcock HR, Kwon S, Riding GH, Fitzpatrick RJ, Bennett JL (1988) Hydrophilic Polyphosphazenes as Hydrogels: Radiation Crosslinking and Hydrogel Characteristics of Poly[bis(methoxyethoxyethoxy)phosphazene]. Biomaterials 19:509–513
46. Greigger PP, Allcock HR (1979) A Spirocyclophosphazene with Iron-Phosphorus Bonds and a P-Fe-Fe Three-Membered Ring. J Am Chem Soc 101:2492
47. Allcock HR, Manners I, Mang MN, Parvez M (1990) Transition Metal Derivatives of Phosphinophosphazenes: X-Ray Crystal Structures of $N_3P_3Cl_4PhPh_2$, $N_3P_3Cl_4PhPPh_4$,Cr $(Co)_5$ and $N_3P_3Cl_4PhPPh_2.Ru_3(CO)_{11}$. Inorg Chem 29:522–529
48. Manners I, Riding GH, Dodge JA, Allcock HR (1989) Role of Ring Strain and Steric Hindrance in a New Method for the Synthesis of Macrocyclic and High Polymeric Phosphazenes. J Am Chem Soc 111:3067–3069
49. Diefenbach U, Cannon AM, Stromberg BE, Olmeijer DL, Allcock HR (2000) Synthesis and Metal Coordination of Thioether-Containing Cyclo- and Poly(organophosphazenes). J Appl Polym Sci 78:650–661
50. Liu X, Breon J, Chen C, Allcock HR (2012) Substituent Exchange Reactions of Linear Oligomeric Aryloxy Phosphazenes with 2,2,2-Trifluoroethoxide. Inorg Chem 51(21):11910–11916
51. Liu X, Breon JP, Chen C, Allcock HR (2012) Substituent Exchange Reactions of Trimeric and Tetrameric Arylocyclophosphazenes with Sodium 2,2,2-Trifluoroethoxide. Roy Soc Chem Dalton Trans 41:2100–2109
52. Allcock HR, Steely L, Kim S, Kang B (2007) Plasma Surface Functionalization of Poly[bis (2,2,2-trifluoroethoxy)phosphazene] Films and Nanofibers. Langmuir 23:8103–8107
53. Tian Z, Chen C, Allcock HR (2014) Ethoxyphosphazene Polymers and their Hydrolytic Behavior. Macromolecules 47:1065–1072

54. Gleria M, De Jaeger R (eds) (2003) Applicative aspects of poly(organophosphazenes). Nova Publishers, New York
55. Chhour P, Gallo N, Cheheltani R, Williams D, Al-Zaki A, Paik T, Nichol JL, Tian Z, Naha PC, Allcock HR, Murray CB, Sourkas TA, Cormode DP (2014) Nano-Disco Balls: Control over Surface versus Core Loading of Active Nanocrystals into Polymer Nanoparticles. ACS Nano 8(9):9143–9153
56. Liu X, Tian Z, Chen C, Allcock HR (2013) UV-Cleavable Unimolecular Micelles: Synthesis and Characterization Toward Photocontrolled Drug Release Carriers. Polym Chem 4:1114–1125
57. Liu X, Zhang H, Tian Z, Sen A, Allcock HR (2012) Preparation of Quaternized Organic-Inorganic Hybrid Brush Polyphosphazene-co-poly-[2-(dimethylamino)ethylmethacrylate] Electrospun Fibers and their Antibacterial Properties. Polym Chem 3:2082–2091
58. Tian Z, Zhang Y, Chen C, Guiltinan MJ, Allcock HR (2013) Biodegradable Polyphosphazenes Containing Antibiotics: Synthesis, Characterization, and Hydrolytic Release Behavior. Polymer 4:1826–1835
59. Allcock HR, Kwon S (1989) An Ionically-Crosslinkable Polyphosphazene: Poly[di (carboxylatophenoxy)phosphazene] and its Hydrogels and Membranes. Macromolecules 22:75–79
60. Peach MS, James R, Toti US, Deng M, Morozowich NL, Allcock HR, Laurencin CT, Kumbar SG (2012) Polyphosphazene Functionalized Polyester Fiber Matrices for Tendon Tissue Engineering: In Vitro Evaluations of Human Mesenchymal Stem Cells. Biomed Mater 7:1–13
61. Nichol JL, Morozowich NL, Allcock HR (2013) Biodegradable Alanine and Phenylalanine Alkyl Ester Polyphosphazenes as Potential Ligament and Tendon Tissue Scaffolds. RSC Polym Chem 4:600–606
62. Peach MS, Kumbar SG, James R, Toti US, Balasubramaniam D, Deng M, Ulery B, Maxxocca AD, McCarthy MB, Morozowich NL, Allcock HR, Laurencin CT (2012) Design and Optimization of Polyphosphazenes Functionalized Fiber Matrices for Soft Tissue Regeneration. J Biomed Nanotechnol 8:107–124
63. Deng M, Kumbar SG, Nair LS, Weikel AL, Allcock HR, Laurencin CT (2011) Biomimetic Structures: Biological Implications of Dipeptide-substituted Polyphosphazene-Polyester Blend Nanofiber Matrices for Load-Bearing Bone Regeneration. Adv Funct Mater 21:2641–2651
64. Deng M, Kumbar SG, Wan Y, Toti US, Allcock HR, Laurencin CT (2010) Polyphosphazene Polymers for Tissue Engineering: An Analysis of Materials Synthesis, Characterization, and Applications. Soft Matter 6:3119–3132
65. Gettleman L (2003) Ch. 2. In: Gleria M, De Jaeger R (eds) Applicative aspects of poly (organophosphazenes). Nova Publishers, New York
66. Celonova Bio-Sciences (2010) Peachtree, Georgia, promotional material (2017) U.S. Patent US7922764
67. Fei S-T, Lee S-HA, Pursel SM, Bashem J, Hess A, Grimes CA, Horn MW, Mallouk TE, Allcock HR (2011) Electrolyte Infiltration in Phosphazene-Based Dye-Sensitized Solar Cells. J Power Sources 196:5223–5230
68. Fei S-T, Allcock HR (2010) Methoxyethoxyethoxyphosphazenes as Ionic Fire Retardant Additives for Lithium Battery Systems. Power Sources 195(7):2082–2088
69. Fei S-T, Allcock HR (2009) Recent Progress with Ethyleneoxy Phosphazenes as Lithium Battery Electrolytes. Mater Res Soc Symp. 1127-T01-05
70. Steely LB, Li Q, Badding JV, Allcock HR (2008) Foam Formation with Fluorinated Polyphosphazenes by Liquid CO_2 Processing. Polym Sci Eng 48:683–686
71. Mukunoki Y, Kubota T (1992) Fuji Photo Film, Japan, U.S. Patent 5,135,846
72. Fukuwatari N, Ueda E, Kurachi Y (1998) Konica Corp., Japan, U.S. Patent 5,840,471
73. Ishikawa W, Fukuwatari N. Konica Corp., Japan, European Patent Application
74. Reed CA, Taylor JP, Guigley KS, Kully KS, Bernheim KA, Coleman MM, Allcock HR (2000) Polyurethane/Poly[bis(carboxylato-phenoxy)phosphazene] Blends and their Potential as Flame Retardant Materials. Report to U.S. Federal Aviation Administration

75. Olshavsky M, Allcock HR (1997) Polyphosphazenes with High Refractive Indices: Optical Dispersion and Molar Reactivity. Macromolecules 30:4179–4183
76. Allcock HR, Bender JD, Chang Y, McKenzie M, Fone MM (2003) Controlled Refractive Index Polymers; Polyphosphazenes with Chlorinated- and Fluoroinated-Aryloxy and Alkoxy Side Groups. Chem Mater 15:473–477
77. Allcock HR, Chang Y, Stone DA (2006) Control of the Conjugation Length and Solubility in Electroluminescent Polymers. J Polym Sci 44:69–76
78. Stewart FF, Luther TA, Harrup MK, Orme CJ (2003) In: Gleria M, De Jaegar R (eds) Applicative aspects of poly(organophosphazenes). Nova Publishers, New York. Chapter 10
79. Venna SR, Spore A, Tian Z, Marti AM, Albenze EJ, Nulwala HB, Rosi NL, Luebke DR, Hopkinson DP, Allcock HR (2017) Polyphosphazene Polymer Development for Mixed Matrix Membranes Using SIFSIX-Cu-2i as Performance Enhancement Filler Particles. J Membr Sci 535:103–112

Chapter 11
Electronic Structure of Element-Block Material As_4S_6 with Cage Shape Toward "Soft Electride"

Kazuyoshi Tanaka and Hiroyuki Fueno

Abstract Electrides are the materials having functions which can draw and release electrons rather freely in general. In this chapter we like to describe theoretical examination on As_4S_6 molecule with cage shape as an element block. It has been found that this molecule can stably accommodate an excessive electron. This fact may open a novel electride material in which As_4S_6 molecules are incorporated into polymer skeleton as element blocks.

Keywords Electride · Cage structure · Electron affinity · Arsenic sulfide

11.1 Introduction

Solid-state materials in which liberated and isolated electrons can exist are called electrides in general. Naming of electride is somewhat similar to fluoride or chloride in that electrons are separated from the counter atoms to have become cations. For instance, alkaline metal crystal such as Na becomes transparent and causes metal-insulating transition under pressure of ca. 200 GPa. This situation is explained by transformation into electride in which liberated electrons from each Na atom exit at the interstitial site of the crystalline lattice [1]. Similar phenomenon has also been reported to appear in Li crystal. Moreover, a theoretical model in which a liberated electron is captured in the virtual quantized energy level at the interstitial site within the crystalline lattice has also been proposed [2].

K. Tanaka (✉)
Fukui Institute for Fundamental Chemistry, Kyoto University, Kyoto, Japan
e-mail: ktanaka@moleng.kyoto-u.ac.jp

H. Fueno
Department of Molecular Engineering, Graduate School of Engineering, Kyoto University, Kyoto, Japan
e-mail: fueno@moleng.kyoto-u.ac.jp

Y. Chujo (ed.), *New Polymeric Materials Based on Element-Blocks*,
https://doi.org/10.1007/978-981-13-2889-3_11

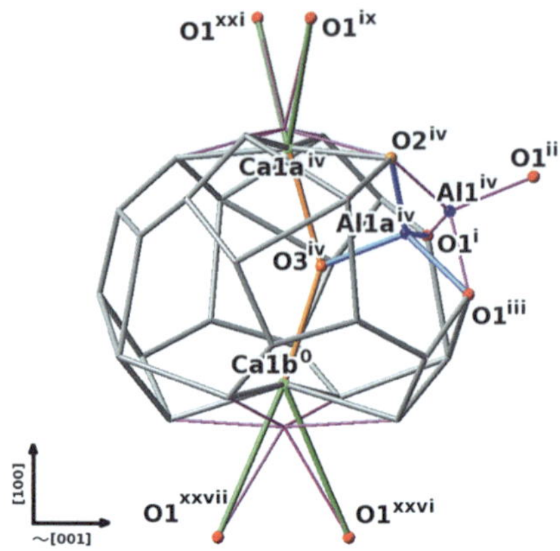

Fig. 11.1 A local cluster shape in $12CaO{\cdot}7Al_2O_3$ crystal. An O_2^- ion designated as O_3 is located at the position close to the center of this cluster cage and possibly supplies electrons to afford electride. (Sakakura et al. [19]; reproduced with permission of the International Union of Crystallography (https://journals.iucr.org/))

There is an organic electride such as Cs^+(18-crown-6)$_2$(e^-) that can exist in inert atmosphere under −40° centigrade [3]. The electride having the component $[Ca_{24}Al_{28}O_{64}]^{4+}(4e^-)$ can exist under ambient condition [4]. This electride contains the liberated electrons in the cage composed of Ca-O-Al as shown in Fig. 11.1. It is of interest to note that this cage-like electride affords considerably high electric conductivity since liberated electrons migrate within the solid as the conduction carriers.

Meanwhile, development of element-block materials has been actively accumulated toward their incorporation into polymers [5]. These materials afford effective strategies not only for structural modification but also for intriguing and useful functions such as peculiar reactivity and electric, electronic, photophysical, and magnetic properties. It will be useful to combine the idea of this element block toward preparation of electride polymer as a soft material.

In this chapter, considering these situations, we like to propose utilization of an As_4S_6 (tetraarsenic hexasulfide) molecule with cage shape (Fig. 11.2a) as an element block to fabricate the electride polymer based on theoretical examination. This molecule or molecular unit has been suggested to be a local electron-capturing center in amorphous As_2S_3 material by one of the authors [6] in relation with the specific photoinduced ESR center [7, 8]. There have been also other reports on As_4S_6 as plausible local structure in this material [9, 10]. Similar cage-shape molecules in Fig. 11.2b, c are compared as the references to As_4S_6.

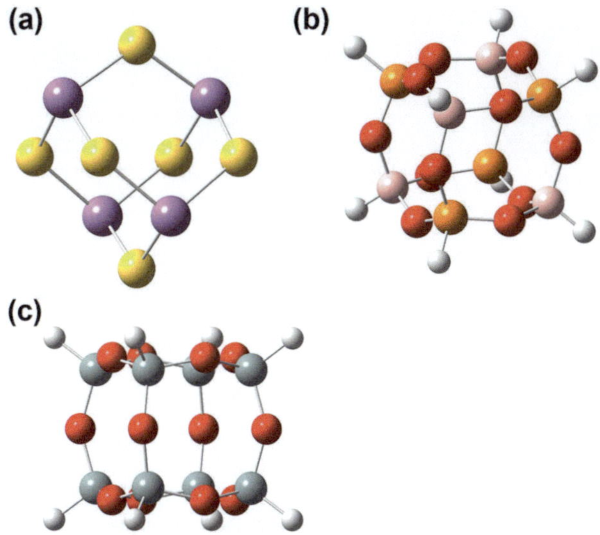

Fig. 11.2 Cage-shape molecules: (**a**) As_4S_6 (As, purple; S, yellow), (**b**) POB (P, orange; O, red; B, peach; H, light gray), and (**c**) POSS (Si, gray; O, red; H, light gray)

11.2 Method

An electron-capturing capability of an As_4S_6 molecule was examined with its rationalization. Calculations were based on the density functional theory (DFT) under the B3LYP method with the 6-31+G** basis set. For the open-shell system such as $As_4S_6^-$ was employed the unrestricted MO scheme. The Gaussian 09 software package was utilized to perform the calculations [11]. More explicitly, structural optimizations were performed with respect to the neutral and monoanionic states of each cage-shape molecule selected here, and energetical stabilization of those was deduced by the comparison of their energies. The reference molecules were dealt with the same method.

11.3 Results and Discussion

First we examine the energetical change between the neutral and the monoanionic states of an As_4S_6 molecule. Molecular structures in the both states are structurally optimized to find that symmetry in the monoanionic state (C_s) is lower than in the neutral one (T_d), the LUMO and the LUMO+1 of the latter being doubly and triply degenerated, respectively. This symmetry lowering is caused by usual Jahn-Teller distortion accompanied by structural relaxation after the injection of an electron to the neutral state.

It is noted that the LUMO level of As_4S_6 is -3.027 eV as seen in Fig. 11.3 and as listed in Table 11.1 showing that this molecule has electron affinity in itself. The

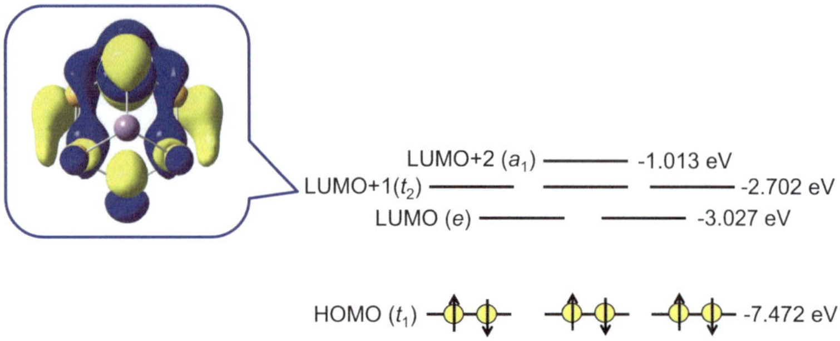

Fig. 11.3 MO diagram of a neutral As_4S_6 (T_d symmetry) and a lobe giving the α-SOMO of As_4S_6 monoanion (see also Fig. 11.4). This lobe is indicated after release of the triply degenerate LUMO +1 by a very slight change from T_d to C_s symmetry. Note that α-SOMO stands for the singly occupied MO occupied with an α-spin electron

Table 11.1 Stability of the monoanionic state along with LUMO energy of the neutral state

Molecule	LUMO energy (in eV)	Energy difference[a] between the neutral and the monoanionic states[b] (in kcal/mol)
As_4S_6	−3.027	−46.17
POB: $[HPO_3BH]_4$	−0.502	−11.32
POSS: $(HSiO_{1.5})_8$	−0.157	11.51

[a]Negative value signifies the stabilization of the monoanionic state
[b]The both states in their optimized structures

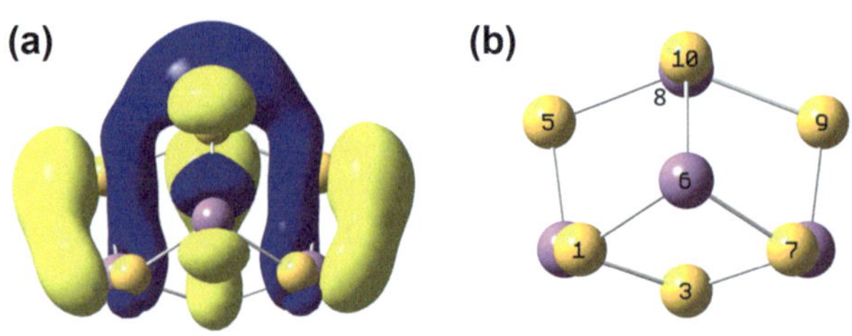

Fig. 11.4 (**a**) α-SOMO (a' symmetry) lobe of $As_4S_6^-$ (C_s symmetry) and (**b**) its molecular structure (As, purple; S, yellow) from the same angle with atomic numbers used in Table 11.2. Note that ^{8}As is under ^{10}S

energy difference between the optimized As_4S_6 and $As_4S_6^-$ seen in Table 11.1 (−46.17 kcal/mol) indicates $As_4S_6^-$ is the more stable with a considerable degree. This result signifies that an injected electron can stay in the As_4S_6 cage in a stable manner. It is of rather interest to point out that the actual α-SOMO of $As_4S_6^-$ comes from one of the triply degenerate LUMO+1, not from the doubly degenerate LUMO, of As_4S_6 as can be seen in the orbital patterns in Figs. 11.3 and 11.4. Hence there must be a certain rearrangement of the electronic structure during the Jahn-Teller distortion from As_4S_6 to $As_4S_6^-$.

On the other hand, the α-SOMOs of both POB^- and $POSS^-$ (Figs. 11.5 and 11.6) simply come from the LUMOs of their neutral species [12, 13]. It is also noted that the POB^- is energetically more stable than POB but that $POSS^-$ less stable than POSS as seen in Table 11.1. In particular, although POSS has a positive value of electron affinity, it will not be able to accommodate an excessive electron in a stable manner.

It is of further interest that the α-SOMO pattern of $As_4S_6^-$ mainly consists of spσ* bonds between ^{8}As-^{5}S and ^{8}As-^{9}S with some lone-pair nature of ^{8}As (see Fig. 11.4b as to the atomic numbers). This pattern is rather different and deviated

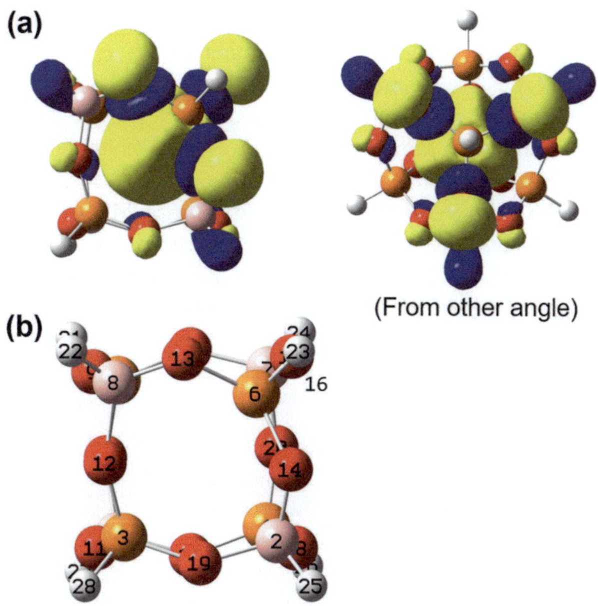

Fig. 11.5 (**a**) α-SOMO (*a* symmetry) lobe of POB^- (C_3 symmetry) and (**b**) its molecular structure (P, orange; O, red; B, peach; H, light gray) from the same angle with atomic numbers used in Table 11.2. Note that ^{16}O is under ^{23}H

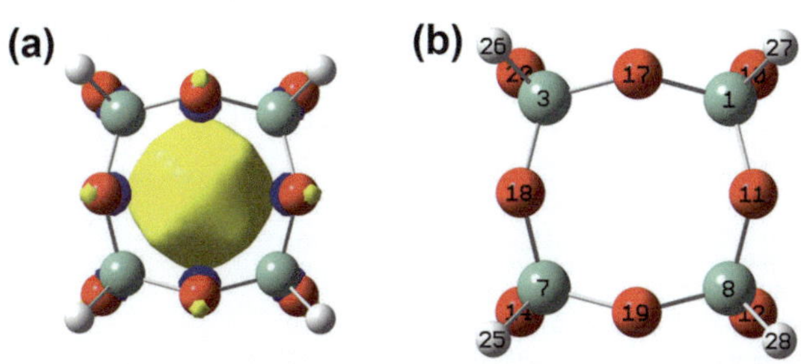

Fig. 11.6 (**a**) α-SOMO (a_{1g} symmetry) lobe of $POSS^-$ (O_h symmetry) and (**b**) its molecular structure (Si, gray; O, red; H, light gray) from the same angle with atomic numbers used in Table 11.2

compared with those of POB^- and $POSS^-$ in Figs. 11.5a and 11.6a, respectively. The α-SOMO pattern of $POSS^-$ is indeed purely spherical [14] and evenly consists of all the Si-O spσ* bonds, but that of POB^- is a bit deviated and consists of three P-O (^{6}P-^{13}O, ^{6}P-^{14}O, and ^{6}P-^{16}O) spσ* bonds (see Fig. 11.5b as to the atomic numbers). These are obviously associated with each molecular symmetry of the concerning monoanions. That is, in $As_4S_6^-$, the Jahn-Teller distortion makes the symmetry far down to C_s, and only one of the triply degenerate LUMO+1 of As_4S_6 becomes the α-SOMO of $As_4S_6^-$, which eventually makes remarkable spatial deviation of the α-SOMO in Fig. 11.4a.

The α-SOMO of POB^-, however, approaches a bit to ^{6}P atom, and the degree of deviation is not so large, which just reflects the C_3 symmetry of this species. The above analyses are in accordance with the fact that ^{8}As in $As_4S_6^-$ and ^{6}P in POB^- accommodate the excessive electron as is understood from Table 11.2.

On the other hand, all the Si atoms are still completely equivalent in $POSS^-$ as seen in Fig. 11.5, and these atoms mainly and evenly accommodate the excessive electron maintaining its spherical shape as indicated in Table 11.2. It is of interest to point out that $POSS^-$ has a completely symmetrical α-SOMO (a_{1g}) whereas POB^- and $As_4S_6^-$ not. Based on the energy difference listed in Table 11.1, $POSS^-$ is energetically less stable than POSS, which might indicate that energetical stability of POB^- and $As_4S_6^-$ is gained at the sacrifice of lowering of molecular symmetry in the neutral state.

Table 11.2 Atomic net charges[a] of the neutral and the monoanionic states

Molecule	Neutral state	Monoanionic state[b]
As_4S_6	As: 0.6189	^{8}As: 0.3390
		^{2,4,6}As: from 0.5541 to 0.5543
	S: −0.4126	^{3}S: −0.5318
		^{5,9}S: −0.5275
		^{1,7,10}S: from −0.4616 to −0.4797
POB: $[HPO_3BH]_4$	P: 2.3036	^{6}P: 1.6745
		^{3,4,5}P: from 2.2968 to 2.2973
	O: −1.0367	^{13,14,16}O: from −1.0216 to −1.0224
		^{9,10,11,12,15,17,18,19,20}O: from −1.0404 to −1.0468
	B: 0.9228	^{1}B: 0.9235
		^{2,7,8}B: from 0.9170 to 0.9178
POSS[c]: $(HSiO_{1.5})_8$	Si: 2.1930	Si: 2.1112
	O: −1.3012	O: −1.3066
	H: −0.2412	H: −0.2763

[a]Obtained by natural population analysis (NPA) method
[b]Atomic numberings are indicated in Figs. 11.4b and 11.5b
[c]The Si, O, and H atoms in $POSS^-$ are all equivalent as those in the neutral POSS due to high symmetry (O_h)

11.4 Further Prospect

Stability of anionic state of As_4S_6 will open possibility of a novel electride when this cage-shape molecule is incorporated into polymer chain as illustrated in Fig. 11.7. From the energetical stability of its monoanionic species, As_4S_6 can capture an excessive electron in a stable manner, which might be applicable to practical use as soft-electride material showing properties derived from, for instance, color centers, magnetic centers, ferroelectric origins, and so on. There can be at least two kinds of mode for incorporation of As_4S_6 cage as element block by two kinds of manners, for instance, as illustrated in Fig. 11.7a, b.

In order to construct these polymers, it is necessary for As_4S_6 to have one or two functional groups on its surface. In Fig. 11.8 is shown a possible example of cycloaddition reaction to attach two ethylene molecules onto the surface of As_4S_6 cage. The HOMO of As_4S_6 moiety shown is one of the triply degenerate, the other two having essentially similar lobes at different S atoms. Since an excessive electron in $As_4S_6^-$ is mainly accommodated at one of As atoms, this cycloaddition reaction will not severely affect the generation of this monoanion. Although, to the authors' knowledge, there have been no reports dealing with such reaction for As_4S_6, it is noted that similar cycloaddition reactions of two ethylene derivatives to an S_4N_4 (tetrasulfur tetranitride) molecule having a kind of cage structure have been reported [15–18].

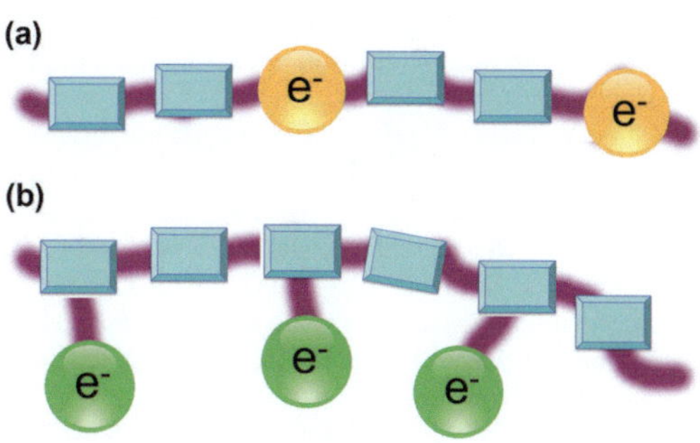

Fig. 11.7 Concept of polymer structures incorporated with (**a**) catenation of the electride clusters (indicated by circles) and (**b**) pendants of those

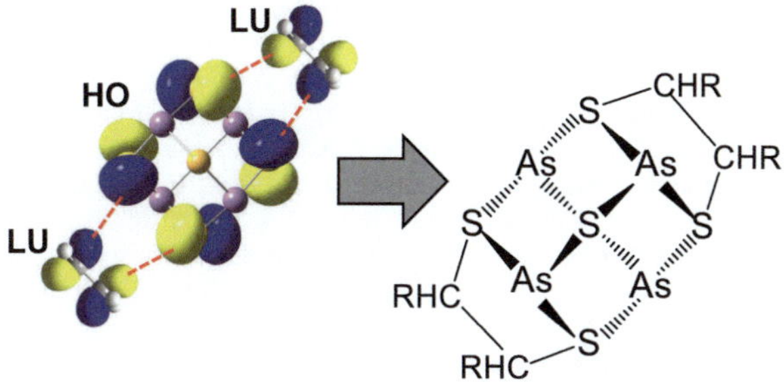

Fig. 11.8 Orbital interaction analysis toward possible cycloaddition reaction to form two five-membered rings at S atoms. Note that the view of As_4S_6 is from the other angle compared with that of Fig. 11.3, and there is another S atom down below the center S atom

11.5 Conclusion

It has theoretically been clarified that an As_4S_6 molecule with cage shape has considerably large electron affinity and that its monoanionic species is energetically more stable than the neutral one. This kind of information would support more utilization of various kinds of inorganic elements in element-block polymers with intriguing electronic properties.

Acknowledgments One of the authors (KT) is grateful to fruitful discussions with Professor Roald Hoffmann for electrides. This work was supported by a Grant-in-Aid for Scientific Research on Innovative Areas "New Polymeric Materials Based on Element-Blocks (No. 2401)" (JSPS KAKENHI Grant Number JP24102014).

References

1. Ma Y, Eremets M, Oganov AR, Xie Y, Trojan I, Medvedev S, Lyakhov AO, Valle M, Prakapenka V (2009) Transparent dense sodium. Nature 458:182–185
2. Miao M-S, Hoffmann R (2014) High pressure electrides: a predictive chemical and physical theory. Acc Chem Res 47:1311–1317
3. Ellaboudy A, Dye JL (1983) Cesium 18-crown-6 compounds. A crystalline ceside and a crystalline electride. J Am Chem Soc 105:6490–6491
4. Matsuishi S, Toda Y, Miyakawa M, Hayashi K, Kamiya T, Hirano M, Tanaka I, Hosono H (2003) High-density electron anions in nanoporous single crystal: $[Ca_{24}Al_{28}O_{64}]^{4+}(4e^-)$. Science 301:626–629
5. Chujo Y, Tanaka K (2015b) New polymeric materials based on element-blocks. Bull Chem Soc Jpn 88:633–643
6. Tachibana A, Yamabe T, Miyake M, Tanaka K, Kato H, Fukui K (1978) Electronic behavior of amorphous chalcogenide models. J Phys Chem 82:272–276
7. Bishop SG, Strom U, Taylor PC (1975) Optically induced localized paramagnetic states in chalcogenide glasses. Phys Rev Lett 34:1346–1350
8. Bishop SG, Strom U, Taylor PC (1977) Optically induced metastable paramagnetic states in amorphous semiconductors. Phys Rev B 15:2278–2294
9. de Neufville JP, Moss SC, Ovshinsky SR (1974) Photostructural transformations in amorphous AS_2Se_3 and AS_2S_3 films. J Non-Cryst Solids 13:191–223
10. Hammam M, Santiago JJ (1986) Evidence for AS_4S_6 molecule as a structural model for amorphous arsenic sulfide from mass spectrometric analysis. Solid State Commun 59:725–727
11. Frisch MJ, Trucks GW, Schlegel HB, Scuseria GE, Robb MA, Cheeseman JR, Scalmani G, Barone V, Mennucci B, Petersson GA, Nakatsuji H, Caricato M, Li X, Hratchian HP, Izmaylov AF, Bloino J, Zheng G, Sonnenberg JL, Hada M, Ehara M, Toyota K, Fukuda R, Hasegawa J, Ishida M, Nakajima T, Honda Y, Kitao O, Nakai H, Vreven T, Montgomery JA Jr, Peralta JE, Ogliaro F, Bearpark M, Heyd JJ, Brothers E, Kudin KN, Staroverov VN, Kobayashi R, Normand J, Raghavachari K, Rendell A, Burant JC, Iyengar SS, Tomasi J, Cossi M, Rega N, Millam JM, Klene M, Knox JE, Cross JB, Bakken V, Adamo C, Jaramillo J, Gomperts R, Stratmann RE, Yazyev O, Austin AJ, Cammi R, Pomelli C, Ochterski JW, Martin RL, Morokuma K, Zakrzewski VG, Voth GA, Salvador P, Dannenberg JJ, Dapprich S, Daniels AD, Farkas Ö, Foresman JB, Ortiz JV, Cioslowski J, Fox DJ (2009) Gaussian 09, revision D.01. Gaussian, Wallingford

12. Zapico J, Shirai M, Sugiura R, Idota N, Fueno H, Tanaka K, Sugahara Y (2017) Borophosphonate cages as element-blocks: ab initio calculation of the electronic structure of a simple borophosphonate, $[HPO_3BH]_4$, and synthesis of two novel borophosphonate cages with polymerizable groups. Chem Lett 46:181–184
13. Tanaka K, Fueno H, Naka K (2014) Toward design of magnetic polymer including POSS block. Polym Prep Japan 63:4546–4547
14. Laine RM, Sulaiman S, Brick C, Roll M, Tamaki R, Asuncion MZ, Neurock M, Filhol J–S, Lee C–Y, Zhang J, Goodson T III, Ronchi M, Pizzotti M, Rand SC, Li Y (2010) Synthesis and photophysical properties of stilbeneoctasilsesquioxanes. Emission behavior coupled with theoretical modeling studies suggest a 3-D excited state involving the silica core. J Am Chem Soc 132:3708–3722
15. Becke-Goehring M, Schläfer D (1968) Tetraschwefel-tetranitrid, ein anorganisches dien. Z Anorg Allg Chem 356:234–238
16. Griffin AM, Sheldrick GM (1975) Tetrasulphur tetranitride-bis(norbornadiene). Acta Crystallogr B31:895–896
17. Mock WL, Mehrotra I (1976) Tetrasulphur tetranitride-bis(norbornadiene). J Chem Soc Chem Commun:123–124
18. Yamabe T, Tanaka K, Tachibana A, Fukui K, Kato H (1979) Orbital interaction in the cycloaddition reactions of tetrasulfur tetranitride, tetraarsenic tetrasulfide, and tetraarsenic tetraselenide. J Phys Chem 83:767–769
19. Sakakura H, Tanaka K, Takenaka Y, Matsuishi S, Hosono H, Kishimoto S (2011) Determination of the local structure of a cage with an oxygen ion in $Ca_{12}Al_{14}O_{33}$. Acta Crystallogr B67:193–204

Chapter 12
Metalloxane Cage Compounds as an Element-Block

Takahiro Gunji and Satoru Tsukada

Abstract A titanium phosphonate cluster with a formula of $[Ti_4(\mu_3\text{-}O)(O^iPr)_5(\mu\text{-}O^iPr)_3(PhPO_3)_3]\cdot$thf was synthesized by the reaction of titanium tetraisopropoxide with phenylphosphonic acid in tetrahydrofuran and the following hydrolysis. The titanium cluster phosphonate was mixed with poly(dimethylsiloxane) (PDMS), poly(methylsilsesquioxane), poly(ethoxysilsesquioxane), poly(methyl methacrylate) (PMMA), poly(vinyl alcohol) (PVA), poly(4-vinylphenol), poly(styrene-*co*-allyl alcohol), or poly(bisphenol A-*co*-epichlorohydrin) to form a hybrid film. The mechanical strengths and strains of PDMS hybrids were very low. The tensile strengths and elongations of PMMA hybrids increased with the increase in the titanium cluster concentration. The tensile strengths and elongations of PVA hybrids were highest when the titanium cluster concentration was 10 wt%.

Keywords Element-block · Titanium phosphonate cluster · Organic-inorganic hybrid · Poly(methyl methacrylate) · Poly(vinyl alcohol)

12.1 Introduction

Organic-inorganic hybrids containing element-blocks, a structural unit consisting of various groups of elements, are promising materials by virtue of their strong performance characteristics, such as high mechanical strength, thermal stability, gas permeability, light emission, and electron conductivity [1]. These characteristics

T. Gunji (✉)
Department of Pure and Applied Chemistry, Faculty of Science and Technology, Tokyo University of Science, Noda, Chiba, Japan
e-mail: gunji@rs.noda.tus.ac.jp

S. Tsukada
Advanced Automotive Research Collaborative Laboratory, Graduate School of Engineering, Hiroshima University, Higashi-Hiroshima City, Hiroshima, Japan

Y. Chujo (ed.), *New Polymeric Materials Based on Element-Blocks*,
https://doi.org/10.1007/978-981-13-2889-3_12

appeared by the hybridization of organic polymer with inorganic material. Indeed, organic-inorganic hybrid materials containing polyhedral octasilsesquioxane (POSS) are reported to show strong gas separation [2] and refractive index [3] qualities.

Metal oxide clusters are unique compounds composed of metal-oxygen bonds as a main-chain bearing side chain, such as organic groups or hydrides, halogens. Polysilsesquioxane cages [4] are famous compounds having silicon and oxygen atoms in a main framework. Oxo-titanium clusters [5] composed of titanium and oxygen atoms are also examples of well-studied metal oxide clusters. Recently, organic-inorganic hybrid polymers prepared by using these clusters were found to improve both thermostability and wear resistivity when compared to the corresponding organic polymers [6–9]. Organic-inorganic hybrid materials are divided into two classes on the basis of bonding type. Class I: the polymers are mixed by weak interactions such as van der Waals force and hydrogen bonding. Class II: the polymers are mixed by strong interactions such as covalent bonding or ionic bonding between organic and inorganic components [10]. For example, Chujo reported that the hybrid materials of polysulfide-bridged POSS with poly(methyl methacrylate) or polystyrene (class I) showed high thermal stability and high refractive indices compared to the calculated values [3]. Schubert reported that poly(methyl methacrylate) cross-linked with $Ti_4O_2(O^iPr)_6(OMc)_6$ (OMc = methacrylate) (class II) has high thermal stability [6]. Therefore, organic-inorganic hybrid materials are very interesting because we can expect the improvement of their physical properties to open new application in materials science.

Clusters containing three components are being reported these days. Titanium phosphonate clusters formed by titanium, oxygen, and phosphorus have the potential for high chemical and thermostable properties bearing Ti-O-P bonds. The interactions among organophosphonate groups are easily controlled by changing the organic groups on titanium or phosphorus atoms. Organic-inorganic hybrid materials having titanium phosphonate clusters would be prepared by an alcohol exchange reaction or the sol-gel process [11] due to the alkoxy group on titanium. The synthesis of titanium phosphonate clusters has been reported [12–17], however, that of organic-inorganic hybrid materials containing these clusters has not been investigated. One of the problems for preparing hybrid materials using previous clusters is dimethyl sulfoxide as coordinating solvent. Dimethyl sulfoxide has strong coordinating properties compared with tetrahydrofuran (THF) that is one of the common coordinating solvents.

12.2 Experimental

12.2.1 Measurements

Nuclear magnetic resonance (NMR) spectra were recorded using a JEOL Resonance JNM-ECP 500 spectrometer (^{1}H at 500.16 MHz, ^{13}C at 125.77 MHz, ^{31}P at

202.46 MHz) at 24 °C. The chemical shifts were reported in ppm with reference to chloroform-*d* ($CDCl_3$) as an internal standard (for ^1H: 7.24 ppm in residual chloroform, ^{13}C: 77.00 ppm) and phosphoric acid (H_3PO_4) as an external standard (for ^{31}P: 0.00 ppm). Infrared (IR) spectra were recorded on a JASCO FT/IR-6100 FT-IR spectrophotometer using attenuated total reflectance (ATR, ZnSe prism, JASCO ATR PRO 0450-S). Thermogravimetric-differential thermal analysis (TG-DTA) was performed with a NETZSCH JAPAN TG-DTA 2000SE analyzer. The samples were heated to 1000 °C under air flow at a rate of 10 °C/min. Transmittance spectra were recorded on a JASCO V-670 spectrophotometer equipped with an integrating-sphere photometer (JASCO ISN-470 type) in the range of 200–800 nm. Refractive indices were determined using an Otsuka Electronics FE-3000 refractive film thickness monitor. Gel permeation chromatography (GPC) was performed using a high-performance liquid chromatography system (LC-6AD, Shimadzu, Kyoto, Japan) attached to a Polymer Laboratory gel 5 μm Mixed-D column. Tetrahydrofuran was used as the eluent (1 mL min^{-1}). RID-10A was used as the detector. The molecular weight was calculated based on polystyrene standards. Transmittance spectra were recorded using a JASCO V-670 spectrophotometer equipped with an integrating-sphere photometer (JASCO ISN-470 type) in the 200–800 nm wavelength range. The tensile strength was recorded using a Shimadzu Autograph AG-50 kN Xplus at the rate of 2 mm/min. The size of the test samples is shown below: height 25 mm, width 5 mm, thickness 50–100 nm.

12.2.2 X-Ray Structure Analyses

For structural determination, crystal data were collected using a Bruker AXS SMART APEX CCD X-ray diffractometer equipped with a rotating-anode X-ray generator emitting graphite-monochromatic Mo-Kα radiation ($\lambda = 0.71073$ Å) at 77 K. Empirical absorption corrections using equivalent reflections and Lorentzian polarization correction were performed using the SADABS program [18]. All data were collected with SMART and Bruker SAINTPLUS (Version 6.45) software packages. The structures were solved using the SHELXS-97 program [19] and refined against F^2 using SHELEXL-97 [19]. CCDC 1487366 contains the supplementary crystallographic data for $[Ti_4(\mu_3\text{-}O)(O^iPr)_5(\mu\text{-}O^iPr)_3(PhPO_3)_3]\cdot thf$. These data can be obtained free of charge via http://www.ccdc.cam.ac.uk/conts/retrieving.html, or from the Cambridge Crystallographic Data Center, 12 Union Road, Cambridge CB2 1EZ, UK; fax: (+44) 1223 336 033; or e-mail: deposit@ccdc.cam.ac.uk.

12.2.3 Materials

The structure and abbreviation of compounds are summarized in Scheme 12.1. All solvents were purified by a standard process [20] and stored over activated molecular sieves. Titanium tetraisopropoxide ($Ti(O^iPr)_4$) and PVA (degree of polymerization, 500) and tetraethylammonium hydroxide (Et_4NOH 20% aqueous solution) were purchased from Wako Pure Chemical Industries. Et_4NOH was concentrated by removal of the water under vacuum. Phenylphosphonic acid ($PhPO_3H_2$) was purchased from Tokyo Chemical Industry. Poly(vinyl alcohol) (PVP) ($M_w = 25{,}000$ g/mol), poly(styrene-*co*-allyl alcohol) (PSA) ($M_w = 2200$ g/mol, allyl alcohol 40 mol %), poly(bisphenol A-*co*-epichlorohydrin) (PBE) ($M_w = 40{,}000$ g/mol), and poly (methyl methacrylate) (PMMA) ($M_w = 997{,}000$ g/mol) were purchased from Sigma-Aldrich (Tokyo, Japan) and used as received. Trimethoxy(methyl)silane (MTMS), tetraethoxysilane (TEOS), and chloro(trimethyl)silane (TMSCl) were purchased from Tokyo Chemical Industry, Tokyo, Japan, and purified by distillation before use. Octamethylcyclotetrasiloxane was purchased from Shin-Etsu Chemical Co., Tokyo, Japan and purified by distillation before use.

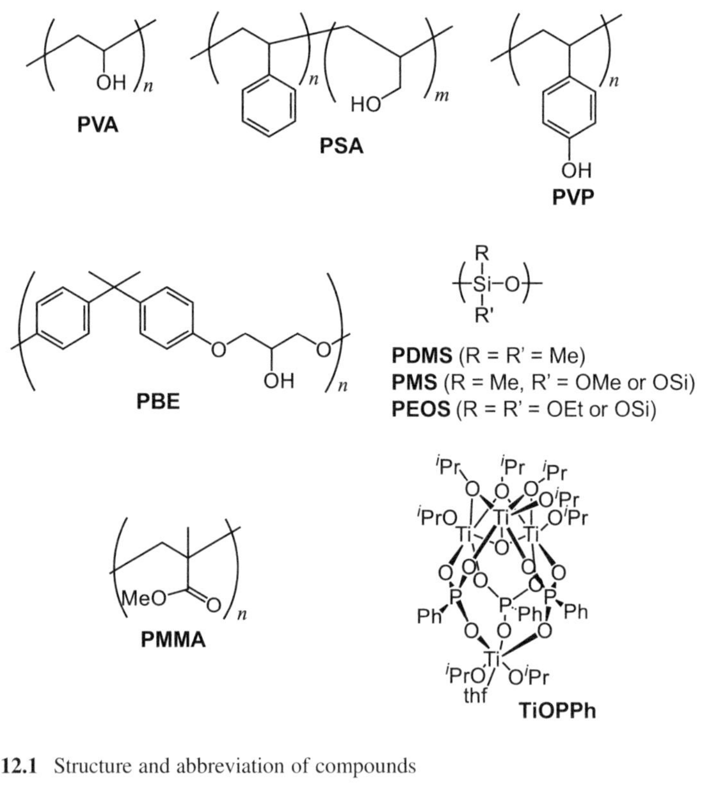

Scheme 12.1 Structure and abbreviation of compounds

12.2.4 Synthesis of $[Ti_4(\mu_3\text{-}O)(O^iPr)_5(\mu\text{-}O^iPr)_3(PhPO_3)_3]\cdot thf$ (TiOPPh)

The reaction was carried out in an argon atmosphere. $Ti(O^iPr)_4$ (36 mL, 0.12 mol) was added to $PhPO_3H_2$ (9.56 g, 60.5 mmol) and H_2O (0.36 mL, 20 mmol) in 48 mL THF at room temperature, after which white precipitation appeared quickly. After several hours of stirring, the mixture became a colorless solution, whereupon stirring was stopped. Colorless black crystals from the solution appeared after several weeks and were then filtrated and dried under reduced pressure. White block crystals were obtained in a yield of 8.25 g (34%).

^1H NMR (500 MHz, $CDCl_3$/7.24 ppm): $\delta = 1.06$ (d, $J = 5.5$ Hz, 12H), 1.37 (d, $J = 6.0$ Hz, 18H), 1.44 (d, $J = 6.0$ Hz, 18H), 1.75–1.81 (m, 6H), 3.79–3.82 (m, 6H), 4.69 (sept, $J = 6.0$ Hz, 3H), 4.94 (br-s, 2H), 5.03 (sept, $J = 6.0$ Hz, 3H), 7.30–7.38 (m, 9H), 7.88 (dd, $J = 7.5$ Hz, $^3J_{P\text{-}H} = 12.0$ Hz, 6H).

$^{13}C\{^1H\}$ NMR (126 MHz, $CDCl_3$/77.0 ppm): $\delta = 24.51$, 24.83, 25.36, 68.84, 78.45, 79.14, 79.40, 127.31 (d, $^2J_{P\text{-}C} = 14.4$ Hz), 129.72 (d, $^4J_{P\text{-}C} = 2.9$ Hz), 131.05 (d, $^3J_{P\text{-}C} = 9.6$ Hz), 134.44 (d, $^1J_{P\text{-}C} = 203.9$ Hz) $^{31}P\{^1H\}$ NMR (202 MHz, $CDCl_3$/ppm): $\delta = 9.22$. Ceramic yield: 40.4% (calcd. for $Ti_4P_3O_{16}$ 41.7%).

12.2.5 Preparation of Poly(dimethylsiloxane) (PDMS)

Octamethylcyclotetrasiloxane (30.0 g, 0.10 mol) and Et_4NOH (0.60 g, 4 mmol) were placed into a 100 mL four-necked flask equipped with an Allihn condenser and a mechanical stirrer. The mixture was stirred at 100 °C for 3 h at 200 rpm. The viscous liquid was extracted with dichloromethane. The organic layer was washed with brine and dried with anhydrous magnesium sulfate. After filtration, the solution was concentrated to remove vapor compounds under vacuum at 60 °C. PDMS was obtained as a colorless viscous liquid (27.2 g, 91%).

^1H NMR (300 MHz, $CDCl_3$/ppm) δ 0.08 (brs, Si-CH_3). ^{29}Si NMR (60 MHz, $CDCl_3$/ppm) δ −21.55. $M_w = 93{,}300$ g/mol, $M_w/M_n = 1.9$.

12.2.6 Preparation of Poly(methylsilsesquioxane) (PMS)

MTMS (27.2 g, 0.20 mol) and MeOH (13.3 g, 0.42 mol) were placed into a 200 mL four-necked flask equipped with nitrogen inlet and outlet tubes and a mechanical stirrer. The mixture was then cooled in an ice bath for 10 min. Water and hydrochloric acid (molar ratios; H_2O/MTMS = 1.0, HCl/MTMS = 0.105) were added. The mixture was stirred in an ice bath for 10 min and then at room temperature for 10 min, followed by heating at 70 °C for 3 h at 150 rpm with a 360 mL/min nitrogen flow. PMS was obtained as colorless liquid (16.3 g).

^{1}H NMR (300 MHz, $CDCl_3$/ppm) δ 0.12 (brs, Si-CH_3), 3.51 (brs, Si-OCH_3). ^{29}Si NMR (60 MHz, $CDCl_3$/ppm) δ −49.35 (T^1, content ratio of 6%), −58.87 (T^2, that of 50%), −67.17 (T^3, that of 44%). M_w = 3800 g/mol, M_w/M_n = 1.8.

12.2.7 Preparation of Poly(ethoxysilsesquioxane) (PEOS)

TEOS (34.8 g, 0.17 mol) and EtOH (15.9 g, 0.35 mol) were placed into a 300 mL four-necked flask equipped with nitrogen inlet and outlet tubes and a mechanical stirrer. The mixture was then cooled in an ice bath for 10 min. Water and hydrochloric acid (molar ratios; H_2O/TEOS = 1.7, HCl/TEOS = 0.105) were added. The mixture was stirred in an ice bath for 10 min and then at room temperature for 10 min, followed by heating at 80 °C for 2 h at 150 rpm with a 360 mL/min nitrogen flow. The resulting product was dissolved into 40 mL THF, then 2 mL TMSCl was added, followed by stirring for 1 day. The mixture was concentrated to remove vapor compounds, and PEOS was obtained (15.7 g).

^{1}H NMR (300 MHz, $CDCl_3$/ppm) δ 0.16 (brs, Si-CH_3), 1.25 (brs, Si-OCH_2CH_3), 3.89 (brs, Si-O$CH_2$$CH_3$). ^{29}Si NMR (60 MHz, $CDCl_3$/ppm) δ 12.76 (M, content ratio of 8%), −95.86 (Q^2, that of 12%), −102.78 (Q^3, that of 80%). M_w = 7200 g/mol, M_w/M_n = 2.7.

12.2.8 Preparation of Freestanding Hybrid Films

Polymer solution of THF or toluene was added to TiOPPh and stirred for 3 h at room temperature. The mixture was poured into a 50 mmϕ Teflon petri dish followed by curing at 50 °C for 1 day and then at 120 °C for 1 day.

12.2.9 Preparation of Hybrid Thin Films

A solution of 0.12 g of PVA in 5 mL of deionized water was added to TiOPPh in 5 mL of THF, and the mixture was stirred at room temperature for overnight. Hybrid thin films were prepared by spin-coating of the solution on silicon wafer at 1000 rpm for 20 s and then heating at 120 °C for 2 min in air.

12.3 Results and Discussion

12.3.1 Characterization of TiOPPh

TiOPPh was synthesized by the reaction of $Ti(O^iPr)_4$ with $PhPO_3H_2$ and water in THF. The formation of this cluster was determined by various NMR spectra, FT-IR spectra, and single-crystal X-ray structural analysis as shown in Fig. 12.1 and Table 12.1. From the single-crystal X-ray structural diffraction, TiOPPh was proved to be built up from four titanium atoms and three RPO_3, five terminal O^iPr, three bridged O^iPr, one μ_3-O, and one THF molecules. The main framework of TiOPPh consisted of three bridged RPO_3 between one Ti atom coordinated by THF and one Ti_3O containing a μ_3-O atom. Three phosphonates are arranged in one direction, one on top of the other, and are capped by a THF-coordinated Ti atom. This structure is very similar to that of $[Ti_4(\mu_3\text{-}O)(O^iPr)_5(\mu\text{-}O^iPr)_3(PhPO_3)_3]\cdot$dmso [10]. In the ^1H NMR spectrum, three doublet signals attributable to CH_3 in the isopropoxy group were found at 1.06, 1.37, and 1.44 ppm with respective integration ratios of 12H, 18H, and 18H. These signals are assigned to the terminal isopropoxy group on the Ti coordinated by THF, the terminal isopropoxy group on the Ti_3O unit, and the bridged isopropoxy group, respectively. CH in the isopropoxy group show signals

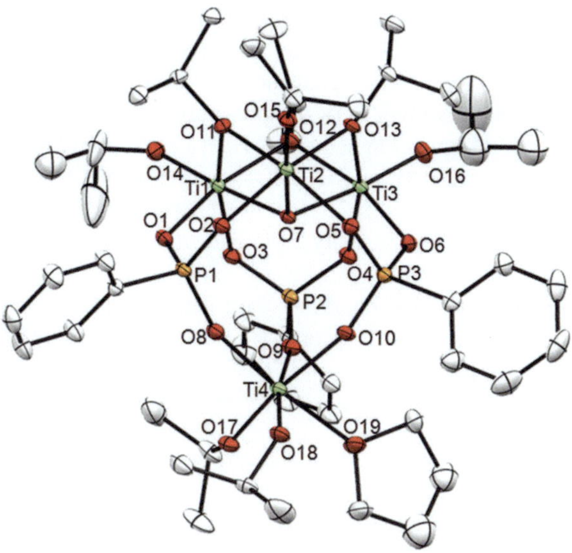

Fig. 12.1 ORTEP drawing of TiOPPh with thermal ellipsoids at the 50% probability level. Hydrogen atoms are omitted for clarity. Selected interatomic distances (Å) and bond angles (°) are as follows: Ti(1)–O(1) 1.972(6), Ti(1)–O(7) 1.962(4), Ti(1)–O(11) 2.010(5), Ti(1)–O(14) 1.779 (5), Ti(4)–O(8) 1.924(6), Ti(4)–O(17) 1.820(7), Ti(4)–O(19) 2.162(6), P(1)–O(1) 1.536(4); O(1)–Ti(1)-O(3) 90.7(2), O(1)–Ti(1)-O(11) 87.5(2), O(7)-Ti(1)-O(14) 175.4(2), O(8)-Ti(4)-O(9) 94.5(2), O(8)-Ti(4)-O(19) 174.3(3), O(1)-P(1)-O(2) 111.4(3), O(1)-P(1)-O(8) 110.9(3)

Table 12.1 Crystal date for TiOPPh

Empirical formula	$C_{46}H_{79}O_{19}P_3Ti_4$
Formula weight	1220.6
Crystal system	Triclinic
Space group	P-1
a (Å)	12.3418 (11)
b (Å)	13.7751 (12)
c (Å)	20.2975 (17)
α (°)	90.957 (1)
β (°)	105.730 (1)
γ (°)	114.034 (1)
Z	2
V (Å^3)	3001.7 (5)
$D_{calc.}$ (g/cm^3)	1.35
$R1$ ($I > 2\sigma(I)$)	0.0707
$wR2$	0.2199
S	1.062

at 4.69, 4.94, and 5.03 ppm. The signals derived from the phenyl group shown at 7.30–7.38 ppm as a multiplet are assigned to *m*- and *p*-CH in the phenyl group with overlapping, and a signal at 7.88 ppm in the double doublet assigned to *o*-CH in the phenyl group was observed. ^{13}C NMR spectrum shows signals due to CH_3 at 24.51 and 24.83 ppm, signals due to THF at 25.36 and 68.84 ppm, and signals due to CH at 78.45, 79.14, and 79.40 ppm. The signals derived from the phenyl group appear in doublet because of the coupling between phosphorus and carbon atoms. Signals were observed at 127.31 ppm assigned to *o*-CH, at 129.72 ppm assigned to *p*-CH, at 131.05 ppm assigned to *m*-CH, and at 134.44 ppm assigned to *ipso*-CH. The ^{31}P NMR spectrum shows signals at 9.22 ppm in $CDCl_3$; in C_6D_6, it shows signals at 6.82 ppm with an integrated intensity of 2 and at 7.16 ppm with an integrated intensity of 1.

The FT-IR spectrum of TiOPPh shows the following absorption bands: $\nu_{C\text{-}H}$ in phenyl at 3056 cm^{-1}, $\nu_{C\text{-}H}$ in the isopropoxy group at 2970 and 2928 cm^{-1}, $\nu_{C\text{-}H}$ in THF at 2862 cm^{-1}, $\nu_{C=C}$ in phenyl at 1439 cm^{-1}, $\delta_{C\text{-}H}$ at 1375 and 1362 cm^{-1}, $\nu_{C\text{-}C}$ at 1156 and 1132 cm^{-1}, $\nu_{C\text{-}O}$ and $\nu_{P\text{-}O}$ at 989–949 cm^{-1} overlapping with $\delta_{C=C}$ at 754 and 696 cm^{-1} assigned to out-of-plane of phenyl, and $\nu_{Ti\text{-}O}$ at 614–558 cm^{-1}. The TG-DTA of TiOPPh shows weight losses at 80–250 °C (41%) assigned to the decomposition of the isopropoxy group, THF, and carbons in phenyl, and at 450 °C (8%) and at 800 °C (10%).

12.3.2 Preparation of Freestanding Hybrid Films

PMMA hybrid films were prepared from toluene solution, and PVA hybrid films were prepared from DMSO and THF solution. A photograph of the hybrid films is

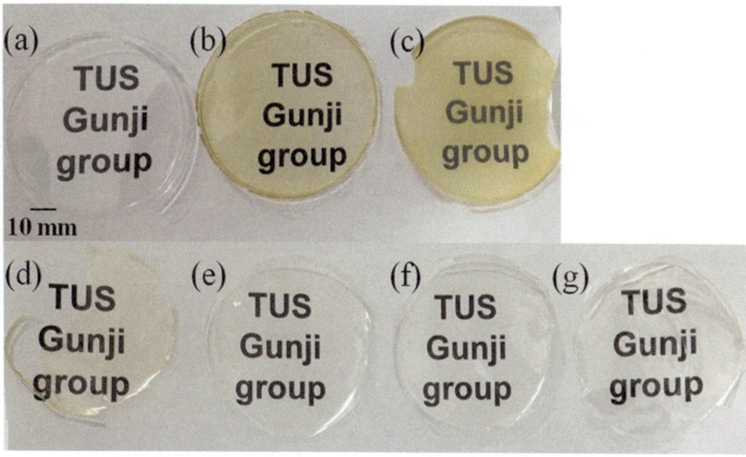

Fig. 12.2 Photograph of hybrid films: (**a**) only PMMA, (**b**) PMMA-10 wt% TiOPPh, (**c**) PMMA-20 wt% TiOPPh, (**d**) only PVA, (**e**) PVA-10 wt% TiOPPh, (**f**) PVA-20 wt% TiOPPh and (**g**) PVA-40 wt% TiOPPh

shown in Fig. 12.2 The PMMA films changed from colorless to yellow as the concentration of TiOPPh increased. On the other hand, colorless and transparent films were prepared by using PVA. We note that we have tried to prepare hybrid films containing 30 wt% of PMMA and 50 wt% of PVA. Unfortunately, these films were not prepared because they were too rigid to retain the form of freestanding films.

FT-IR spectra are shown in Fig. 12.3. All PMMA hybrid films show $\nu_{C\text{-}H}$ at 3000–2950 cm^{-1}, $\nu_{C\text{-}H}$ at 1723 cm^{-1}, and $\nu_{C\text{-}O\text{-}C}$ at 1190 and 1144 cm^{-1}. PMMA-TiOPPh hybrid films show new bands due to $\nu_{P=O}$ at 1040 cm^{-1} and $\nu_{Ti\text{-}O}$ at around 550 cm^{-1}. However, the band of $\nu_{P\text{-}O}$ could be observed only for the 20 wt% TiOPPh. These bands derived from PMMA were very similar, especially $\nu_{C=O}$ at 1723 cm^{-1}. These results indicated that the carbonyl groups of PMMA did not coordinate to Ti atom of TiOPPh via an exchange reaction between THF and the carbonyl groups. However, $\nu_{C\text{-}O\text{-}C}$ shifted from 1144 cm^{-1} (PMMA) to 1142 cm^{-1} (PMMA-TiOPPh hybrid). Therefore, PMMA and TiOPPh can be blended by using weak interaction between methoxy group of PMMA and TiOPPh. On the other hand, all PVA hybrid films showed $\nu_{O\text{-}H}$ at 3300 cm^{-1}, $\nu_{C\text{-}H}$ at 2911 cm^{-1}, and $\nu_{C\text{-}C}$ at 1421 cm^{-1}. The signal intensity due to $\nu_{P\text{-}O}$ at 1000 cm^{-1} was increased, and that due to $\nu_{O\text{-}H}$ decreased when TiOPPh was increased. From these results, an alcohol-exchange reaction between the hydroxy group in PVA and the isopropoxy group in TiOPPh.

Transmittance spectra are shown in Fig. 12.4 and the data are summarized in Table 12.2. The transparency of PMMA hybrid films significantly decreased in the visible region as the TiOPPh concentration increased. Moreover, these TiOPPh-PMMA hybrid films were yellow. On the other hand, the transparency of the PVA hybrid films was high compared with that of the PVA films. The origin of the yellow

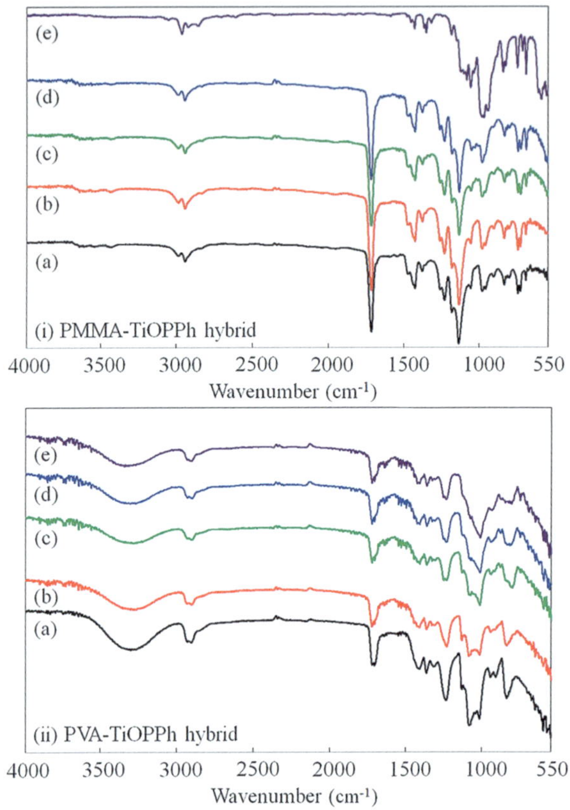

Fig. 12.3 FT-IR spectra of (i) PMMA-TiOPPh hybrid films (a) only PMMA, (b) PMMA-2.5 wt% TiOPPh, (**c**) PMMA-10 wt % TiOPPh, (d) PMMA-20 wt% TiOPPh and (e) only TiOPPh and (ii) PVA-TiOPPh hybrid films (a) only PVA, (b) PVA-2.5 wt% TiOPPh, (**c**) PVA-10 wt% TiOPPh, (d) PVA-20 wt% TiOPPh and (e) PVA-40 wt% TiOPPh

color in PMMA films is the aggregation and/or polymerization of TiOPPh caused by water in air. In contrast, TiOPPh in PVA seems to be highly dispersed. Therefore, PVA hybrid films were obtained as colorless film. These ideas are also supported by the results of IR spectroscopy.

TG-DTA thermograms of hybrid films are shown in Fig. 12.5, and the data are summarized in Table 12.2. The temperatures of 10% weight loss (T_{d10}) were 279 °C and 278 °C for PMMA and PVA, respectively. The values of T_{d10} for PMMA-TiOPPh hybrid films were about 310 °C and increased 30 °C compared to that of PMMA. TiOPPh improves the thermal stability of PMMA such that the weight loss is 40 wt% at 310 °C. The improvement of thermal stability was as same as other blend or hybrid PMMA materials in previously reported papers which contain polysulfide-bridged POSS [3], zirconia nanocrystals modified with 3-(methacryloxy)propyl-trimethoxysilane [21], and titania modified with 2-hydroxyethyl methacrylate [19]. On the other hand, the T_{d10} of the PVA-TiOPPh hybrid films decreased as the concentration of TiOPPh increased, such that PVA-30 wt% of the TiOPPh hybrid film was decreased at 45 °C compared

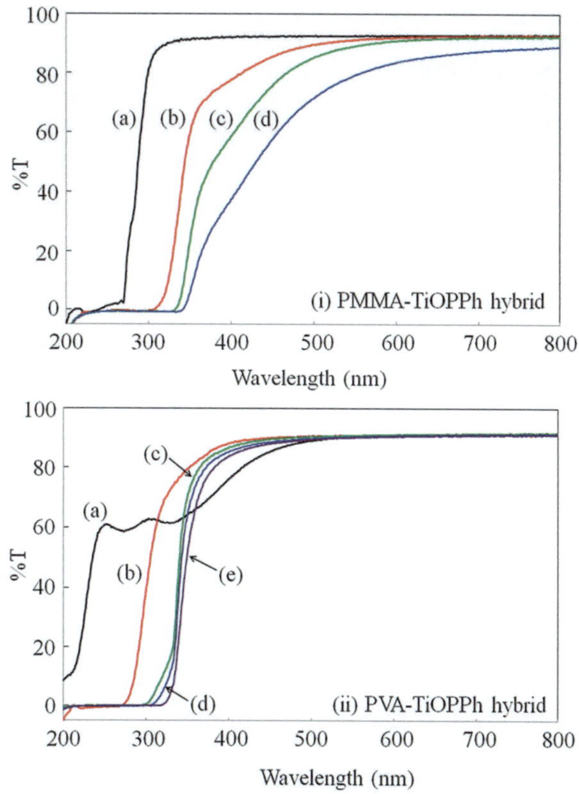

Fig. 12.4 UV-Vis transmission spectra of (i) PMMA-TiOPPh hybrid films (a) only PMMA, (b) PMMA-2.5 wt% TiOPPh, (c) PMMA-10 wt % TiOPPh and (d) PMMA-20 wt% TiOPPh and (ii) PVA-TiOPPh hybrid films (a) only PVA, (b) PVA-2.5 wt% TiOPPh, (c) PVA-10 wt% TiOPPh, (d) PVA-20 wt% TiOPPh and (e) PVA-40 wt% TiOPPh

Table 12.2 Optical and thermal properties of TiOPPh hybrid films

	Transmittance		Thermal analysis	
Sample	T_{550} (%)[a]	T_{420} (%)[b]	T_{d10} (°C)[c]	Ceramic yield (wt%)[d]
Only PMMA	92.3	91.7	279	1.4
PMMA-2.5 wt% TiOPPh	90.9	79.6	312	1.4
PMMA-10 wt% TiOPPh	89.2	66.9	311	5.2
PMMA-20 wt% TiOPPh	78.6	42.9	308	10.3
Only PVA	90.4	81.0	278	0.4
PVA-2.5 wt% TiOPPh	90.9	89.6	270	1.3
PVA-10 wt% TiOPPh	91.0	88.4	264	5.9
PVA-20 wt% TiOPPh	90.6	87.3	259	8.4
PVA-40 wt% TiOPPh	90.3	86.4	249	13.9

[a]Measured by UV-Vis spectroscopy at 550 nm
[b]Measured by UV-Vis spectroscopy at 420 nm
[c]Determined from the decomposition temperature with 10% weight loss
[d]Weight percent of the residue at 1000 °C

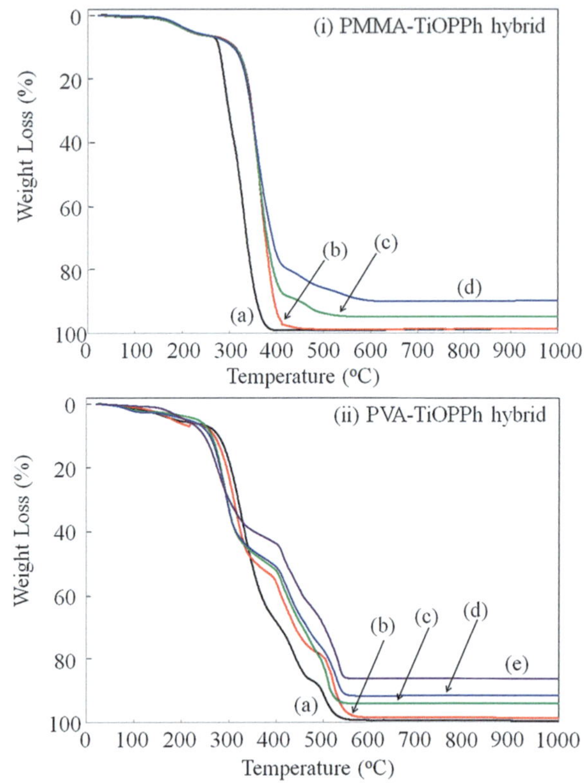

Fig. 12.5 TGA curves of (i) PMMA-TiOPPh hybrid films (a) only PMMA, (b) PMMA-2.5 wt% TiOPPh, (c) PMMA-10 wt % TiOPPh and (d) PMMA-20 wt% TiOPPh and (ii) PVA-TiOPPh hybrid films (a) only PVA, (b) PVA-2.5 wt% TiOPPh, (c) PVA-10 wt% TiOPPh, (d) PVA-20 wt% TiOPPh and (e) PVA-40 wt% TiOPPh

to PVA. The decrease in the value of T_{d10} is attributable to the reaction of the isopropoxy group in TiOPPh with the residual hydroxy group in PVA.

The refractive indices of PVA-hybrid thin films at 633 nm were measured. The refractive index of only PVA was 1.488, and those of PVA-2.5 wt% TiOPPh and PVA-10 wt% TiOPPh were 1.500 and 1.501, respectively. The increase of refractive index supports the formation of hybrids.

12.3.3 Preparation and Properties of Silicone Polymers/TiOPPh Hybrid

The silicone/TiOPPh hybrids were prepared by mixing silicone polymers with TiOPPh in toluene (Table 12.3). Freestanding PDMS/TiOPPh films were prepared in the concentration of 20–40 wt% TiOPPh, and the obtained films were highly flexible. When the concentration was less than 20 wt%, the hybrids were oily, viscous, and adhesive. On the other hand, when the concentration was more than 40 wt%, the hybrids were obtained as glassy solids. In the IR spectra of the hybrids,

Table 12.3 Results of the formation of TiOPPh containing freestanding films

Polymer	Formation of freestanding films[a] at the wt concentration 0	10	20	30	40
PDMS[b]	–	–	++	++	++
PMS[b]	–	++	++	++	++
PEOS[b]	+	+	+	+	+
PMMA[b]	++	++	++		
PVA[c]	++	++	++	++	++
PVP[d]	+	+	+	+	+
PSA[d]	+	+	+	+	+
PBE[d]	++	++	++	++	++

[a]++: film formation of crack-free;
+: film formation with many cracks;
–: no film formation
[b]Polymer (0.25 g) was dissolved in toluene 5 mL and mixed with TiOPPh
[c]PVA (0.13 g) was dissolved in DMSO 10 mL and mixed with TiOPPh
[d]Polymer (0.13 g) was dissolved in THF 5 mL and mixed with TiOPPh

no absorption band assigned to $\nu_{\text{Si-O-Ti}}$ (around 950 cm^{-1}) [22] was observed. The transparency of PDMS hybrid was higher than 89% (at 420 nm) even at a concentration of 40 wt%. This transparency is higher than the PMMA composite (43% at 420 nm, due to the aggregation of TiOPPh in PMMA) [23]. The hybrids using PDMS were, therefore, composed of segregated polymer domains with uniform dispersions of TiOPPh in PDMS. The thermogravimetric traces for PDMS hybrids are shown in Table 12.2. The temperatures at which 10% weight loss occurred (T_{d10}) were 414 °C (PDMS), 281 °C (TiOPPh 20 wt%), and 264 °C (TiOPPh 40 wt%). The decrease of T_{d10} was very similar to the tendency of PDMS/metal oxide composite that was reported to accelerate the oxidation of methyl groups and the depolymerization with formation of cyclosiloxanes [24]. The char yield of PDMS only and TiOPPh only at 1000 °C in air were 28.2% and 40.4%, respectively. Thus, the thermal decomposition of PDMS mainly occurs by the depolymerization because the char yield would be 81.0% if the thermal decomposition of PDMS was proceed to oxidize all of the methyl groups. The char yields of PDMS/20 wt%, 30 wt%, and 40 wt% TiOPPh were 28.9%, 31.2%, and 37.2%, respectively, where we can expect 30.6% (for 20 wt%), 31.9% (for 30 wt%), and 33.1% (for 40 wt%). The depolymerization of PDMS would be more accelerated when the content of TiOPPh is low, and the oxidation of PDMS would be slightly accelerated at high concentration.

The freestanding PMS/TiOPPh films were prepared in the concentration of 10–40 wt% TiOPPh, but the films were brittle and easily broken. The PMS hybrid containing 40 wt% TiOPPh was yellow and 67% transparent at 420 nm. Other films were colorless with 93% transparent at 420 nm. In the IR spectra of 40 wt%, a new small absorption band was observed at 912 cm^{-1}, which was assigned to $\nu_{\text{Si-O-Ti}}$. Unfortunately, this band was so weak that this band was not observed when the concentration was less than 40 wt%. The T_{d10} decreased when the concentration of

TiOPPh was increased as 573 °C for PMS, 557 °C for 20 wt% TiOPPh, and 452 °C for 40 wt% TiOPPh. TiOPPh might accelerate the oxidation of methyl group on PMS as with thermal decomposition of PDMS.

Freestanding PEOS hybrid films were barely formed because the hybrids were rigid and brittle. In the IR spectrum of the hybrid containing 40 wt% TiOPPh, a new absorption band assigned to $\nu_{\text{Si-O-Ti}}$ (at 908 cm^{-1}) was observed. The T_{d10} of the PEOS hybrid decreased slightly compared to that of only PEOS. TiOPPh maybe act as a catalyst of hydrolytic polycondensation of alkoxysilane such as metal acetylacetonate complexes [25].

12.3.4 Preparation and Properties of Organic Polymers/TiOPPh Hybrids

The organic polymers/TiOPPh hybrids were prepared by mixing organic polymers with TiOPPh in THF (Table 12.3). When TiOPPh was added to PVP solution, the solution immediately changed from colorless to red, and the coloration increased with the concentration of TiOPPh. This phenomenon was similar to that observed in the synthesis of $[Ti_2(\mu\text{-}OPh)_2(OPh)_6(HOPh)_2]$ by the reaction of $Ti(O\textit{i}Pr)_4$ with phenol [26]. Thus, the alcohol exchange reaction is expected to proceed between PVP and TiOPPh. These freestanding films were rigid. The coating films were prepared on a silicon wafer by spin-coating, which were colored in orange. The intensity of the hydroxyl group at 3200–3500 cm^{-1} in the FT-IR spectra decreased with the increase in TiOPPh. The T_{d10} were 242 °C (PVP) and 406 °C (50 wt% concentration), showing that TiOPPh acts as a good cross-linker to PVP matrix.

Freestanding PSA hybrid films were barely formed because of the low molecular weight (2200 g/mol) of PSA. In the IR spectra, the intensity of the hydroxyl group decreased as the concentration of TiOPPh increased. The T_{d10} is summarized in Table 12.4 as 332 °C (PSA), 241 °C (PSA 20 wt%), and 279 °C (PSA 40 wt%).

Freestanding PBE hybrid films were prepared at concentrations of less than 50 wt % TiOPPh, and the obtained films were orange. In the IR spectra of PBE hybrids, the intensity of ν_{OH} decreased with the increase in TiOPPh content. Thus, TiOPPh was reacted with hydroxyl groups of PBE. Also, the top of the absorption bands assigned to ether groups ($\nu_{\text{C-O-C}}$) shifted from 1036 cm^{-1} (PBE) to 1032 cm^{-1} (PEB hybrids). In the case of Zn crown ether-type complex coordinated from the oxygen atoms to zinc atom, the band of $\nu_{\text{C-O-C}}$ was shifted to decrease 6 cm^{-1} less than only crown ether ligand [12]. Therefore, the PBE hybrids suggest the formation of chelate by the coordination of the oxygen atom to titanium. The transmittance of freestanding films is summarized in Table 12.4 as 89% (PBE), 55% (10 wt% TiOPPh), and 40% (50 wt % TiOPPh) at 420 nm. The T_{d10} values were 397 °C (PBE), 350 °C (20 wt% TiOPPh), and 341 °C (50 wt% TiOPPh). The thermal stabilities of the PSA and PBE hybrids were lower than those of PSA and PBE polymers and PVA hybrid as reported before [23].

Table 12.4 Temperature for 10% weight loss occurred and transmittance at 420 nm

Polymer	Content of TiOPPh (wt%)	Temperature of 10% weight loss (T_{d10}) (°C)[a]	Transmittance (%) at 420 nm[b]
PDMS	0	414	–
	20	281	92
	40	264	90
PMS	0	573	–
	20	557	93
	40	452	67
PEOS	0	235	–
	10	234	–
	20	217	–
PVP	0	242	–
	20	259	–
	50	406	–
PSA	0	332	–
	20	241	–
	40	279	–
PBE	0	397	89
	20	350	44
	50	341	40

[a]Measured by DTA-TG analysis
[b]Measured by UV-Vis spectrometry

12.3.5 Tensile Strengths of Freestanding Films

The measured tensile strengths of freestanding films of PDMS and PVA hybrids and PMMA composites are summarized in Table 12.5. The freestanding PDMS hybrid films show very low tensile strengths: 0.6 MPa (PDMS 20 wt% TiOPPh) and 0.2 MPa (30 wt% TiOPPh). The tensile strengths of PDMS-TEOS hybrid materials containing a titanium cross-linker were reported to be increased by the increase in the titanium content [27]. Moreover, the tensile strength of PDMS materials is known to depend strongly on the cross-linked network [28]. Therefore, the molecular interaction between PDMS and TiOPPh would be weak.

The tensile strengths of PMMA composite films are shown in Fig. 12.3 (i). Stress and strain increased with the increase in the concentration of TiOPPh for 0, 2.5, and 10 wt%. The tensile strength was increased about threefold from only PMMA film by mixing TiOPPh in 10 wt %. This improvement is higher than that in PMMA–montmorillonite composite material, which showed an improvement of ca. 1.1-fold [29]. The strain was also increased with the increase in the TiOPPh concentration, suggesting mixing PMMA with TiOPPh will yield hard and brittle composites.

The tensile strengths of PVA hybrid films are shown in Fig. 12.6 (ii). The strain increased with the similar manner with PMMA hybrids to 10 wt% cluster contain. On the other hand, the strain decreased when the content was increased from 20 to

Table 12.5 Tensile strength, strain, and Young's moduli of freestanding films

Polymer	Content of TiOPPh (wt %)	Tensile strength (MPa)	Strain (%)	Young's modulus (MPa)
PDMS	20	0.8	5.9	23
	30	0.3	2.3	19
PMMA	0	18.3	2.9	391
	2.5	44.1	3.7	693
	10	56.5	6.1	614
PVA	0	15.1	36.5	185
	2.5	18.6	55.3	338
	10	28	84.8	476
	20	19.2	60.2	227
	30	17.9	31.4	271

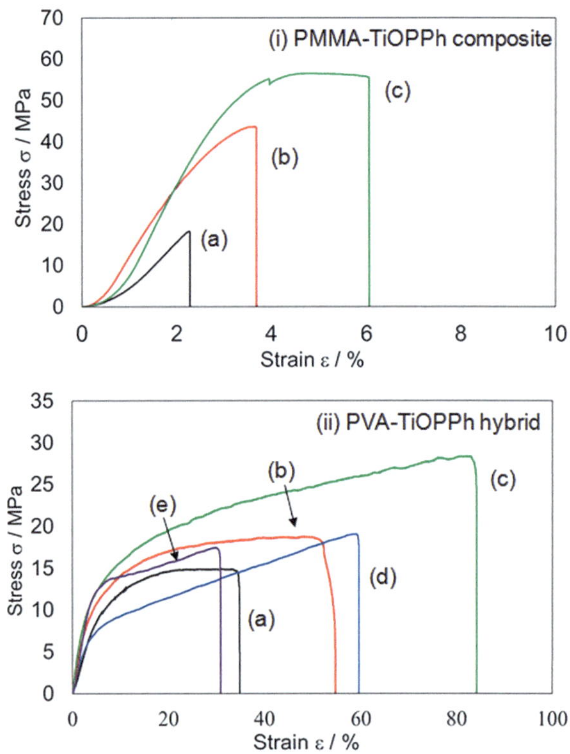

Fig. 12.6 Stress-strain curves of (i) PMMA (PMMA (a), TiOPPh 2.5 wt % (b), and TiOPPh 10 wt% (c), and (ii) PVA (PVA (a), TiOPPh 2.5 wt% (b), TiOPPh 10 wt% (c), TiOPPh 20 wt% (d), and TiOPPh 30 wt% (e)

30 wt %. The Young's modulus was evaluated for the stress and strain values of the initial stage. The value was increased by the TiOPPh content that suggests the formation of brittle and hard hybrids due to the cross-linking by TiOPPh. Maximum strain and stress were observed when the TiOPPh content was 10 wt % because TiOPPh acts as a good cross-linker to form hard and brittle hybrids.

12.4 Conclusion

A titanium phosphonate cluster with a formula of $[Ti_4(\mu_3\text{-}O)(O^iPr)_5(\mu\text{-}O^iPr)_3(PhPO_3)_3]\cdot thf$ was synthesized by the reaction of titanium tetraisopropoxide and phenylphosphonic acid in water and tetrahydrofuran. Titanium phosphonate was mixed with PMMA to form a yellow hybrid film at a concentration of less than 30 wt%. The degradation temperature increased about 30 °C when the titanium phosphonate cluster was hybridized with PMMA. On the other hand, titanium phosphonate was mixed with PVA to form a hybrid film. The resulting content was less than 50 wt% and colorless. Isopropyl alcohol was detected after the formation of the hybrid films, which confirms the reaction of PVA with the isopropoxy group in titanium phosphonate cluster. The application of titanium phosphonate clusters for preparing hybrid materials is expected to be useful for developing new organic-inorganic hybrid materials including various reactive positions in a molecule.

Organic-inorganic hybrids containing TiOPPh as a new element-block were prepared by using silicon polymers such as PDMS, PMS, and PEOS and organic polymers such as PMMA, PVA, PVP, PSA, and PBE. The concentration of TiOPPh was increased to 40 wt% to form a transparent film for PDMS, PMS, PVA, and PBE, whereas the concentration was 20 wt% to form such a film for PMMA. The temperature at which the 10 wt% weight loss occurred was increased for the hybrid using PVP because the alcohol exchange reaction between TiOPPh and PVP was expected to form a rigid network, whereas the 10 wt% weight loss temperature decreased with the increase in TiOPPh concentration for PSA, PBE, PDMS, PMS, and PEOS. The PDMS hybrids showed very low tensile strengths and elongations. The tensile strengths and elongations of PMMA hybrids increased with the increase in the TiOPPh concentration. The tensile strengths and elongations of PVA hybrid were highest when the concentration of TiOPPh was 10 wt%. As a result, TiOPPh was found to be a good cross-linker to form hard and brittle hybrids.

Acknowledgments This work was supported by a Grant-in-Aid for Scientific Research on Innovative Areas "New Polymeric Materials Based on Element-Blocks (No.2401)" (JSPS KAKENHI Grant Number JP24102008). This work was supported by JSPS KAKENHI Grant Number JP16K17951.

Shinji Ogihara, Ryuta Kitamura, and Ryosuke Matsuzaki are greatly acknowledged for their technical assistance in the tensile strength measurement.

References

1. Chujo Y, Tanaka K (2015) New polymer materials based on element–blocks. Bull Chem Soc Jpn 88:633–643
2. Kanezashi M, Shioda T, Gunji T, Tsuru T (2012) Gas permeation properties of silica membranes with uniform pore sizes derived from polyhedral oligomeric silsesquioxane. AICHE J 58:1733–1743

3. Tanaka K, Yamane H, Mitamura K, Watase S, Matsukawa K, Chujo Y (2014) Transformation of sulfur to organic–inorganic hybrids employed by networks and their application for the modulation of refractive indices. J Polym Sci Part A Polym Chem 52:2588–2595
4. Cordes DB, Lickiss PD, Rataboul F (2010) Recent developments in the chemistry of cubic polyhedral oligosilsesquioxanes. Chem Rev 110:2081–2173
5. Bradley DC, Mehrotra RC, Rothwell IP, Singh A (2001) Alkoxo and aryloxo derivatives of metals. Academic, San Diego
6. Schubert U (2004) Organofunctional metal oxide clusters as building blocks for inorganic-organic hybrid materials. J Sol-Gel Sci Techn 31:19–24
7. Rozes L, Steunou N, Fornasieri G, Sanchez C (2006) Titanium-oxo clusters, versatile nanobuilding blocks for the design of advanced hybrid materials. Monatsh Chem 137:501–528
8. Gross S (2011) Oxocluster-reinforced organic–inorganic hybrid materials: effect of transition metal oxoclusters on structural and functional properties. J Mater Chem 21:15853–15861
9. Schubert U (2011) Cluster-based inorganic–organic hybrid materials. Chem Soc Rev 40:575–582
10. Kickelbick G (2006) Hybrid materials. Wiley VCH, Weinheim
11. Guerrero G, Mutin PH, Vioux A (2000) Mixed nonhydrolytic/hydrolytic sol−gel routes to novel metal oxide/phosphonate hybrids. Chem Mater 12:1268–1272
12. Guerrero G, Mehring M, Mutin PH, Dahan F, Vioux A (1999) Syntheses and single-crystal structures of novel soluble phosphonate- and phosphinato-bridged titanium oxo alkoxides. J Chem Soc Dalton Trans:1537–1538
13. Mehring M, Guerrero G, Dahan F, Mutin PH, Vioux A (2000) Syntheses, characterizations, and single-crystal x-ray structures of soluble titanium alkoxide phosphonates. Inorg Chem 39:3325–3332
14. Czakler M, Artner C, Schubert U (2013) Influence of the phosphonate ligand on the structure of phosphonate-substituted titanium oxo clusters. Eur J Inorg Chem:5790–5796
15. Czakler M, Artner C, Schubert U (2014) Acetic acid mediated synthesis of phosphonate-substituted titanium Oxo clusters. Eur J Inorg Chem 2014:2038–2045
16. Czakler M, Artner C, Schubert U (2015) Titanium oxo/alkoxo clusters with both phosphonate and methacrylate ligands. Monatsh Chem 146:1249–1256
17. Kalita L, Kalita AC, Murugavel R (2014) Organotitanium phosphates with free P–OH groups: synthesis, spectroscopy and solid state structures. J Organomet Chem 751:555–562
18. Sheldrick GM (1996) SADABS, program for Siemens area detector absorption correction. University of Göttingen, Germany
19. Sheldrick GM (1997) SHELXS-97, program for crystal structure solution. University of Göttingen, Germany
20. Armarego WLF, Chai C (2012) Purification of laboratory chemicals, 7th edn. Elsevier, Oxford
21. Otsuka T, Chujo Y (2010) Poly(methyl methacrylate) (PMMA)-based hybrid materials with reactive zirconium oxide nanocrystals. Polym J 42:58–65
22. Julián B, Gervais C, Cordoncillo E, Escribano P, Babonneau F, Sanchez C (2003) Synthesis and characterization of transparent PDMS-metal-oxo based organic–inorganic nanocomposites. Chem Mater 15:3026–3034
23. Hayami R, Wada K, Sagawa T, Tsukada S, Watase S, Gunji T (2017) Preparation and properties of organic-inorganic hybrid polymer films using $[Ti_4(\mu_3\text{-}O)(O\textit{i}Pr)_5(\mu\text{-}O\textit{i}Pr)_3(O_3PPh)_3]\cdot thf$. Polym J 49:223–228
24. Guńko VM, Borysenko MV, Pissis P, Spanoudaki A, Shinyashiki N, Sulim IY, Kulik TV, Palyanytsya BB (2007) Polydimethylsiloxane at the interfaces of fumed silica and zirconia/fumed silica. Appl Surf Sci 253:7143–7156
25. Adachi K, Hirano T (2008) Good linear relationship between logarithms of Eigen's water exchange constants for several divalent metal ions and activation energies of corresponding metal-catalyzed alkoxysilane hydrolysis in ethylene-propylene copolymer system. Eur Polym J 44:542–549

26. Svetich GW, Voge AA (1972) The crystal and molecular structure of *sym-trans*-di-μ-phenoxyhexaphenoxydiphenolatodititanium (IV). Acta Crystallogr B 28:1760
27. Glaser RH, Wilkes GL (1988) Structure property behavior of polydimethylsiloxane and poly (tetramethylene oxide) modified TEOS based sol-gel materials V. effect of titaniumisopropoxide incorporation. Polym Bull 19:51–57
28. Liu M, Sun J, Sun Y, Bock C, Chen Q (2009) Thickness-dependent mechanical properties of polydimethylsiloxane membranes. J Micromech Microeng 19:35028–35031. https://doi.org/10.1088/0960-1317/19/3/035028
29. Lee DC, Jang LW (1996) Preparation and characterization of PMMA–clay hybrid composite by emulsion polymerization. J Appl Polym Sci 61:1117–1122

Chapter 13
Preparation of Element-Block Materials Using Inorganic Nanostructures and Their Applications

Naokazu Idota and Yoshiyuki Sugahara

Abstract The evolution of organic-inorganic hybrids is highly desirable for the further acquisition of functionalities not achievable with conventional polymer materials in terms of mechanical, electronic, optical, and magnetic properties. Element-blocks, which are heterogeneous structures consisting of organic and inorganic components mixed at the element level, and their highly ordered polymeric derivatives, element-block polymers, are highly useful for overcoming a number of difficult problems. Among the various approaches to establishing element-blocks, this review focuses on surface modification of inorganic nanostructures with organic molecules to control interactions at the interfaces between organic and inorganic components in the organic-inorganic hybrids. For the design of surface-modified inorganic-nanostructure-based element-blocks, the dimensional features of inorganic nanostructures and the methods of modifying organic molecules on the surfaces are discussed from the viewpoint of nanomaterials. Finally, various applications using surface-modified inorganic-nanostructure-based element-blocks are introduced in terms of polymer-based hybrids and hierarchal architectures to provide successful examples, which are important to the development of polymeric materials based on element-blocks.

Keywords Nanostructure · Surface modification · Polymer-based hybrid · Hierarchical architecture

N. Idota
Department of Chemical Science and Technology, Faculty of Bioscience and Applied Chemistry, Hosei University, Koganei-shi, Tokyo, Japan

Y. Sugahara (✉)
Department of Applied Chemistry, Faculty of Advanced Science and Engineering, Waseda University, Shinjuku-ku, Tokyo, Japan

Kagami Memorial Research Institute for Materials Science and Technology, Waseda University, Shinjuku-ku, Tokyo, Japan
e-mail: ys6546@waseda.jp

Y. Chujo (ed.), *New Polymeric Materials Based on Element-Blocks*,
https://doi.org/10.1007/978-981-13-2889-3_13

13.1 Introduction

Organic-inorganic hybrid materials have been routinely developed to improve various properties of organic compounds, including their optical, thermal, electrochemical, and mechanical properties, by incorporation of inorganic compounds into organic compounds [1]. This concept is based not only on their complementary abilities to supply each other with lacking properties but also on the generation of new functionality with no loss of the intrinsic characteristics of either of their components. Since the functionalities often appear at the interfaces between organic and inorganic components, increasing the ratio of interfaces by reducing their domain sizes and mixing these components homogeneously is an important issue [2]. Phase separation or aggregation occurs easily, however, with decreases in the sizes of organic and inorganic components due to differences in their intrinsic compatibility, resulting in reduced performance compared with the original materials. Thus, a new technology is highly desirable to break through the concept of traditional organic-inorganic hybrid materials. Recently, heterogeneous structures consisting of organic and inorganic components mixed at the element level have been developed as a new family of functional materials, whose structural units are referred to as element-blocks [3]. These element-blocks and their highly ordered polymeric derivatives, element-block polymers, are expected to encourage the development of functional materials with various properties which are not achievable with conventional strategies for preparation of organic-inorganic hybrid materials.

The development of many types of element-blocks has been proposed using well-defined molecular design of organic and inorganic structures, such as polyhedral oligomeric silsesquioxanes (POSS) [4], size-controlled metal clusters [5], and silicon-bridged bithiophenes [6]. While thc preparation procedures of these element-blocks are bottom-up approaches based on molecular design, another strategy is the use of inorganic nanostructures, which are preprogrammed nanomaterials with size-dependent properties. Inorganic nanostructures have been studied for decades because of their unique size-dependent characteristics, and the preparation methods and functionalities of various inorganic nanostructures, such as metal (oxide) nanoparticles [7], carbon nanotubes [8], and graphene nanosheets [9], have been investigated. This approach to preparing element-blocks is advantageous for creating predetermined functionalities based on the properties of the inorganic nanostructures, and the preparation of highly ordered element-block polymers is possible by introducing polymerizable groups or sites into element-block structures derived from inorganic nanostructures. Since it is difficult to achieve precise control of hierarchical structures using inorganic nanostructures and organic components with different intrinsic properties as mentioned above, a key technology for realizing element-blocks based on inorganic nanostructures is surface modification with organic molecules [10]. Surface properties are among the most essential factors in all organic-inorganic hybrid materials, because their interfaces play a crucial role in determining the microscopic and mesoscopic structures of final organic-inorganic hybrid materials. The use of element-blocks prepared via surface modification of inorganic nanostructures with organic molecules is thus an attractive strategy for adjusting their surface properties to achieve affinity with the conjugated organic components to prevent the phase separation or aggregation. Although a variety of surface coupling reagents have been extensively studied in academic as well as industrial settings, organic coupling molecules for surface modification of inorganic nanostructures should be properly selected to achieve desirable structures for potential applications. Therefore, innovative new element-blocks have been developed through an approach that differs from the molecular bottom-up approach, as demonstrated in Fig. 13.1.

This review focuses on the surface modification of inorganic nanostructures with organic molecules for preparation of element-blocks. The various inorganic nanostructures are classified into categories such as nanoparticles, nanotubes, and nanosheets to understand their dimensional characters. We then review the surface modification techniques used for inorganic nanostructures, especially the nanoparticles and nanosheets of transition metal oxides, employing organic molecules for the design of element-blocks. Based on currently available knowledge, recent possible applications using the modified inorganic nanostructures are discussed, mainly with respect to polymer-based hybrids.

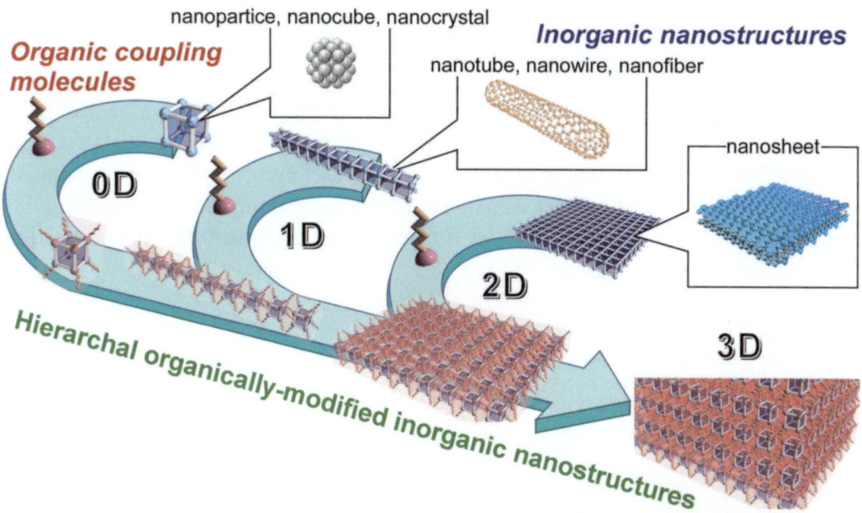

Fig. 13.1 Schematic concept for element-block materials comprising hierarchal organically modified inorganic nanostructures

13.2 Classification of Inorganic Nanostructures

Attention has recently been paid to inorganic nanostructures in a wide range of research fields with attention to the inorganic synthesis and control of the physicochemical properties of inorganic materials. Inorganic nanostructures consist mainly of metals and metal oxides, and the size-dependent properties of inorganic nanostructures are often discussed in terms of classifications according to their shape; nanoparticles, nanotubes/nanofibers, and nanosheets are categorized as zero-dimensional (0D), one-dimensional (1D), and two-dimensional (2D) inorganic nanostructures, respectively (Fig. 13.2) [11]. Among them, one of the most important characteristics for inorganic nanostructures is the quantum size effect. Decreasing metal and semiconductor materials to sizes smaller than a de Broglie wavelength induces significant restrictions on the electron motion by quantum confinement in all three spatial dimensions, resulting in a decrease in the energy level number and widening of the bandgaps. Such changes in the bandgaps provide unique properties compared with conventional bulk materials. For example, semiconductor nanocrystals showed particle-size-dependent control of adsorption and emission wavelengths by the quantum size effect [12]. The quantum size effect can also be obtained by anisotropic decreases in the sizes of inorganic components, including nanotubes and nanosheets. In this section, the characteristics of inorganic nanostructures are introduced based on their dimensional characters.

13.2.1 0D Inorganic Nanostructures

Among the typical inorganic nanostructures exhibiting the quantum size effect are nanoparticles categorized as 0D inorganic nanostructures. A decrease in all the axial lengths of the bulk materials to smaller than the wave number of electrons leads to immobile electrons of any dimension in the materials, resulting in zero dimension for electrons (so-called quantum dots) [13]. Semiconductor quantum dots have been

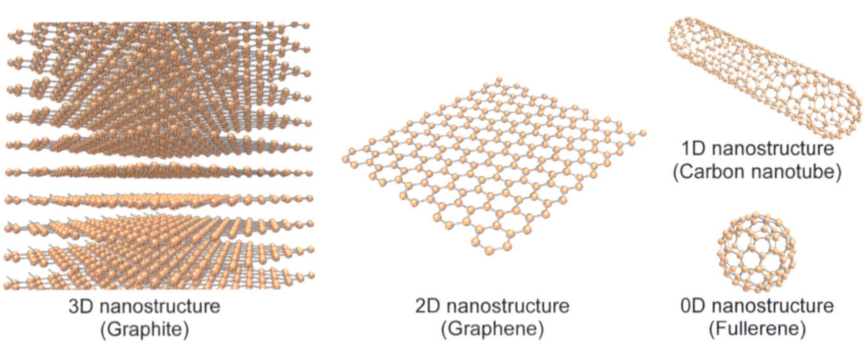

Fig. 13.2 Dimensional classification of nanostructures as an example of carbon compounds

extensively studied in a variety of fields with respect to their numerical theories and physical and chemical features, and a desire has emerged to use them for optical information communication systems and quantum cipher communication devices because of their huge information storage capacity. The quantum dots also possess a number of size-dependent properties, such as high electron density and strong optical absorption (e.g., Au nanoparticles [14]), photoluminescence (e.g., CdSe nanoparticles [15]), phosphorescence (e.g., Y_2O_3 nanoparticles [16]), and magnetic moments (e.g., iron oxide nanoparticles [17]). Procedural techniques for controlling the functions of nanoparticles have advanced over recent decades. The preparation of nanoparticles typically involves either a direct synthetic route of physical (e.g., vapor deposition, laser ablation) or chemical (e.g., sol-gel, micelle, pyrolysis) processes [18]. Reduction methods using metal salts with stabilizers are well known for synthesizing metallic nanoparticles in chemical liquid processes. The size, shape, and polydispersity of metallic nanoparticles can be controlled by varying the reaction conditions, including the concentration, use of reducing agents, and addition of capping ligands, which act as inhibitors to particle growth. In metal oxide nanoparticles, the common methods are coprecipitation, thermal decomposition, and hydrothermal synthesis, which allow synthesis of high-quality nanoparticles [19]. The other route to preparation of nanoparticles is the use of milling to reduce the size of large particles. It has been shown that wet milling using fine ceramic beads below 30 μm in diameter enables a decrease in the particle size to as low as 1–10 nm [20]. The 0D nanoparticle serves as a foundation for various inorganic nanostructures, and higher-ordered inorganic nanostructures can be precisely constructed by self-organization of assembled nanoparticles [21]. The self-assembly of nanoparticles can be controlled by balancing the attraction forces between them, leading to a generation of various highly ordered inorganic nanostructures, including chains, sheets, vesicles, and 3D architectures.

13.2.2 1D Inorganic Nanostructures

Although completely confined electrons are present in quantum dots, as mentioned above, one-dimensionally confined electrons are present in inorganic nanostructures called quantum wires [13], in which electrons can move freely only in one direction. Quantum wires surrounded by a material with a large bandgap confine electron and holes in one dimension (carriers can only move in one dimension) due to the larger bandgap. Since quantum wires exhibit a unique conductivity behavior which does not follow Orme's law, it is desirable to develop their applications for single-electron transfer devices. Carbon nanotubes are considered to be among the typical 1D inorganic nanostructures, and these have been evolving continuously in response to investigation of their synthesis, theory, characterization, and applications [22]. Carbon nanotubes consist of graphitic sheets rolled into cylindrical shapes with nanometer diameters. The properties of carbon nanotubes can be varied as a function of their helicity and tube diameter, and carbon nanotubes behave like metals or semiconductors according to the arrangement of the hexagon rings along their tubular surfaces [23]. In addition to the unique properties derived from the 1D structure, carbon nanotubes exhibit effective release of electrons from their tips, even with low voltages applied, due to the electric field concentration [24], leading to applications as cantilevers for probe microscopy and nanoactuators. Other one-dimensional inorganic nanostructures, such as metal nanorods (e.g., Au, Ag, Si), metal oxide nanowires (e.g., Al_2O_3, ZnO, TiO_2), nitride nanotubes (e.g., BN, AlN), and semiconductor nanowires (e.g., GaAs, InP), have attracted considerable attention [25, 26]. Chemical processes, such as chemical vapor deposition (CVD), precursor decomposition, and solvothermal, hydrothermal, and carbothermal processes, have been employed in the preparation of 1D inorganic nanowires. In these processes, nucleation and nuclear growth are important for controlling the crystallization processes that determine their properties [27]. Sufficiently high concentrations of the raw materials facilitate aggregation into small clusters through homogeneous nucleation, and larger clusters are subsequently formed by further growth of these small clusters. In order to control the crystallization, various methods have been widely investigated, such as the introduction of solid-liquid interfaces, use of templates or capping agents to direct the one-dimensional formation, and hierarchical self-assembly from 0D inorganic nanostructures. Compared to solid nanowires, inorganic nanotubes have more complex hollow structures, and the use of carbon-like hexagonal boron nitride [28] and template techniques for other inorganic materials [29] have been reported to form nanotube structures.

13.2.3 2D Inorganic Nanostructures

In quantum chemistry, two-dimensionally confined electrons are known to be present in quantum wells [13], which are used as semiconductor lasers to develop blue light-emitting diodes for large-capacity information storage. Graphene nanosheets exfoliated from graphite are representative of two-dimensional inorganic nanostructures [30], though their behavior is expected to differ from those of the quantum wells in semiconductor interfaces. The differences are caused by the unique electronic properties of graphene nanosheets, which exhibit electron-hole degeneracy and disappearance of carrier mass near the neutral charge point [31]. Since the structure of graphene nanosheets is a plane of sp^2 carbon atoms, their fundamental properties are similar to those of carbon nanotubes, whose structure consists of graphene nanosheets rolled into cylindrical shapes. On the other hand, the typical preparation procedure for graphene nanosheets (e.g., mechanical or chemical exfoliation from graphite) is easier than that for carbon nanotubes. Other methods of synthesizing graphene nanosheets include CVD, annealing at high temperature, unzipping of nanotubes, and microwave synthesis [32]. Graphene nanosheets exhibit good electrical conductivity, high thermal conductivity, strong mechanical properties, and optical transmittance due to their 2D quantum effects, and their electronic and thermal properties can be controlled by adjusting the number of layers [33].

Other types of inorganic nanosheets can be prepared, meanwhile, by exfoliation from some ion-exchangeable inorganic layered compounds, which accommodate guest species to form intercalation compounds. Typical inorganic layered compounds include clay minerals, such as montmorillonite and kaolinite, but other synthetic compounds, including zirconium phosphates, layered double hydroxides (LDHs), layered metal oxyhalides, layered titanates, layered niobates, and layered perovskites, can also be employed [34]. Although smectic clay minerals exfoliate spontaneously in water because of their low layer charge densities [35], the exfoliation of inorganic layered compounds with high layer charge densities, such as layered titanates and layered perovskites, is mainly induced by intercalation reactions or interlayer surface modification in appropriate combinations between the host and guest species [36]. Among examples, the interlayer distance of ion-exchangeable layered compounds is expanded by substituting the original interlayer ions with bulky organic ones (e.g., tetrabutylammonium cations (TBA^+)). The expansion can decrease the electrostatic interactions at the interlayers, resulting in exfoliation with the aid of mechanical shaking. In neutral layered compounds such as transition metal chalcogenides, intercalation of guest species proceeds after weakly charging the interlayers by reduction [37]. Exfoliated nanosheets can be obtained as dispersions, and applications using deposited nanosheets have been investigated by the Langmuir-Blodgett method and layer-by-layer technique [38].

13.3 Surface Modification of Inorganic Nanostructures with Organic Molecules

Surface modification with organic molecules is an effective technique for controlling the surface properties of inorganic materials that has been used routinely in not only scientific research but also in industrial applications [10]. In inorganic nanostructures, grafting of organic molecules as monolayers is an essential factor in preventing drastic changes in their well-defined size for maintaining the intrinsic properties of the inorganic nanostructures. Appropriate selection of surface coupling reagents based on the shape and quality of the inorganic nanostructures is therefore important. In this section, the focus is on surface modification of inorganic nanoparticles and nanosheets, and the applicability of various surface coupling reagents is discussed.

13.3.1 Surface Modification of Nanoparticles

Since nanoparticles have large specific surface areas compared with bulk materials, interactions between nanoparticles such as the van der Waals force, electrostatic interactions, and hydrogen bonds are significantly strengthened by differences in the surface energies between the nanoparticles and dispersion media, easily resulting in phase separation or aggregation [39]. The dispersion technique is therefore important for maintaining the size-dependent properties of the nanoparticles. Although the use of electric repulsion between surface-charged nanoparticles achieved by adjusting the pH of the dispersion medium is a simple method, surface modification with organic molecules can be an effective technique for controlling the polarity of the nanoparticle surfaces according to solvent properties. Typical coupling reagents for metal nanoparticles are thiol derivatives, which can react with metal atoms to modify them into monolayers [40]. For example, alkanethiols form a commensurate $\sqrt{3} \times \sqrt{3}$ R30° structure on Au(111) surfaces, resulting in all-trans alkyl chain formation with a tilt angle of ca. 30°. The tilt angle depends on the type of metal atoms, as exemplified by $16 \pm 2°$ for palladium, 12° for copper, ca. 10° for silver, and $\approx 0°$ for mercury [41]. The sulfur atoms in the alkanethiols can bind with metal atoms in stable quasi-covalent bonds, and van der Waals forces between their alkyl groups lead to decreases in the surface free energy. Longer alkyl chains in the alkanethiols contribute to restricting their molecular rotations, and the surface modification becomes more stable consequently. Since the headgroups of alkanethiol can be replaced with various functional groups, preparation of modified nanoparticles has been developed using thiol derivatives with hydrophilic and polar headgroups to improve water dispersibility [42].

Silane coupling reagents are used extensively for surface modification of transition metal oxide nanoparticles. Surface modification with silane coupling reagents, or silanization, was initially developed in surface functionalization of silica particles [43]. Common silane coupling reagents bear alkoxysilyl (≡Si-OR) or chlorosilyl groups (≡Si-Cl), which can be hydrolyzed and subsequently reacted with hydroxyl groups on transition metal oxide surfaces through condensation (Fig. 13.3a), and various organic functional groups are available in the other terminated side of silane molecules. Inter- and intramolecular condensation reactions of silyl groups (the so-called homocondensation), on the other hand, also occur during the silanization, resulting in multilayered modification. Thus, the selection of appropriate reaction conditions, particularly with respect to water content, is important for surface modification of inorganic nanostructures while maintaining their original size [44]. Silanization proceeds not only in liquid-phase reactions but also in vapor-phase deposition, and the vaporization enables it to proceed under vacuum conditions at low temperature with the use of silane coupling reagents with large molecular weights [45]. A variety of silane coupling reagents is commercially available, and the dispersibility of modified TiO_2 nanoparticles in organic solvents can be controlled by managing the mixing ratio of silane alkoxides [46].

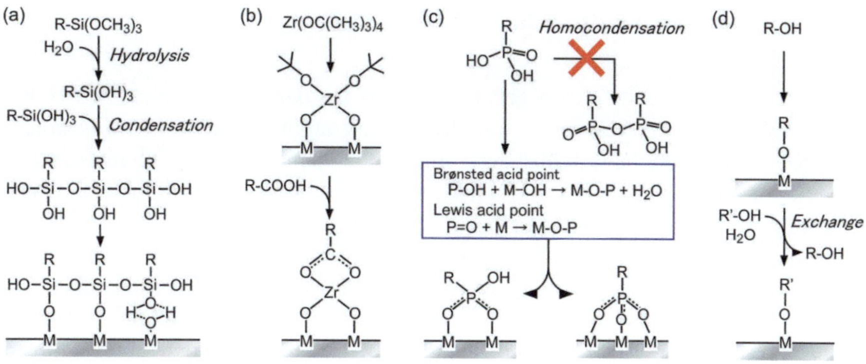

Fig. 13.3 Modification process with organic groups onto metal oxide surfaces. (**a**) Silanization, (**b**) carboxylic acids via zirconium tetra-*tert*-butoxide [47], (**c**) phosphorus coupling, and (**d**) alcohol-exchange-type reactions

Transition metal oxide nanoparticles are covered with hydroxyl groups, which play important roles in solid catalysts and modification with coupling molecules. Since these surface hydroxyl groups can serve as Brønsted acid sites, carboxylates and phosphonates can be modified onto the surfaces through acid-base reactions. Although organocarboxylic acids can easily react with the transition metal oxide nanoparticles, stable surface modification using organocarboxylic acids is often difficult because of the relatively weak and reversible interactions. In order to overcome this issue, the deposition reaction of zirconium tetra-*tert*-butoxide ($Zr(O^tBu)_4$) onto transition metal surfaces has been reported [47]. Organocarboxylic acids can be stably and irreversibly reacted with ZrO_2 surfaces to form strong bidentate bonds, and modification with carboxylic acids is also applicable to native oxide surfaces, such as hydroxylated alumina, by pretreatment with $Zr(O^tBu)_4$ (Fig. 13.3b). TiO_2 nanoparticles were also reported to permit bidentate binding with carboxylic acid moieties in a fashion similar to ZrO_2 [48].

Phosphorus coupling reagents have been attracting attention recently for their ability to modify transition metal oxides stably as self-assembled monolayers. The phosphorus coupling reagents form covalent M-O-P bonds, which are relatively stable with respect to hydrolysis, by reacting with hydroxyl groups of transition metal oxide surfaces (Fig. 13.3c) [49]. It is worth noting that no homocondensation of phosphorous coupling reagents proceeds under mild conditions [50]. Thus, surface modification of phosphorus coupling reagents is a suitable technique for functionalization of transition metal oxide nanoparticles. Phosphoric acid esters are also attractive for use in surface modification because of their high solubility in organic solvents and easy preparation from commercially available phosphoric acid and alcohol [51]. Other organic acids involving sulfonic acids have been used to introduce functional groups onto nanoparticle surfaces [52].

13.3.2 Surface Modification of Nanosheets

A variety of metal oxide nanosheets prepared by the deposition method or exfoliation from layered materials can be modified using coupling reagents in a way similar to nanoparticles. Graphene nanosheets, on the other hand, are inert with respect with organic coupling reagents, since graphene nanosheets consist of two-dimensional sp ^{2}-hybridized carbon atoms arrayed in a honeycomb pattern. The surface modification of graphene nanosheets with organic molecules is mainly achieved by two general routes: free radical or Diels-Alder reaction with C=C bonds of pristine graphene [53] or formation of covalent bonds with oxygen groups of graphene oxides (GO) [54]. Highly reactive free radicals generated by heating a diazonium salt can attack the sp^2 carbon atoms of graphene nanosheets forming a covalent bond, and dienophile derivatives are also bound to these though a cycloaddition reaction. GO is a single graphitic monolayer partially containing oxidized aliphatic regions (sp^3 carbon atoms), such as hydroxyl, epoxy, carbonyl, and carboxyl functional groups. Using these oxidized species, organic molecules can be covalently bound through formation of ester and amide bonds. The deactivation of the instinct properties of graphene nanosheets by surface modification is relatively low, and a chemical reduction of GO can remove the oxygen atoms to recover the pristine graphene, although the recovery is incomplete. Among other modification methods, direct attachment of hydrogen and halogen atoms to graphene surfaces by plasma treatment and heat-induced halogenation has been reported [9].

In preparation of surface-modified inorganic-nanosheet-based element-blocks, the use of ion-exchangeable layered compounds is attractive for simultaneous employment of interlayer surface modification with coupling molecules and weakening of the electrostatic interactions at the interlayer for subsequent exfoliation to surface-modified inorganic nanosheets as well as intercalation. Grafting reactions of the interlayer surfaces of layered perovskites, $HLaNb_2O_7 \cdot xH_2O$ (HLaNb), using various *n*-alcohols via so-called alcohol-exchange-type reactions have been reported (Fig. 13.3d) [55]. In this type of reaction, alkoxy groups that are bound covalently to the $[LaNb_2O_7]$ units can be exchanged with other alkoxy groups upon reaction with alcohols in the presence of a small amount of water via a hydrolysis-reesterification mechanism. The technique is available for subsequent modification with larger alcohols by using layered perovskites modified with smaller alcohols as intermediates. That is to say, alcohol-exchange reactions in less reactive hosts can be employed for grafting reactions with larger *n*-alcohols after direct reaction with methanol. This approach also induces decreases in the electrostatic interactions at the interlayers, and *n*-alkoxy-modified nanosheets can consequently be obtained by an exfoliation process through ultrasonication. The *n*-alkoxy groups on the HLaNb surface form all-trans conformations as bilayers at the interlayers with a tilt angle of ca. 57° [56]. The title angle is independent on the alkyl chain length in the *n*-alkoxy groups, and the increase in the interlayer distance is 0.214 nm per one carbon atom. The title angle and increase ratio of the interlayer distance in the *n*-alkoxy groups depend significantly on the type of host compounds, such as 41° and 0.166 nm for

$HCa_2Nb_3O_{10}{\cdot}xH_2O$ [57] and 75° and 0.244 nm for $H_2La_2Ti_3O_{10}$ [58]. In addition to *n*-alcohols, branched and fluorinated alcohols can also be bound though alcohol-exchange reactions [59, 60]. Although the alkoxy groups on inorganic nanosheets exhibited high thermal stability, hydrolysis occurred under alkaline conditions to remove the *n*-alkoxy groups easily from the surfaces.

In addition to inorganic nanoparticles, phosphorus coupling reagents are also powerful tools for preparation of surface-modified inorganic-nanosheet-based element-blocks from layered compounds of transition metal oxides by forming self-assembled monolayers. The monolayered modification provides clear regularity of the interlayer distances in the host compounds determined by XRD patterns, and the exfoliated products can be used as well-defined element-blocks. Organophosphonic acids can react with the *n*-decoxy derivative of HLaNb as an intermediate [61]. The phosphorus environment of the organophosphonic acid moieties is monodentate with a phosphoryl group and P-OH group remaining, although the phosphorus environment is mainly bidentate or tridentate for transition metal oxide nanoparticles. Since organophosphonic acids are relatively strong acids, on the other hand, the reactions with organophosphonic acids often resulted in dissolution of the host compounds. When kaolinite was reacted with phenyphosphonic acids with the intention of intercalation, for example, lamellar aluminum phenylphosphonate was crystallized after a part of the kaolinite had dissolved [62]. In surface modification of layered compounds with organophosphonic acids, a mechanical exfoliation process such as ultrasonication is also required for obtaining surface-modified nanosheets in a way similar to that of *n*-alkoxy-modified layered compounds. Another approach to preparing surface-modified nanosheets is surface-initiated polymerizations at the interlayers of the host compounds. Such polymerization can be achieved using interlayer modification with organophosphonic acids comprising atom transfer radical polymerization (ATRP) initiators, resulting in spontaneous exfoliation from the organically modified layered compounds in solution by the excluded volume effects of grafted polymer chains [63].

A remarkable layered compound for preparation of interesting nanosheets is potassium hexaniobate, $K_4Nb_6O_{17}{\cdot}3H_2O$, since creation of single- or double-layered nanosheets can be achieved by utilizing its unique structure [64]. $K_4Nb_6O_{17}{\cdot}3H_2O$ has alternately stacked hydrated and anhydrous interlayers with different reactivities. Some large guest molecules can recognize the difference in reactivity and preferentially intercalate into the hydrated interlayers. Based on this structural feature, two types of intercalation compounds can be prepared: intercalation compounds bearing guest species in every other interlayer (A-type) and intercalation compounds whose interlayers are occupied completely by guest species (B-type). It is also possible to create two types of organophosphonate derivatives separately by using two types of intercalation compounds as intermediates (Fig. 13.4). As a consequence, single- or double-layered nanosheets can be obtained by exfoliation upon ultrasonication from the organophosphonate-modified derivatives.

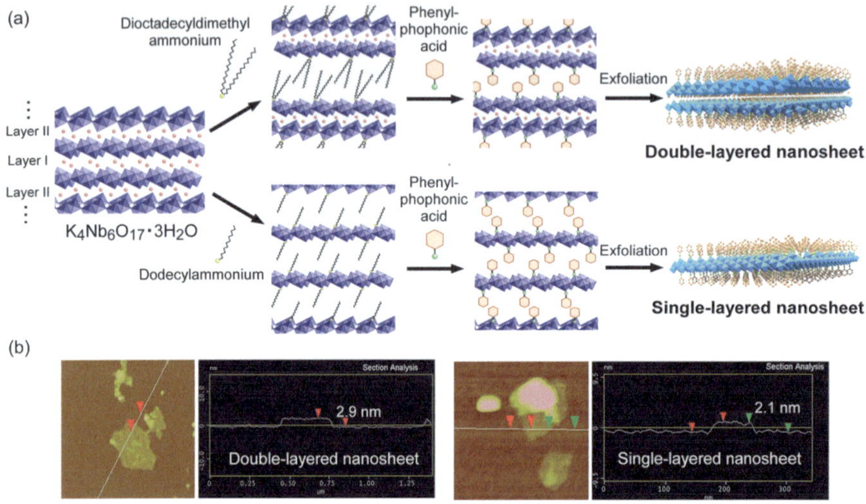

Fig. 13.4 Selective surface grafting of layered potassium hexaniobate, $K_4Nb_6O_{17}\cdot 3H_2O$, for preparation of single- or double-layered nanosheets [64] (© 2014 American Chemical Society). (**a**) $K_4Nb_6O_{17}\cdot 3H_2O$ are intercalated with dioctadecyldimethylammonium ions in every other interlayer (A-type) or occupied completely by dodecylammonium ions (B-type), and single- or double-layered nanosheets can be prepared via selective grafting with phenylphosphonic acids followed by exfoliation by ultrasonication. (**b**) AFM top-view and cross-sectional images for single- (left) and double-layered (right) nanosheets

13.4 Applications Using Surface-Modified Inorganic-Nanostructure-Based Element-Blocks

Functional organic-inorganic hybrid materials derived from inorganic nanostructures via surface modification with organic groups are playing important roles in various applications, such as catalysts, separation materials, electronic materials, and biomedical materials. The unique features of inorganic nanostructures such as a photothermal effect in Au [14] and superparamagnetism in Fe_3O_4 [17] are used in innovative ways, and surface modification with organic groups generates synergistic effects, in particular, for biological processes with the addition of bio-compatibility and stimuli responsibility [7]. Another attractive capability is the control of interactions between organic and inorganic components in organic-inorganic hybrids using the surface-modified inorganic-nanostructure-based element-blocks to improve their performances. Since inorganic nanostructures aggregate easily in organic solvents or polymer matrices, surface modification of inorganic nanostructures with organic groups is an essential technique for well-defined dispersion in organic components to provide the instinct properties of the inorganic nanostructures [2]. Based on the strategic designs of modified organic structures, graft architectures, and certain types of inorganic nanostructures, possible applications using organic-inorganic hybrids with inorganic nanostructures have been demonstrated across a wide range one after another. This section focuses on recent applications of organic-inorganic hybrids using surface-modified inorganic-nanostructure-based element-blocks.

13.4.1 Polymer-Based Hybrids

Inorganic nanostructures have been routinely used for polymer-based hybrids as nanofillers to improve the thermal, mechanical, and optical properties of polymer matrices. It is critical for polymer-based hybrids to maintain the instinct properties of polymer matrices, such as transparency, lightweight, excellent formability, and low cost, even after incorporation of inorganic nanostructures. The aggregation or phase separation of inorganic nanostructures in polymer-based hybrids could, however, cause a reduction of the properties of polymer matrices. Even using small nanoparticles, for an example, Rayleigh scattering at the interfaces between the aggregated inorganic nanoparticles and the polymer matrices occurs when the aggregate size increases to above 40 nm, leading to a decline in the transparency of polymer-based hybrids [65]. Since the aggregation is caused mainly by differences in compatibility between the surface properties of inorganic nanostructures and polymer matrices, surface modification of inorganic nanostructures with appropriate organic groups is important for suppressing their aggregation. Although functionalized polymers have recently been developed through molecular design, demand remains for improvement of the properties of commercially available polymers by incorporating inorganic nanostructure for various industrial applications. The use of inorganic nanoparticles with high refractive indices as nanofillers is attractive for improving the refractive indices of common optical polymers such as poly(methyl methacrylate) (PMMA) and epoxy resins, since they have a restricted range of refractive indices of from 1.3 to 1.7 [66]. Surface modification of rutile-type TiO2 nanoparticles, which have high refractive indices (n = 2.5–2.7), has been demonstrated in the application of transparent polymer-based hybrids with high refractive indices. In these applications, *n*-octylphosphonic acid [67] and oleyl phosphate [68] are covalently bound to the surfaces of TiO_2 nanoparticles to form a monolayer, and the homogeneous dispersion of the nanoparticles in organic solvents and polymer matrices is achieved by varying the surface properties of the nanoparticles. The polymer-based hybrid thin films with surface-modified nanoparticles exhibit excellent transparency, and their refractive indices increase with increases in the TiO_2 content. Such ex situ approaches are suitable for preparing optical hybrid thin films with thicknesses of less than several μm, because of significant effects of the dynamic structural changes occurring during solvent evaporation for thick hybrid films [69]. For preparation of bulk optical hybrids such as photovoltaic devices and ophthalmic lenses, in situ polymerization of monomer dispersions containing surface-modified nanoparticles without solvent evaporation is a useful technique, and grafting polymer chains used in the matrices of the bulk hybrids onto the nanoparticles help to prevent their aggregation during in situ polymerization [70]. In addition, it was reported that modification of the polymer chain lengths on the nanoparticles was critical for dispersion in the polymer matrices (Fig. 13.5a). Surface-initiated living polymerization processes [71] such as ATRP, nitroxide-mediated polymerization (NMP), and reversible addition fragmentation chain transfer (RAFT) are therefore effective in modifying well-defined polymer chains on inorganic nanostructures for preparation of transparent bulk hybrids with high refractive indices.

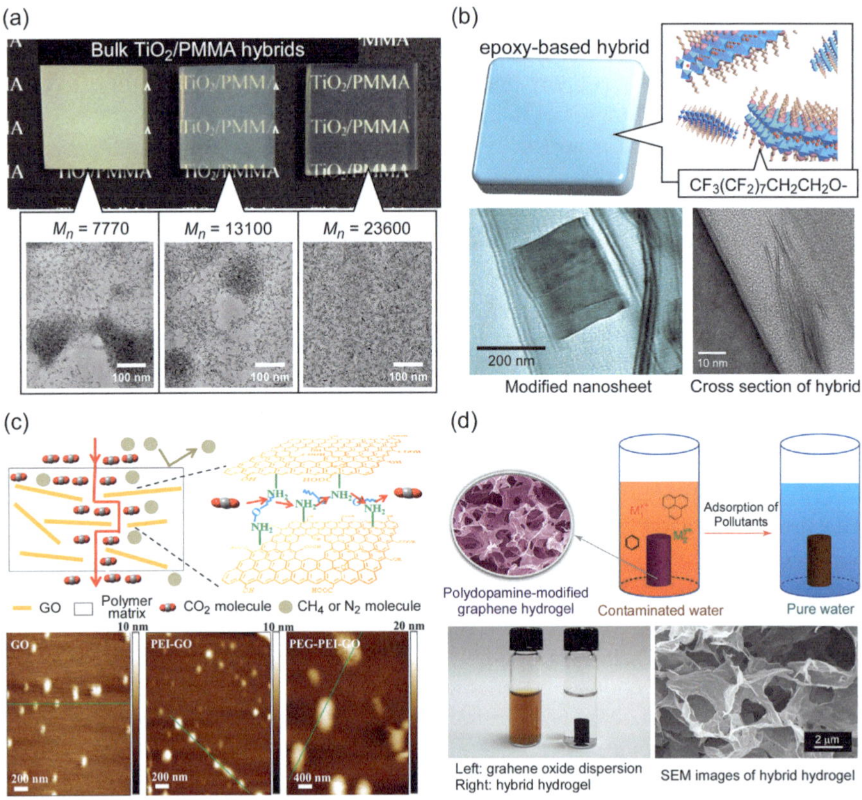

Fig. 13.5 Polymer-based hybrids using organically modified nanostructures. (**a**) Photographs and TEM images of bulk PMMA hybrids containing PMMA-modified TiO_2 nanoparticles with different chain length [70] (© 2016 American Chemical Society), (**b**) overview of epoxy-based hybrids with layered perovskite nanosheets bearing hydrophobic fluorinated alkoxy groups and its microscopic images (left, TEM image of the modified nanosheets; right, cross-section FE-TEM image of the hybrid) [75] (© 2014 Royal Society of Chemistry), (**c**) overview of CO_2 permselective polymer-based hybrid membrane incorporating graphene oxide nanosheets functionalized with PEG and PEI and AFM images of the modified nanosheets [76] (© 2015 American Chemical Society), and (**d**) overview of functional hydrogels containing polydopamine-modified graphene nanosheets as a reusable adsorbent of heavy metal ions for water purification and its images (left, photographs of the products; right, SEM images of the hybrid) [78] (© 2013 American Chemical Society)

Other improvements in the properties of polymer-based hybrids using inorganic nanostructures have been experimentally and theoretically investigated. The mechanical properties of synthetic polymers, including tensile strength, flexural strength, hardness, Young's modulus, and stiffness, can be improved by the incorporation of surface-modified inorganic nanostructures via reinforcement mechanisms [72]. The well-dispersed inorganic nanostructures in the polymer-based hybrids also act as thermal insulators and mass transport barriers to the volatiles generated during decomposition, enabling them to exhibit stable thermal properties [73]. Along with such advantageous properties for polymer materials, inorganic nanosheets have additional superior barrier properties with respect to gas penetration, including penetration by oxygen, carbon dioxide, and organic gases, due to their two-dimensional high aspect ratios. The incorporated nanosheets in polymer matrices create tortuous pathways for the diffusion of gaseous molecules passing through the polymer-based hybrids. Although gaseous molecules can penetrate perpendicularly through a bare polymer film, diffused molecules navigate around impenetrable platelets and through interfacial zones, which have permeability characteristics [74]. These properties are useful for food packaging and flame-retardant materials. These properties of the polymer-based hybrids are also further improved by enhanced dispersibility through surface modification of inorganic nanostructures with organic molecules. In addition, the surface-modified nanosheets allow control of affinities for specific molecules to the polymer-based hybrids according to the type of coupling reagent. Epoxy-based hybrids with layered perovskite nanosheets bearing hydrophobic fluorinated alkoxy groups, $CF_3(CF_2)_7C_2H_4O$, on the surface exhibited not only improved mechanical and thermal properties but also reduced water uptake compared to neat epoxy resins due to the surface-modifying hydrophobic groups (Fig. 13.5b) [75]. CO_2 permselective hybrid membranes can be fabricated by incorporating graphene oxide nanosheets functionalized with polyethylene glycol (PEG) and polyethylenimine (PEI) into commercially available polymers (Fig. 13.5c) [76]. In these modified nanosheets, oxyethylene groups in modified PEG and various types of amine groups in PEI showed excellent affinity and reversible reactivity to CO_2, respectively. The hybrid membranes with modified nanosheets thus had selective properties of catch and release for CO_2. Titanate nanosheets functionalized with sulfonic acids were used as nanofillers to improve the proton conductivity of Nafion® [77]. The sulfonic acid groups on the nanosheets simultaneously assisted the proton transfer and dispersion in the polymer matrices, and the methanol permeability of hybrid membranes, which caused a decline in their proton conductivity, was decreased by the tortuous pathways of dispersed nanosheets in the membranes. Functional hydrogels containing polydopamine-modified graphene nanosheets were reported to act as reusable adsorbents of heavy metal ions for water purification (Fig. 13.5d) [78]. The dopamine residues on the nanosheets play the role of reducing agents for graphene and the active sites for heavy metals ions through electrostatic, bidentate chelating, or hydrogen-bonding interactions. Surface modification with organic molecules can thus provide polymer-based hybrids with additional functionalities and synergistic effects.

13.4.2 Construction of Hierarchical Architectures

A next generation of organic-inorganic hybrids is being developed by constructing hierarchical architectures using organic modification of inorganic nanostructures through self-assembly. Although the aforementioned polymer-based hybrids exhibit various improved and added properties depending on the surface-modifying organic groups and inorganic nanostructures, a paradigm shift to further advances in organic-inorganic hybrid materials is required for assembling modified inorganic nanostructures displaying new properties as a result of interactions between the excitons, magnetic moments, or surface plasmons of individual inorganic nanostructures [21]. It will be important for achieving this purpose to construct hierarchical architectures comprising modified inorganic nanostructures as element-blocks through precise control of the spacing and alignment of individual inorganic nanostructures. The well-defined alignment of inorganic nanostructures has been investigated using self-assembly through a balance between attraction and repulsion forces [79], and surface modification of the inorganic nanostructures leads to binding of their hierarchical architectures. A possible technique for preparing self-assembled high-order architectures is site- or area-selective surface modification of inorganic nanostructures, in which the patterned organic groups act as links between modified inorganic nanostructures. In Au nanoparticle surfaces modified with thiolated molecules, there are two diametrically opposed singularities that are present at the particle poles, at which modified molecules are not optimally stabilized by intermolecular interactions with their neighbors, resulting in replacement with other thiolated molecules through place-exchange reactions [80]. Use of this characteristic enables the reactive organic groups to be introduced at only the two polar positions of the Au nanoparticles, and linear chains comprising the modified nanoparticles can be fabricated by linkage between the reactive organic groups. In addition, Ag nanorods have different reactivities between the {100} and {111} facets in their crystalline planes [81]. Individual surface modification of the Au nanorods with cetyltrimethylammonium bromide (CTAB) and polystyrene was led to the construction of a variety of hierarchal architectures such as chains, rings, spheres, and vesicles by changing the solvent composition (Fig. 13.6a) [82]. Such hierarchical organization based on patterning modification can also be achieved by using Janus nanoparticles, with the surfaces of both hemispheres modified with different organic groups [83].

Homogeneous surface modification of inorganic nanostructures also allows construction of higher-order architectures through control of self-assembly. Inorganic nanoparticles modified with block copolymers comprising hydrophilic and hydrophobic segments spontaneously exhibit vesicular or tubular formation by conformational changes of tethered polymer chains (Fig.13.6b) [84]. This mechanism is mainly based on rearrangements of the polymer conformation for minimizing the interfacial energy through partial hydration of the block copolymers, and the morphologies and geometries of these assemblies can be controlled by the

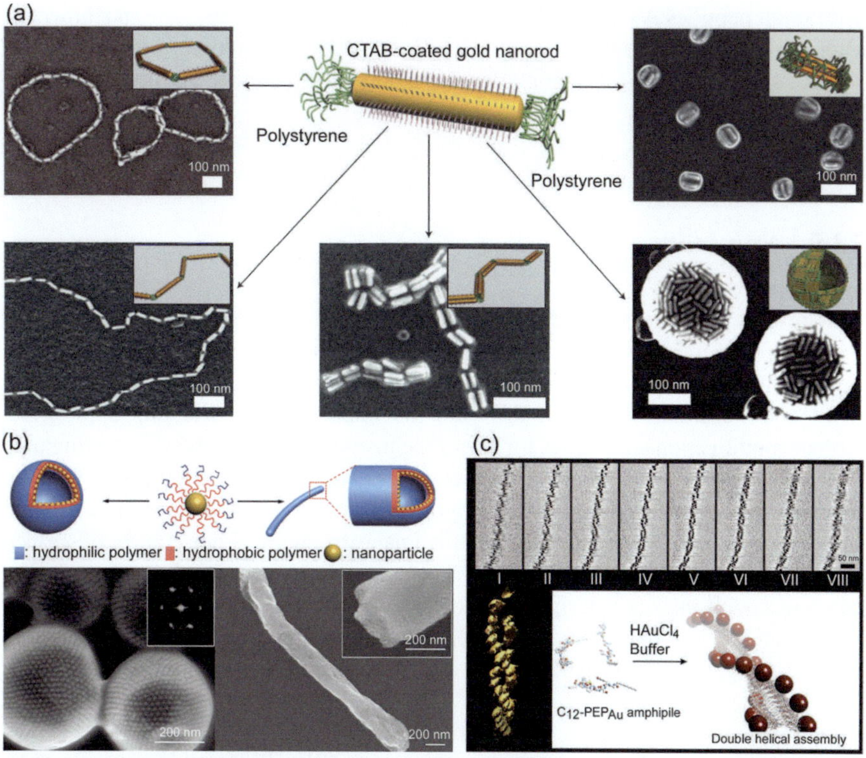

Fig. 13.6 Hierarchical architectures based on organically modified nanostructures. (**a**) Schematic images of self-assembly comprising polymer-modified gold nanorods in selective solvents and its SEM images (insets in the SEM images show corresponding schematic diagrams of the nanorod assemblies) [82] (© 2007 Nature Publishing Group), (**b**) schematic images of amphiphilic block copolymer-assisted self-assembly of nanoparticles into vesicles or tubules and its SEM images (left, vesicles with inset of the FFT pattern; right, tubules with inset of high-resolution image) [84] (© 2012 American Chemical Society), and (**c**) schematic images of forming gold nanoparticle double helices and its electron tomography data (X-Y computational slices (I–VIII) of 3D tomographic volume containing the double helical gold nanoparticle assembly) [86] (© 2008 American Chemical Society)

nanoparticle size and molecular weight of modified block copolymers. Assembly of Ag nanocubes have shown one-dimensional edge-to-edge and face-to-face orientations with modification of the nanocube surfaces with polymers of different chain lengths [85]. During solvent annealing processing of a polystyrene solution containing poly(*N*-vinyl-2-pyrrolidone)-grafted nanocubes, the nanocube-polymer hybrids underwent spontaneous self-organized phase segregation to form nanocube strings according to the balance between the chain lengths and effective surface energy of the Ag nanocubes. Biological self-assembly is a potential technique for building up hierarchical architectures. An interesting report was published on formation of double helices through peptide conjugation using Au nanoparticles with small peptides AYSSGAPPMPPF attached (Fig. 13.6c) [86]. The attachment of aliphatic carbon tails to the N-terminuses of the peptides would promote creation of various multidimensional supramolecular assemblies, such as 1D peptide amphiphilic structures.

13.5 Summary

This review introduces element-blocks prepared by surface modification of inorganic nanostructures with functional organic groups for the development of versatile organic-inorganic hybrid materials. Inorganic nanostructures are classified according to their dimensional characteristics, and the importance of selecting appropriate surface modification techniques for use of the instinct features of both nanostructures and modified organic groups is described. It is further demonstrated that feasible applications such as polymer-based hybrids and hierarchal architectures can be achieved by designing organically modified nanostructures. Since control of surface properties is an essential factor in every field of materials science, the surface design combined with interface design between organic and inorganic components at the molecular level is expected to contribute to advances in not only conventional organic-inorganic hybrid materials but also in element-block materials.

Acknowledgments The research presented in this article was financially supported in part by a Grant-in-Aid for Scientific Research on Innovative Areas, "New Polymeric Materials Based on Element-Blocks (No. 2401)" (JSPS KAKENHI Grant Number JP24102002).

References

1. Sanchez C, Soler-Illia GJAA, Ribot F, Lalot T, Mayer CR, Cabuil V (2001) Designed hybrid organic–inorganic nanocomposites from functional nanobuilding blocks. Chem Mater 13(10):3061–3083. https://doi.org/10.1021/cm011061e
2. Kango S, Kalia S, Celli A, Njuguna J, Habibi Y, Kumar R (2013) Surface modification of inorganic nanoparticles for development of organic–inorganic nanocomposites—a review. Prog Polym Sci 38(8):1232–1261. https://doi.org/10.1016/j.progpolymsci.2013.02.003
3. Chujo Y, Tanaka K (2015) New polymeric materials based on element-blocks. Bull Chem Soc Jpn 88(5):633–643. https://doi.org/10.1246/bcsj.20150081
4. Naka K, Irie Y (2017) Synthesis of single component element-block materials based on siloxane-based cage frameworks. Polym Int 66(2):187–194. https://doi.org/10.1002/pi.5121
5. Niihori Y, Hossain S, Sharma S, Kumar B, Kurashige W, Negishi Y (2017) Understanding and practical use of ligand and metal exchange reactions in thiolate-protected metal clusters to synthesize controlled metal clusters. Chem Rec. https://doi.org/10.1002/tcr.201700002
6. Ohshita J, Nodono M, Kai H, Watanabe T, Kunai A, Komaguchi K, Shiotani M, Adachi A, Okita K, Harima Y, Yamashita K, Ishikawa M (1999) Synthesis and optical, electrochemical, and electron-transporting properties of silicon-bridged bithiophenes. Organometallics 18(8):1453–1459. https://doi.org/10.1021/om980918n

7. Jiang S, Win KY, Liu S, Teng CP, Zheng Y, Han M-Y (2013) Surface-functionalized nanoparticles for biosensing and imaging-guided therapeutics. Nanoscale 5(8):3127. https://doi.org/10.1039/c3nr34005h
8. Baughman RH (2002) Carbon nanotubes – the route toward applications. Science 297(5582):787–792. https://doi.org/10.1126/science.1060928
9. Georgakilas V, Otyepka M, Bourlinos AB, Chandra V, Kim N, Kemp KC, Hobza P, Zboril R, Kim KS (2012) Functionalization of graphene: covalent and non-covalent approaches, derivatives and applications. Chem Rev 112(11):6156–6214. https://doi.org/10.1021/cr3000412
10. Sperling RA, Parak WJ (2010) Surface modification, functionalization and bioconjugation of colloidal inorganic nanoparticles. Philos Trans R Soc Lond Ser A 368(1915):1333–1383. https://doi.org/10.1098/rsta.2009.0273
11. Geim AK, Novoselov KS (2007) The rise of graphene. Nat Mater 6(3):183–191. https://doi.org/10.1038/nmat1849
12. Halperin WP (1986) Quantum size effects in metal particles. Rev Mod Phys 58(3):533–606. https://doi.org/10.1103/RevModPhys.58.533
13. Sun CQ (2007) Size dependence of nanostructures: impact of bond order deficiency. Prog Solid State Chem 35(1):1–159. https://doi.org/10.1016/j.progsolidstchem.2006.03.001
14. Yavuz MS, Cheng Y, Chen J, Cobley CM, Zhang Q, Rycenga M, Xie J, Kim C, Song KH, Schwartz AG, Wang LV, Xia Y (2009) Gold nanocages covered by smart polymers for controlled release with near-infrared light. Nat Mater 8(12):935–939. https://doi.org/10.1038/nmat2564
15. Kulakovich O, Strekal N, Yaroshevich A, Maskevich S, Gaponenko S, Nabiev I, Woggon U, Artemyev M (2002) Enhanced luminescence of CdSe quantum dots on gold colloids. Nano Lett 2(12):1449–1452. https://doi.org/10.1021/nl025819k
16. Sotiriou GA, Schneider M, Pratsinis SE (2012) Green, silica-coated monoclinic Y_2O_3:Tb^{3+} nanophosphors: flame synthesis and characterization. J Phys Chem C 116(7):4493–4499. https://doi.org/10.1021/jp211722z
17. Lu A-H, Salabas EL, Schüth F (2007) Magnetic nanoparticles: synthesis, protection, functionalization, and application. Angew Chem Int Ed 46(8):1222–1244. https://doi.org/10.1002/anie.200602866
18. Masala O, Seshadri R (2004) Synthesis routes for large volumes of nanoparticles. Annu Rev Mater Res 34(1):41–81. https://doi.org/10.1146/annurev.matsci.34.052803.090949
19. Wu W, He Q, Jiang C (2008) Magnetic iron oxide nanoparticles: synthesis and surface functionalization strategies. Nanoscale Res Lett 3(11):397–415. https://doi.org/10.1007/s11671-008-9174-9
20. Klaine SJ, Alvarez PJJ, Batley GE, Fernandes TF, Handy RD, Lyon DY, Mahendra S, McLaughlin MJ, Lead JR (2008) Nanomaterials in the environment: behavior, fate, bioavailability, and effects. Environ Toxicol Chem 27(9):1825. https://doi.org/10.1897/08-090.1
21. Nie Z, Petukhova A, Kumacheva E (2009) Properties and emerging applications of self-assembled structures made from inorganic nanoparticles. Nat Nanotechnol 5(1):15–25. https://doi.org/10.1038/nnano.2009.453
22. Popov V (2004) Carbon nanotubes: properties and application. Mater Sci Eng R 43(3):61–102. https://doi.org/10.1016/j.mser.2003.10.001
23. Ebbesen TW, Lezec HJ, Hiura H, Bennett JW, Ghaemi HF, Thio T (1996) Electrical conductivity of individual carbon nanotubes. Nature 382(6586):54–56. https://doi.org/10.1038/382054a0
24. de Heer WA, Ch telain A, Ugarte D (1995) A carbon nanotube field-emission electron source. Science 270(5239):1179–1180. https://doi.org/10.1126/science.270.5239.1179
25. Rao CNR, Deepak FL, Gundiah G, Govindaraj A (2003) Inorganic nanowires. Prog Solid State Chem 31(1–2):5–147. https://doi.org/10.1016/j.progsolidstchem.2003.08.001
26. Xia Y, Yang P, Sun Y, Wu Y, Mayers B, Gates B, Yin Y, Kim F, Yan H (2003) One-dimensional nanostructures: synthesis, characterization, and applications. Adv Mater 15(5):353–389. https://doi.org/10.1002/adma.200390087

27. Burda C, Chen X, Narayanan R, El-Sayed MA (2005) Chemistry and properties of nanocrystals of different shapes. Chem Rev 105(4):1025–1102. https://doi.org/10.1021/cr030063a
28. Chopra NG, Luyken RJ, Cherrey K, Crespi VH, Cohen ML, Louie SG, Zettl A (1995) Boron nitride nanotubes. Science 269(5226):966–967. https://doi.org/10.1126/science.269.5226.966
29. Xiong Y, Mayers BT, Xia Y (2005) Some recent developments in the chemical synthesis of inorganic nanotubes. Chem Commun (40):5013. https://doi.org/10.1039/b509946c
30. Geim AK (2009) Graphene: status and prospects. Science 324(5934):1530–1534. https://doi.org/10.1126/science.1158877
31. Zhang Y, Tan Y-W, Stormer HL, Kim P (2005) Experimental observation of the quantum Hall effect and Berry's phase in graphene. Nature 438(7065):201–204. https://doi.org/10.1038/nature04235
32. Bhuyan MSA, Uddin MN, Islam MM, Bipasha FA, Hossain SS (2016) Synthesis of graphene. Int Nano Lett 6(2):65–83. https://doi.org/10.1007/s40089-015-0176-1
33. Nagashio K, Nishimura T, Kita K, Toriumi A (2010) Systematic investigation of the intrinsic channel properties and contact resistance of monolayer and multilayer graphene field-effect transistor. Jpn J Appl Phys 49(5):051304. https://doi.org/10.1143/jjap.49.051304
34. Nicolosi V, Chhowalla M, Kanatzidis MG, Strano MS, Coleman JN (2013) Liquid exfoliation of layered materials. Science 340(6139):1226419–1226419. https://doi.org/10.1126/science.1226419
35. Nadeau PH, Wilson MJ, McHardy WJ, Tait JM (1984) Interstratified clays as fundamental particles. Science 225(4665):923–925. https://doi.org/10.1126/science.225.4665.923
36. Osada M, Sasaki T (2009) Exfoliated oxide nanosheets: new solution to nanoelectronics. J Mater Chem 19(17):2503. https://doi.org/10.1039/b820160a
37. Joensen P, Frindt RF, Morrison SR (1986) Single-layer MoS_2. Mater Res Bull 21(4):457–461. https://doi.org/10.1016/0025-5408(86)90011-5
38. Osada M, Sasaki T (2012) Two-dimensional dielectric nanosheets: novel nanoelectronics from nanocrystal building blocks. Adv Mater 24(2):210–228. https://doi.org/10.1002/adma.201103241
39. Caseri W (2000) Nanocomposites of polymers and metals or semiconductors: historical background and optical properties. Macromol Rapid Commun 21(11):705–722. https://doi.org/10.1002/1521-3927(20000701)21:11<705::aid-marc705>3.0.co;2-3
40. Ulman A (1996) Formation and structure of self-assembled monolayers. Chem Rev 96(4):1533–1554. https://doi.org/10.1021/cr9502357
41. Love JC, Estroff LA, Kriebel JK, Nuzzo RG, Whitesides GM (2005) Self-assembled monolayers of thiolates on metals as a form of nanotechnology. Chem Rev 105(4):1103–1170. https://doi.org/10.1021/cr0300789
42. Xia Y, Whitesides GM (1998) Soft lithography. Angew Chem Int Ed 37(5):550–575. https://doi.org/10.1002/(sici)1521-3773(19980316)37:5<550::aid-anie550>3.0.co;2-g
43. Brzoska JB, Azouz IB, Rondelez F (1994) Silanization of solid substrates: a step toward reproducibility. Langmuir 10(11):4367–4373. https://doi.org/10.1021/la00023a072
44. Mutin PH, Guerrero G, Vioux A (2003) Organic–inorganic hybrid materials based on organophosphorus coupling molecules: from metal phosphonates to surface modification of oxides. C R Chim 6(8–10):1153–1164. https://doi.org/10.1016/j.crci.2003.07.006
45. Okusa H, Kurihara K, Kunitake T (1994) Chemical modification of molecularly smooth mica surface and protein attachment. Langmuir 10(10):3577–3581. https://doi.org/10.1021/la00022a034
46. Iijima M, Takenouchi S, Lenggoro IW, Kamiya H (2011) Effect of additive ratio of mixed silane alkoxides on reactivity with TiO_2 nanoparticle surface and their stability in organic solvents. Adv Powder Technol 22(5):663–668. https://doi.org/10.1016/j.apt.2010.09.015
47. Aronoff YG, Chen B, Lu G, Seto C, Schwartz J, Bernasek S (1997) Stabilization of self-assembled monolayers of carboxylic acids on native oxides of metals. J Am Chem Soc 119(2):259–262. https://doi.org/10.1021/ja953848+

48. Khaled SM, Sui R, Charpentier PA, Rizkalla AS (2007) Synthesis of TiO_2–PMMA nanocomposite: using methacrylic acid as a coupling agent. Langmuir 23(7):3988–3995. https://doi.org/10.1021/la062879n
49. Marcinko S, Fadeev AY (2004) Hydrolytic stability of organic monolayers supported on TiO_2 and ZrO_2. Langmuir 20(6):2270–2273. https://doi.org/10.1021/la034914l
50. Gao W, Dickinson L, Grozinger C, Morin FG, Reven L (1996) Self-assembled monolayers of alkylphosphonic acids on metal oxides. Langmuir 12(26):6429–6435. https://doi.org/10.1021/la9607621
51. Freedman LD, Doak GO (1957) The preparation and properties of phosphonic acids. Chem Rev 57(3):479–523. https://doi.org/10.1021/cr50015a003
52. Imai Y, Terahara A, Hakuta Y, Matsui K, Hayashi H, Ueno N (2009) Transparent poly (bisphenol A carbonate)-based nanocomposites with high refractive index nanoparticles. Eur Polym J 45(3):630–638. https://doi.org/10.1016/j.eurpolymj.2008.12.031
53. Sarkar S, Bekyarova E, Niyogi S, Haddon RC (2011) Diels–Alder chemistry of graphite and graphene: graphene as diene and dienophile. J Am Chem Soc 133(10):3324–3327. https://doi.org/10.1021/ja200118b
54. Dreyer DR, Park S, Bielawski CW, Ruoff RS (2010) The chemistry of graphene oxide. Chem Soc Rev 39(1):228–240. https://doi.org/10.1039/b917103g
55. Sugahara Y (2014) Chemical processes employing inorganic layered compounds for inorganic and inorganic-organic hybrid materials. J Ceram Soc Jpn 122(1427):523–529. https://doi.org/10.2109/jcersj2.122.523
56. Takahashi S, Nakato T, Hayashi S, Sugahara Y, Kuroda K (1995) Formation of methoxy-modified interlayer surface via the reaction between methanol and layered perovskite $HLaNb_2O_7 \cdot xH_2O$. Inorg Chem 34(20):5065–5069. https://doi.org/10.1021/ic00124a023
57. Tahara S, Sugahara Y (2003) Interlayer surface modification of the protonated triple-layered perovskite $HCa_2Nb_3O_{10} \cdot xH_2O$ with *n*-alcohols. Langmuir 19(22):9473–9478. https://doi.org/10.1021/la0343876
58. Tahara S, Ichikawa T, Kajiwara G, Sugahara Y (2007) Reactivity of the Ruddlesden–Popper phase $H_2La_2Ti_3O_{10}$ with organic compounds: intercalation and grafting reactions. Chem Mater 19(9):2352–2358. https://doi.org/10.1021/cm0623662
59. Suzuki H, Notsu K, Takeda Y, Sugimoto W, Sugahara Y (2003) Reactions of alkoxyl derivatives of a layered perovskite with alcohols: substitution reactions on the interlayer surface of a layered perovskite. Chem Mater 15(3):636–641. https://doi.org/10.1021/cm0200902
60. Takeda Y, Suzuki H, Notsu K, Sugimoto W, Sugahara Y (2006) Preparation of a novel organic derivative of the layered perovskite bearing $HLaNb_2O_7 \cdot nH_2O$ interlayer surface trifluoroacetate groups. Mater Res Bull 41(4):834–841. https://doi.org/10.1016/j.materresbull.2005.10.004
61. Shimada A, Yoneyama Y, Tahara S, Mutin PH, Sugahara Y (2009) Interlayer surface modification of the protonated ion-exchangeable layered perovskite $HLaNb_2O_7 \bullet xH_2O$ with organophosphonic acids. Chem Mater 21(18):4155–4162. https://doi.org/10.1021/cm900228c
62. Gardolinski JEFC, Lagaly G, Czank M (2004) On the destruction of kaolinite and gibbsite by phenylphosphonic, phenylphosphinic and phenylarsonic acids: evidence for the formation of new Al compounds. Clay Miner 39(4):391–404. https://doi.org/10.1180/0009855043940142
63. Idota N, Fukuda S, Tsukahara T, Sugahara Y (2015) Preparation of thermoresponsive nanosheets exhibiting phase transitions in water via surface modification of layered perovskite nanosheets with poly(*N*-isopropylacrylamide) (PNIPAAm). Chem Lett 44(2):203–205. https://doi.org/10.1246/cl.140956
64. Kimura N, Kato Y, Suzuki R, Shimada A, Tahara S, Nakato T, Matsukawa K, Mutin PH, Sugahara Y (2014) Single- and double-layered organically modified nanosheets by selective interlayer grafting and exfoliation of layered potassium hexaniobate. Langmuir 30(4):1169–1175. https://doi.org/10.1021/la404223x
65. Althues H, Henle J, Kaskel S (2007) Functional inorganic nanofillers for transparent polymers. Chem Soc Rev 36(9):1454. https://doi.org/10.1039/b608177k

66. Lü C, Yang B (2009) High refractive index organic–inorganic nanocomposites: design, synthesis and application. J Mater Chem 19(19):2884. https://doi.org/10.1039/b816254a
67. Kobayashi M, Saito H, Boury B, Matsukawa K, Sugahara Y (2013) Epoxy-based hybrids using TiO_2 nanoparticles prepared via a non-hydrolytic sol-gel route. Appl Organomet Chem 27(11):673–677. https://doi.org/10.1002/aoc.3027
68. Fujita M, Idota N, Matsukawa K, Sugahara Y (2015) Preparation of oleyl phosphate-modified TiO_2/poly(methyl methacrylate) hybrid thin films for investigation of their optical properties. J Nanomater 2015:1–7. https://doi.org/10.1155/2015/297197
69. Tao P, Li Y, Rungta A, Viswanath A, Gao J, Benicewicz BC, Siegel RW, Schadler LS (2011) TiO_2 nanocomposites with high refractive index and transparency. J Mater Chem 21(46):18623. https://doi.org/10.1039/c1jm13093e
70. Maeda S, Fujita M, Idota N, Matsukawa K, Sugahara Y (2016) Preparation of transparent bulk TiO_2/PMMA hybrids with improved refractive indices via an in situ polymerization process using TiO_2 nanoparticles bearing PMMA chains grown by surface-initiated atom transfer radical polymerization. ACS Appl Mater Interfaces 8(50):34762–34769. https://doi.org/10.1021/acsami.6b10427
71. Barbey R, Lavanant L, Paripovic D, Schüwer N, Sugnaux C, Tugulu S, Klok H-A (2009) Polymer brushes via surface-initiated controlled radical polymerization: synthesis, characterization, properties, and applications. Chem Rev 109(11):5437–5527. https://doi.org/10.1021/cr900045a
72. Wang K, Chen L, Wu J, Toh ML, He C, Yee AF (2005) Epoxy nanocomposites with highly exfoliated clay: mechanical properties and fracture mechanisms. Macromolecules 38(3):788–800. https://doi.org/10.1021/ma048465n
73. Sinha Ray S, Okamoto M (2003) Polymer/layered silicate nanocomposites: a review from preparation to processing. Prog Polym Sci 28(11):1539–1641. https://doi.org/10.1016/j.progpolymsci.2003.08.002
74. Cui Y, Kundalwal SI, Kumar S (2016) Gas barrier performance of graphene/polymer nanocomposites. Carbon 98:313–333. https://doi.org/10.1016/j.carbon.2015.11.018
75. Asai Y, Ariake Y, Saito H, Idota N, Matsukawa K, Nishino T, Sugahara Y (2014) Layered perovskite nanosheets bearing fluoroalkoxy groups: their preparation and application in epoxy-based hybrids. RSC Adv 4(51):26932. https://doi.org/10.1039/c4ra01777c
76. Li X, Cheng Y, Zhang H, Wang S, Jiang Z, Guo R, Wu H (2015) Efficient CO_2 capture by functionalized graphene oxide nanosheets as fillers to fabricate multi-permselective mixed matrix membranes. ACS Appl Mater Interfaces 7(9):5528–5537. https://doi.org/10.1021/acsami.5b00106
77. Rhee CH, Kim Y, Lee JS, Kim HK, Chang H (2006) Nanocomposite membranes of surface-sulfonated titanate and Nafion® for direct methanol fuel cells. J Power Sources 159(2):1015–1024. https://doi.org/10.1016/j.jpowsour.2005.12.006
78. Gao H, Sun Y, Zhou J, Xu R, Duan H (2013) Mussel-inspired synthesis of polydopamine-functionalized graphene hydrogel as reusable adsorbents for water purification. ACS Appl Mater Interfaces 5(2):425–432. https://doi.org/10.1021/am302500v
79. Bishop KJM, Wilmer CE, Soh S, Grzybowski BA (2009) Nanoscale forces and their uses in self-assembly. Small 5(14):1600–1630. https://doi.org/10.1002/smll.200900358
80. DeVries GA, Brunnbauer M, Hu Y, Jackson AM, Long B, Neltner BT, Uzun O, Wunsch BH, Stellacci F (2007) Divalent metal nanoparticles. Science 315(5810):358–361. https://doi.org/10.1126/science.1133162
81. Liu K, Nie Z, Zhao N, Li W, Rubinstein M, Kumacheva E (2010) Step-growth polymerization of inorganic nanoparticles. Science 329(5988):197–200. https://doi.org/10.1126/science.1189457
82. Nie Z, Fava D, Kumacheva E, Zou S, Walker GC, Rubinstein M (2007) Self-assembly of metal–polymer analogues of amphiphilic triblock copolymers. Nat Mater 6(8):609–614. https://doi.org/10.1038/nmat1954

83. Sardar R, Shumaker-Parry JS (2008) Asymmetrically functionalized gold nanoparticles organized in one-dimensional chains. Nano Lett 8(2):731–736. https://doi.org/10.1021/nl073154m
84. He J, Liu Y, Babu T, Wei Z, Nie Z (2012) Self-assembly of inorganic nanoparticle vesicles and tubules driven by tethered linear block copolymers. J Am Chem Soc 134(28):11342–11345. https://doi.org/10.1021/ja3032295
85. Gao B, Arya G, Tao AR (2012) Self-orienting nanocubes for the assembly of plasmonic nanojunctions. Nat Nanotechnol 7(7):433–437. https://doi.org/10.1038/nnano.2012.83
86. Chen C-L, Zhang P, Rosi NL (2008) A new peptide-based method for the design and synthesis of nanoparticle superstructures: construction of highly ordered gold nanoparticle double helices. J Am Chem Soc 130(41):13555–13557. https://doi.org/10.1021/ja805683r

Chapter 14
Design of Element Blocks for Photoresponsive Organosiloxane-Based Materials

Sufang Guo, Kazuyuki Kuroda, and Atsushi Shimojima

Abstract In this chapter, preparation, structures, and photoresponsive properties of various azobenzene–siloxane hybrid nanomaterials based on the design of element blocks are described. Hydrolysis and polycondensation reactions accompanied by self-assembly of different types of azobenzene-functionalized alkoxysilane precursors yield hybrids with ordered mesostructures. Lamellar films show reversible *d*-spacing changes and macroscopic bending–unbending motion by partial *trans–cis* photoisomerization of the azobenzene groups. Further, incorporation of a cage oligosiloxane into the precursor leads to the formation of cylindrical assemblies, in which efficient photoisomerization of azobenzene is achieved. These findings will contribute to the creation of novel photoresponsive materials.

Keywords Photoresponsive materials · Azobenzene · Siloxane · Self-assembly · Sol–gel processing

S. Guo
Kagami Memorial Research Institute for Materials Science and Technology, Waseda University, Shinjuku-ku, Tokyo, Japan

Present address: School of Materials Science and Technology, China University of Geosciences, Haidian District, Beijing, China

K. Kuroda
Kagami Memorial Research Institute for Materials Science and Technology, Waseda University, Shinjuku-ku, Tokyo, Japan

Department of Applied Chemistry, Waseda University, Shinjuku-ku, Tokyo, Japan

A. Shimojima (✉)
Department of Applied Chemistry, Waseda University, Shinjuku-ku, Tokyo, Japan
e-mail: shimojima@waseda.jp

Y. Chujo (ed.), *New Polymeric Materials Based on Element-Blocks*,
https://doi.org/10.1007/978-981-13-2889-3_14

14.1 Introduction

Photoresponsive materials are attracting increasing attention because of their wide range of potential applications [1–3]. Azobenzene is one of the most widely used chromophores for the preparation of photoresponsive materials. Azobenzene undergoes quick and reversible isomerization between the *trans* and *cis* states upon UV and visible light irradiation, which is accompanied by large changes in molecular shape, size, and polarity. Azobenzene-containing polymer particles, liquid crystal elastomers, and liquid crystal gels displaying macroscopic size/shape changes and bending–unbending motions upon photoirradiation have been reported. More recently, photoresponsive materials capable of motions such as oscillation, rolling, inchworm movement, and wavelike-movement have been prepared [4–7]. These materials have potential applications in smart sensors, motors, and actuators. To achieve such unique photoresponses, free volume and mobility should be guaranteed to realize efficient photoisomerization of azobenzene. Additionally, azobenzene groups should be regularly arranged or selectively excited with polarized light so that the molecular-level shape change or motions manifest at the macroscopic scale. The aforementioned soft polymer materials have satisfied these requirements. For example, in liquid crystalline elastomers, azobenzene groups can be unidirectionally aligned in the film by a rubbing procedure, with the flexibility of the polymer chains ensuring adequate free volume and mobility. Furthermore, photoresponsive azobenzene–inorganic nanocomposites have also been investigated. The inorganic moiety can act as scaffolds, controlling the spatial arrangement of the azobenzene moiety, in addition to enhancing the mechanical property and thermal stability. Organosiloxane-based materials, consisting of Si–O–Si and Si–C bonds, are particularly important for producing photoresponsive materials because of their high stability, transparency, and excellent controllability of their structures.

14.2 General Approach to Nanostructured Azobenzene–Siloxane Hybrids

Organosiloxane materials are easily obtained by hydrolysis and polycondensation of organoalkoxysilanes through sol–gel processes. By controlling the reaction conditions, organosiloxanes with various structures and morphologies can be obtained. The use of azobenzene-containing organosilane precursors allows the introduction of azobenzene groups into siloxane networks at the molecular level. Controlling the nanostructure is the key to realizing novel photoresponsive properties. Two self-assembly approaches are commonly used to obtain organosiloxane materials with ordered nanostructures.

Ordered mesoporous organosiloxane-based materials can be prepared by a surfactant-directed process. Amphiphilic molecules and block copolymers interact with hydrolyzed alkoxysilane species and self-assemble into ordered nanocomposites with various mesostructures such as two-dimensional (2D) hexagonal and three-dimensional cubic structures. Mesopores are generated after the removal of organic moieties by calcination or extraction. Moreover, organically functionalized mesoporous materials can be prepared by co-condensation of organoalkoxysilanes and tetraalkoxysilanes [8]. This method has been employed for designing photoresponsive azobenzene-functionalized mesoporous materials. The azobenzene groups are grafted onto the pore walls by using precursors having a pendant azobenzene group, $RSi(OR')_3$ (R = azobenzene group, R′ = Me, Et, etc.), while they are embedded in the siloxane network by using precursors with a bridging azobenzene group, $(R'O)_3Si\text{-}R\text{-}Si(OR')_3$ [9–15]. Photoinduced reversible changes in pore sizes and photo-controlled release of guest molecules have been achieved.

In another approach, ordered organosiloxanes are formed by the surfactant-free self-assembly of organosilane precursors through weak intermolecular interactions such as Van der Waals force, H-bonding, and π–π interaction [16, 17]. It is expected that the organic groups would be distributed more homogeneously and with a higher density compared to those in surfactant-templated materials. Liu et al. reported surfactant-free self-assembly of 4,4′-bis(3-triethoxysilylpropylureido)azobenzene into a lamellar structure by intermolecular H-bonding [18]. However, this material exhibited almost no photoresponse possibly due to the strong intermolecular H-bonding interactions derived from the bis-ureido groups that inhibited the photoisomerization of the bridging azobenzene groups.

14.3 Photoresponsive Lamellar Azobenzene–Siloxane Hybrids by Self-Assembly

In order to improve the photoresponsive properties of the self-assembled azobenzene–siloxane hybrids, we designed and synthesized several azobenzene-functionalized alkoxysilane precursors through hydrosilylation reactions [19, 20]. To avoid strong intermolecular interactions, the Si atoms were connected with azobenzene by simple $-O(CH_2)_3-$ chains. The structures of the precursors with pendant azobenzene groups (**P1**–**P2**) and those with bridging azobenzene groups (**P3**–**P5**) are shown in Fig. 14.1. All the precursors show quick and reversible *trans*–*cis* photoisomerization in ethanol solutions upon UV–Vis irradiations.

When the precursors **P1** and **P2** were hydrolyzed in an ethanol solvent under an acidic condition, platelike particles (Fig. 14.2) were precipitated [19]. X-ray diffraction (XRD) analysis revealed the formation of ordered mesostructures through self-assembly. When THF was used as the solvent, thin films (**H1** and **H2**) were obtained by spin coating the hydrolyzed solutions on glass plates. Their XRD patterns (Fig. 14.3a) revealed that lamellar structures (Fig. 14.1) were obtained. The large difference in the *d*-spacings of **H1** and **H2** (3.20 nm and 2.37 nm, respectively) can be attributed to the different tilt angles of the azobenzene groups on the siloxane layers, suggesting that the arrangement of azobenzene was influenced by the type of silyl

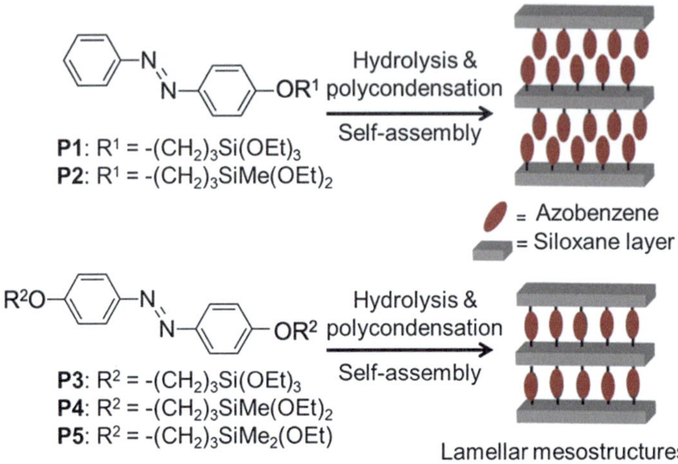

Fig. 14.1 Structures of precursors **P1**–**P5** and their formation of lamellar structures

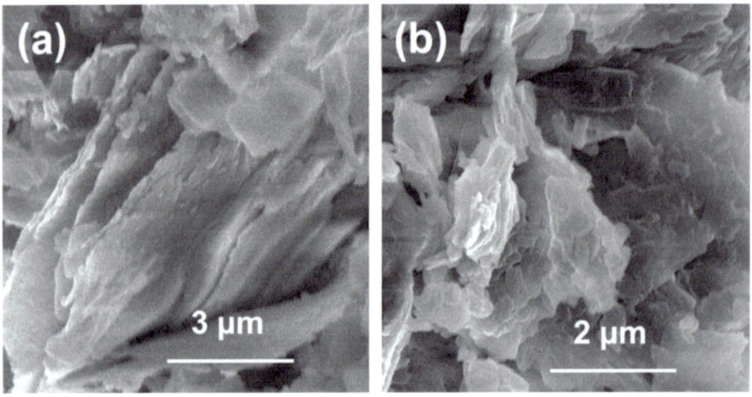

Fig. 14.2 Scanning electron microscopy (SEM) images of the precipitates formed from (**a**) **P1** and (**b**) **P2.** (Adapted from [19] with permission from The Royal Society of Chemistry)

groups. Our previous research showed that alkylalkoxysilane with a $SiMe(OMe)_2$ group did not self-assemble into ordered structures [16]. In the case of **P2**, the interaction between the azobenzene groups (π–π interactions) might play a role in the formation of ordered structures. Solid-state ^{29}Si MAS NMR analysis showed that **H1** had a siloxane network consisting of R**Si**$(OSi)_2(OH)$ and R**Si**$(OSi)_3$ units (R = azobenzene group), whereas **H2** was an assembly of dimers (R(HO)MeSi-O-SiMe (OH)R).

We investigated the photoresponsive properties of **H1** and **H2**. Upon UV (340 nm) and subsequent Vis (>420 nm) irradiations, reversible and partial *trans*–*cis* photoisomerization was observed by UV–Vis spectroscopic analysis. Meanwhile, reversible changes of the lamellar periodicity were observed by XRD

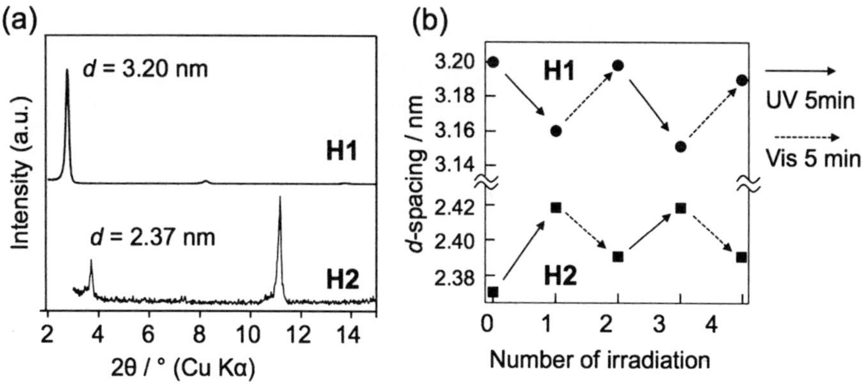

Fig. 14.3 (**a**) XRD patterns of **H1** and **H2** and (**b**) variations in the *d*-spacings upon UV and Vis irradiations. (Adapted from [19] with permission from The Royal Society of Chemistry)

(Fig. 14.3b). After UV irradiation, a slight decrease (ca. 0.04 nm) in *d*-spacing was observed for **H1**, while a slight increase (ca. 0.03 nm) was observed for **H2**. For both films, subsequent Vis irradiation recovered the *d*-spacings, and no fatigue was observed after several cycles of UV and Vis irradiations. It is likely that the different tilt angles of the azobenzene groups in **H1** and **H2** caused the opposite changes in the *d*-spacings. This is the first report that demonstrated the photoinduced structural variation in the self-assembled azobenzene–siloxane hybrids, although similar phenomena have been reported in azobenzene-intercalated layered silicate composites, where azobenzene is not covalently attached to the silicate surface [21, 22].

Lamellar hybrid films **H3**, **H4,** and **H5** were also obtained in a similar manner from the precursors **P3**, **P4,** and **P5** with bridging azobenzene groups [20]. Hydrolyzed **P3** demonstrated a strong self-assembling ability due to the large number of Si–OH groups compared to the other two precursors. However, no photoisomerization and *d*-spacing changes were observed for **H3** upon UV–Vis irradiations, which is probably because both the ends of azobenzene were immobilized on the siloxane networks. **H4** and **H5** showed partial photoisomerization; however, no reversible structural change was observed.

14.4 Photoinduced Macroscopic Bending of Lamellar Azobenzene–Siloxane Hybrids

As described in the previous section, hydrolyzed organoalkoxysilanes with either pendant or bridging azobenzene groups possess self-assembling ability to form lamellar mesostructures without strong intermolecular interactions such as H-bonding. To achieve more dynamic photoresponses of the azobenzene–siloxane hybrids, we synthesized new precursors, **P6** (pendant type) and **P7** (bridge type),

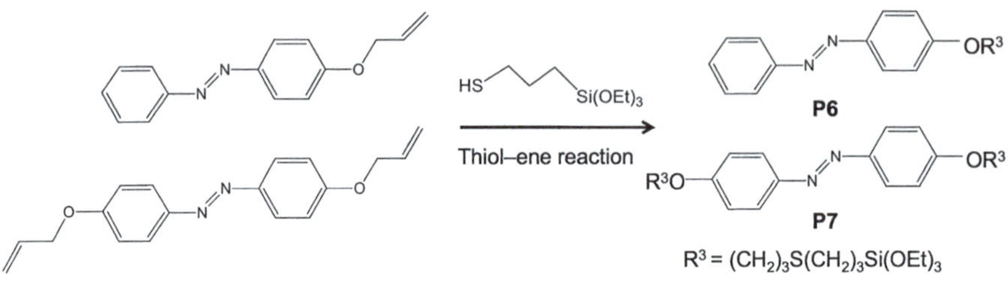

Fig. 14.4 Synthesis of precursors **P6** and **P7** through thiol–ene reactions

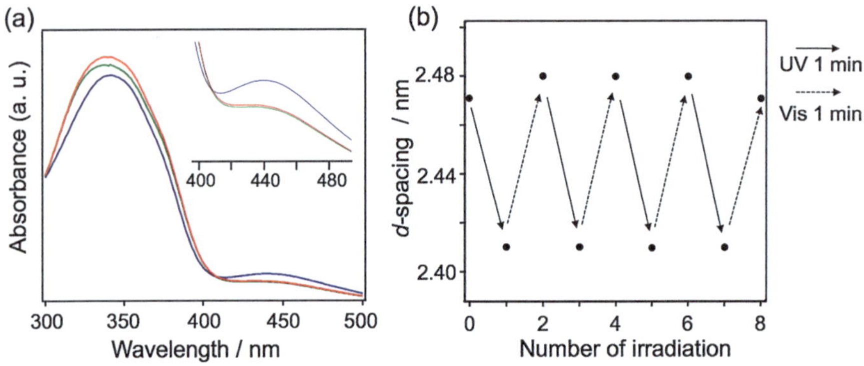

Fig. 14.5 (**a**) UV–Vis spectra of **H6** (green, before irradiation; blue, after UV irradiation (1 min); red, after subsequent Vis irradiation (1 min)) and (**b**) variation in the *d*-spacing of **H6** upon UV and Vis irradiations

through thiol–ene reactions (Fig. 14.4) [23]. Compared with **P1** and **P3**, the organic linkers between azobenzene and Si atom are longer and more flexible.

Lamellar hybrid films **H6** and **H7** were obtained from **P6** and **P7**, respectively, through sol–gel processes. **H6** showed reversible and partial *trans–cis* photoisomerization (Fig. 14.5a) and, compared with **H1**, displayed much quicker (10 s) and larger magnitude (0.06 nm) of reversible *d*-spacing changes (Fig. 14.5b). Reversible photoisomerization of the bridging azobenzene groups was also observed in **H7**, although there was no change in lamellar periodicity. These results suggest the improved mobility of the azobenzene groups owing to the increased length and higher flexibility of the organic linkers (cf. **H1** and **H3**).

A free-standing film could be prepared from a mixture of **P6** and **P7**. The film was flexible, had several microns in thickness, and possessed an ordered lamellar structure. Upon UV irradiation, the film bent away from the UV source quickly with a very large bending angle (180°), as shown in Fig. 14.6. The film recovered the original shape upon subsequent Vis irradiation. This reversible bending–unbending can be repeated without any detectable fatigue. The bending was attributed to the partial *trans–cis* isomerization of the film (Fig. 14.6). Due to the high concentration of azobenzenes and large thickness of the film, UV irradiation can induce *trans*-to-

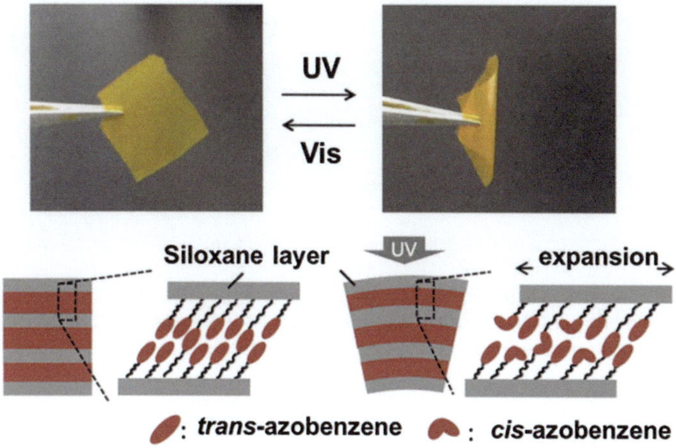

Fig. 14.6 Photoinduced bending of the free-standing film and the proposed mechanism. (Adapted with permission from [23]. Copyright 2015, American Chemical Society)

cis isomerization only in the surface region of the film. Because the bent *cis* isomers possess larger sizes in the horizontal plane than the *trans* isomers do, expansion occurred in the horizontal direction and was transduced to the inner part of the film through the bridged azobenzene groups, leading to the bending of the film away from UV source.

The free-standing hybrid film also exhibited excellent thermal and mechanical properties. Its thermogravimetry (TG) curves in both air and N_2 showed almost no weight loss up to 200 °C. Besides, in the differential scanning calorimetry (DSC) curve, no phase transition was observed in the temperature range of 0–150 °C. The dynamic mechanical analysis of the film at room temperature suggested that the storage modulus was ca. 0.2 GPa, which is much larger than that of rubbery liquid crystal elastomers (<1 MPa). The reversible photoinduced bending–unbending of the film was also observed in a wide temperature range from room temperature up to 130 °C.

14.5 Controlling the Mesostructure of Azobenzene–Siloxane Hybrid

Although some unique photoresponsive properties have been observed for the aforementioned lamellar azobenzene–siloxane hybrids, the isomerization degrees of azobenzene were very low. This is probably due to the close packing of the azobenzene groups in the lamellar structure. To address this issue, we introduced a bulky cage-type oligosiloxane instead of the alkoxysilyl groups to adjust the arrangement of the azobenzene groups.

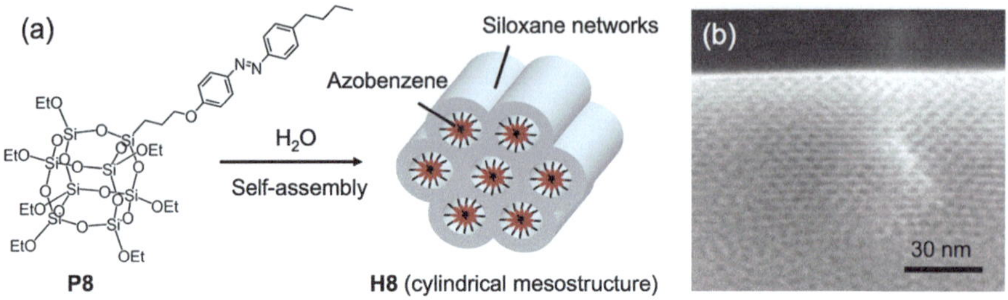

Fig. 14.7 (**a**) Schematic depicting the self-assembly of **P8** into a mesostructure consisting of cylindrical assemblies and (**b**) HAADF-STEM image of the cross-section of the **H8** film. (Adapted from [24] with permission from The Chemical Society of Japan)

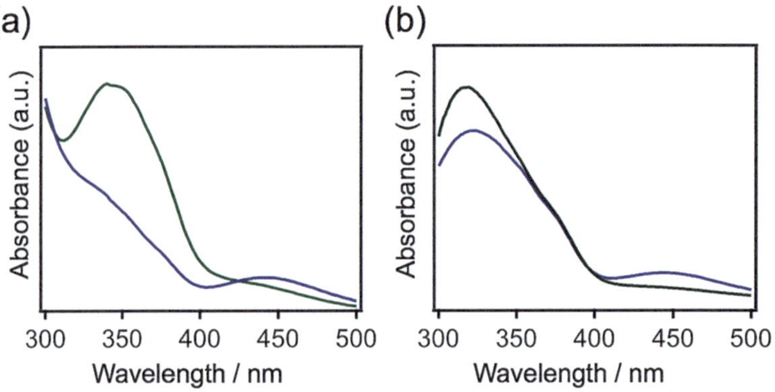

Fig. 14.8 UV–Vis spectra of (**a**) **H8** and (**b**) **H9**: green, before irradiation; blue, after UV irradiation (3 min)

We synthesized an azobenzene-containing cage-like siloxane precursor **P8** (Fig. 14.7a) [24]. The siloxane cage possesses seven OEt groups and one azobenzene group having flexible alkyl chains at both ends. A yellow transparent thin film (**H8**) was obtained by spin coating the THF solution of hydrolyzed **P8**.

The XRD pattern of **H8** indicated an ordered structure with a *d*-spacing of 3.52 nm. In addition, the 2D XRD pattern and the high-angle annular dark-field (HAADF)-scanning transmission microscope (STEM) image (Fig. 14.7b) clearly revealed a distorted 2D hexagonal structure. The mesostructure was considered to form through the self-assembly of the amphiphilic hydrolyzed **P8** into cylinders, with the azobenzene chains inside and the siloxane cages outside. We have also synthesized a similar precursor **P9** possessing a single $Si(OEt)_3$ group instead of a siloxane cage. A hybrid film (**H9**) with a lamellar structure was obtained. This indicated that the bulky siloxane cage contributed to the formation of the cylindrical assemblies.

Compared with the lamellar hybrid film (**H9**), a much larger extent of reversible *trans–cis* isomerization of azobenzene was observed for **H8** (Fig. 14.8). Note that when azobenzene is doped in silica gel matrix or covalently attached on mesoporous silica, photoisomerization is severely inhibited or requires a considerably longer time to occur [8, 25–27]. Thus, the incorporation of siloxane cages probably provided a larger free volume for isomerization.

In another work, we prepared a 3D network consisting of cage siloxanes and azobenzene linkers by hydrosilylation reactions of 4, 4′-diallyloxy-azobenzene and octahydridosilsesquioxane ($H_8Si_8O_{12}$) [28]. The reaction degree of H-POSS was ca. 70%. The obtained gel shrinks after drying in air and swells when immersed in organic solvents such as toluene and hexane. In the swelling state, a very large extent of photoinduced *trans–cis* isomerization of azobenzenes was observed. Such photoresponsive property of the gel is of great significance in host–guest applications.

14.6 Conclusions

In this chapter, our recent efforts in the synthesis of a new class of photoresponsive azobenzene–siloxane hybrids based on the concept of element blocks were introduced. Various structures such as lamellar structures, a distorted 2D hexagonal structure, and a 3D network structure have been obtained from different types of precursors. Most importantly, molecular design of the azobenzene-functionalized alkoxysilanes led to the formation of self-assembled lamellar hybrids with unique properties, i.e., reversible variation of the lamellar periodicity and bending–unbending motions under photoirradiation. Furthermore, incorporation of a cage siloxane unit allowed controlling the mesostructure to improve the efficiency of photoisomerization of azobenzene. These materials are promising for many applications. Further design of the precursors will ensure the creation of azobenzene–siloxane hybrids with novel structures and photoresponsive properties.

References

1. Ercole F, Davis TP, Evans RA (2010) Photo-responsive systems and biomaterials: photochromic polymers, light-triggered self-assembly, surface modification, fluorescence modulation and beyond. Polym Chem 1:37–54
2. Yagai S, Kitamura A (2008) Recent advances in photoresponsive supramolecular self-assemblies. Chem Soc Rev 37:1520–1529
3. Bandarab HMD, Burdette SC (2012) Photoisomerization in different classes of azobenzene. Chem Soc Rev 41:1809–1825
4. White TJ, Tabiryan NV, Serak SV et al (2008) A high frequency photodriven polymer oscillator. Soft Matter 4:1796–1798
5. Yamada M, Kondo M, Mamiya J et al (2008) Photomobile polymer materials: toward light-driven motors. Angew Chem Int Ed 47:4986–4988
6. Yamada M, Kondo M, Miyasato R et al (2009) Photomobile polymer materials: various three-dimensional movements. J Mater Chem 19:60–62

7. Gelebart AH, Mulder DJ, Varga M et al (2017) Making waves in a photoactive polymer film. Nature 546(29):632–647
8. Hoffmann F, Cornelius M, Morell J et al (2006) Silica-based mesoporous organic-inorganic hybrid materials. Angew Chem Int Ed 45:3216–3251
9. Liu N, Chen Z, Dunphy DR et al (2003) Photoresponsive nanocomposites formed by self-assembly of an azobenzene-modified silane. Angew Chem Int Ed 42:1731–1734
10. Liu N, Dunphy DR, Atanassov P et al (2003) Photoregulation of mass transport through a photoresponsive azobenzene modified nanoporous membrane. Nano Lett 4(4):551–554
11. Angelos S, Choi E, Vogtle F et al (2007) Photo-driven exclusion of molecules from mesostructured silica nanoparticles. J Phys Chem C 111(18):6589–6592
12. Angelos S, Yang YW, Khashab NM et al (2009) Dual controlled nanoparticles exhibiting AND logic. J Am Chem Soc 131(32):11344–11346
13. Lu J, Choi E, Tamanoi F et al (2008) Light-activated nanoimpeller-controlled drug release in cancer cells. Small 4(4):421–426
14. Alvaro M, Benitez M, Das D et al (2005) Reversible porosity changes in photoresponsive azobenzene-containing periodic mesoporous silicas. Chem Mater 17(20):4958–4964
15. Besson E, Mehdi V, Lerner DA et al (2005) Photoresponsive ordered hybrid materials containing a bridged azobenzene group. J Mater Chem 15:803–809
16. Shimojima A, Kuroda K (2006) Designed synthesis of nanostructured siloxane–organic hybrids from amphiphilic silicon-based precursors. Chem Rec 6:53–63
17. Chemtob A, Ni L, Croutxé-Barghorn C et al (2014) Ordered hybrids from template-free organosilane self-assembly. Chem Eur J 20:1790–1806
18. Liu N, Yu K, Smarsly B et al (2002) Self-directed assembly of photoactive hybrid silicates derived from an azobenzene-bridged silsesquioxane. J Am Chem Soc 124(49):14540–14541
19. Guo S, Sugawara-Narutaki A, Okubo T et al (2013) Synthesis of ordered photoresponsive azobenzene–siloxane hybrids by self-assembly. J Mater Chem C 1:6989–6995
20. Guo S, Chaikittisilp W, Okubo T et al (2014) Azobenzene–siloxane hybrids with lamellar structures from bridged-type alkoxysilyl precursors. RSC Adv 4:25319–25325
21. Ogawa M, Ishii T, Miyamoto N et al (2001) Photocontrol of the basal spacing of azobenzene-magadiite intercalation compound. Adv Mater 13(14):1107–1109
22. Nabetani Y, Takamura H, Masui D et al (2011) A photoactivated artificial muscle model unit: reversible, photoinduced sliding of nanosheets. J Am Chem Soc 133:17130–17133
23. Guo S, Matsukawa K, Miyata T et al (2015) Photoinduced bending of self-assembled azobenzene–siloxane hybrid. J Am Chem Soc 137:15434–15440
24. Guo S, Sasaki J, Tsujiuchi S et al (2017) Role of cubic siloxane cages in mesostructure formation and photoisomerization of azobenzene–siloxane hybrid. Chem Lett 46(8):1237–1239
25. Ide T, Ozama Y, Matsui K (2011) Photochemistry of azobenezene in sol-gel systems. J Non-Cryst Solids 357:100–104
26. Ueda M, Kim HB, Ichimura K (1994) Photochemical and thermal isomerization of azobenzene derivatives in sol-gel bulk materials. Chem Mater 6(10):1771–1775
27. Ueda M, Kim HB, Ikeda T et al (1993) Photoisomerizability of an azobenzene covalently attached to silica-gel matrix. J Non-Cryst Solids 163:125–132
28. Guo S, Okubo T, Kuroda K et al (2016) A photoresponsive azobenzene-bridged cubic silsesquioxane network. J Sol-Gel Sci Technol 79:262–269

Chapter 15
How Can We Control the "Element-Blocks" in Transition Metal Oxide Crystals?

Katsuhisa Tanaka and Koji Fujita

Abstract Crystalline transition metal oxides show intriguing properties and important functionalities relevant to the electrical conduction, dielectricity, magnetism, and optical phenomenon. It seems that the concept of "element-blocks" and "element-block polymers" suitable to the organic polymers and organic-inorganic hybrid materials is *not* adequate *at all* for the consideration of structure and properties of transition metal oxide crystals in which ionic bonds between cations and oxide ions are predominant. However, it is possible to consider the oxygen polyhedron, at the center of which the transition metal ion is set, to be an element-block and to regard the whole crystal structure or a part of the structure as an element-block polymer. Also, one can modify the structure of element-block polymers inside the transition metal oxide crystals so as to change the electrical, magnetic, and optical properties drastically, although the process to realize the modification is not so sophisticated as organic polymers and organic-inorganic hybrid materials, for which a variety of chemical reactions are effectively utilized. We exemplify some transition metal oxide crystals for which the control of element-blocks is possible to achieve a drastic change in the magnetic, dielectric, and optical properties. We present three topics: (1) ferrimagnetism induced in spinel-type $ZnFe_2O_4$ by exchange of cations; (2) ferromagnetism caused by the change of cell volume or crystal system in perovskite-type $EuTiO_3$, $EuZrO_3$, and $EuHfO_3$; and (3) piezoelectricity as well as optical second-order nonlinearity realized by oxygen octahedral rotation in Ruddlesden-Popper phases, Na*R*TiO_4 (*R*: rare-earth element). The methods to control the element-blocks mentioned in the three topics are to change the way by which element-blocks are connected with each other, to change the chemical interaction or the overlap of atomic orbitals between element-blocks, and to make local displacement (rotation) of element-blocks to alter drastically the overall symmetry of the long-range structure, respectively.

K. Tanaka (✉) · K. Fujita
Department of Material Chemistry, Graduate School of Engineering, Kyoto University, Kyoto, Japan
e-mail: tanaka@dipole7.kuic.kyoto-u.ac.jp

Y. Chujo (ed.), *New Polymeric Materials Based on Element-Blocks*,
https://doi.org/10.1007/978-981-13-2889-3_15

Keywords Transition metal oxide · Magnetic properties · Dielectric properties · Electronic structure

15.1 Introduction

The concept of an "element-block polymer" is an extended version of a traditional organic polymer composed of carbon atoms. Although some inorganic polymers such as polythiazyl and polyphosphazene have been found, the organic polymers are much more predominant; there exist a huge number of organic polymers based on carbon atoms. This is because not only the carbon atom can form single, double, and triple bonds but also the catenation can occur much more easily for carbon atom than for other elements. Nonetheless, the number of polymers comprising a variety of elements other than carbon, i.e., so-called hetero elements often referred to in the organic chemistry, is increasing rapidly. As for the solid-state oxide dealt with in the present chapter, a polyoxoanion, a polymerized state of oxoanions, each of which contains a metal ion at the center of its coordination polyhedron, is a prototype of inorganic element-block polymer. The polyoxoanion is formed for transition elements including V, Nb, Mo, Ta, and W as well as metalloid elements like B, Si, P, S, Se, and Te. As mentioned below, we can find an element-block polymer composed of transition elements and oxygen even in ionic crystals of transition metal oxides.

The transition metal oxides have attracted considerable attention for the last several decades because of their peculiar electronic structures; intriguing physical properties including electrical conduction, dielectricity, magnetism, and optical phenomenon; and functionalities of significant benefits to many practical applications. The so-called high-T_c (critical temperature) superconductors based on cuprate, like $(La,Ba)_2CuO_4$ [1], $YBa_2Cu_3O_7$ [2], $Bi_2Ca_2Sr_2Cu_3O_{10}$ [3], and $HgBa_2CaCu_2O_{6+\delta}$ [4], discovered at the end of the 1980s to the beginning of the 1990s are an archetypal example, where the doping of the two-dimensional CuO_4 layers with positive holes leads to the formation of Cooper pair which is stable even at 40–130 K beyond the BCS (Bardeen-Cooper-Schrieffer) theory.

Also, interesting and important dielectric and optical properties such as ferroelectricity, pyroelectricity, piezoelectricity, and optical second-order nonlinearity accompanying those dielectric properties have been often found in many transition metal oxides. In particular, not only perovskite oxides such as $BaTiO_3$, $Pb(Zr,Ti)O_3$, $(Pb,La)(Zr,Ti)O_3$, $KNbO_3$, and $SrTiO_3$ but lithium niobate-type (or extremely distorted rhombohedral perovskite-type) oxides like $LiNbO_3$ and $LiTaO_3$ are important materials from a point of view of practical applications as well as fundamental physics and chemistry of condensed matter. $BaTiO_3$ is a main constituent of commercially available capacitors. $Pb(Zr,Ti)O_3$, often referred to as PZT, is broadly utilized as an actuator. $LiNbO_3$ and $LiTaO_3$ find their applications as nonlinear optical materials and surface acoustic wave (SAW) devices. $SrTiO_3$ adopts an ideal perovskite structure, for which the so-called tolerance factor defined as

$$t = \frac{r_A + r_O}{\sqrt{2}(r_B + r_O)}, \tag{15.1}$$

where r_A, r_B, and r_O denote the ionic radii of icosahedral A-site cation surrounded by 12 oxide ions, octahedral B-site cation, and O^{2-} ion, respectively, is almost unity. Here, it should be noted that the coordination number of A-site cation is reduced from 12 in the perovskite structure as the tolerance factor becomes smaller than unity. This fact is very important for the achievement of piezoelectricity by means of the rotation of oxygen octahedra, as described in Sect. 15.4. $SrTiO_3$ shows interesting electrical properties from a fundamental viewpoint; the pure compound is a quantum paraelectric compound at low temperatures and is converted into a superconductor when oxygen vacancy is introduced or cations are replaced with La^{3+} (for Sr^{2+}) and Nb^{5+} (for Ti^{4+}) [5–8]. A recent study has disclosed a curious phenomenon that the two-dimensional electron gas is generated at the interface between $SrTiO_3$ and $LaAlO_3$ and that the interface manifests superconducting transition at 200 mK although both compounds are insulators [9].

A variety of spin correlations among transition metal or rare-earth ions in oxides give rise to plentiful magnetic properties which have drawn considerable interest from both fundamental and practical viewpoints. Many types of ferrites, i.e., iron oxide-based compounds, have been effectively utilized in many applications thus far. Spinel-type ferrites such as Fe_3O_4, γ-Fe_2O_3, $(Mn,Zn)Fe_2O_4$, $(Ni,Zn)Fe_2O_4$, $(Cu,Zn)Fe_2O_4$, and so forth have been used as a magnetic head, a magnetic core in the high frequency range, a recording medium, and a magnetic fluid. Barium ferrite, $BaFe_{12}O_{19}$, one kind of magnetoplumbite-type ferrites, is a main component of a permanent magnet. Garnet-type ferrites such as $Y_3Fe_5O_{12}$, $(Gd,Bi)_3Fe_5O_{12}$, and so forth can be applied to electronic and photonic devices such as an isolator and a modulator for optical telecommunications and microwave circuits. Also, a recent study has revealed an interesting phenomenon that an electrical signal can be transferred through $Y_3Fe_5O_{12}$ although the compound is an insulator. This has been accomplished by the spin Hall effect at the platinum electrode and the efficient generation of spin waves in $Y_3Fe_5O_{12}$ [10, 11]. All of those ferrites form ferrimagnetic structures and have large magnetizations as well as transition temperatures higher than room temperature thanks to the strong superexchange interactions among the transition metal ions, in particular, Fe^{3+} ions. There exist transition metal oxides other than ferrites which display curious magnetic properties. It is well known that perovskite-type $(La,Sr)MnO_3$ and $(La,Sr)CoO_3$ show both ferromagnetism and metallic conduction due to the double exchange interaction among transition metal ions with mixed valence states, i.e., Mn^{3+}/Mn^{4+} and Co^{3+}/Co^{4+} combinations. Not only $(La,Sr)MnO_3$ but similar perovskite-type manganese oxides such as (La,Ca)MnO_3, (La,Ba)MnO_3, and (Pr,Ca)MnO_3 exhibit giant magnetoresistive (GMR) or colossal magnetoresistive (CMR) effects [12–17]. Since the spin-polarized carrier can be transferred in those ferromagnetic oxides, the compounds find potential applications in the field of spintronics. In other transition metal oxides, magnetically and dielectrically ordered structures coexist, as observed in $BiMnO_3$ [18], $BiFeO_3$ [19], $TbMnO_3$ [20], and so forth. These compounds are categorized into materials called multiferroics [21, 22]. In particular, compounds which possess a strong coupling between spin and phonon mode are promising as a spin transistor and a multiple state memory, because in such a material, one can control the magnetization by electric field and at the same time modify the dielectric polarization by magnetic field. Thus, the combination of magnetism with other properties such as electrical conduction and ferroelectricity is of great importance to develop novel and advanced electronic devices.

In the present chapter, we discuss the magnetic, dielectric, and optical properties of several transition metal oxides, investigated by our research group thus far, in terms of the concept of "element-block" and "element-block polymer." We show three examples: (1) ferrimagnetism with high magnetization and high magnetic transition temperature of disordered spinel-type zinc ferrite, which stems from the change of the way to connect two types of element-blocks; (2) ferromagnetism of perovskite oxides containing divalent europium, which is induced by the alternation of orbital overlap between element-blocks; and (3) piezoelectricity and optical second-order nonlinearity accomplished by the rotation of element-blocks in a series of oxides called Ruddlesden-Popper phases.

15.2 Changing the Way to Connect Element-Blocks: Ferrimagnetism of $ZnFe_2O_4$ with Disordered Spinel Structure

It is known that the zinc ferrite, i.e., $ZnFe_2O_4$, crystallizes in a normal spinel-type structure at ambient temperature and pressure, where all the divalent Zn^{2+} ions occupy the tetrahedral sites and all the trivalent Fe^{3+} ions are present at the octahedral sites [23]. In other words, the compound comprises two types of element-blocks – $Zn^{2+}O_4$ tetrahedron and $Fe^{3+}O_6$ octahedron – which are connected with each other via their corners and edges to construct the spinel-type structure. Figure 15.1 illustrates a schematic of normal spinel-type structure. It is found in the figure that the $Fe^{3+}O_6$ octahedra are connected to each other by sharing their edges, so that they form a linear chain running along the <110> directions of the cubic spinel-type structure. This one-dimensional chain structure can be regarded as an element-block polymer, although one chain is connected to others running along different directions. In the normal spinel-type structure of $ZnFe_2O_4$, the negative superexchange interaction works between the nearest neighboring $Fe^{3+}O_6$ octahedra. Strictly speaking, the magnetic frustration emerges because the Fe^{3+} ions are placed on the four corners of a tetrahedron formed at the intersection of two $Fe^{3+}O_6$ one-dimensional

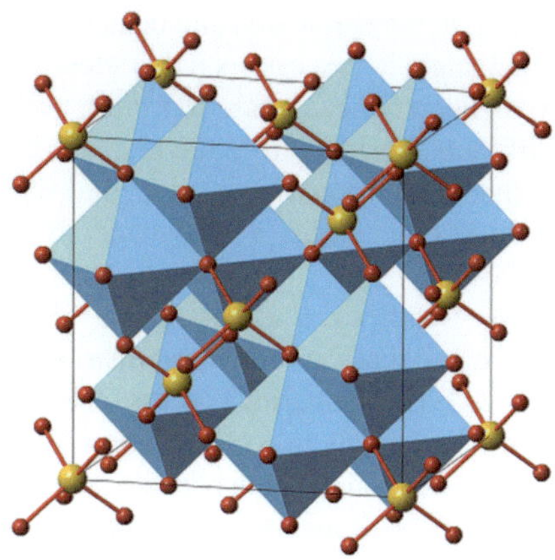

Fig. 15.1 A schematic illustration of normal spinel structure observed for $ZnFe_2O_4$ at ambient temperature and pressure. The structure is composed of cations at tetrahedral (ocher circles) and octahedral (pale blue octahedra) sites and oxide ions (red circles)

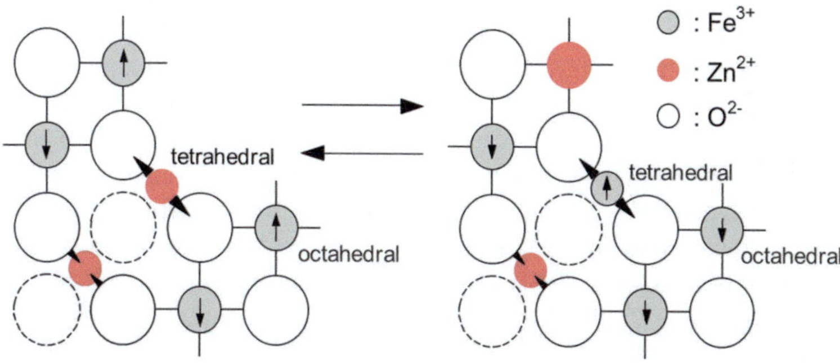

Fig. 15.2 Schematic illustrations of {100} planes of normal (left panel) and disordered (right panel) spinel structures for $ZnFe_2O_4$. The gray and pale red circles denote Fe^{3+} and Zn^{2+} ions, respectively. The arrows on the gray circles stand for magnetic moments of Fe^{3+} ions

chains. Nonetheless, the ordered magnetic structure observed in the normal spinel $ZnFe_2O_4$ can be approximately regarded as antiferromagnetic. Besides, since the bond angle of Fe^{3+}-O^{2-}-Fe^{3+} in the chain structure is 90°, the superexchange interaction is rather weak. Consequently, the Néel temperature is as low as about 10 K. In other words, the normal spinel-type $ZnFe_2O_4$ is paramagnetic at room temperature, and its magnetization is very small.

On the other hand, when some of the Zn^{2+} ions at the tetrahedral sites are exchanged with the same number of Fe^{3+} ions at the octahedral sites, magnetic structure is drastically altered. Such a crystal structure is called disordered spinel-type structure because Zn^{2+} and Fe^{3+} ions are randomly distributed at the tetrahedral and octahedral sites. The difference in a local structure or a distribution of cations can be clearly found in Fig. 15.2, in which {100} planes of the normal (left panel) and disordered (right panel) structures are depicted. Looking at the phenomenon from a different angle, the exchange of Zn^{2+} with Fe^{3+} is equivalent to the operation that one $Fe^{3+}O_6$ chain is cut and connected to the "side" of the other chain via $Fe^{3+}O_4$ tetrahedron. In such a structure, the negative superexchange interaction between Fe

$^{3+}$ ions at the tetrahedral and octahedral sites is dominant because the Fe^{3+}-O^{2-}-Fe^{3+} bond angle in this case is rather close to 180°. This situation is expected to give rise to higher magnetic phase transition temperature compared to the normal spinel-type $ZnFe_2O_4$. Besides, since the number of Fe^{3+} ions at the tetrahedral sites is not equal to that of Fe^{3+} ions at the octahedral sites, the magnetic structure becomes ferrimagnetic. In other words, a large magnetization is expected unlike the normal spinel-type $ZnFe_2O_4$.

We synthesized $ZnFe_2O_4$ thin film on a glass substrate by using a sputtering method so that the disordered distribution of Fe^{3+} and Zn^{2+} ions would be achieved [24–28]. The sputtering process, one kind of the gas phase synthesis methods of thin films, enables us to realize a disordered structure preferred at higher temperatures (gas phase) and to quench it as a metastable state. Figure 15.3a shows magnetization as a function of magnetic field measured at 300 K for as-deposited $ZnFe_2O_4$ thin film [25]. The magnetization is apt to be saturated even at low magnetic fields, and the saturation magnetization reaches 32 emu/g, which is about one third of the saturation magnetization of magnetite, Fe_3O_4. The inset of Fig. 15.3a depicts a magnified view of the low magnetic field range. A clear hysteresis loop is seen although it is not very large. These facts evidently indicate that the as-deposited $ZnFe_2O_4$ thin film has an ordered magnetic structure (in this case, ferrimagnetic structure as mentioned above) even at room temperature, which is in sharp contrast to the magnetic structure of normal spinel-type $ZnFe_2O_4$, that is, antiferromagnetic state with the Néel temperature much lower than room temperature. Figure 15.3b illustrates field-cooled (FC) and zero-field-cooled (ZFC) magnetizations as a function of temperature for the as-deposited $ZnFe_2O_4$ thin film. The variation of field-cooled magnetization with

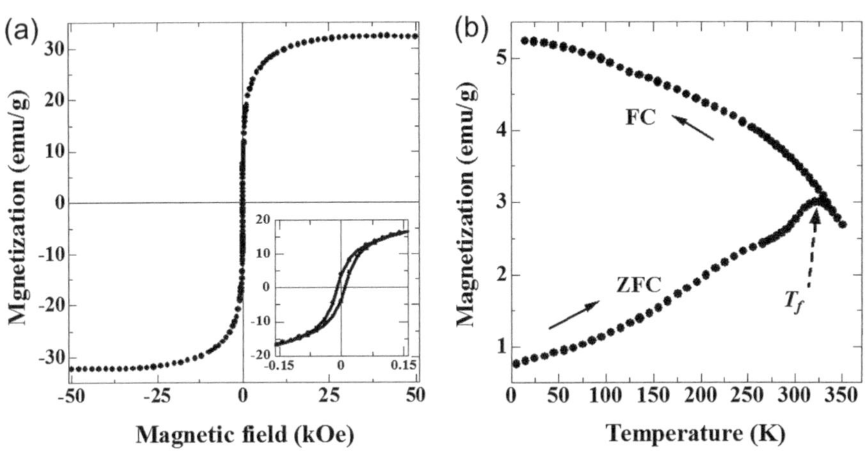

Fig. 15.3 Dependence of magnetization on (**a**) magnetic field at 300 K and (**b**) temperature at 20 Oe for as-deposited $ZnFe_2O_4$ thin film on glass substrate. The inset of (**a**) is a magnified view of low magnetic field region. For the temperature dependence of magnetization shown in (**b**), field-cooling (FC) and zero-field-cooling (ZFC) processes were performed. T_f indicates the cluster spin glass transition temperature. (Reproduced with permission from Figs. 2 and 3 in [25]. Copyright 2005 by the Institute of Physics)

temperature is a characteristic of a ferrimagnet and suggests that the Curie temperature is higher than room temperature. In contrast, a maximum is observed at around 325 K in the temperature dependence of zero-field-cooled magnetization; the temperature variations of FC and ZFC magnetizations manifest rather different features from each other below around 325 K. The phenomena are explainable in terms of the concept of cluster spin glass; the temperature of maximum magnetization in the ZFC process, indicated by T_f in Fig. 15.3b, is the cluster spin glass transition temperature. In other words, the ferrimagnetic order is restricted within each of the clusters or element-block polymers composed of $Fe^{3+}O_6$ and $Fe^{3+}O_4$ polyhedra.

The disordered distribution of Fe^{3+} and Zn^{2+} ions over the tetrahedral and octahedral sites, i.e., the random connection of $Fe^{3+}O_6$ chains via $Fe^{3+}O_4$ tetrahedron, was confirmed by Zn K-edge XANES (X-ray absorption near edge structure) spectroscopy and first principle calculations using density functional theory [26, 27]. Both the experiments and calculations proved that Zn^{2+} ions occupy octahedral as well as tetrahedral sites in the spinel structure and suggested that about 60% of Zn^{2+} ions are present at the octahedral sites in the as-deposited $ZnFe_2O_4$ thin film.

It was also clarified that not only the as-deposited $ZnFe_2O_4$ thin film but thin films annealed at low temperatures like 300 and 400 °C exhibit a large Faraday effect in a short wavelength range [28]. In particular, the $ZnFe_2O_4$ thin film annealed at 300 °C shows the largest Faraday effect among the present $ZnFe_2O_4$ thin films prepared under different conditions; the magnitude of Faraday rotation angle reaches 1.65×10^{4}°/cm at a wavelength of 386 nm. This value is comparable to those of Fe_3O_4 (3.9×10^{4}°/cm at 632.8 nm) [29], $Gd_3Fe_5O_{12}$ (5×10^{3}°/cm at 520 nm) [30], $Gd_{1.6}Bi_{1.4}Fe_5O_{12}$ (1.2×10^{5}°/cm at 520 nm) [30], and $BaFe_{10.42}Co_{0.78}Ti_{0.8}O_{19}$ (7.5×10^{3}°/cm at 780 nm) [31]. Thus, the change of connectivity of oxygen polyhedra containing Fe^{3+} results in a drastic change of magneto-optical as well as magnetic properties of spinel-type $ZnFe_2O_4$.

15.3 Changing the Overlap Between Atomic Orbitals of Element-Blocks: Ferromagnetism of Metastable Perovskite Oxides Containing Divalent Europium

Compounds having the chemical formula of EuMO$_3$, where europium takes the divalent state and M stands for the group 4 elements, crystallize in a perovskite structure at ambient temperature and pressure. The crystal structure of $EuTiO_3$ is cubic at room temperature and atmospheric pressure, as shown schematically in Fig. 15.4a, whereas the stable phase of $EuZrO_3$ and $EuHfO_3$ adopts an orthorhombic structure [32–34]. Here, the crystal structure in Fig. 15.4a is visualized using VESTA code [35]. For the cubic $EuTiO_3$, the crystal structure can be regarded as a result of three-dimensional polymerization of two types of element-blocks, i.e., $Eu^{2+}O_{12}$ icosahedron and $Ti^{4+}O_6$ octahedron. On the other hand, the orthorhombic

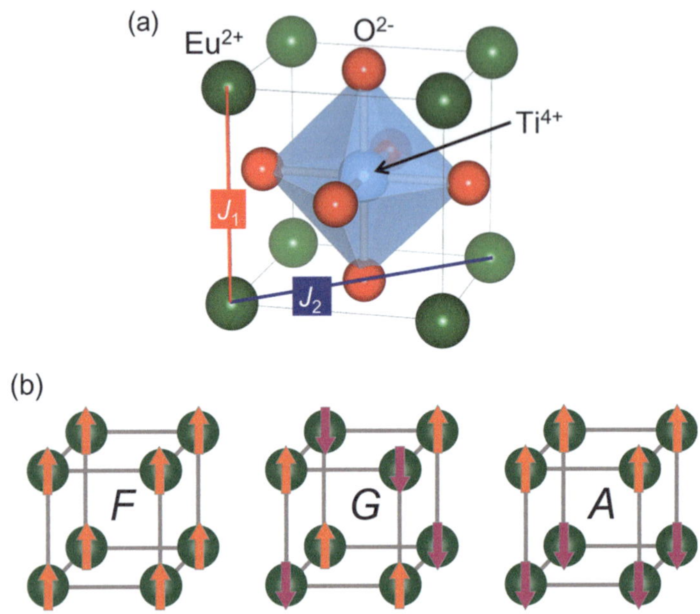

Fig. 15.4 (**a**) A schematic illustration of $EuTiO_3$ unit cell along with the exchange coupling constants between nearest-neighboring (J_1) and next-nearest-neighboring (J_2) Eu^{2+} ions. (**b**) Ferromagnetic (F), G-type antiferromagnetic (G), and A-type antiferromagnetic (A) structures assumed for $EuTiO_3$. The arrows denote the magnetic moments of Eu^{2+} ions. (Reproduced with permission from Fig. 1c in [45]. Copyright 2011 by the American Physical Society)

structure results from the rotation of octahedra, which is caused due to the relatively large ionic radii of Zr^{4+} and Hf^{4+}; the rotation of octahedron in perovskite structure takes place more easily as the tolerance factor defined as Eq. (15.1) becomes smaller. The rotation of octahedron commonly occurs in perovskite-related structures as discussed in detail in the following section, Sect. 15.4. It is known that all of the three compounds undergo the antiferromagnetic transition at low temperatures; the Néel temperature of $EuTiO_3$, $EuZrO_3$, and $EuHfO_3$ is 5.3, 4.1, and 3.9 K, respectively [33, 34, 36, 37]. For $EuTiO_3$ and $EuZrO_3$, the permittivity as a function of temperature manifests a drastic change at the Néel temperature [38, 39]. Namely, those oxides show the magnetodielectric effect, indicating that they are multiferroics. As mentioned below in more detail, the antiferromagnetic structure stems from the superexchange interaction between Eu^{2+} ions via group 4 cations, and we can switch the macroscopic magnetic structure between antiferromagnetic and ferromagnetic states by changing the degree of orbital overlap between Eu^{2+} and group 4 cations.

Both experimental and theoretical approaches have been carried out to examine the effect of stress and/or strain on the magnetic and dielectric properties of $EuTiO_3$. We synthesized $EuTiO_3$ epitaxial thin film on (001) plane of single-crystalline $SrTiO_3$ as a substrate by using a pulsed laser deposition method [40]. We accidentally found that the lattice constant of the thin film was elongated by 2.4% in a

Table 15.1 Difference in lattice constants and cell volume between bulk $EuTiO_3$ and $EuTiO_3$ thin films epitaxially grown on $LaAlO_3$, $SrTiO_3$, and $DyScO_3$ substrates. Δl_1 and Δl_2 are relative changes of lattice constants in directions parallel and perpendicular to the surface of the thin film, respectively. ΔV is the change of cell volume relative to the bulk $EuTiO_3$

Substrate	Δl_1 (%)	Δl_2 (%)	ΔV (%)
$LaAlO_3$	−0.3	+1.89	+1.3
$SrTiO_3$	0.0	+2.41	+2.41
$DyScO_3$	+1.05	+1.77	+3.92

direction perpendicular to the film surface and that the as-deposited $EuTiO_3$ thin film manifested ferromagnetic transition at around 2 K. We also revealed that by post-annealing of the thin film, the lattice constant was decreased to become the value reported for bulk $EuTiO_3$ and the resultant $EuTiO_3$ thin film exhibited antiferromagnetic transition at 5.1 K, as expected for the stable phase of $EuTiO_3$. Furthermore, we deposited $EuTiO_3$ epitaxial thin films on $LaAlO_3$, $SrTiO_3$, and $DyScO_3$ substrates which have lattice constants different from each other and explored the effect of cell volume of $EuTiO_3$ thin film on the magnetic properties [41]. The lattice constants of $EuTiO_3$, $LaAlO_3$, $SrTiO_3$, and $DyScO_3$ are 0.3905, 0.3790, 0.3905, and 0.3944 nm, respectively. Table 15.1 summarizes the difference in lattice constants (those in directions perpendicular and parallel to the thin film surface) and cell volume between the $EuTiO_3$ thin films and bulk $EuTiO_3$ [41]. These values were evaluated from the reciprocal space mappings obtained by X-ray diffraction measurements for $EuTiO_3$ thin films deposited on the three types of substrates. The change in lattice constants and cell volume depends on the difference in lattice constant between $EuTiO_3$ and the substrate materials. The cell volume is increased in all the $EuTiO_3$ thin films. Also, the increase in cell volume is the largest for the $DyScO_3$ substrate which has the largest lattice constant among the four compounds including $EuTiO_3$, and the increase in cell volume is the smallest for the $LaAlO_3$ which has the smallest lattice constant. Namely, we can modify the separation between Eu^{2+} ions and hence the degree of overlap between atomic orbitals of the element-blocks, i.e., oxygen polyhedra of Eu^{2+} and Ti^{4+}.

Our measurements of magnetization as a function of temperature and magnetic field suggested that the $EuTiO_3$ thin films on $LaAlO_3$ and $DyScO_3$ as well as $SrTiO_3$ substrates are ferromagnetic at low temperatures, in sharp contrast to the antiferromagnetic transition observed for the stable phase of $EuTiO_3$ [41]. Besides, a comparison of magnetization at low temperatures among the thin films deposited on the different kinds of substrates indicated that the magnetizations of $EuTiO_3$ thin films on $DyScO_3$ and $SrTiO_3$ substrates are about four and two times larger than the thin film on a $LaAlO_3$ substrate, respectively. In other words, the magnetization of $EuTiO_3$ thin film monotonically increases with an increase in the cell volume.

Theoretical calculations have been performed to clarify the relation between the change in size of $EuTiO_3$ lattice and magnetic properties. Fennie and Rabe [42] predicted that ferromagnetic and ferroelectric states are stabilized when a biaxial compressive strain above 1.2% is induced in the (001) plane parallel to the surface of $EuTiO_3$ thin film. In the calculations, the cell volume is kept constant, so that the compressive strain in the surface of the $EuTiO_3$ thin film leads to elongation of the lattice in a direction perpendicular to the surface. Ranjan et al. [43] calculated the energy for some magnetic structures by using LDA (local density approximation) + U (on-site Coulomb repulsion) approach and revealed that the ferromagnetic state becomes more stable as either the cell volume of $EuTiO_3$ or the on-site Coulomb repulsion for a 4f electron of Eu^{2+} ion is increased. Lee et al. [44] carried out spin-polarized GGA (generalized gradient approximation) + U calculations for $EuTiO_3$ thin film and disclosed that tensile as well as compressive strain leads to the stabilization of ferromagnetic and ferroelectric states. They also experimentally ascertained that $EuTiO_3$ thin film under tensile strain becomes both ferromagnetic and ferroelectric as expected from the theoretical calculations. Akamatsu et al. [45] performed hybrid Hartree-Fock density functional calculations for G-type antiferromagnetic, A-type antiferromagnetic, and ferromagnetic $EuTiO_3$ with the cell volume being isotropically varied. Those magnetic structures are depicted in Fig. 15.4b, where the arrangement of magnetic moments of Eu^{2+} ions is shown in the unit cell of $EuTiO_3$. Here, it should be noted that the stable phase of $EuTiO_3$ is experimentally found to be G-type antiferromagnetic below the Néel temperature. The magnetic structure of $EuTiO_3$ can be expressed in terms of the spin Hamiltonian for Heisenberg system:

$$H_{\text{spin}} = -2\sum_{i>j} J_{\text{ij}} S_i \cdot S_j. \tag{15.2}$$

The energy derived from Eq. (15.2) is

$$E_F = E_0 + 2S(S+1)(-12J_1 - 24J_2) \tag{15.3}$$

and

$$E_G = E_0 + 2S(S+1)(12J_1 - 24J_2) \tag{15.4}$$

for ferromagnetic and G-type antiferromagnetic spin configurations, respectively, where J_1 and J_2 are the exchange coupling constant for nearest-neighboring and next-nearest-neighboring Eu^{2+} ions, respectively, as illustrated in Fig. 15.4a. Figure 15.5a shows the cell volume dependence of energy difference among G-type antiferromagnetic, A-type antiferromagnetic, and ferromagnetic structures [45]. It is found that the energy of the G-type antiferromagnetic state is the lowest among the three types of magnetic structures at the equilibrium cell volume, i.e., at $\Delta V = 0$, corresponding to the stable phase of $EuTiO_3$. This is in good agreement with the experimental result. The difference in energy between ferromagnetic and G-type antiferromagnetic structures is determined only by the sign of J_1, as indicated by Eqs. (15.3) and (15.4). Figure 15.5b depicts the variation of the exchange coupling constants with the cell volume. It is clear that the value of J_1 monotonically increases with an increase in the cell volume of $EuTiO_3$ and that J_1 changes its sign from negative to positive when the increase in cell volume becomes above about 5%,

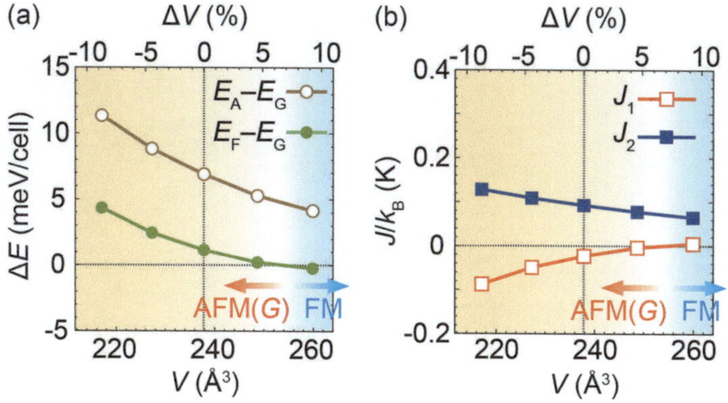

Fig. 15.5 Cell volume dependence of (**a**) energy difference among ferromagnetic (the energy of which is denoted by E_F), G-type antiferromagnetic (E_G), and A-type antiferromagnetic (E_A) structures and (**b**) exchange coupling constants between nearest-neighboring (J_1) and next-nearest-neighboring (J_2) Eu^{2+} ions. ΔV is the relative change of cell volume. $\Delta V = 0$ corresponds to the stable phase of $EuTiO_3$. (Reproduced with permission from Fig. 3 in [45]. Copyright 2011 by the American Physical Society)

suggesting that the ferromagnetic spin configuration becomes more stable compared to the G-type antiferromagnetic structure when the cell volume exceeds this critical value. The cell volume variation of J_1 derived from the calculations is qualitatively coincident with the abovementioned experimental result that the ferromagnetic magnetization of $EuTiO_3$ thin film increases with an increase in the cell volume.

Figure 15.6 illustrates charge-density isosurfaces obtained by the hybrid Hartree-Fock density functional calculations for an Eu 4f state at the Γ point at 0.003 $Å^{-3}$ for G-type antiferromagnetic structure of $EuTiO_3$ with different cell volumes [45]. It is clear that the degree of overlap between the 4f orbitals of Eu^{2+} and the 3d orbitals of Ti^{4+} becomes larger as the cell volume is decreased. Such an orbital overlap can lead to negative superexchange interactions among the nearest-neighboring Eu^{2+} ions, because Eu 4f and Ti 3d orbitals are non-orthogonal to each other [46]. In addition, the indirect exchange interaction works between 4f spins of nearest-neighboring Eu^{2+} ions via vacant 5d states of themselves, leading to the ferromagnetic coupling between the magnetic moments of the Eu^{2+} ions [37, 47–49]. It is considered that in $EuTiO_3$ the cell volume of which is enlarged, the indirect exchange interaction between Eu^{2+} ions to stabilize ferromagnetic structure overcomes the antiferromagnetic interaction via the 3d orbital of Ti^{4+} because the overlap between Eu^{2+} 4f and Ti^{4+} 3d orbitals is not so significant in such a structure. The importance of overlap between Eu^{2+} 4f orbitals and d orbitals of transition metal cations at the octahedral sites was pointed out for orthorhombic $EuZrO_3$ and $EuHfO_3$ as well [50]. As described above, the orthorhombic perovskite structure of the two oxides originates in the rotation of octahedral sites. According to the calculations, such a rotation gives rise to an increased overlap between Eu^{2+} 4f orbitals and nd ($n = 4$ and 5) orbitals of Zr^{4+} and Hf^{4+}, and consequently, antiferromagnetic superexchange interaction becomes dominant as experimentally observed. On the other hand, the calculations

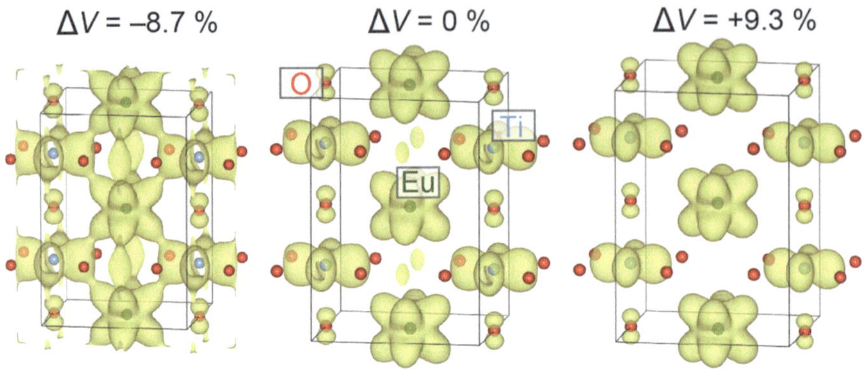

Fig. 15.6 Charge-density isosurfaces (0.003 $Å^{-3}$) calculated for an Eu 4f state at the Γ point for the *G*-type antiferromagnetic $EuTiO_3$. Three types of cell volumes, $\Delta V = -8.7$, 0, and +9.3%, are assumed for the calculations. The crystal structures and isosurfaces are visualized using VESTA code [35]. (Reproduced with permission from Fig. 4f in [45]. Copyright 2011 by the American Physical Society)

predict that the exchange coupling constant for nearest-neighboring Eu^{2+} ions (J_1) is positive when cubic perovskite structure is assumed for $EuZrO_3$ and $EuHfO_3$. Namely, cubic perovskite-type $EuZrO_3$ and $EuHfO_3$ are expected to be ferromagnetic, although those compounds have been never synthesized. The rotation of $Zr^{4+}O_6$ and $Hf^{4+}O_6$ octahedra is suppressed in the cubic perovskite structure, so that the overlap between Eu^{2+} 4f and Zr^{4+} 4d or Hf^{4+} 5d orbitals becomes less significant and the antiferromagnetic interaction between Eu^{2+} ions is weakened. Consequently, the ferromagnetic structure becomes more stable.

In general, it is rather difficult to change the degree of overlap between atomic orbitals of element-blocks so as to modify the electronic properties drastically in the ionic crystals when compared to the organic polymers. However, it is not impossible to do so as exemplified by the conversion of antiferromagnetic spin configuration into a ferromagnetic one in the strained or metastable phases of $EuTiO_3$, $EuZrO_3$, and $EuHfO_3$. To synthesize ferromagnetic $EuZrO_3$ and $EuHfO_3$, which were predicted to exist by the theoretical calculations, remains to be realized.

15.4 Rotating Element-Blocks: Piezoelectricity and Optical Second-Order Nonlinearity of Ruddlesden-Popper Phases

The abovementioned phenomenon that the rotation of oxygen octahedron in perovskite structure drastically changes properties of the oxides is not restricted to the magnetic properties. In this section, we demonstrate that such a rotation leads to piezoelectric and second-order nonlinear optical properties in layered perovskite oxides called Ruddlesden-Popper phase. Recent theoretical studies have suggested that it is possible to break the inversion symmetry by means of the rotation of oxygen octahedron in layered perovskites involving the Ruddlesden-Popper phase [51–57]. A series of Ruddlesden-Popper phases possess layered structures in which perovskite and rock-salt layers are stacked along the *c*-axis in an ordered way, and the chemical formula of the phases is denoted as $(ABO_3)_n(AO)$, where *A* and *B* stand for the *A*- and *B*-site cations, respectively, in the perovskite layer and *n* takes an integer. Here, we focus on one kind of $n = 1$ phases, $NaRTiO_4$ [58], where *R* denotes the rare-earth element, but the following discussion can be applied to other layered perovskite oxides involving the Ruddlesden-Popper phases with *n* besides 1.

A schematic illustration of the crystal structure of $n = 1$ Ruddlesden-Popper phase, the formula of which is A_2BO_4, is shown in Fig. 15.7a. In the figure, *A*-site cations and oxygen octahedra containing *B*-site cations, which are important element-blocks in the structure, are depicted. It should be noted that the space group of the crystal structure is *I*4/*mmm*, which has an inversion symmetry. In Fig. 15.7a, the crosses stand for the inversion centers. In other words, neither piezoelectricity nor optical second-order nonlinearity can be observed in A_2BO_4 with the structure shown in the figure. The inversion centers at the *B*-sites can be eliminated when two different types of cations occupy the *A*-sites and the two types of the *A*-site cations, i.e., *A* and *A*', are arranged in an ordered way along the *c*-axis as illustrated in Fig. 15.7b. The structure in Fig. 15.7b belongs to the space group *P*4/*nmm*, which still has inversion centers as indicated by the crosses at the midpoints between the same types of *A*-site cations. If one can induce a rotation of octahedral sites as shown in Fig. 15.7c, it is possible to remove all the inversion centers from the

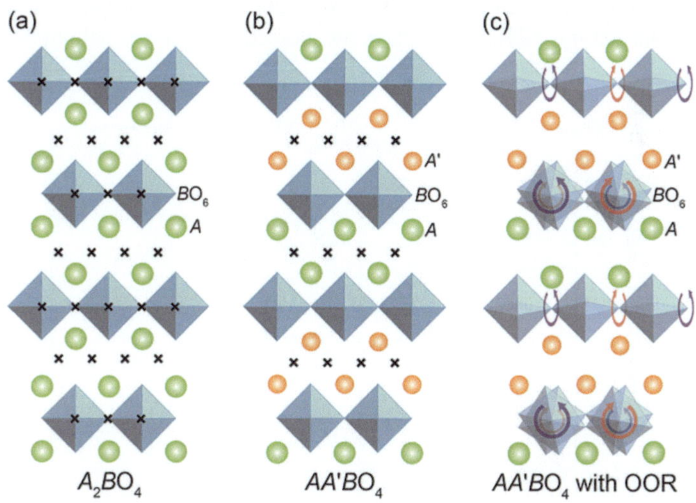

Fig. 15.7 Schematic illustrations of structures for $n = 1$ Ruddlesden-Popper phases: (**a**) A_2BO_4, the structure of which belongs to the space group *I*4/*mmm*; (**b**) $AA'BO_4$, in which *A*-sites are occupied by two types of cations in an ordered way along the *c*-axis, the space group is *P*4/*nmm*; and (**c**) $AA'BO_4$ with rotations of oxygen octahedra as depicted in the figure, the space group of the structure is $P\bar{4}2_1m$. The crosses in (**a**) and (**b**) denote the inversion centers. (Reproduced with permission from Fig. 1 in [60]. Copyright 2014 by the American Physical Society)

structure, so that the oxides adopting such a structure are expected to exhibit piezoelectric and second-order nonlinear optical properties. In Fig. 15.7c, the nearest-neighboring octahedra are rotated in opposite directions to each other in each of the layers, and the axis of rotation in one layer is just perpendicular to the one in the nearest-neighboring layer. The crystal structure in Fig. 15.7c belongs to the space group $P\bar{4}2_1m$ without the inversion symmetry.

As described in Sect. 15.3, the rotation of *B*-site octahedron in perovskite structure occurs more readily as the size of *A*-site cation relative to the *B*-site cation becomes smaller. For the case of $NaRTiO_4$, it is expected that *P*4/*nmm* and $P\bar{4}2_1m$ structures are preferred for larger rare-earth ion such as La^{3+} and smaller one such as Y^{3+}, respectively. Figure 15.8 shows the phonon band structure in a range from Γ to M calculated with the phonopy code [59] for $NaLaTiO_4$ (Fig. 15.8a) and $NaYTiO_4$ (Fig. 15.8b) by assuming *P*4/*nmm* structure for both compounds [60]. In the phonon band structure for $NaYTiO_4$, imaginary frequencies appear at M point, while no unusual phenomenon is observed for the band structure of $NaLaTiO_4$. This indicates that the *P*4/*nmm* structure is suitable to $NaLaTiO_4$ but a soft phonon mode must be generated in $NaYTiO_4$ to induce structural phase transition from the *P*4/*nmm* structure. Actually, calculations of total energies of some possible structures involving *P*4/*nmm* and $P\bar{4}2_1m$ for $NaYTiO_4$ reveal that $P\bar{4}2_1m$ is the most stable structure.

The result of theoretical calculations that the stable phase of $NaYTiO_4$ adopts the $P\bar{4}2_1m$ structure without inversion symmetry can be experimentally demonstrated by utilizing optical second-harmonic generation (SHG), one of the second-order nonlinear optical phenomena. We prepared polycrystalline specimens of $NaRTiO_4$ with *R* = La, Nd, Sm, Eu, Gd, Dy, Ho, and Y and measured second-harmonic intensity as a function of temperature [60]. We obtained the following results: (1) all the specimens exhibit SHG when they are cooled to low temperatures, and the

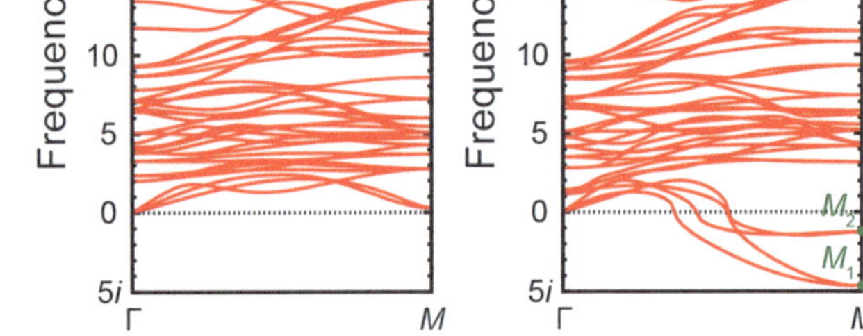

Fig. 15.8 Phonon band structures between Γ and M points calculated for (**a**) $NaLaTiO_4$ and (**b**) $NaYTiO_4$ by assuming that the structure of both compounds is *P*4/*nmm*. Imaginary frequencies are observed at M point for $NaYTiO_4$. (Reproduced with permission from Fig. 2a in [60]. Copyright 2014 by the American Physical Society)

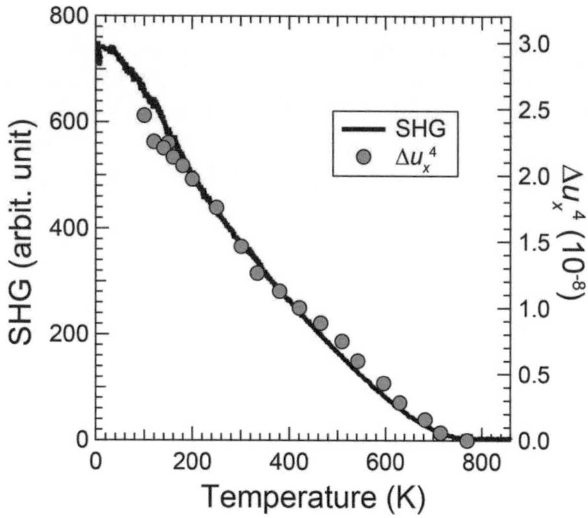

Fig. 15.9 Temperature variation of second-harmonic intensity and the fourth power of displacement of rare-earth ion, Δu_x, for $NaSmTiO_4$. The data indicate that Eq. (15.6) holds. (Reproduced with permission from Fig. 4c in [60]. Copyright 2014 by the American Physical Society)

second-harmonic intensity decreases with an increase in temperature, and (2) the temperature at which the second-harmonic intensity becomes zero as the temperature is increased is higher for the smaller rare-earth ion. The temperature referred to in (2) is the structural phase transition temperature at which the transition takes place between structures with and without inversion symmetry. As an example, the temperature dependence of second-harmonic intensity is shown for $NaSmTiO_4$ in Fig. 15.9 [60]. It is seen that the second-harmonic intensity monotonically decreases and eventually disappears at around 760 K as the temperature is increased. The result of SHG measurements indicates that the structure without inversion symmetry grows more stable as the ionic radius of *A*-site cation becomes smaller.

The order parameter, η, which is relevant to the M_1-mode in the phonon band structure (Fig. 15.8b) and governs the phase transition between $P4/nmm$ and $P\bar{4}2_1m$ structures, is represented by the displacement of rare-earth ion, Δu_x, a component along the *a*-axis of the vibration within the *ab*-plane of $\sqrt{2} \times \sqrt{2} \times 1$ supercell of $P4/nmm$ lattice [60]. Calculations based on the Landau theory lead to the result that the component of second-harmonic polarization, P_i, is proportional to the square of the order parameter:

$$P_i \propto \eta^2 E_j E_k \tag{15.5}$$

where E_j and E_k are the components of electric field of fundamental wave. Since the second-harmonic intensity, I, is proportional to the square of the second-order polarization, the relation

$$I \propto \Delta u_x^4 \tag{15.6}$$

is derived. Figure 15.9 illustrates the temperature variation of $\Delta u_x{}^4$ experimentally obtained from the temperature-dependent structures of $NaSmTiO_4$, in addition to the temperature dependence of second-harmonic intensity [60]. The temperature variations of these two quantities are in good agreement with each other. Furthermore, the polycrystalline specimens of Na*R*TiO_4 with $R = $ Dy and Ho proved to be piezoelectric at room temperature by using piezoresponse force microscopy [60]. The effective piezoelectric constant tensor, d_{33}^{eff}, is about 20 and 30 pC/N for the polycrystalline $NaDyTiO_4$ and $NaHoTiO_4$, respectively. The results of theoretical calculations and experiments demonstrate that the rotation of *B*-site octahedra illustrated in Fig. 15.7c surely contributes to the piezoelectricity as well as the optical second-order nonlinearity of Na*R*TiO_4.

Similar phenomena have been observed for other $n = 1$ Ruddlesden-Popper phases such as Li*R*TiO_4 and H*R*TiO_4 with *R* being the rare-earth element [61, 62]. Here, we can regard the perovskite layer composed of the *B*-site octahedron as a two-dimensional element-block polymer, and the crystal structure itself is considered to be equivalent to the three-dimensional integration of the two-dimensional element-block polymers. The phenomena observed in the Ruddlesden-Popper phases are rephrased as follows: one can control the geometrical structure of the element-block polymers to cause a drastic change in electrical and optical properties of the integrated structure.

15.5 Conclusion

The concept of element-block polymer science includes the synthesis of element-blocks which correspond to the monomers for the traditional organic polymers but contain elements other than carbon, polymerization of the element-blocks via sophisticated chemical reactions, construction of higher-order structures, and forming of hierarchical interfaces so that one can derive curious properties and important functionalities never reached so far. Hence, strictly speaking, the transition metal oxides described in the present chapter may be outside the purview of the concept of element-block polymers. Nonetheless, as exemplified above, a variety of intriguing electrical, magnetic, and optical properties observed in a huge number of transition metal oxides essentially originate in the valence state, nature of chemical bond, and structure of oxygen polyhedron of a transition element; one-, two-, and three-dimensional connectivity of the polyhedra composed of transition elements; and the whole crystal structure. Therefore, it seems to be reasonable in some cases to interpret the properties of transition metal oxides in terms of the concept of element-block polymers. Also, the structural analysis and clarification of the physical and chemical properties of the inorganic crystals involving the oxides may be useful for the further development of novel element-block polymers.

Acknowledgment We would like to thank many collaborators: Dr. Seisuke Nakashima (currently Shizuoka University); Prof. Hirofumi Akamatsu (currently Kyushu University); Dr. Yanhua Zong; Naoki Wakasugi; Yuya Maruyama; Toshihiro Kuge; Dr. Yoshiro Kususe; Hideo Murakami; Dr. Takahiro Kawamoto; Koji Iwata; Dr. Shunsuke Murai; Prof. Kazuyuki Hirao of the Department of Material Chemistry, Kyoto University; Prof. Isao Tanaka; Prof. Tomoyuki Yamamoto (currently Waseda University); Prof. Fumiyasu Oba (currently Tokyo Institute of Technology); Dr. Yu Kumagai; Dr. Hiroyuki Hayashi; Prof. Atsushi Togo of the Department of Materials Science and Engineering, Kyoto University; Prof. Venkatraman Gopalan; Prof. Long-Qing Chen; Dr. Arnab Sen Gupta; Dr. Shiming Lei; Dr. Fei Xue; Dr. Greg Stone of Pennsylvania State University; and Prof. James M. Rondinelli of Drexel University. Especially, we are indebted to Prof. Isao Tanaka and his research group members as well as Prof. Hirofumi Akamatsu for the theoretical calculations. Also, we thank Suguru Yoshida of the Department of Material Chemistry, Kyoto University, for the preparation of Fig. 15.1.

References

1. Bednorz JG, Müller KA (1986) Possible high T_c superconductivity in the Ba–La–Cu–O system. Z Phys B 64(2):189–193
2. Wu MK, Ashburn JR, Torng CJ, Hor PH, Meng RL, Gao L, Huang ZJ, Wang YQ, Chu CW (1987) Superconductivity at 93 K in a new mixed-phase Y-Ba-Cu-O compound system at ambient pressure. Phys Rev Lett 58(9):908–910
3. Maeda H, Tanaka Y, Fukutomi M, Asano T (1988) A new high-T_c oxide superconductor without a rare earth element. Jpn J Appl Phys 27:L209–L210
4. Schilling A, Cantoni M, Guo JD, Ott HR (1993) Superconductivity above 130 K in the Hg–Ba–Ca–Cu–O system. Nature 363:56–58
5. Schooley JF, Hosler WR, Ambler E, Becker JH, Cohen ML, Koonce CS (1965) Dependence of the superconducting transition temperature on carrier concentration in semiconducting $SrTiO_3$. Phys Rev Lett 14:305–307
6. Baratoff A, Binnig G (1981) Mechanism of superconductivity in $SrTiO_3$. Phys B+C 108:1335–1336
7. Leitner A, Rogers CT, Price JC, Rudman DA, Herman DR (1998) Pulsed laser deposition of superconducting Nb-doped strontium titanate thin films. Appl Phys Lett 72:3065–3067
8. Olaya D, Pan F, Rogers CT, Price JC (2004) Superconductivity in La-doped strontium titanate thin films. Appl Phys Lett 84:4020–4022
9. Reyren N, Thiel S, Caviglia AD, Fitting Kourkoutis L, Hammerl G, Richter C, Schneider CW, Kopp T, Rüetschi A-S, Jaccard D, Gabay M, Muller DA, Triscone J-M, Mannhart J (2007) Superconducting interfaces between insulating oxides. Science 317:1196–1199
10. Kajiwara Y, Harii K, Takahashi S, Ohe J, Uchida K, Mizuguchi M, Umezawa H, Kawai H, Ando K, Takanashi K, Maekawa S, Saitoh E (2010) Transmission of electrical signals by spin-wave interconversion in a magnetic insulator. Nature 464:262–266
11. Serga AA, Chumak AV, Hillebrands B (2010) YIG magnonics. J Phys D 43:264002
12. Jin S, Tiefel TH, McCormack M, Fastnacht RA, Ramesh R, Chen LH (1994) Thousandfold change in resistivity in magnetoresistive La-Ca-Mn-O films. Science 264:413–415
13. Chahara K, Ohno T, Kasai M, Kozono Y (1993) Magnetoresistance in magnetic manganese oxide with intrinsic antiferromagnetic spin structure. Appl Phys Lett 63:1990–1992
14. Tokura Y, Urushibara A, Moritomo Y, Arima T, Asamitsu A, Kido G, Furukawa N (1994) Giant magnetotransport phenomena in filling-controlled Kondo lattice system $La_{1-x}Sr_xMnO_3$. J Phys Soc Jpn 63:3931
15. Urushibara A, Moritomo Y, Arima T, Asamitsu A, Kido G, Tokura Y (1995) Insulator-metal transition and giant magnetoresistance in $La_{1-x}Sr_xMnO_3$. Phys Rev B 51:14103
16. von Helmolt R, Meeker HB, Schultz L, Samwer K (1993) Giant negative magnetoresistance in perovskitelike $La_{2/3}Ba_{1/3}MnO_x$ ferromagnetic films. Phys Rev Lett 71:2331
17. Tomioka Y, Asamitsu A, Moritomo Y, Tokura Y (1995) Anomalous magnetotransport properties of $Pr_{1-x}Ca_xMnO_3$. J Phys Soc Jpn 64:3626
18. Kimura T, Kawamoto S, Yamada I, Azuma M, Takano M, Tokura Y (2003a) Magnetocapacitance effect in multiferroic $BiMnO_3$. Phys Rev B 67:180401

19. Wang J, Neaton JB, Zheng H, Nagarajan V, Ogale SB, Liu B, Viehland D, Vaithyanathan V, Schlom DG, Waghmare UV, Spaldin NA, Rabe KM, Wuttig M, Ramesh R (2003) Epitaxial $BiFeO_3$ multiferroic thin film heterostructures. Science 299:1719
20. Kimura T, Goto T, Shintani H, Ishizaka K, Arima T, Tokura Y (2003b) Magnetic control of ferroelectric polarization. Nature 426:55–58
21. Cheong SW, Mostovoy M (2007) Multiferroics: a magnetic twist for ferroelectricity. Nat Mater 6(1):13–20
22. Tokura Y (2006) Multiferroics as quantum electromagnets. Science 312:1481–1482
23. Chikazumi S (1978) Physics of ferromagnetism, vol 1. Syokabo, Tokyo, p 216 [in Japanese]
24. Tanaka K, Nakashima S, Fujita K, Hirao K (2003) High magnetization and the Faraday effect for ferrimagnetic zinc ferrite thin film. J Phys Condens Matter 15:L469–L474
25. Nakashima S, Fujita K, Tanaka K, Hirao K (2005) High magnetization and the high-temperature superparamagnetic transition with intercluster interaction in disordered zinc ferrite thin film. J Phys Condens Matter 17(1):137–149
26. Nakashima S, Fujita K, Tanaka K, Hirao K, Yamamoto T, Tanaka I (2007a) Thermal annealing effect on magnetism and cation distribution in disordered $ZnFe_2O_4$ thin films deposited on glass substrates. J Magn Magn Mater 310:2543–2545
27. Nakashima S, Fujita K, Tanaka K, Hirao K, Yamamoto T, Tanaka I (2007b) First-principles XANES simulations of spinel zinc ferrite with a disordered cation distribution. Phys Rev B 75:174443–174443
28. Tanaka K, Nakashima S, Fujita K, Hirao K (2006) Large Faraday effect in a short wavelength range for disordered zinc ferrite thin films. J Appl Phys 99:106103
29. Borrelli NF, Murphy JA (1971) Magnetooptic properties of magnetite films. J Appl Phys 42:1120–1123
30. Takeuchi H (1975) The Faraday effect of bismuth substituted rare-earth iron garnet. Jpn J Appl Phys 14:1903–1910
31. Nakamura H, Ohmi F, Kaneko Y, Sawada Y, Watada A, Machida H (1987) Cobalt-titanium substituted barium ferrite films for magneto-optical memory. J Appl Phys 61:3346–3348
32. Viallet V, Marucco J-F, Saint J, Herbst-Ghysel M, Dragoe N (2008) Structural, magnetic and electrical properties of a perovskite containing divalent europium $EuZrO_3$. J Alloys Compd 461:346
33. Zong Y, Fujita K, Akamatsu H, Murai S, Tanaka K (2010) Antiferromagnetism of perovskite $EuZrO_3$. J Solid State Chem 183:168–172
34. Akamatsu H, Fujita K, Hayashi H, Kawamoto T, Kumagai Y, Zong Y, Iwata K, Oba F, Tanaka I, Tanaka K (2012) Crystal and electronic structure and magnetic properties of divalent europium perovskite oxides $EuMO_3$ (M=Ti, Zr, and Hf): experimental and first-principles approaches. Inorg Chem 51:4560–4567
35. Momma K, Izumi F (2008) VESTA: a three-dimensional visualization system for electronic and structural analysis. J Appl Crystallogr 41:653–658
36. McGuire TR, Shafer MW, Joenk RJ, Alperin HA, Pickart SJ (1966) Magnetic structure of $EuTiO_3$. J Appl Phys 37:981–982
37. Chien C-L, DeBenedetti S, De F, Barros S (1974) Magnetic properties of $EuTiO_3$, Eu_2TiO_4, and $Eu_3Ti_2O_7$. Phys Rev B 10:3913
38. Katsufuji T, Takagi H (2001) Coupling between magnetism and dielectric properties in quantum paraelectric $EuTiO_3$. Phys Rev B 64:054415
39. Kolodiazhnyi T, Fujita K, Wang L, Zong Y, Tanaka K, Sakka Y, Takayama-Muromachi E (2010) Magnetodielectric effect in $EuZrO_3$. Appl Phys Lett 96:252901
40. Fujita K, Wakasugi N, Murai S, Zong Y, Tanaka K (2009) High-quality antiferromagnetic $EuTiO_3$ epitaxial thin films on $SrTiO_3$ prepared by pulsed laser deposition and post-annealing. Appl Phys Lett 94:062512
41. Tanaka K, Fujita K, Maruyama Y, Kususe Y, Murakami H, Akamatsu H, Zong Y, Murai S (2013) Ferromagnetism induced by lattice volume expansion and amorphization in $EuTiO_3$ thin films. J Mater Res 28:1031
42. Fennie CJ, Rabe KM (2006) Magnetic and electric phase control in epitaxial $EuTiO_3$ from first principles. Phys Rev Lett 97:267602

43. Ranjan R, Nabi HS, Pentcheva R (2007) Electronic structure and magnetism of $EuTiO_3$: a first-principles study. J Phys Condens Matter 19:406217
44. Lee JH, Fang L, Vlahos E, Ke X, Jung YW, Fitting Kourkoutis L, Kim J-W, Ryan PJ, Heeg T, Roeckerath M, Goian V, Bernhagen M, Uecker R, Hammel PC, Rabe KM, Kamba S, Schubert J, Freeland JW, Muller DA, Fennie CJ, Schiffer P, Gopalan V, Johnston-Halperin E, Schlom DG (2010) A strong ferroelectric ferromagnet created by means of spin–lattice coupling. Nature 466:954
45. Akamatsu H, Kumagai Y, Oba F, Fujita K, Murakami H, Tanaka K, Tanaka I (2011) Antiferromagnetic superexchange via 3d states of titanium in $EuTiO_3$ as seen from hybrid Hartree-Fock density functional calculations. Phys Rev B 83:214421
46. Anderson PW (1959) New approach to the theory of superexchange interactions. Phys Rev 115:2–13
47. Shafer MW (1965) Preparation and crystal chemistry of divalent europium compounds. J Appl Phys 36:1145–1152
48. Kunes J, Ku W, Pickett WE (2005) Exchange coupling in Eu monochalcogenides from first principles. J Phys Soc Jpn 74:1408–1411
49. Souza-Neto NM, Haskel D, Tseng Y-C, Lapertot G (2009) Pressure-induced electronic mixing and enhancement of ferromagnetic ordering in Eu*X* (*X*=Te, se, S, O) magnetic semiconductors. Phys Rev Lett 102:057206
50. Akamatsu H, Kumagai Y, Oba F, Fujita K, Tanaka K, Tanaka I (2013) Strong correlation between magnetic interaction and oxygen octahedral rotation in divalent europium perovskites. Adv Funct Mater 23:1864
51. Bousquet E, Dawber M, Stucki N, Lichtensteiger C, Hermet P, Gariglio S, Triscone J-M, Ghosez P (2008) Improper ferroelectricity in perovskite oxide artificial superlattices. Nature 452:732–736
52. Benedek NA, Fennie CJ (2011) Hybrid improper ferroelectricity: a mechanism for controllable polarization-magnetization coupling. Phys Rev Lett 106(10):107204
53. Benedek NA, Mulder AT, Fennie CJ (2012) Polar octahedral rotations: a path to new multifunctional materials. J Solid State Chem 195:11–20
54. Rondinelli JM, Fennie CJ (2012) Octahedral rotation-induced ferroelectricity in cation ordered perovskites. Adv Mater 24:1961–1968
55. Mulder AT, Benedek NA, Rondinelli JM, Fennie CJ (2013) Turning ABO_3 antiferroelectrics into ferroelectrics: design rules for practical rotation-driven ferroelectricity in double perovskites and $A_3B_2O_7$ Ruddlesden-popper compounds. Adv Funct Mater 23:4810
56. Balachandran PV, Puggioni D, Rondinelli JM (2014) Crystal-chemistry guidelines for noncentrosymmetric A_2BO_4 Ruddlesden-popper oxides. Inorg Chem 53:336
57. Fukushima T, Stroppa A, Picozzi S, Perez-Mato JM (2011) Large ferroelectric polarization in the new double perovskite $NaLaMnWO_6$ induced by non-polar instabilities. Phys Chem Chem Phys 13:12186
58. Toda K, Kameo Y, Kurita S, Sato M (1996) Crystal structure determination and ionic conductivity of layered perovskite compounds $NaLnTiO_4$ (ln = rare earth). J Alloys Compd 234:19–25
59. Togo A, Oba F, Tanaka I (2008) First-principles calculations of the ferroelastic transition between rutile-type and $CaCl_2$ -type SiO_2 at high pressures. Phys Rev B 78:134106
60. Akamatsu H, Fujita K, Kuge T, Gupta AS, Togo A, Lei S, Xue F, Stone G, Rondinelli JM, Chen L-Q, Tanaka I, Gopalan V, Tanaka K (2014) Inversion symmetry breaking by oxygen octahedral rotations in Ruddlesden-popper $NaRTiO_4$ family. Phys Rev Lett 112:187602
61. Gupta AS, Akamatsu H, Strayer ME, Lei S, Kuge T, Fujita K, dela Cruz C, Togo A, Tanaka I, Tanaka K, Mallouk TE, Gopalan V (2015) Improper inversion symmetry breaking and piezoelectricity through oxygen octahedral rotations in layered Perovskite family, $LiRTiO_4$ (R=rare earths). Adv Electron Mater 1:1500196
62. Gupta AS, Akamatsu H, Brown F, Nguyen M, Strayer M, Lapidus S, Yoshida S, Fujita K, Tanaka K, Tanaka I, Mallouk T, Gopalan V (2017) Competing structural instabilities in the Ruddlesden-popper derivatives $HRTiO_4$ (R = rare earths): oxygen octahedral rotations inducing noncentrosymmetricity and layer sliding retaining centrosymmetricity. Chem Mater 29:656–665

Part II
Applications of Element-Blocks as New Functional Materials

Chapter 16
Organic-Inorganic Hybrid Material with Surface-Modified Zirconia Nanoparticles as Element Blocks

Kimihiro Matsukawa, Koji Mitamura, and Seiji Watase

Abstract Organic-inorganic hybrid materials by dispersing inorganic nanoparticles in organic polymers are expected as functional substances with synergistic properties. Recently, high refractive index thin films are most important materials for optical devices such as flat panel display, touch panel, etc. These hybrids with dispersion of surface-modified zirconia nanoparticles can provide the controlled refractive index materials. It is necessary for the preparation of transparent hybrid materials to disperse zirconia nanoparticles in organic solvent or polymer solution by a beads mill process. Effective surface modifications of zirconia nanoparticles were successfully developed by two treatments of (1) with silane coupling agent and isocyanate compound and (2) with dual-site silane coupling agent. These dispersions can be applied to prepare the high refractive index thin films.

Keywords Nanoparticles · High refractive index · Organic-inorganic hybrids · Thin films

16.1 Introduction

Organic polymers have excellent properties in flexibility and moldability; on the other hand, inorganic nano-materials produce in durability, hardness, and optical properties. The organic-inorganic hybrids from these substances are possible candidates for the practical materials in versatile industrial fields. In order to prepare the organic-inorganic hybrid materials, inorganic nano-materials were dispersed in organic polymers by interactions such as covalent bonds, hydrogen bonds, coordination bonds, π-π interactions, ionic interactions, hydrophobic interactions, and so

K. Matsukawa (✉)
Kyoto Institute of Technology, Kyoto, Japan
e-mail: kmatsu@kit.ac.jp

K. Mitamura · S. Watase
Osaka Research Institute of Industrial Science and Technology, Osaka, Japan

Y. Chujo (ed.), *New Polymeric Materials Based on Element-Blocks*,
https://doi.org/10.1007/978-981-13-2889-3_16

on [1–5]. Inorganic substances are typically prepared by a sol-gel method of metal alkoxides, but these have some disadvantages such as large curing shrinkage during gelation, alcohol production due to hydrolysis of metal alkoxides, and a long reaction time of curing sol in a polymer solution. However, polysilsesquioxanes are one of possible materials preventing from these disadvantages, because the hydrolysis and condensation in a sol-gel method was completed before mixing with polymers [6–8].

The preparation of organic-inorganic hybrids with metal oxide nanoparticles instead of sol-gel materials is seemed to be an effective procedure reducing the defect of the sol-gel method. For example, a transparent organic-inorganic hybrid can be produced by uniformly dispersing colloidal silica of nanometer size in an organic polymer. However, the colloidal silica dispersed in water has high hydrophilic surface that cannot be dispersed in an organic polymer, so it is necessary to change the surface property of colloidal silica to be hydrophobic. We have succeeded in obtaining a low refractive index thin film from organic-inorganic hybrid consisting of polymethylphenylsilane (PMPS) and phenyl-modified silica nanoparticles [9–11]. The thin film containing nanovoids between these silica particles resulted in low refractive index.

It is well known that high refractive index material can be prepared by introducing aromatic rings, halogen elements except fluorine, sulfur elements, and metal oxides like titania or zirconia. In this point of view, zirconia nanoparticles play a large role as a factor to control the refractive index in the hybrid material, and it is possible to design a material capable of controlling the optical properties. As titania has photocatalytic activity, there is a risk of affecting the organic polymer matrix under UV light. However, the bandgap energy of zirconia (5.0 eV) is much larger than titania, and it exhibits photocatalytic activity only in the ultraviolet range of less than 250 nm. Furthermore, zirconia has a refractive index of 2.17, which is relatively high among metal oxides, and organic-inorganic hybrids containing zirconia nanoparticles can be considered to be high refractive index material excellent in light stability. For the purpose of preparing high refractive index hybrid materials, it is indispensable to develop the dispersion of zirconia nanoparticles in organic solvents by the surface treatment with a beads mill process. In this chapter, the preparation of zirconia nanoparticle dispersion and the application to high refractive index organic-inorganic hybrid thin films will be explained.

16.2 Problem on Preparation of Zirconia Nanoparticle Dispersion Using Silane Coupling Agents by Beads Mill Process

Dispersions of nanoparticles in organic solvents were prepared from the aggregated nanoparticles by a top-down method using a beads mill with media beads as shown in Fig. 16.1 [12–14]. These nanoparticles in dispersions indicated almost similar

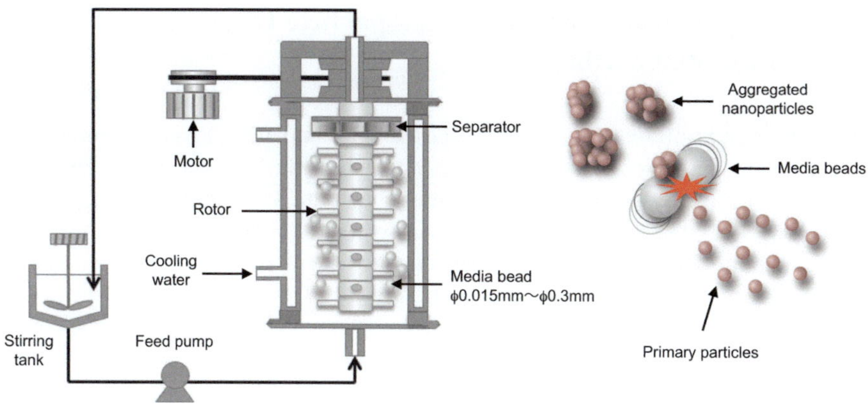

Fig. 16.1 Apparatus illustration of beads mill and disintegration of aggregated particles

particle size obtained by a sol-gel method. Using dispersants are indispensable for dispersing nanoparticles, and some silane coupling agents are commonly added to the beads mill process. The silanols generated by hydrolysis of silane coupling agents can be reacted with hydroxyl groups on the surface of zirconia nanoparticle to form Zr-O-Si bonds, and the surface is changed to be hydrophobic, which this process is theoretically simple and easy. However, the strong mechanical forces such as shearing, friction, and compression generated between the media beads during a beads mill process generated on the surface of the nanoparticles. This reaction condition is quite different from the usual simple stirring or static reaction with silane coupling agents. As all of three silanols hydrolyzed from silane coupling agent cannot bond to the surface of nanoparticles, unreacted silanols may form siloxane bonds with silanols on other nanoparticles in such a strong reaction field of beads mill process. Therefore, it is presumed that silanols on certain amount of nanoparticles are covalently bonded and finally caused reaggregation of nanoparticles.

This phenomenon is a very serious problem in practical process of dispersion preparation by beads mill. In order to resolve this problem, it is intended to reduce the amount of the silane coupling agent to be used. But it is difficult to obtain an good nanoparticle dispersions in organic solvents by merely reducing silane coupling agent, so that it is important to develop the processes using new dispersants. We have been studying about two procedures of (1) two-step method using silane coupling agent and isocyanate compound and (2) surface modification using dual-site silane coupling agent.

16.3 Preparation of Zirconia Nanoparticle Dispersion by Two-Step Method

A small amount of silane coupling agent cannot completely cover the surface of zirconia nanoparticles, so it does not perform an effective dispersion although reaggregation hardly occurs. In this point of view, the preparation of nanoparticle dispersion was investigated by the two-step method using firstly half amount of silane coupling agent, and another dispersant reacted with remained hydroxyl groups on the surface of zirconia nanoparticles [15, 16].

As shown in Fig. 16.2, aggregated zirconia nanoparticles were treated with vinyltrimethoxysilane (VTMS), which was half amount of whole treatment for the surface of nanoparticle, in methyl ethyl ketone (MEK) by bead mill, then 2-methacryloxyoxyethyl isocyanate (MOI) was added to the solution, and the stable dispersion of zirconia nanoparticle (ZrO_2-MOI) containing methacrylic groups on the surface of particle was obtained after ultrasonic treatment. MOI is a good dispersing agent in the second step, because isocyanate group can form a urethane bond with Zr-OH of zirconia surface but not covalent bond between MOIs. The formation of urethane bond was confirmed by a FT-IR measurement, in which the peak of isocyanate at near 2270 cm^{-1} decreased and the formation of urethane bond appeared after ultrasonic treatment as shown in Fig. 16.3. The reaction of MOI to zirconia was relatively fast, because it seemed that urethane formation from isocyanate was accelerated by the Lewis acidity of zirconia. However, in the case of reaction of MOI to silica nanoparticles without any catalyst, it is known that urethane bonds do not generate at all [17, 18]. Furthermore, it was confirmed that silanols, which were hydrolyzed from silane coupling agent, and isocyanate do not react without catalyst. From these results, MOI is not only a good second dispersant for zirconia nanoparticles but also an advantage that no by-product with the silane coupling agent is generated. However, when the bead mill treatment was performed using only MOI as a dispersant, reaggregation of zirconia nanoparticles occurred. The reason for reaggregation is presumed that urethane bonds formed on the surface of zirconia nanoparticles may cause hydrogen bonding between nanoparticles.

Figure 16.4 indicates that DLS average particle diameter of nanoparticles prepared by the two-step method using each 2 wt% of both VTMS and MOI was 8.5 nm at the cumulative 50% volume. And this particle size of dispersion is approximately equal to the primary particle diameter. On the other hand, the average particle diameter of dispersion obtained by using 4 wt% of only VMTS was 332.7 nm, and it was found to occur a considerable aggregate after 2 weeks standing. It is assumed that the reaggregation occurred due to formation of siloxane bonds between the silanol of VTMS on the nanoparticle surface. As the zirconia nanoparticle dispersion by this two-step method has methacrylic groups derived from MOI, it is easy to form a hybrid thin film by a photoradical polymerization with multifunctional acrylate monomers. The photocured hybrid thin films are quite transparent as shown in Fig. 16.5, and even when the content of zirconia nanoparticles was increased up to 60 wt%, a transmittance of the hybrid thin film was same level as that of only DPHA thin film.

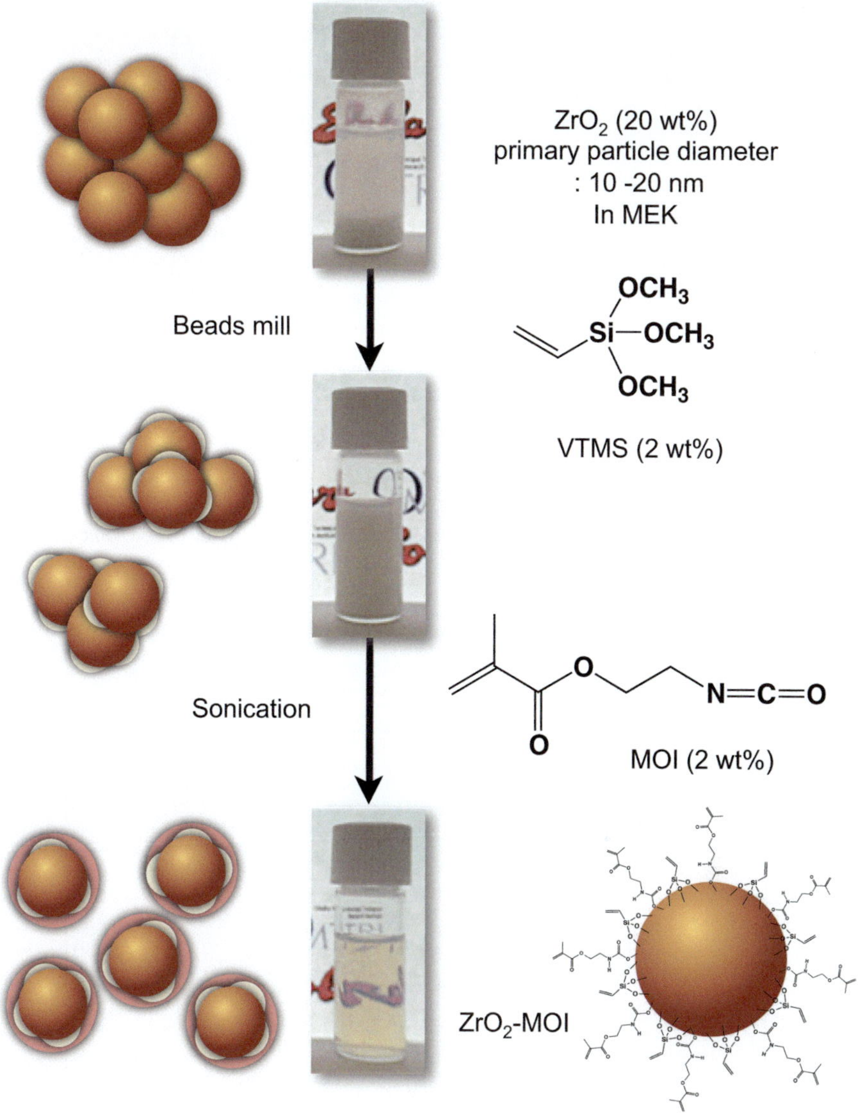

Fig. 16.2 Preparation of zirconia nanoparticle dispersion by two-step method

Furthermore, as shown in Fig. 16.6, the refractive index of hybrid thin film was increased by the addition of zirconia nanoparticles to DPHA. For example, the refractive index of hybrid thin film containing 80 wt% of zirconia nanoparticles was 1.68; on the other, refractive index of DPHA is 1.52. From this result, it is concluded that the refractive index of thin films can be controlled easily by the composition ratio of zirconia nanoparticles and applied to various optical materials.

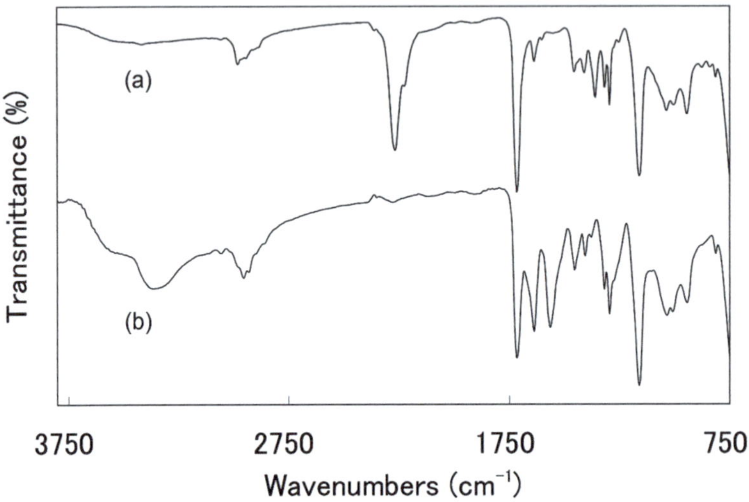

Fig. 16.3 FT-IR spectra of zirconia nanoparticles with addition of MOI; (**a**) before sonication, (**b**) after sonication

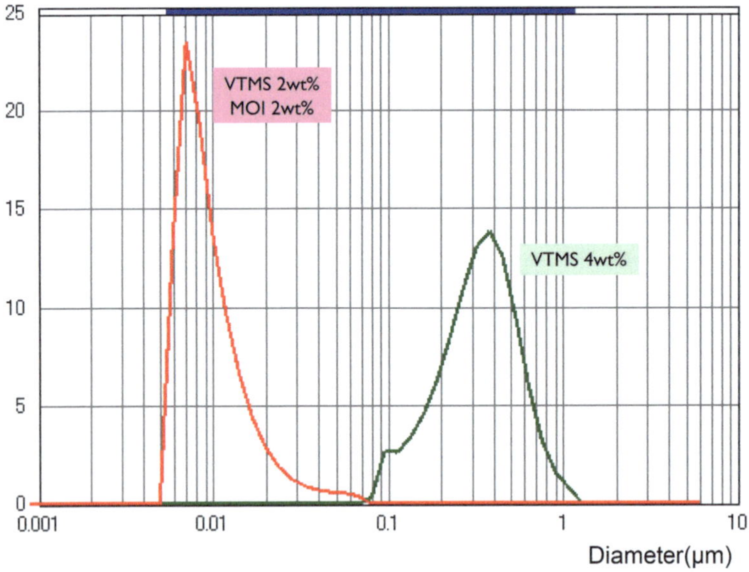

Fig. 16.4 DLS average particle diameter of zirconia nanoparticles due to different dispersant

Fig. 16.5 UV-Vis transmittance of photocured DPHA-hybrid thin films with different composition ratio containing zirconia nanoparticles dispersed by two-step method

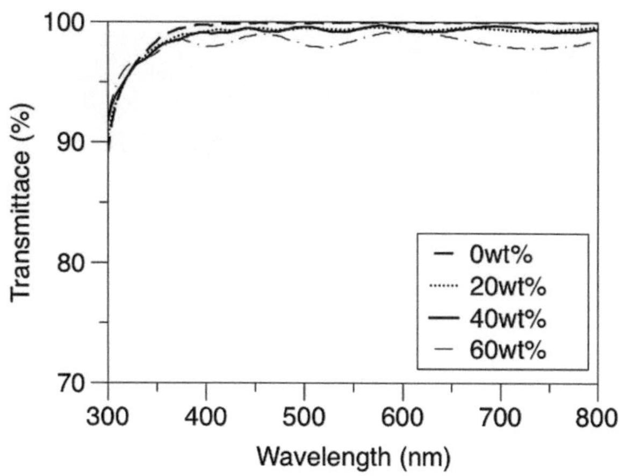

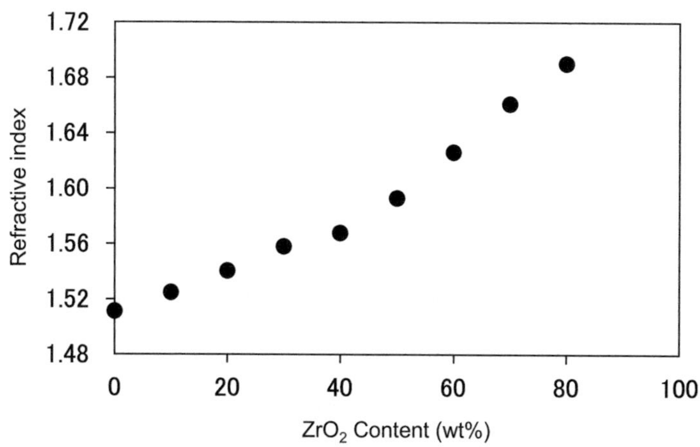

Fig. 16.6 Refractive index change of photocured DPHA-hybrid thin films with different composition ratio containing zirconia nanoparticles dispersed by two-step method

16.4 Preparation of Zirconia Nanoparticle Dispersions by Dual-Site Silane Coupling Agent

A novel silane coupling agent, which is preventing from a condensation reaction between unreacted silanols in beads mill process, is seemed to be an essential dispersant without lowering the reactivity for the nanoparticle surface. In order to accomplish this target, we developed a dual-site-type silane coupling agent, which has two silane coupling agent units in one direction of a relatively large organic group and forms a unique bidentate structure on the surface of the zirconia nanoparticle indicated in Fig. 16.7; hence, this structure is expected to suppress the formation of siloxane bonds with silane coupling agents on the surface and unreacted

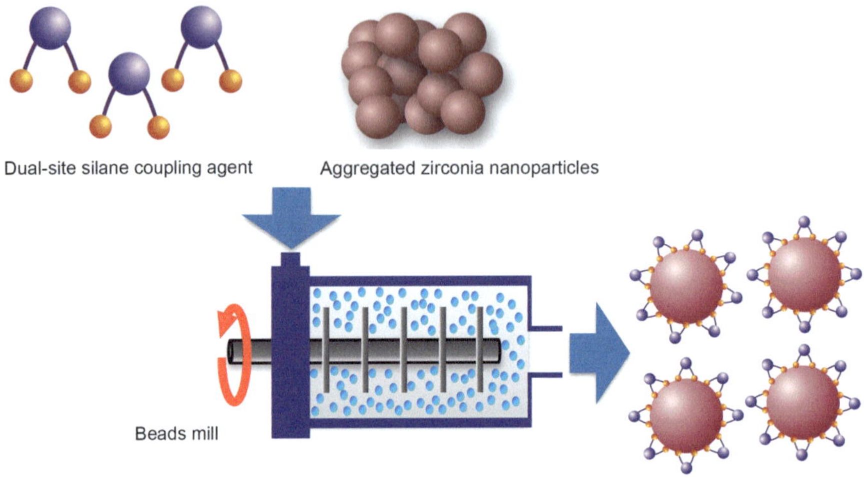

Fig. 16.7 Preparation of zirconia nanoparticle dispersion by using dual-site silane coupling agent

ones. Herein, a dual-site silane coupling agent was synthesized from bisphenylfluorene or o-phthalate ester derivatives.

16.4.1 Dual-Site-Type Silane Coupling Agent from Bisphenylfluorene Derivatives

It is known that bisphenylfluorene is a characteristic aromatic compound with a cardo structure formed by equivalent four aromatic rings connecting on one carbon, and it has properties of high refractive index and low birefringence. We have reported that high refractive index hybrid thin films could be prepared by using epoxy containing polysilsesquioxane and bisphenylfluorene derivatives [6, 7].

A structure of dual-site silane coupling agent based on bisphenylfluorene is illustrated in Fig. 16.8, which two trialkoxysilane groups are attached at the end of side chain from two phenyl groups. This silane coupling agent was synthesized from thiol-ene reaction of bis(3-methylphenyl)fluorene allyl ether (BCF-allyl) and mercaptopropyltrimethoxysilane (MPTMS) [19]. The thiol-ene reaction used here is a radical addition reaction of thiyl radical to a carbon-carbon double bond, which is known as one of click reactions [20, 21]. As shown in Fig. 16.9, a dual-site silane coupling agent (BSF) can be synthesized quantitatively by the ultraviolet irradiation for a very short time in the presence of a photoradical initiator, and it has been confirmed the existence of alkoxysilyl groups by FT-IR and NMR analysis. The refractive index of BSF is 1.57, which is higher than that of usual dispersants; therefore, it is suitable for the purpose of attaining a high refractive index of hybrid materials.

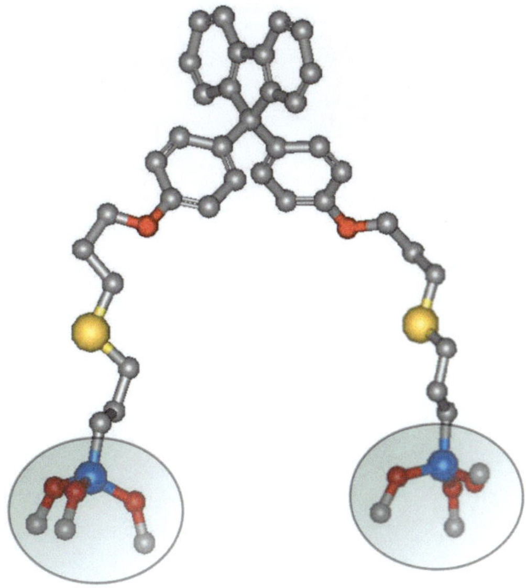

Fig. 16.8 Chemical structure of dual-site silane coupling agent based on bisphenylfluorene

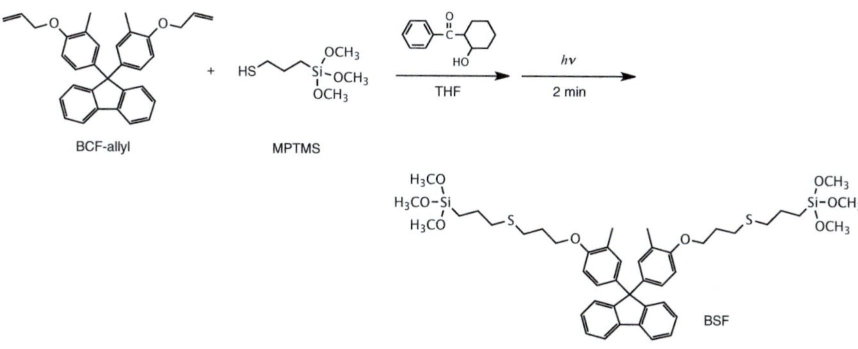

Fig. 16.9 Synthesis of dual-site silane coupling agent (BSF) from BCF-allyl and MPTMS

The bisphenylfluorene-modified zirconia nanoparticle dispersion (ZrO_2/BSF) could be prepared by surface treatment of zirconia particles (primary particle size: 10–20 nm) with BSF in MEK. The dual-site silane coupling agent formed a bidentatae structure on the surface of the zirconia nanoparticle during beads mill process as shown in Fig. 16.10 and a hydrophobic surface, so it can be presumed that it is easy to disperse in the organic solvent. The obtained nanoparticle dispersion is 14.6 nm at the lative 50% particle size, and it has high dispersibility and transparency.

A photoradical initiator (Irgacure 184) was added to the solution of bis-phenoxyethanol fluorene diacrylate (BPEFA) and the prepared zirconia nanoparticle dispersion. After spin coating of the mixture on a glass substrate, a photocuring hybrid thin film was prepared by ultraviolet irradiation. Even when

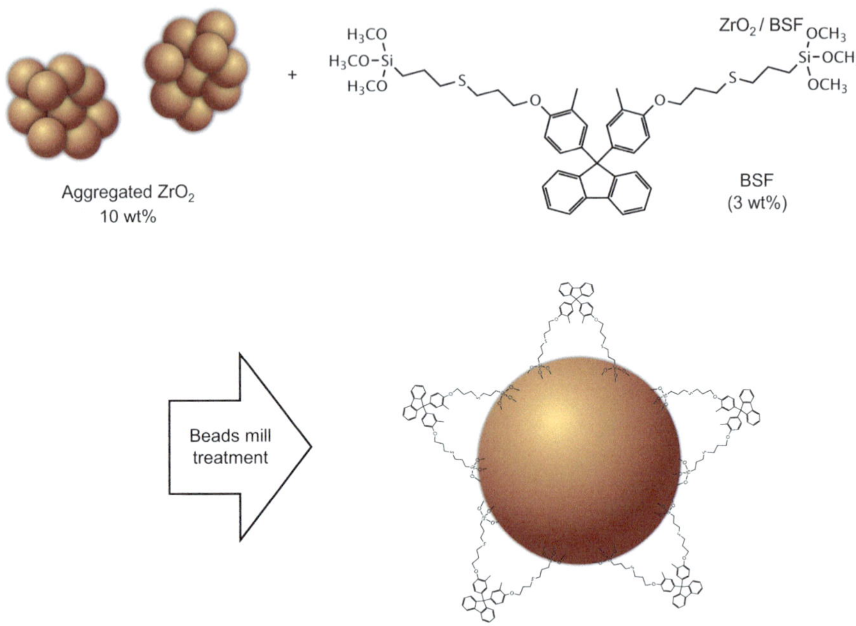

Fig. 16.10 Preparation of zirconia nanoparticle dispersion by using BSF

Fig. 16.11 Refractive index change of polymer hybrid thin films with different composition ratio containing BSF-treated zirconia nanoparticles

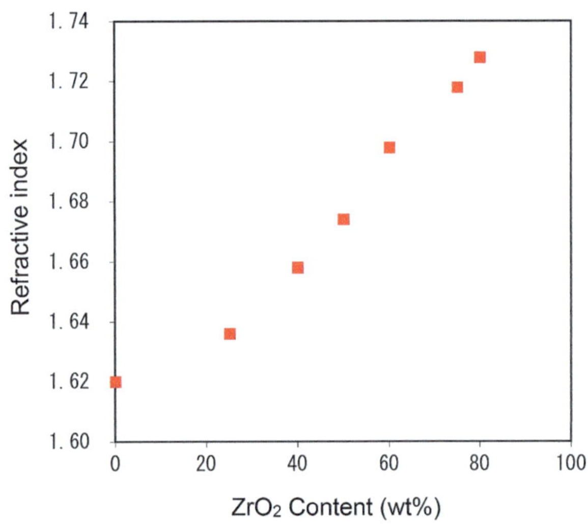

containing zirconia nanoparticles at a high concentration, the transparency is quite high, and the refractive index can be controlled according to the content of nanoparticles in the hybrid material, which is from 1.62 to 1.73 as shown in Fig. 16.11. It was concluded to provide the controllable refractive index of hybrids in a wide range.

16.4.2 Dual-Site Silane Coupling Agent from Diallyl Phthalate

When two trialkoxysilyl groups bond in one side of a nonequivalent organic molecule, the dual-site silane coupling agent can be designed to exhibit its function. From the synthetic point of view, o-diallyl phthalate (DAP) is a possible starting compound for thiol-ene reaction with MPTMS. As shown in Fig. 16.12, the dual-site silane coupling agent (o-DAP-Si) was prepared quantitatively by thiol-ene reaction to the terminal allyl groups under ultraviolet irradiation. The preparation of zirconia nanoparticle dispersion using o-DAP-Si as a dispersant was examined by a beads mill [22]. Then, TEM photographs indicate that the dispersion was obtained in good condition. It can be seen to be difficult to disperse zirconia particles without surface modification, and they were considerably aggregated (Fig. 16.13a). On the other, zirconia particles with surface modification using o-DAP-Si showed the excellent dispersibility corresponding to primary particles (Fig. 16.13b), which the cumulative 50% particle diameter of the dispersion by DLS was 27.5 nm. In order to confirm the ability of dual-site silane coupling agent for dispersing effect, it was examined to compare with the single-site silane coupling agent having an organic molecular structure similar to DAP. Hence the thiol-ene reaction of allyl benzoate (AB) and MPTMS was carried out to form the silane coupling agent (AB-Si). As the result of

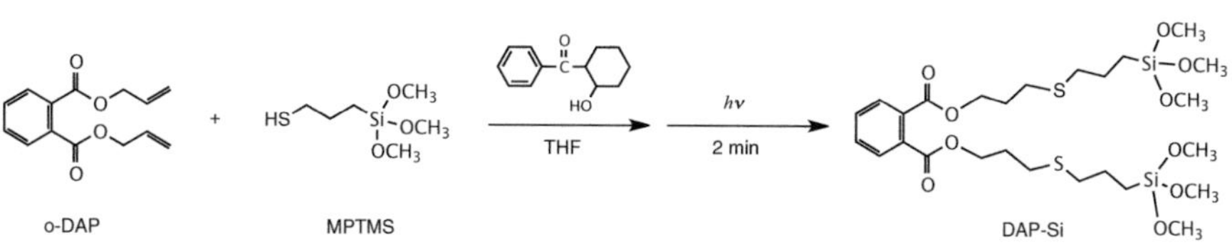

Fig. 16.12 Synthesis of dual-site silane coupling agent (DAP-Si) from o-DAP and MPTMS

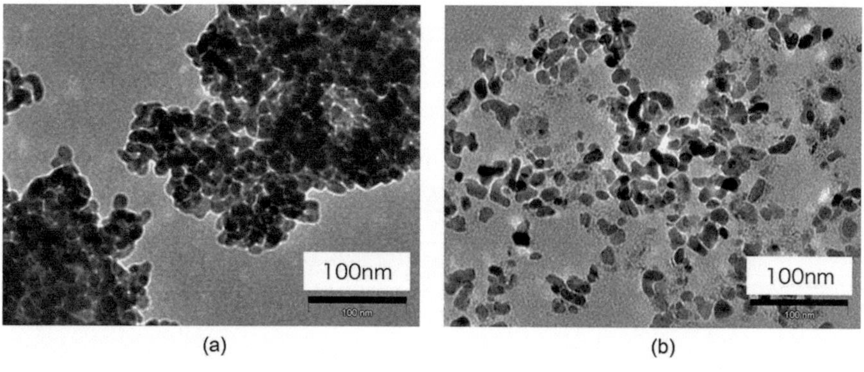

Fig. 16.13 TEM images of zirconia nanoparticles after beads mill treatment; (**a**) without dispersant, (**b**) modified with DAP-Si

comparison, it was found that AB-Si showed not only poor dispersibility than DAP-Si but also reaggregation of the zirconia nanoparticles under the excessive addition. This phenomenon is commonly observed from the case of dispersing process with ordinary silane coupling agents, so it is considered that reaggregation occurred by the condensation reaction of silanols between unreacted silane coupling agents. However, in the case of excessive addition of dual-site silane coupling agent, the reaggregation did not occur at all, even if unreacted silanols were present on the surface of nanoparticles. As it is inferred that the second reaction of DAP-Si is extremely difficult to occur on the surface of zirconia nanoparticles, the dual-site silane coupling agent is seemed to be a very effective dispersant for preparing a stable dispersion.

On the other hand, since the phthalate ester due to the basic skeleton of DAP-Si has the chemical structure of plasticizer for polymers, it seemed to be excellent compatibility with various polymers. Therefore, since the zirconia nanoparticles treated by DAP-Si can be regarded as having a function of plasticizer, there is a great advantage for producing a hybrid material. Thus obtained dispersion has a good compatibility to thermoplastics, such as polymethyl methacrylate (PMMA), polystyrene (PS), etc., and multifunctional acrylate monomers, which would generate transparent hybrid materials. Optical properties of these hybrid thin films were measured from transmittance, haze value, and refractive index.

When PMMA hybrid was prepared with 40 wt% of untreated zirconia nanoparticle dispersion, the transmittance and haze value of the hybrid thin film were, respectively, 88.5% and 44%. And also the photocured thin film prepared from trimethylolpropane triacrylate (TMPTA) and 50 wt% of untreated zirconia nanoparticle dispersion showed a transmittance of 82.8% and a haze value of 38% similar to PMMA hybrid. These hybrid thin films were turbid; accordingly it is not suitable for the optical application.

On the contrary, the hybrid thin film containing DAP-Si-treated zirconia nanoparticle dispersions indicated the same transmittance as the polymers as shown in Table 16.1. Furthermore, the haze value of hybrid thin films was summarized in

Table 16.1 Total light transmission of polymer hybrid thin films containing DAP-Si-treated zirconia nanoparticles

ZrO_2 content (wt%)	PMMA	PS	DPHA	TMPTA	PETA
0	92.6	90.6	91.4	90.5	92.0
10	92.3	88.0	91.9	93.4	91.4
20	92.3	90.6	92.1	90.9	92.9
30	89.1	90.0	93.5	92.0	91.3
40	89.9	90.0	91.7	93.4	92.7
50	88.4	89.2	93.1	91.7	91.7
60	90.0	91.3	92.6	91.0	91.3
70	89.2	91.9	91.9	91.2	91.1
80	88.8	88.8	89.0	86.8	90.6
90	86.0	–	90.0	87.0	88.6

Table 16.2 Haze value of polymer hybrid thin films containing DAP-Si-treated zirconia nanoparticles

ZrO_2 content (wt%)	PMMA	PS	DPHA	TMPTA	PETA
0	0.1	0.3	0.3	0.2	0.1
10	0.1	0.3	0.1	0.1	0.2
20	0.1	0.3	0.2	0.1	0.1
30	0.2	0.3	0.1	0.3	0.2
40	0.3	0.6	0.1	0.2	0.2
50	0.2	0.7	0.2	0.1	0.1
60	0.2	0.8	0.2	0.4	0.1
70	0.3	0.7	0.1	0.1	0.2
80	0.2	0.5	0.2	0.2	0.3
90	0.1	–	0.1	0.3	0.1

Table 16.2. It was found that the thin films were very transparent and clear, even when 90 wt% of zirconia nanoparticles was added to the polymer components. These results suggest good compatibility and dispersibility of the zirconia nanoparticle dispersion treated with DAP-Si in some polymers. The refractive index of hybrid thin films with various composition ratio of zirconia nanoparticles and polymer was summarized in Fig. 16.14. It was concluded that the high refractive index hybrid thin films based on PMMA or photocured polymers can be controlled from 1.5 to over 1.7.

16.5 Conclusions

A controlled refractive index thin film is a useful indispensable material for an information equipment such as flat panel displays, smartphones, tablet PCs, and so on. In particular, these thin films are applied to antireflection films and refractive index matching layers for ITO, which contribute greatly to improve the visibility of device. Therefore this optical property is important and necessary for versatile devices in the future.

In order to introduce the zirconia nanoparticles of high refractive index to polymer matrices, the surface treatment of zirconia nanoparticles was studied about the preparation of dispersion using new systems, which can be regarded the new zirconia dispersion as the element blocks. We developed two practical methods using beads mill, which were used by (1) two-step method with silane coupling agent and isocyanate and (2) dual-site silane coupling agents. Both methods can provide the effective zirconia nanoparticle dispersions for high refractive index organic-inorganic hybrids. The transparent thin films of high refractive index were created by photopolymerization of multifunctional acrylate monomers containing zirconia nanoparticle dispersions. The refractive index of these thin films was controlled from 1.5 to over 1.7 by changing the composition ratio.

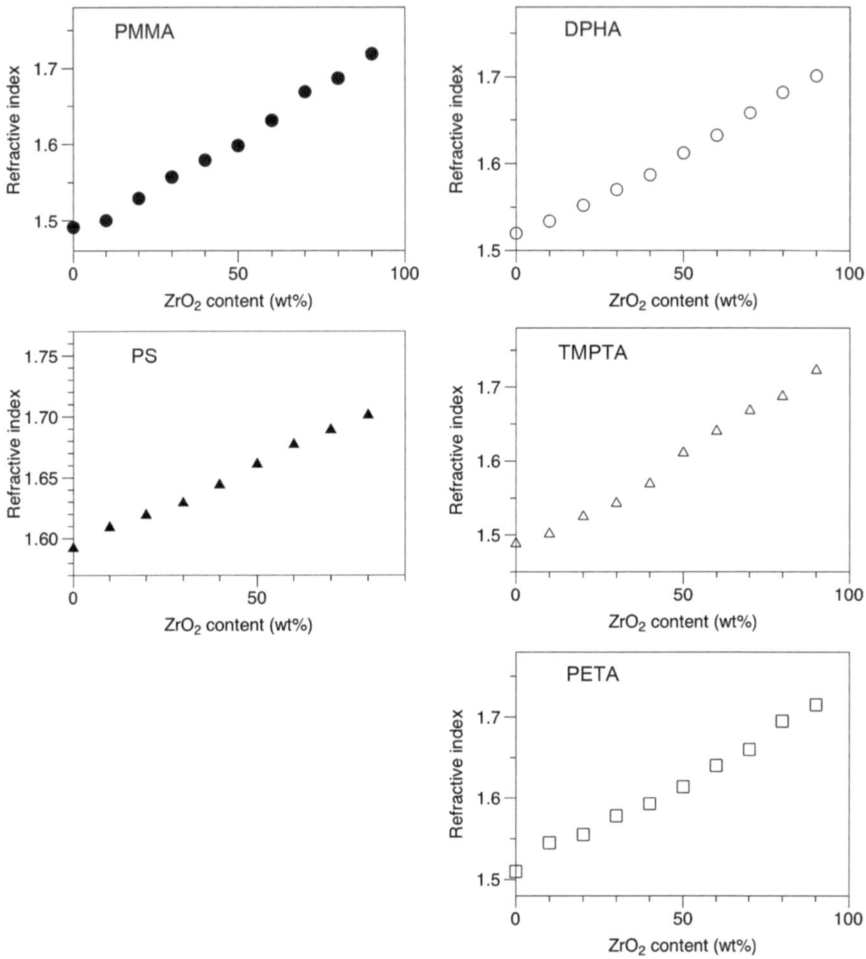

Fig. 16.14 Refractive index change of polymer hybrid thin films with different composition ratio containing DAP-Si-treated zirconia nanoparticles

From this investigation, the organic-inorganic hybrid materials containing the element blocks of surface-modified zirconia nanoparticles are capable of controlling the refractive index, and it is expected that this function will continue to lead to the development of new optical materials.

Acknowledgment We thank Dr. Yuki Minami (Mikuni Color Ltd.), Mr. Eiichiro Takeuchi (Osaka Institute of Technology), Mr. Takashi Fukuda, and Mr. Atsushi Sasaki (Osaka Tele-communication University). This work was supported in part by a Grant-in-Aid for Scientific Research on Innovative Areas "New Polymeric Materials Based on Element-Blocks (No. 2401)" provided by the Ministry of Education, Culture, Sports, Science and Technology, Japan.

References

1. Chujo Y (1996) Polym Mater Encycl 6:4793
2. Wen J, Wilkes GL (1996) Chem Mater 8:1667
3. Gomez-Romero P (2001) Adv Mater 13:161
4. Sanchez C, Julian B, Belleville P, Popall M (2005) J Mater Chem 15:3559
5. Gomez-Romero P, Sanchez C (2006) Functional hybrid materials. Wiley
6. Matsukawa K, Matsuura Y, Nakamura A, Nishioka N, Motokawa T, Murase H (2006) J Photopolym Sci Technol 19:89
7. Matsukawa K, Matsuura Y, Nakamura A, Nishioka N, Murase H, Kawasaki S (2007) J Photopolym Sci Technol 20:307
8. Matsukawa K, Fukuda T, Watase S, Goda H (2010) J Photopolym Sci Technol 23:115
9. Matsukawa K, Katada K, Nishioka N, Matsuura Y, Inoue H, Photopolym J (2004) Sci Technol 17:51–52
10. Matsukawa K, Matsuura Y, Katayama K, Nishioka N, Chikaraishi M, Naito H (2008) Kobunshi Ronbunsyu 65:440
11. Matsukawa K, Matsuura Y, Michiwaki Y, Chikaraishi M, Naito H (2009) J Photopolym Sci Technol 22:307
12. Murata K, Minami Y, Yomoyasu H, Matsukawa K (2008) J Adhes Soc Jpn 44:438
13. Joni M, Purwanto A, Iskandar F, Okuyama K (2009) Ind Eng Chem Res 48:6916
14. Iskandar F (2009) Adv Powder Technol 20:283
15. Minami Y, Murata K, Watase S, Matsukawa K (2010) Kobunshi Ronbunsyu 67:397
16. Matsukawa K, Minami Y, Watase S (2012) IDW/AD' 12 Proc 965
17. Sugimoto H, Nakanishi E, Yamauchi K, Daimatsu K, Yasumura T, Inomata K (2004) Polym Bull 52:209
18. Sugimoto H, Daimatsu K, Nakanishi E, Ogasawara Y, Yasumura T, Inomata K (2006) Polymer 47:3754
19. Minami Y, Murata K, Watase S, Matsumoto A, Matsukawa K (2013) J Photopolym Sci Technol 26:491
20. Hoyle CE, Lee TY, Roper T (2004) J Polym Sci. Part A: Polym Chem 5301:42
21. Hoyle CE, Bowman CN (2010) Angew Chem Int Ed 49:1540
22. Minami Y, Murata K, Watase S, Matsukawa K (2014) Photopolym J Sci Technol 27:261–262

Chapter 17
Network Formation Conditions Control Water Drop Adhesion for VK100 and a Model Pt-Cured Silicone

Jennie B. Lumen, Rebecca M. Jarrell, Sithara S. Nair, Chenyu Wang, Ashraf M. Kayesh, and Kenneth J. Wynne

Abstract Unexpected wetting behavior is reported for silicone elastomers platinum cured at 37 °C in water or saline. These conditions were prompted as a way to mimic cure under physiologically relevant conditions for VK100, a Pt-cured silicone used for vertebral augmentation. Water contact angles (CAs) were determined by the drop addition/withdrawal method. Network formation in air, water, or saline gave high advancing CAs (θ_A). However, compared to 74° for air cure, network formation in water (56°) or saline (46°) gave low receding CAs (θ_R). Thus, water drop adhesion to VK100 and a model Pt-cured silicone depends on whether network formation is carried out in water or saline ("sticky") or in air ("slippery"). For cure in water or saline, autoxidation (Si-H → Si-OH) and near-surface entrapment of cross-linking chains containing –Si-OH are proposed to account for low receding CAs. The origin of the low θ_R and high contact angle hysteresis (54–72°) is correlated with the theory of Johnson and Dettre by which a small area fraction of polar groups impedes retraction of a receding water drop. These results are of interest given the importance of polar interactions at interfaces that favor adhesion to bone and influence on biofouling, adhesion of proteins, and interactions with human cells.

Keywords Contact angles · Platinum-cured silicones · Cure · Network formation · PDMS

J. B. Lumen · R. M. Jarrell · S. S. Nair · C. Wang · A. M. Kayesh · K. J. Wynne (✉)
Chemical and Life Science Engineering, Virginia Commonwealth University, Richmond, VA, USA
e-mail: kjwynne@vcu.edu

Y. Chujo (ed.), *New Polymeric Materials Based on Element-Blocks*,
https://doi.org/10.1007/978-981-13-2889-3_17

17.1 Introduction

Plasma or other high-energy treatments of poly(dimethylsiloxane) elastomers such as Sylgard 184 are often used to create a hydrophilic surface [1, 2]. Because a siliceous film forms, *both* advancing and receding contact angles (CAs) decrease. McCarthy reported a 30 s oxygen plasma treatment for Sylgard 184 reduced advancing and receding contact angles (θ_A/θ_R) from $117 \pm 8°/104 \pm 4°$ to $10 \pm 3°/5 \pm 3°$ with better stability to hydrophobic recovery.

In contrast to decreasing both θ_A and θ_R, we reported that network formation for platinum-cured Sylgard 184 at 60 °C or lower resulted in retention of high advancing contact angles (~115°) but low receding CAs [3]. Water drop adhesion can be approximated by w_p, the practical work of adhesion [4] which is a function of receding CAs ($w_p \propto 1 + \cos\theta_R$) [5]. Thus, decoupling θ_A from θ_R resulted in water drops running off Sylgard 184 cured at 100 °C but sticking to coatings cured at $\leq$60 °C. A laboratory-prepared analog without filler (Pt-PDMS) showed similar behavior. That is, both Sylgard 184 and Pt-PDMS had low receding contact angles when cured at $\leq$60 °C (sticky) but high θ_R (slippery) when network formation was carried out at 100 °C.

Achieving a low receding CA for Pt-cured silicone required a lengthy cure time in air for Sylgard 184 and Pt-PDMS at $\leq$60 °C. In contrast, we now report that in 10–15 min, cure for two platinum-cured silicones in water or saline at 37 °C results in low receding CAs and "sticky" surfaces for water drops.

The present investigation began with examining wetting behavior of VK100, a two-part platinum-cured silicone manufactured by BONWRx LLC [6, 7]. This silicone elastomer is formulated for vertebral augmentation by which the solidifying resin is injected into the periosteum of targeted vertebral pedicles to generate a bone void filler and relieve pain associated with vertebral compression fractures.

To characterize wetting behavior, VK100 coatings were formed on glass microscope slides and cured in water or saline at 37 °C to mimic physiological conditions. A laboratory-prepared Pt-cured silicone designated D^VD^H (experimental) was also prepared without filler, silicone oil, or $BaSO_4$ radiopacifier. Without an inhibitor, VK100 and D^VD^H cure rapidly (10–15 min). High advancing CAs were observed for cure in air (108–118°). However, in contrast to the receding CAs for cure in air (74°), both VK100 and D^VD^H gave receding CAs of 46° in saline and 56° in water. Because adhesion is a function of the receding CA, water drop adhesion to VK100 and the model D^VD^H was "sticky" for cure in water or saline and "slippery" for cure in air.

While the hydrophobicity of platinum-cured silicones in air is well known [3, 8–10], to our knowledge wetting behavior after cure in an aqueous medium has not been investigated. Contact angles generated for cure in air, water, or saline are discussed in terms of network formation by hydrosilylation and autoxidation of cross-linker Si-H to Si-OH. The low θ_R and high contact angle hysteresis is correlated with the theory of Johnson and Dettre by which a small area fraction of polar groups impedes retraction of a receding water drop. These results are of interest given the importance of polar interactions at interfaces that favor adhesion to bone and influence biofouling, adhesion of proteins, and interactions with human cells.

17.2 Experimental

17.2.1 Materials

VK100 was obtained from BONWRx LLC. VK100 is comprised of two components that when mixed form a viscous but flowable composition [7, 11]. One component consists of vinyldimethyl-terminated polydimethylsiloxane (D^{Vi}-PDMS) and trimethylsiloxy-terminated polydimethylsiloxane in ~1:1 ratio by weight, amorphous silica, barium sulfate powder, and a platinum catalyst. The second component is similar but instead of Pt catalyst contains a poly(methylhydro-*co*-dimethylsiloxane), D^{H}-PDMS, cross-linking agent [7, 11]. For the model silicone, components were D^{Vi}-PDMS (MW 28 kD, Gelest DMS-V31R), D^{H}-PDMS (MW 1 kD, 50–55 mole % MeHSiO, HMS-501), and platinum-divinyltetramethyldisiloxane complex solution (3% in vinyl-terminated PDMS, Gelest SIP6830.3).

17.2.2 Sample Preparation

VK100 The two components were mixed and extruded with the plastic device provided with the kit in conjunction with the procedural steps recommended in the VK100 BONWRx LLC pamphlet. A typical run employed 1–2 g of VK100 precursor. The mixture was spread quickly on pre-cleaned glass microscope slides and cured at 37 °C. Gelation and solidification were noted after 10 min, but samples were allowed to remain in water or saline for 1 h. Saline was laboratory-prepared (0.9% NaCl/L; 154 mEq/L). Three sterilized Kimble Chase glass petri dishes were used to hold water or saline during the cure. Conventional cure in air was used as a reference for surface analysis. An image of VK100 cured in water at 37 °C is shown in Fig. 17.1. The coating is white due to the presence of $BaSO_4$ radiopacifier.

Model Divinyl Polydimethylsiloxane ($D^{Vi}D^{H}$) Elastomer Pt-cured elastomer $D^{Vi}D^{H}$ is similar to VK100 but without a siliceous filler, $BaSO_4$, or silicone oil. Preparation followed a procedure similar to that described by McCarthy [2] and Wang [3]. In a

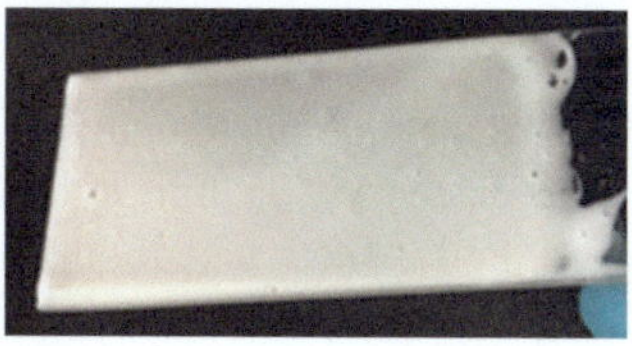

Fig. 17.1 VK100 after cure in water at 37 °C

typical preparation, 20 g divinyl-terminated polydimethylsiloxane (D^{Vi}-PDMS) and platinum-divinyltetramethyldisiloxane complex solution (40 μL, 1 wt% in hexane) were hand-mixed. Trimethylsiloxy-terminated methylhydrosiloxane-dimethylsiloxane (D^{H}-PDMS) copolymer (2 g) was added and hand-mixed for 30 s. An optically transparent, viscous resin was obtained which was spread on microscope slides and left in air for 5 min for viscosity buildup. Subsequently, coated slides were placed in a preheated oven (37 °C) or in a preheated saline solution (37 °C). Gelation and cure were evident after 10 min. After 4 h the samples were removed for CA measurements.

17.2.3 Contact Angles

Deionized water (~18.2 MΩ, 20 μL drops) was used as the probe liquid. Contact angle measurements were carried out with a Rame-Hart goniometer equipped with an LCD camera. Captured images were analyzed, and water CAs were obtained using Dropview image software version 1.4.11. Average values were obtained from three observations. Sessile drops were used to obtain static contact angles, while advancing and receding drops provided (θ_A) and (θ_R) measurements, respectively. Negligible change in CAs was found over the course of ~ 1 min.

17.2.4 AFM Imaging

Morphological and nanomechanical investigations were carried out using a Bruker Dimension ICON (Digital Instruments, CA) atomic force microscope (AFM) with a NanoScope V controller. Tapping mode imaging was performed in air using microfabricated silicone cantilevers (Tap300E-G, resonant frequency, 300 kHz; force constant, 40 N/m, budget sensors). Unless otherwise noted in figures, tapping force corresponded to a set point ratio r_{sp} of 0.8 where $r_{sp} = A_{exp}/A_o$, A_o is free oscillation amplitude and A_{exp} is the experimental oscillation amplitude. Images were analyzed by using NanoScope v815r3 software. For comparisons 3D height images were also generated.

17.2.5 SEM Imaging

Images were obtained using a Hitachi SU-70 scanning electron microscope. Samples were placed on 25 mm cylindrical aluminum mounts using carbon adhesive tabs (Ted Pella, Inc.) and then coated with light carbon sputtering at 5×10^{-2} torr. Samples were then analyzed using 15.0 kV beam energy with short exposure times to prevent excessive charging. Sample working distance was set to 15 mm and no tilt was used during analysis.

17.3 Results and Discussion

VK100 is a two-part silicone product formulated for vertebral augmentation. The coatings are white due to incorporation of $BaSO_4$ for radiopacity (Fig. 17.1). Initial insight into near-surface morphology was obtained by scanning electron microscopy (SEM) and tapping mode atomic force microscopy (TM-AFM). Cure in air provided a convenient reference as platinum-catalyzed hydrosilylation for network formation in air is a conventional process. Cure in water or saline provided a connection with the physiological environment after injection of VK100 into the periosteum of the targeted vertebral pedicles. Preliminary CA measurements for VK100 and the model divinyl polydimethylsiloxane ($D^{Vi}D^{H}$) elastomer after cure at 37 °C in water or saline revealed surprising differences compared to cure in air and prompted a detailed study of surface morphology and wetting behavior.

17.3.1 *Scanning Electron Microscopy*

An SEM image for VK100 extruded and cured in air is shown in Fig. 17.2a. Uniformly distributed 40–50 nm features in Fig. 17.2a are attributed to silica nanoparticles [6, 7]. Similar images were obtained for VK100 cured in water and in saline.

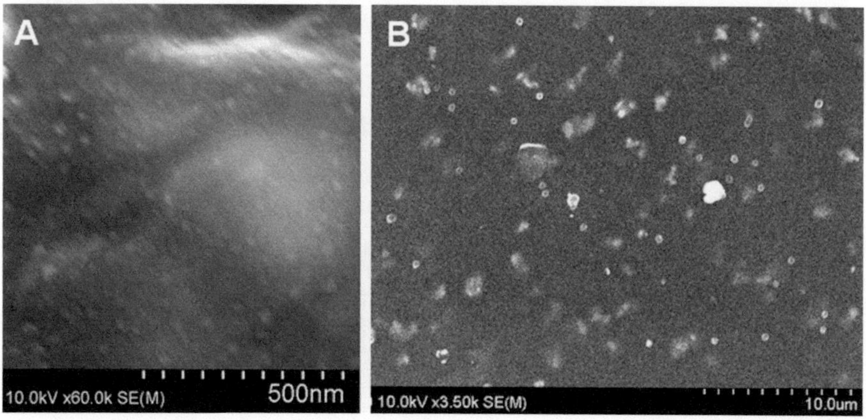

Fig. 17.2 SEM image of VK100 extruded and cured in air: (**a**) 10 kV, 60 k magnification and (**b**) 10 kV, 3.5 k magnification

A special compositional feature of VK100 is $BaSO_4$ (10–40 wt%), which is a well-known radiopacifier [6, 7]. This is essential in vertebral augmentation, as the surgeon uses continuous X-ray imaging to guide the location of the injected material. In the image shown in Fig. 17.2b for VK100 cured in air, the light-colored features were identified by EDX (energy-dispersive X-ray spectroscopy) as $BaSO_4$ particles (0.5–1 μm). Similar SEM images were seen for VK100 cured in water. These particles lead to roughness as discussed in the next section on AFM imaging.

17.3.2 Tapping Mode Atomic Force Microscopy

Figure 17.3 shows TM-AFM images for VK100 cured in air and water. AFM imaging was difficult because of surface roughness. That is, finding an area to

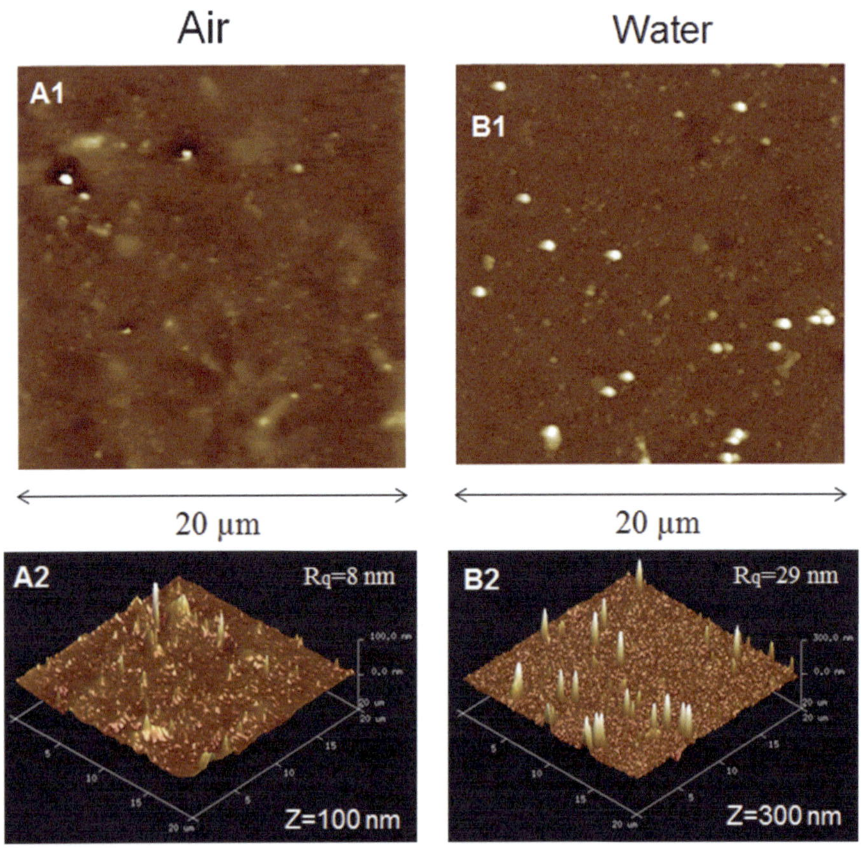

Fig. 17.3 TM-AFM images (20 × 20 μm) of VK100 cured in air and water: **A1** and **B1** are 2D images; **A2** and **B2** are 3D images

image was challenging because of height features (Fig. 17.3b) that caused the tip to retract. Therefore, it should be borne in mind that the images shown are selected from a number of trials.

In false-color imaging, light-colored features are attributed to hard (high modulus) entities [12]. Therefore, prominent surface and near-surface features (~1 μm) are attributed to $BaSO_4$ particles. For the 20 × 20 μm^2 areas shown, roughness (Rq) that ranges from 8 to 29 nm is also attributed to $BaSO_4$ particles. Because images are from areas selected for lack of large features, roughness determined by Rq is not representative.

17.3.3 Network Formation and Surface Chemistry

Figure 17.4 shows two chemical processes that influence the presence of near-surface nonpolar and polar groups [34, 35]. Hydrosilylation (Fig. 17.4, Eq. 17.1) brings about rapid network formation via the formation of nonpolar C-H and Si-C bonds. Due to the absence of inhibitor for hydrosilylation, buildup of viscosity and gelation for VK100 occurs in a few minutes. This relatively short "working time" is desirable for vertebral augmentation in order to minimize migration away from the target vertebral area after injection [13]. Figure 17.4 and Eq. 17.2 illustrate autoxidation that converts Si-H to Si-OH. Recently, autoxidation was proposed to occur contemporaneously with hydrosilylation resulting in near-surface Si-OH [3]. After considering CA measurements for VK100 and $D^{Vi}D^{H}$, near-surface chemistry described by Eqs. 17.1 and 17.2 is revisited.

17.3.4 Contact Angle Dependence on VK100 Processing Conditions

Air Cure at 37 °C Advancing (θ_A) and receding (θ_R) CAs were measured by water drop addition/withdrawal using a Rame-Hart goniometer and drop imaging system (Fig. 17.5 and Table 17.1) [14]. Curing VK100 in air resulted in $\theta_A = 108°$ and $\theta_R = 74°$, giving a 34° contact angle hysteresis (CAH = $\theta_\Delta = \theta_A - \theta_R$). VK100 contact angles after cure in air are in the range found for Sylgard 184 cured in air at 60 °C (θ_A = ~120° and θ_R = 60–70°) [3]. Without inhibitor, platinum cure in air results in rapid network formation that entrains surface-concentrated PDMS chains (Fig. 17.4, Eq. 17.1). Other components, particularly $BaSO_4$, do not greatly perturb the contact angles from those typically observed for platinum-cured silicone [8–10].

Cure in Saline 37 °C The robust nature of commercially available hydrosilylation-cured "biomedical grade" silicone after immersion in water at elevated temperatures was reported by Chaudhury [15]. In testimony to chemical stability, negligible

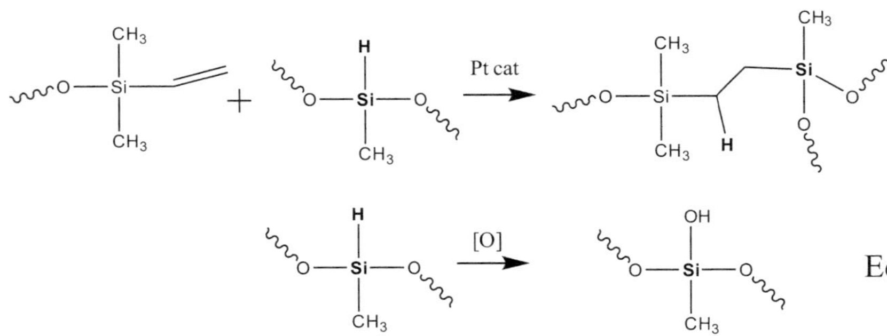

Eq 17.1

Eq 17.2

Fig. 17.4 Network formation reaction (Eq. 17.1) and Si-H autoxidation (Eq. 17.2)

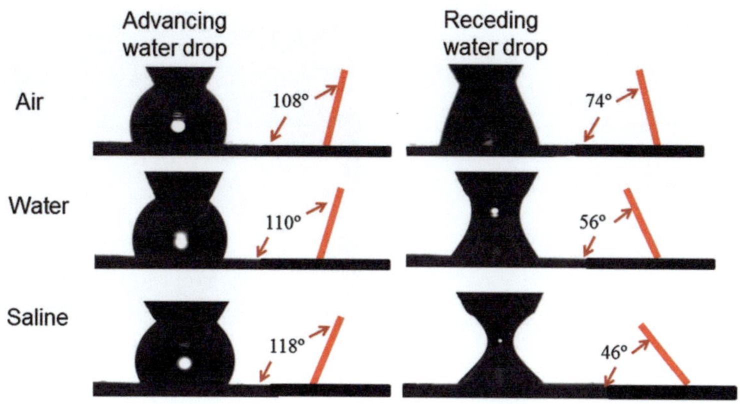

Fig. 17.5 Images of advancing and receding water drops on VK100 cured under the conditions indicated

Table 17.1 Contact angles (in degrees) after cure of VK100 in air, water, and saline

Cure medium	Average adv (°)	Std dev	Average rec (°)	Std dev	CAH (°)
Air	108	1.0	74	3.2	34
Water	110	3.2	56	2.6	54
Saline	118	2.9	46	2.7	71

a. Contact angle hysteresis, CAH or $\theta_\Delta = \theta_A - \theta_R$

changes in CAs were observed after heating a plaque at 70 °C in saline for 40 h (θ_A = 118° and θ_R = 70°).

CAs provided in Table 17.1 show that curing VK100 in water has negligible impact on advancing CAs (111–118°). Considering the stability of contact angles at elevated temperatures in water [15] and in contrast to cure in air, curing VK100 in aqueous media at 37 °C has a striking effect on receding contact angles (Fig. 17.5 and Table 17.1). Our discussion focuses on VK100 cure in saline (0.9% NaCl/L; 154 mEq/L), which was chosen to mimic a physiological medium [16, 17]. The salt concentration of saline approximates that found in cerebral spinal fluid (140–145 mEq/L) [18]. Compared to cure in air (74°), the receding CA for VK100 cured in saline is 46° (28° lower).

The low receding CA for network formation in saline is reminiscent of low receding CAs found for ambient temperature cure in air of Sylgard 184, which is also a platinum-cured silicone [3]. However, without inhibitor, cure takes only a few minutes compared with hours/days for Sylgard 184. Another characteristic of low receding CAs for VK100 cured in saline that is shared with Sylgard 184 cured at ambient temperature is the stability of receding CAs in air for at least 1 month. This stability stands in contrast to plasma-treated silicone coatings that become hydrophilic but undergo hydrophobic recovery over hours in air [2].

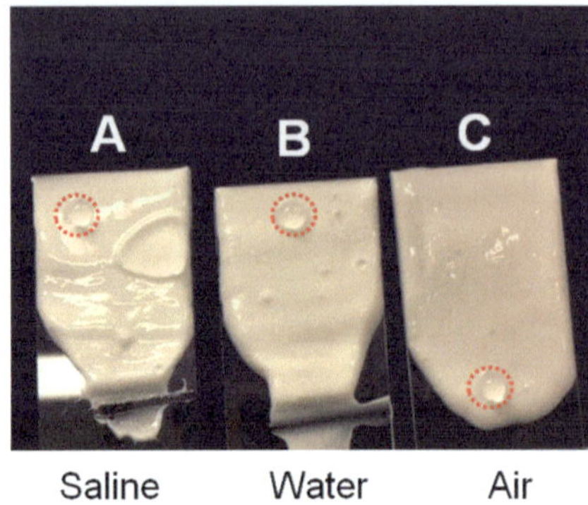

Fig. 17.6 A frame from a video showing water drop motion starting at tilt of 65°: "slippery" on VK100 cured in air (**c**) but "sticky" on VK100 cured in saline (**a**) or water (**b**). Droplets circled in red

Cure in Water and Growth Medium at 37 °C In contrast to the ~10° span of advancing CAs, receding CAs for water (56°) and saline (46°) result in contact angle hysteresis of 54 and 71°, respectively (Table 17.1). Surprisingly, cure in saline results in a 10° lower receding CA than cure in water. This difference cannot be due to a difference in the surface tension of water (72 dyne/cm) and saline (73 dyne/cm) [19]. A surface model after cure in water or saline is suggested below.

17.3.5 Impact of Water Droplet Adhesion

A frame from a video of a 25 μL water droplet flowing on VK100 cured in air or fixed on slides cured in water or saline is shown in Fig. 17.6. Water drops are "stuck" for VK100 cured in water or saline but slip easily off the surface cured in air. The effect of low receding contact angles is seen in "stickiness," while water begins to slide off VK100 cured in air (receding CA >70°), at a tilt angle of >65°. To investigate the origin of the difference and to provide data that would lead to a model for the sharply different wetting behavior, a model Pt-cured silicone, $D^{Vi}D^{H}$, was prepared. Contact angles for this model system are discussed in the next section.

17.3.6 Model Divinyl Polydimethylsiloxane ($D^{Vi}D^{H}$) Elastomer

Low receding contact angles for VK100 cured in water or saline led to an investigation of a model system similar without $BaSO_4$, silica nanoparticles, or silicone oil. A broad investigation of compositional and processing variations for model systems can be envisaged, but the present study focuses only on cure in air and saline, which are conditions relevant to cure of VK100 in vertebral augmentation.

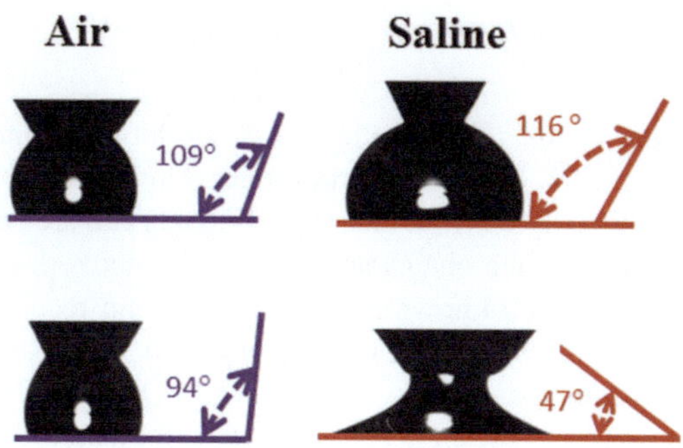

Fig. 17.7 Contact angles for model D^VD^H silicone on coverslips cured at 37 °C in air and saline

To generate Pt-cured elastomers, $D^{Vi}D^H$ samples were made in a manner analogous to that described by McCarthy [2] and Wang [3]. In brief, D^{Vi}-PDMS and platinum-divinyltetramethyldisiloxane complex were hand-mixed. D^H-PDMS copolymer was added and hand-mixed for 30 s. Microscope slides were drip coated, held in air briefly for viscosity buildup, and placed either in an oven (37 °C) or in saline (37 °C). Gelation and solidification were evident in 10 min. Samples were removed after 4 h for CA measurements.

Images for advancing and receding water droplets are shown in Fig. 17.7 along with contact angles that are the average of three measurements. Interestingly, after cure of the $D^{Vi}D^H$ model system in saline, CAs are virtually identical to those for VK100. These results help elucidate the differences in outermost surface chemistry for network formation in air and in saline. That is, the presence of $BaSO_4$ and associated roughness, silicone oil, and silica nanoparticles in VK100 does not have a measurable influence on wetting behavior of the outermost surface.

In prior work, we proposed that low receding CAs for Sylgard 184 cured at ambient temperature resulted from autoxidation of Si-H to Si-OH on the cross-linker (Fig. 17.4, Eq. 17.2) [3]. With an inhibitor for hydrosilylation, network formation for Pt-cured Sylgard 184 was slow (24–48 h). Slow cure at ambient temperature for a $D^{Vi}D^H$ model system resulted in receding CAs that were similar to Sylgard 184 (≤50°). Because the process for cure in these systems was slow, it seemed that autoxidation (Fig. 17.4, Eq. 17.1) was also slow. In any event, the notion of near-surface autoxidation of Si-H to Si-OH was attractive as high PDMS chain flexibility could account for high CA hysteresis.

The $D^{Vi}D^H$ model system for VK100 is similar to the model system used for Sylgard 184, but there is no inhibitor for hydrosilylation cross-linking. Like VK100, the $D^{Vi}D^H$ model system gelled and set up to form a non-tacky solid in ~10–15 min. The low and stable receding CA for both VK100 and the $D^{Vi}D^H$ model system after only 10–15 min cure at 37 °C in water or saline was unexpected. To account for the 94° receding CA for $D^{Vi}D^H$ cured in air and 47° for cure in saline (Fig. 17.7), a more detailed model is suggested.

17.3.7 Dynamic Wettability (DW)

Cure in Air Though the rate is not yet determined, it is assumed that autoxidation (Fig. 17.4, Eq. 17.2) occurs in air. However, nonpolar PDMS chains are favored at the air-polymer interface (Fig. 17.8a-1) and are entrapped upon gelation, which is the point at which chains are confined by network linkages that prevent flow [20]. The result is well-known high advancing and receding CAs (Fig. 17.8a-2, Table 17.1) [8–10]. That is, weak interactions such as that of siloxane oxygen with water, which are depicted in Fig. 17.8a-2 with O highlighted in blue, account for low CA hysteresis [21, 22]. As far as CA measurements are concerned, the presence of Si-OH from autoxidation is masked by entrapment in the bulk via cross-linking. For CA measurements, "bulk" may be as little as 2–3 nm from the outermost surface with "depth" defined by the root-mean-square end-to-end distance of PDMS chains.

DW After Cure in Saline As depicted in Fig. 17.8b-1, the outermost surface after cure in saline is dominated by -SiO$(CH_3)_2$- originating from the divinyl resin and from the $D^{Vi}D^{H}$ copolymer. Thus the advancing CA is similar to that for VK100 cure in air. For cure in saline, autoxidation (Si-H $\rightarrow$ Si-OH, Fig. 17.4, Eq. 17.2) is proposed to occur within the time for gelation (10–15 min). In contrast to cure in air, polar Si-OH moieties formed in this short time are favored at the water/polymer interface via enthalpically driven hydrogen bonding. Rapid gelation due to hydrosilylation entraps chains containing -Si-OH moieties at the outermost surface. Strong hydrogen bonding with water (blue dashed line) is depicted in Fig. 17.8b-2 by OH groups with dark blue circles. A combination of high chain flexibility, strong hydrogen bonding, and the sensitivity of CA measurements to outermost surface functionality accounts for low receding contact angles.

The proposed dynamic wetting model (Fig. 17.8b1–2) for CA hysteresis raises a question about the area fraction of near-surface Si-OH groups that give rise to low

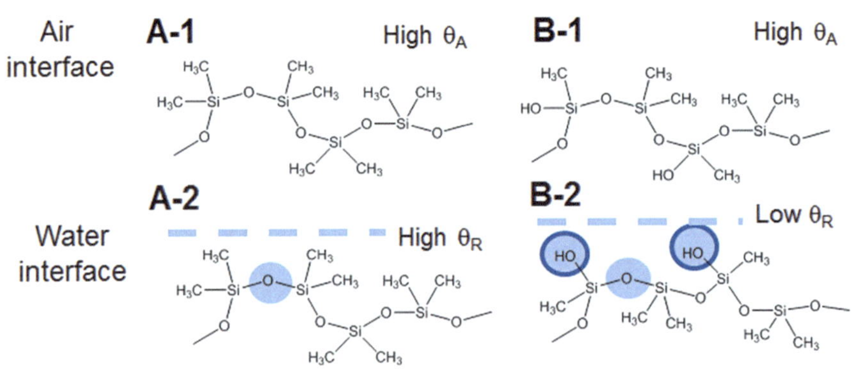

Fig. 17.8 Depiction of the outermost $D^{Vi}D^{H}$ PDMS surface during CA measurements: (**A**-1,2) network formed in air and (**B**-1,2) network formed in water or saline. Similar CAs were obtained for VK100. (See text for details)

receding CAs. For the $D^{Vi}D^{H}$ model system, the Si-OH area fraction may be estimated based on stoichiometry, molecular weights, and geometric considerations. The feed D^{Vi}/D^{H} weight ratio is 10:1 giving a mole ratio of $SiO(CH_3)_2$ to SiO(H) CH_3 of 19:1 and a mole ratio of -$SiO(CH_3)_2CH{=}CH_2$ to $SiO(H)CH_3$ of 1:10. From this analysis, ~10% Si-H moieties are used in hydrosilylation so that vinyl content is consumed leaving excess -$SiO(H)CH_3$-. Assuming complete autoxidation of Si-H, the mole ratio of $SiO(CH_3)_2$ to $SiO(OH)CH_3$ segments is ~21:1. Johnson and Dettre and others found receding CAs were highly variable at low area fractions of polar moieties [23–25]. Keeping in mind high PDMS chain flexibility due to an exceptionally low T_g (~−120 °C), a 10% area fraction of near-surface Si-OH reasonably accounts for a dynamic wetting and receding CAs that are stable and less than 50°.

Given the sensitivity of receding CAs to area fraction of near-surface polar moieties [25, 26], variables such as the scale and distribution of hydrophilic heterogeneity are likely of importance in pinning water at the three-phase contact line. Factors that affect the proposed autoxidation of Si-H to Si-OH and near-surface concentration of Si-OH groups during the initial stages of network formation are largely unknown. Further studies will be necessary to confirm the influence of components such as salt in aqueous media. Spectroscopic investigations and the sensitivity of receding CAs to near-surface Si-OH will facilitate these studies.

17.3.8 Adhesion to Glass

Enhanced polarity for VK100-cured water or saline was evidenced by strong adhesion of VK100 coatings to the glass substrate. We propose that this adhesion is due in part to interaction of Si-OH moieties formed by autoxidation and the glass surface. Such behavior is not unprecedented, as Bartell reasoned that molecular orientation of cellulose polymers against glass was responsible in making the surface more hydrophilic (as evidenced by low receding contact angles) [23]. By analogy, the adhesion of VK100 to bone is expected to be enhanced as well by low receding contact angles and attendant polarity as the hydrophilic nature of the bone surface is attributed to the presence of collagen fibers [27].

17.4 Conclusions

The present investigation began with characterization of VK100 a BONWRx two-part silicone bone cement formulated for vertebral augmentation. Contact angle measurements were carried out to determine whether cure under simulated physiological conditions would affect wetting behavior. A high θ_A was found which is typical for silicones [8–10]. However, cure in saline or water at 37 °C resulted in receding CAs of 46 and 56°, respectively (Table 17.1). The unexpectedly low receding CAs led to an investigation of a model system similar to VK100 but without silica nanoparticles, $BaSO_4$, or silicone oil. Cure of a model divinyl polydimethylsiloxane ($D^{Vi}D^{H}$) elastomer focused only on cure in air and saline. Interestingly, after cure of the $D^{Vi}D^{H}$ model system in saline, contact angles were virtually identical to those for VK100.

The impact of low receding CAs can be seen in a frame from a video showing slides tilted (Fig. 17.6). Water drops are "stuck" for VK100 cured in aqueous media but slip easily off the surface cured in air starting at a tilt angle of ~65°. These observations are correlated with work of adhesion for water drops (w_a) which is proportional to $1 + \cos\theta_R$ [3].

For cure in water or saline, autoxidation (Si-H → Si-OH) and near-surface entrapment of cross-linking chains containing –Si-OH are proposed to account for low receding CAs. The origin of the low θ_R and high contact angle hysteresis is correlated with the theory of Johnson and Dettre by which a small area fraction of polar groups impedes retraction of a receding water drop [24, 25].

In summary, the impact of this surface chemistry is twofold. One, VK100 presents a hydrophilic surface under physiological conditions, indicating likely biocompatibility. Secondly, its excellent adhesion to glass suggests similar adhesion to bone due to the nature of surface interactions.

Surprisingly, we have been unable to locate publications or patents on platinum-catalyzed hydrosilylation cure of divinyl silicones in an aqueous medium with which to compare our results. Given the importance of hydrophilic character of PDMS to resistance against biofouling [28, 29], adhesion of proteins [30, 31], and interactions with human cells [32, 33], experiments are planned to elucidate further factors affecting wetting behavior of Pt-cured polydimethylsiloxane elastomers.

Acknowledgment J. Lumen thanks BONWRx LLC for a summer fellowship and materials. K.J.W. thanks the National Science Foundation, Division of Materials Research, Polymers Program (DMR-1206259) and Polymers/Biomaterials Programs (DMR-1608022), and the School of Engineering Foundation for the support of this research.

References

1. Efimenko K, Wallace WE, Genzer J (2002) Surface modification of Sylgard-184 poly(dimethyl siloxane) networks by ultraviolet and ultraviolet/ozone treatment. J Colloid Interface Sci 254:306
2. Lien N, Hang M, Wang W, Tian Y, Wang L, McCarthy TJ, Chen W (2014) Simple and improved approaches to long-lasting, hydrophilic silicones derived from commercially available precursors. ACS Appl Mater Interfaces 6:22876
3. Wang C, Nair SS, Veeravalli S, Moseh P, Wynne KJ (2016) Sticky or slippery wetting: network formation conditions can provide a one-way street for water flow on platinum-cured silicone. ACS Appl Mater Interfaces 8:14252
4. Mittal KL (ed) (1978) ASTM special technical publication, vol. 640: adhesion measurement of thin films, thick films, and bulk coatings. ASTM, Philadelphia
5. Gao LC, McCarthy TJ (2009) Wetting 101 degrees. Langmuir 25:14105

6. Seaton JP, Carmichael R (2008) Materials and apparatus for in-situ bone repair. WO2008039807A2 28pp
7. Seaton JP, Carmichael R (2009) Materials and apparatus for in-situ bone repair. WO2009064541A1 32pp
8. Patel SK, Malone S, Cohen C, Gillmor JR, Colby RH (1992) Elastic-modulus and equilibrium swelling of poly(dimethylsiloxane) networkS. Macromolecules 25:5241
9. Perutz S, Kramer EJ, Baney J, Hui CY (1997) Adhesion between hydrolyzed surfaces of poly (dimethylsiloxane) networks. Macromolecules 30:7964
10. Uilk JM, Mera AE, Fox RB, Wynne KJ (2003) Hydrosilation-cured poly(dimethylsiloxane) networks: intrinsic contact angles via dynamic contact angle analysis. Macromolecules 36:3689
11. Seaton JP, Trebing LM (2007) Injectable compositions containing curable polysiloxane elastic materials for repair and reconstruction of intervertebral discs and other reconstructive surgery. WO2007062082A2 24pp
12. Magonov SN, Elings V, Whangbo MH (1997) Phase imaging and stiffness in tapping-mode atomic force microscopy. Surf Sci 375:L385
13. Gasbarrini A, Ghermandi R, Girolami M, Boriani S, Akman YE (2017) Elastoplasty as a promising novel technique: vertebral augmentation with an elastic silicone-based polymer. Acta Orthop Traumatol Turc 51:209–214
14. Mackel MJ, Sanchez S, Kornfield JA (2007) Humidity-dependent wetting properties of high hysteresis surfaces. Langmuir 23:3
15. Kennan JJ, Peters YA, Swarthout DE, Owen MJ, Namkanisorn A, Chaudhury MK (1997) Effect of saline exposure on the surface and bulk properties of medical grade silicone elastomers. J Biomed Mater Res 36:487
16. Di Terlizzi R, Platt S (2006) The function, composition and analysis of cerebrospinal fluid in companion animals: part I – function and composition. Vet J 172:422
17. Spector R, Snodgrass SR, Johanson CE (2015) A balanced view of the cerebrospinal fluid composition and functions: focus on adult humans. Exp Neurol 273:57
18. Hooshfar S, Basiri B, Bartlett MG (2016) Development of a surrogate matrix for cerebral spinal fluid for liquid chromatography/mass spectrometry based analytical methods. Rapid Commun Mass Spectrom 30:854
19. Heller W, Cheng MH, Greene BW (1966) Surface tension measurements by means of microcone tensiometer. J Colloid Interface Sci 22:179
20. Chanda M (2000) Advanced polymer chemistry. Marcel Dekker, New York
21. Owen MJ (1981) Why silicones behave funny. Chem Tech 11:288
22. Owen MJ (1990) In: Zeigler JM, Fearon FW (eds) Siloxane surface activity, vol 224. American Chemical Society, Washington, DC
23. Bartell FE, Ray BR (1952) Wetting characteristics of cellulose derivatives. I. Contact angles formed by water and by organic liquids. J Am Chem Soc 74:778
24. Johnson RE Jr, Dettre RH (1964) Contact angle hysteresis. III. Study of an idealized heterogeneous surface. J Phys Chem 68:1744
25. Dettre RH, Johnson RE (1965) Contact angle hysteresis. 4. Contact angle measurements on heterogeneous surfaces. J Phys Chem 69:1507
26. Pease DM (1945) The significance of the contact angle in relation to the solid surface. J Phys Chem 49:107
27. Erli HJ, Marx R, Paar O, Niethard FU, Weber M, Wirtz DC (2003) Surface pretreatments for medical application of adhesion. Biomed Eng Online 2:15
28. Hawkins ML, Fay F, Rehel K, Linossier I, Grunlan MA (2014) Bacteria and diatom resistance of silicones modified with PEO-silane amphiphiles. Biofouling 30:247
29. Wenning BM, Martinelli E, Mieszkin S, Finlay JA, Fischer D, Callow JA, Callow ME, Leonardi AK, Ober CK, Galli G (2017) Model amphiphilic block copolymers with tailored molecular weight and composition in PDMS-based films to limit soft biofouling. ACS Appl Mater Interfaces 9:16505

30. Elwing H, Welin S, Askendal A, Nilsson U, Lundstrom I (1987) A wettability gradient-method for studies of macromolecular interactions at the liquid solid interface. J Colloid Interface Sci 119:203
31. Lin SY, Parasuraman VR, Mekuria SL, Peng S, Tsai HC, Hsiue GH (2017) Plasma initiated graft polymerization of 2-methacryloyloxyethyl phosphorylcholine on silicone elastomer surfaces to enhance bio(hemo)compatibility. Surf Coat Technol 315:342
32. Liu PS, Chen Q, Yuan B, Chen MZ, Wu SS, Lin SC, Shen J (2013) Facile surface modification of silicone rubber with zwitterionic polymers for improving blood compatibility. Mater Sci Eng C-Mater Biol Appl 33:3865
33. Pedraza E, Brady AC, Fraker CA, Stabler CL (2013) Synthesis of macroporous poly (dimethylsiloxane) scaffolds for tissue engineering applications. J Biomater Sci-Polym Ed 24:1041
34. Kurian P, Kennedy JP (2002) Novel tricontinuous hydrophilic-lipophilic-oxyphilic membranes: synthesis and characterization. J Polym Sci A Polym Chem 40(9):1209–1217
35. Simpson TRE, Tabatabaian Z, Jeynes C, Parbhoo B, Keddie JL (2004) Influence of interfaces on the rates of crosslinking in poly(dimethyl siloxane) coatings. J Polym Sci A Polym Chem 42(6):1421–1431

Chapter 18
Block Copolymers with Element Blocks: The Metal-Bisterpyridine Linkage

Andreas Winter and Ulrich S. Schubert

Abstract In classical block copolymers, the constituent blocks are linked via a covalent bond. In supramolecular copolymers, this connection is replaced by a more labile, in some cases reversible, one – hydrogen bonding, host-guest interaction, and metal-to-ligand complexation represent the most prominent and versatile examples in this respect. The structural modification has a profound impact on the overall material's properties: The characteristics of the polymeric subunits are retained and combined with the special features of the non-covalent linkage. With respect to block copolymers bridged by a cationic transition metal ion complex, the newly gained properties can go far beyond a (reversible) chemical linkage and might include, e.g., a photo- or magnetochemical behavior. Moreover, the charged nature of the linking complex can change the physical characteristics of the overall block copolymer when compared to a classical covalent counterpart – this may hold true in solution, in the melt as well as in the solid state. These features together allow considering the metal-complex linkage in such assemblies not to be "innocent" but rather to be a very short block or segment on its own. In this chapter, the metal-bisterpyridine linkage within linear copolymer architectures will be highlighted exemplarily. By this, the modularity of metallo-supramacromolecular chemistry, referred to as playing LEGO™ with macromolecules or as using the connection as element block, will be shown.

Keywords Block copolymers · Metallopolymers · Self-assembly · Supramolecular polymerization · Terpyridine complexes

A. Winter · U. S. Schubert (✉)
Laboratory for Organic and Macromolecular Chemistry (IOMC), Friedrich Schiller University Jena, Jena, Germany

Jena Center for Soft Matter (JCSM), Jena, Germany

Center for Energy and Environmental Chemistry Jena (CEEC Jena), Jena, Germany
e-mail: ulrich.schubert@uni-jena.de

Y. Chujo (ed.), *New Polymeric Materials Based on Element-Blocks*,
https://doi.org/10.1007/978-981-13-2889-3_18

18.1 Introduction

Block copolymers represent one of the central structural motifs in polymer science with the field of applications ranging from materials since to (opto-)electronic devices and drug delivery systems. Supramolecular block copolymers are obtained when the covalent linkage between two blocks is replaced by a non-covalent, i.e., supramolecular, one [1]. For example, the supramolecular interactions can be employed to connect two different homopolymer chains or even to link different block copolymer architectures. In such systems, the intrinsic properties of classical block copolymers are not only retained (e.g., microphase separation) but combined with the special features arising from the non-covalent linkage of the subunits (e.g., reversibility and addressability of the supramolecular bonds) [2]. In order to establish a supramolecular analog to a classical copolymer, two main criteria have to be met: a sufficiently high binding strength of the non-covalent interaction as well as a high directionality thereof. In this regard, mainly three types of supramolecular systems are in the focus of interest: hydrogen bonding [3, 4], host-guest interaction [5–7], and metal-to-ligand complexation [1, 2, 8]. Though multiple hydrogen-bonding motifs within linear arrays have been optimized to reach high association constants in common organic solvents, still their lability in aqueous media represents a major drawback. In contrast, many assemblies relying on metal-to-ligand coordination have been found to be stable even in water [2, 9, 10]. As an extra benefit, the incorporated metal complex might not only function as plain linkage between the polymer chains but might introduce special properties far beyond reversibility/addressability, such as redox activity or optoelectronic and magnetic features. Due to the chemical/physical properties being significantly different from the attached polymer chains, the metal complexes located in the main chain of the (block co) polymer architecture can be even defined as a short block or segment on its own. From all the systems commonly used in the field of metallo-supramolecular polymers, only a few assemblies are of practical use when aiming for the formation of defined (block co)polymers based on metal-to-ligand coordination. This is due to the fact that one additional property needs to be considered which is heterolepticity, i.e., the ability to selectively assemble a block copolymer of the generalized formula A-[M]-B, while the two possible "homopolymers," A-[M]-A and B-[M]-B, are only obtained in negligible amounts (here, A-[and B-[represent two different polymer chains, each equipped with a vacant metal coordination site; M denotes a transition metal ion). In order to prepare such materials, a directed two-step assembly strategy is typically followed (Fig. 18.1). Taking all the abovementioned criteria, the number of applicable metal/ligand combinations is limited down to two: bisterpyridine and pincer complexes [10]. Within the scope of this chapter, the focus will be on the former ones.

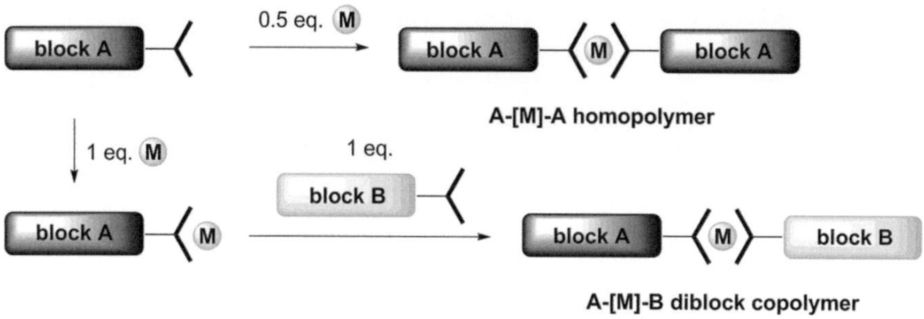

Fig. 18.1 Schematic representation of the formation of homopolymers and diblock copolymers via a directed stepwise metal-to-ligand coordination

18.2 Bisterpyridine Complexes

Before detailing the various types of (block co)polymers based on bisterpyridine complexes, a short insight into the chemistry of selective terpyridine complexation has to be given. Bisterpyridine complexes with transition metal ions have already been known since the initial report on 2,2′:6′,2″-terpyridine (tpy) itself [11]. The major characteristic of the pseudo-octahedral complexes with the generalized formula $[M(tpy)_2]^{2+}$ is the strength of the coordinative bonds (K) which is strongly related on the nature of the incorporated metal center. The kinetics of the stepwise formation of the biscomplexes were determined by isothermal titration calorimetry (ITC) [12] and were found to be in good agreement with the Irving-Williams series established for complexes of the first-row transition metal ions [13]: $Mn^{II} < Fe^{II} < Co^{II} < Ni^{II} < Cu^{II} > Zn^{II}$.

Among other analytical methods, mass spectrometry (MS) has also been applied for the characterization of bisterpyridine complexes, and, in particular, matrix-assisted laser desorption/ionization time-of-flight (MALDI-TOF) MS appeared to be an appropriate tool to conclude on the complex stability, though in a rather qualitative fashion [14]: The degree of fragmentation of the $[M(tpy)_2]^{2+}$ cation depended on the applied laser energy, and, by this, the binding strength could be estimated. The stability of bisterpyridine complexes represents the crucial factor when solution-based characterization tools have to be utilized, e.g., size exclusion chromatography (SEC) in the case of polymers [10, 15]. Often the solvent and temperature dependency of the stability constant cannot be neglected, and, consequently, the K value might also depend on the concentration. In other words, too small binding constants yield kinetically labile complexes only that are prone to dissociate under the typical SEC conditions. However, when K is sufficiently high

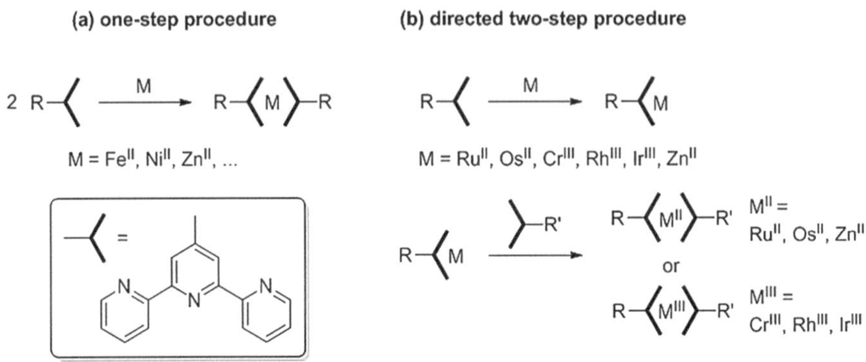

Fig. 18.2 The one-step and directed two-step strategies toward homoleptic (**a**) or heteroleptic bisterpyridine complexes (**b**)

(as for the so-called kinetically inert complexes), SEC can well be applied to study metallo-supramolecular polymers with bisterpyridine complexes. So far, this is only applicable to Ni^{II}, Co^{III}, Ru^{II}, Os^{II}, and Ir^{III} ions (in very special cases, also Fe^{II}-/Co^{II} - [16] and Zn^{II}-containing systems [17] are compatible with a SEC analysis). Worth mentioning, the rather labile Co^{II} complexes can be oxidized to the +III state, thus forming kinetically stable bisterpyridine complexes [18].

When aiming for homoleptic complexes, the ligand R-tpy, where R denotes any substituent, is reacted with an appropriate metal ion (e.g., Zn^{II}, Co^{II}, Cu^{II}, Ni^{II}, Fe^{II}, Ru^{II}, Ir^{III}, etc.) in a 2:1 metal-to-ligand ratio (Fig. 18.2); the isolation typically involves an exchange of the counterions in order to guarantee solubility in organic solvents, such as acetone or acetonitrile. The addition of a metal salt to a mixture of two different tpy ligands will always yield a statistical mixture of both homoleptic complexes besides the heteroleptic one. Consequently, a directed two-step procedure has to be applied for the selective preparation of heteroleptic complexes. Due to the inherent stability of their monoterpyridine complexes, Ru^{II} and Os^{II} are particularly suited for this purpose (numerous protocols varying in the reaction conditions and the employed precursor complex are known from literature [15]). Basically, also heteroleptic Rh^{III} and Ir^{III} bisterpyridine complexes can be obtained, though under rather drastic conditions only [19]. Gohy and co-workers widened the range of accessible heteroleptic systems by preparing kinetically stable Ni^{II} and Co^{III} bisterpyridine via a directed two-step synthesis [18]. Remarkably, also heteroleptic Cr^{III} bisterpyridine complexes have been found to be stable, at least in the absence of coordinating solvents [20]. Very recently, the Chan group reported on heteroleptic Zn^{II} bisterpyridine complexes that were stabilized by π-π interaction of the coordinated ligands, the formation of the respective homoleptic species was suppressed by steric hindrance and kinetic lability (Fig. 18.3) [17].

Having considered the complexation of tpy ligands, either in a homoleptic or a heteroleptic fashion, a short comment on the synthesis of the most relevant tpy building block in this field, which is the 4′-chloro derivative, has to be made. Today, the majority of tpy compounds, at least those having aromatic substituents in

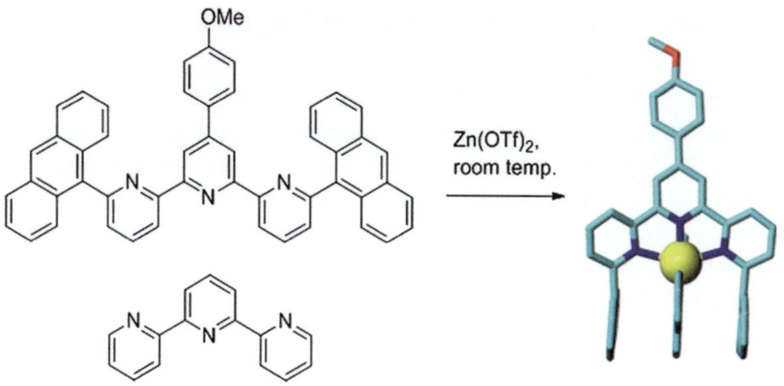

Fig. 18.3 Synthesis of a heteroleptic Zn^{II} bisterpyridine complex. (Figure reproduced with permission from [17])

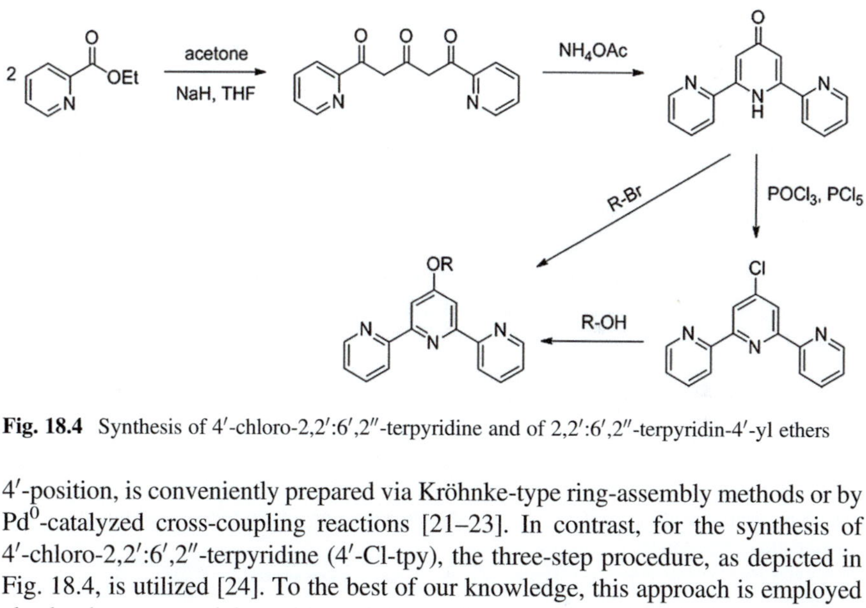

Fig. 18.4 Synthesis of 4′-chloro-2,2′:6′,2″-terpyridine and of 2,2′:6′,2″-terpyridin-4′-yl ethers

4′-position, is conveniently prepared via Kröhnke-type ring-assembly methods or by Pd^0-catalyzed cross-coupling reactions [21–23]. In contrast, for the synthesis of 4′-chloro-2,2′:6′,2″-terpyridine (4′-Cl-tpy), the three-step procedure, as depicted in Fig. 18.4, is utilized [24]. To the best of our knowledge, this approach is employed also by the commercial providers of this particular compound [25]. 4′-Cl-tpy and its precursor, 2,6-di(pyridin-2-yl)pyridin-4(1*H*)-one in its tautomeric enol form, represent the most common substrates for the synthesis of 4′-terpyridinoxy derivatives, including polymers, by aromatic or S_N2-type nucleophilic substitution, respectively. An overview over the scope and limitations of these transformations has already been given elsewhere [15, 21, 22, 26–28]. The accessibility to appropriately functionalized tpy derivatives paved the way to investigated tpy-containing polymers – though with a delay by about 30 years when comparing to the famous bidentate analog 2,2′-bipyridine (bpy) [27].

18.3 Synthesis of tpy Functionalized Polymers

In the context of this chapter, the focus is on polymers that bear a tpy moiety on (at least) one chain end. Only from these materials linear copolymer architectures, for instance, A-[M]-B assemblies, can be obtained by metal ion complexation (see Fig. 18.1). It was emphasized in the previous section that only a handful of transition metal ions are suited for the formation of such supramolecular polymers, and a literature survey shows that practically almost exclusively Ru^{II} ions are employed for this purpose [10, 15]. Over the last years, the scope was widened to some extend by utilizing also Ni^{II}, Co^{III} [18], and, very recently, also Zn^{II} ions [17]. In comparison, the range of useable transition metal ions for the metallo-supramolecular polymerization of telechelic]-A[molecules is much broader and covers the late first-row divalent transition metal ions (i.e., Fe^{II}, Co^{II}, Ni^{II}, Cu^{II}, and Zn^{II}) as well as Cd^{II} and Ru^{II}. The resulting polymers, in which A represents either a π-conjugated rigid chromophore or a flexible unit (including short alkyl chains and even polymers), are beyond the scope of this chapter, and the reader is referred to the available review articles published elsewhere [10, 21, 29–32].

Basically, there are two synthetic strategies for introducing tpy units at the end (s) of a polymer chain: end-functionalizing of a preformed polymer or using a tpy-bearing initiator (Fig. 18.5). Though these approaches can, in principle, also be applied to bisterpyridine complexes, only scattered examples for this can be found in literature. Recently, Jäger et al. explored this methodology starting from heteroleptic Ru^{II} biscomplexes of tpy's relative 2,6-di(quinolin-8-yl)pyridine (dqp) in order to prepared A-[Ru]-B assemblies in which photo-induced charge separation between the donor-type block A and the acceptor-type block B could be observed [33–36].

The concept of using transition metal complexes or the corresponding metal-free ligands as initiators in living/controlled polymerization reactions was pioneered by Fraser and her co-workers [37]. All the protocols that have been established for the

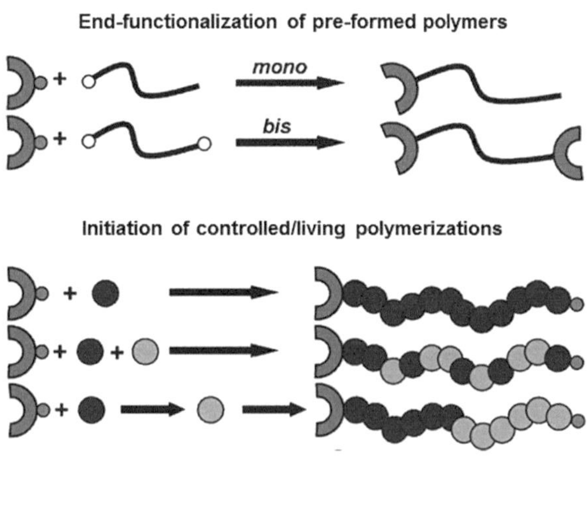

Fig. 18.5 Synthesis of ligand-equipped polymers via the end-functionalization and the initiator route. (Figure reproduced with permission from [37])

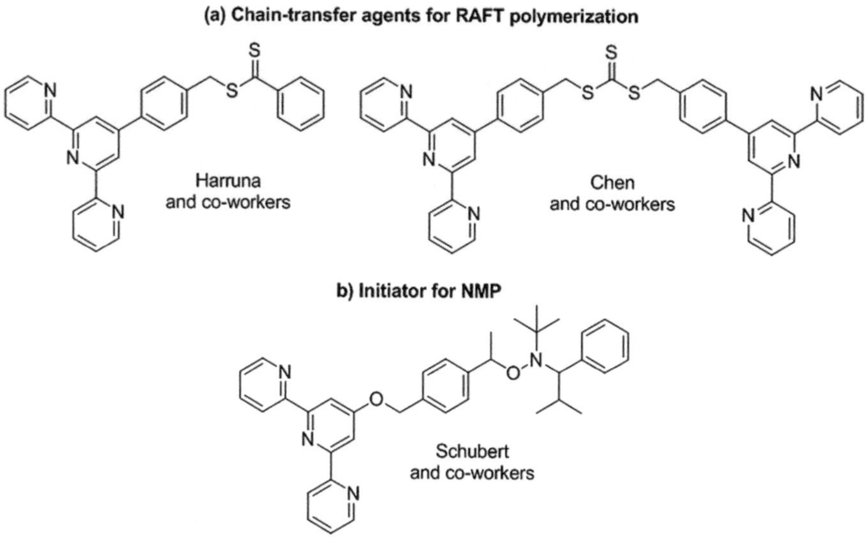

Fig. 18.6 Schematic representation of tpy-equipped chain-transfer agents for RAFT polymerization and initiators for NMP

bpy-containing systems (i.e., ligands and complexes) could be transferred also to their tpy-based analogs. In this respect, the ring-opening polymerization (ROP) of cyclic esters, such as ε-caprolactone or L-lactic anhydride, shall be named first. In these cases, OH-functionalized terpyridine derivatives have been employed as initiators in combination with Lewis-acidic catalysts (e.g., Sn^{II} laurate or $AlEt_3$) [38–41]. Moreover, the three most relevant types of controlled radical polymerization (CRP) methods have been initiated by tpy derivatives [10, 26, 37]: atom transfer radical polymerization (ATRP), reversible addition-fragmentation polymerization (RAFT), and nitroxide-mediated polymerization (NMP). Pefkianakis et al. demonstrated that bromomethylated tpy derivatives can readily be used for the polymerization of styrenes under ATRP conditions (though, some supporting bpy ligand had to be added to accelerate the reaction) [42]. The RAFT polymerization of a range of monomers (i.e., styrenes, acrylates, and acryl amides) has been performed independently by the Harruna [43] and Chen [44] groups; the chain-transfer agents used in these studies are depicted in Fig. 18.6a. In 2004, Schubert and co-workers introduced the first high-performance NMP initiator featuring a designated binding site for metal ion complexation (Fig. 18.6b) [45]. This initiator, derived from Hawker's universal TIPNO-type initiator (TIPNO: 2,2,5-trimethyl-4-phenyl-3-azahexane nitroxide) [46], is today the most popular and versatile derivative in this field allowing to polymerize a wide range of monomers into homo- and block copolymers with good control over the molar masses, and, in most cases, narrow dispersities are obtained [18, 45, 47–53].

4′-Chloro-2,2′:6′,2″-terpyridine, as already introduced in Sect. 18.2, is the most commonly used building block in order to graft tpy units onto preformed polymers

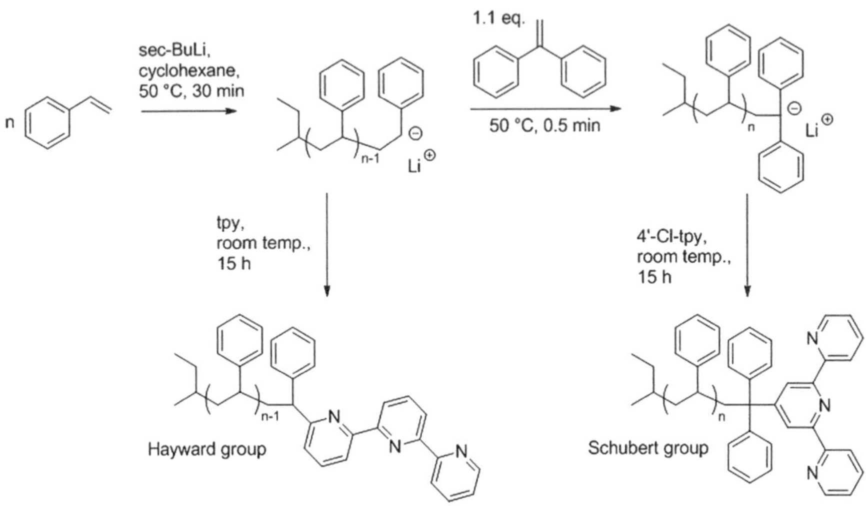

Fig. 18.7 Synthesis of tpy-functionalized macroligands by anionic polymerization

with OH groups at the chain end(s) or the pending side chains [15]. Thereby, polymers with a chemically robust backbone structure can be modified in a straightforward nucleophilic substitution reaction; however, the particular grafting conditions need to be modified in each case to meet the requirements of the polymer dealt with (i.e., solubility and reactivity). The polymer blocks that have thusly been modified include poly(ethylene glycol)s (PEGs) [54–58], polystyrenes (PSs) [54, 59], poly(vinylpyridine)s [60], poly(ethylene-*co*-butylene)s (PEBs) [54], poly (ex-tetrahydrofuran)s (PTHFs) [61], poly(2-oxazoline)s [61], poly (dimethylsiloxane)s [62], and Ploronics™ [61]. With increasing reactivity/instability of the polymer backbone under basic conditions, much milder coupling procedures and different tpy derivatives for the functionalization have to be chosen [63, 64], one particular example in this respect is the Cu^{I}-catalyzed azide-alkyne cycloaddition (CuAAc) reaction between appropriately functionalize polymers and tpy derivatives [65, 66]. As discussed in more detail elsewhere [15], the anionic polymerization, in particular of styrene-type monomers, represents one further important method to obtain tpy-functionalized polymers simply by quenching the "living" anionic polymer chain ends with a suitable tpy-based terminating agent (two examples for this are given in Fig. 18.7) [67–71]. In comparison to the other end-capping approaches, two main advantages of the experimentally more demanding anionic polymerization have to be named [67]: (i) excellent control over the obtained molar masses in combination with very narrow dispersities and (ii) formation of a chemically inert linkage between the polymer chain and the metal-binding site (i.e., C–C single bond). The latter one is of relevance when Ru^{II} or Os^{II} ions have to be coordinated to the macroligands – more sensitive macroligands might decompose under the required harsh reaction conditions.

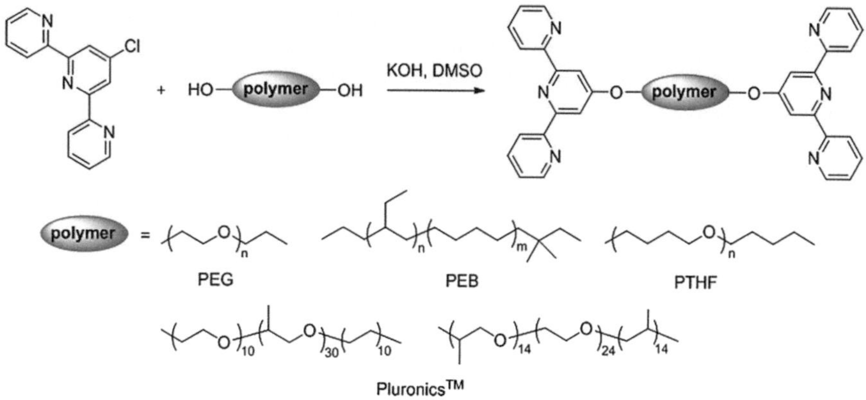

Fig. 18.8 Telechelic bisterpyridine polymers prepared from end-group modification of α,ω-dihydroxy-functionalized polymers

As for the polymers with a tpy moiety at one chain end, also telechelic polymers – those carrying tpy units at both termini – are conveniently prepared from α,ω-dihydroxy-functionalized polymers and 4′Cl-tpy [26]. Schubert and co-workers synthesized a range of such telechelic bisterpyridines polymers featuring different macroscopic properties (e.g., hydrophilic or hydrophobic); some examples are summarized in Fig. 18.8 [57, 61, 72, 73].

18.4 (Block Co)polymers with One Bisterpyridine Complex in the Main Chain

In this section, polymers of the general types A-[M]-A and A-[M]-B are evaluated with respect to their preparation and characterization as well as the potential fields of application. To be in line with the relevant literature, the A-[M]-A systems are simply referred to as homopolymers in the following (thereby neglecting the central metal complex "block"); consequently, the materials of an A-[-M-]-B structure are regarded as block copolymers. The modular combination of macroligands by metal ion complexation into more advanced polymer architecture has been compared to playing LEGO™, but with macromolecules [74].

18.4.1 A-[M]-A Homopolymers

The A-[M-]-A homopolymers based on various divalent first-row transition metal ions (i.e., Fe^{II}, Co^{II}, Ni^{II}, Cu^{II}, Zn^{II}, and Cd^{II}) and Ru^{II} have been assembled and investigated in detail, mainly to establish characterization methods that could be applied later for more advanced structures. Thus, these materials can be regarded as model polymers, and the basic lessons of the metallo-supramolecular "playing LEGO™" concept can be learned from this homoleptic complexation [26]. In particular, tpy-functionalized PEG monomethyl ethers (MeOPEG-tpy) have been studied by SEC, MALDI-TOF MS, UV-vis spectroscopy, and, in case of diamagnetic complexes, also ^{1}H NMR spectroscopy [56, 75]. Regarding the mass spectrometric characterization of the MeOPEG-[M]-PEGOMe, only the homopolymers with Co^{II} or Ru^{II} ions revealed a peak corresponding to the unfragmented assembly (besides a peak assigned to the metal-free macroligand) [14]. With increasing laser intensity, the degree of fragmentation increased, eventually allowing to conclude on the relative binding strength of the polymeric homoleptic complexes under MALDI-TOF MS conditions.

From the state-of-the-art techniques commonly applied in analyzing polymer samples, SEC has maybe to be named first. Being a relative method only, reliable reference materials are required in order to assure quantitative results – for all types of metal-containing polymers, this is not the case. Moreover, non-specific interaction of charged (metallo-)polymers with the column material makes SEC much more demanding as if dealing with "classical" polymers. Meier et al. faced this problem and established an eluent system (i.e., DMF containing 5.5 mM NH_4PF_6) that enabled the separation of metal-containing polymers only by size (thus, excluding all chromatographic effects) [76]. In this study, MeOPEG-tpy, MeOPEG-typ$RuCl_3$, and $[Ru(tpy\text{-}PEG\text{-}OMe)_2](PF_6)_2$ were used as analytes; the SEC traces of these polymers are depicted in Fig. 18.9a. Furthermore, the coupling of an in-line photodiode array (PDA) detector to the SEC system afforded an insight into the structural integrity of a UV/vis-absorbing polymer as well as its purity (in this case from the characteristic MLCT absorption band of the Ru^{II} bisterpyridine units at ca. 490 nm). As pointed out in Sect. 18.2, the stability of bisterpyridine complexes is decreasing in the order $Ni^{II} > Co^{II} > Fe^{II}$ – this holds also true when performing SEC measurements. Accordingly, polymers containing Ni^{II} bisterpyridine complexes

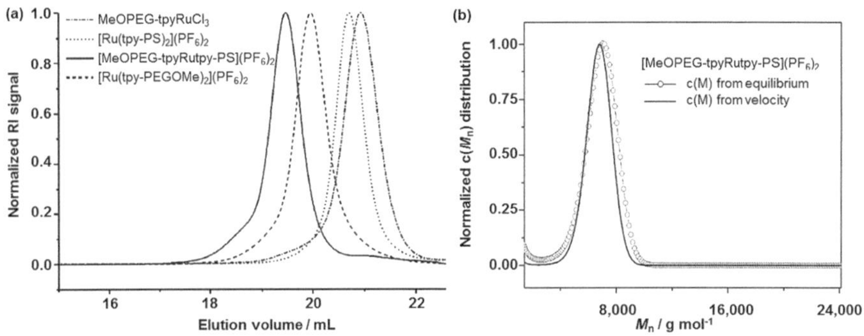

Fig. 18.9 (**a**) SEC traces of various Ru^{II}-containing polymers (eluent: DMF with 5.5 mM NH_4PF_6). (**b**) Normalized $c(M_n)$ distributions according to AUC measurements of an A-[Ru]-B diblock copolymer. (Figure reproduced with permission from [76, 77], respectively)

can be investigated by SEC (note that ^{1}H NMR spectroscopy cannot be applied for these paramagnetic materials) [16].

Analytical ultracentrifugation (AUC) was originally developed to analyze the hydrodynamic properties of proteins and nucleic acids; this analytical tool was adopted later also for synthetic polymers. Schubert et al. utilized the technique to investigate various types of metallo-supramolecular polymers containing bisterpyridine complexes of Fe^{II} or Ru^{II} ions, and no dissociation of the polymers was observed during the experiments (an example is shown in Figure 18.9b) [77, 78]. Performing sedimentation equilibrium and velocity measurements allowed to determine various hydrodynamic parameters, such as the average friction ratio (f/f_0), the sedimentation coefficient (S_n), and the average diffusion coefficient (D) as well as the molar mass (M_n). In total, four independent characterization techniques are now available to estimate the molar mass of metallo-supramolecular (block co) polymer architectures: SEC, ^{1}H NMR spectroscopy, MS, and AUC. The latter, as maybe the most sophisticated one, is believed to be the most accurate and circumvents severe limitations, such as dissociation or fragmentation (as in case of SEC and MS, respectively) [79]. In addition to the abovementioned methods, capillary zone electrophoresis (CZE) was successfully used to analyze a series of A-[M]-A systems ($M = Ru^{II}$ or Ni^{II}, A = PEG or PS). Thereby, the electrophoretic mobility was found to be independent from the nature of the transition metal ion, and, based on reference samples, the dispersities could be investigated [80].

A special feature of metal-containing polymers is the stimuli responsiveness, i.e., the ability to undergo a (reversible) transformation when they addressed by a change in the environment (e.g., pH value, temperature, light, concentration, redox chemistry, competing ligand). In this respect, one emerging application of metallopolymers with bisterpyridine (or related) complexes is the field of self-healing materials [81–83]. The A-[M]-A homopolymers obtained from PEG-based macroligands with Fe^{II}, Co^{II}, Zn^{II}, or Cd^{II} ions experienced fast decomplexation at high and low pH values (i.e., at pH values of 13 and 1, respectively), whereas the kinetically more stable Cu^{II}-based polymer dissociated slowly over days at a low pH value. In contrast, when using Ru^{II} or Ni^{II} ions instead, the homopolymers resisted to changes over the entire pH regime from 0 to 14 [56]. Thus, for several transition metal ions, reversible complexation/decomplexation of the A-[M]-A polymers, as a function of the applied pH value can be achieved. The most robust homopolymers, i.e., those containing Ru^{II} ions, could only be opened under forcing condition by either adding a strong competitive ligand, such as the sodium salt of *N*-hydroxyethylethylenediamine triacetic acid (HEEDTA) [84, 85], or by oxidation of the metal center to the +III state using Ce^{IV} salts [86].

In more advanced systems, not only the linking metal complex but also the polymer chain can be addressed separately ("multi-responsiveness"). One example in this respect is the A-[M-]-A homopolymer formed by the self-assembly of a tpy-functionalized PNIPAM macroligand with Fe^{II} or Zn^{II} ions (PNIPAM: poly(*N*-isopropylacrylamide), Fig. 18.10) [64]. Thermoresponsive polymers, such as PNIPAM, are known for their remarkable physical behavior that is characterized by a reversible transition from soluble to collapsed precipitated chains upon heating in aqueous solution [87]. Basically, the temperature at which precipitation occurs is

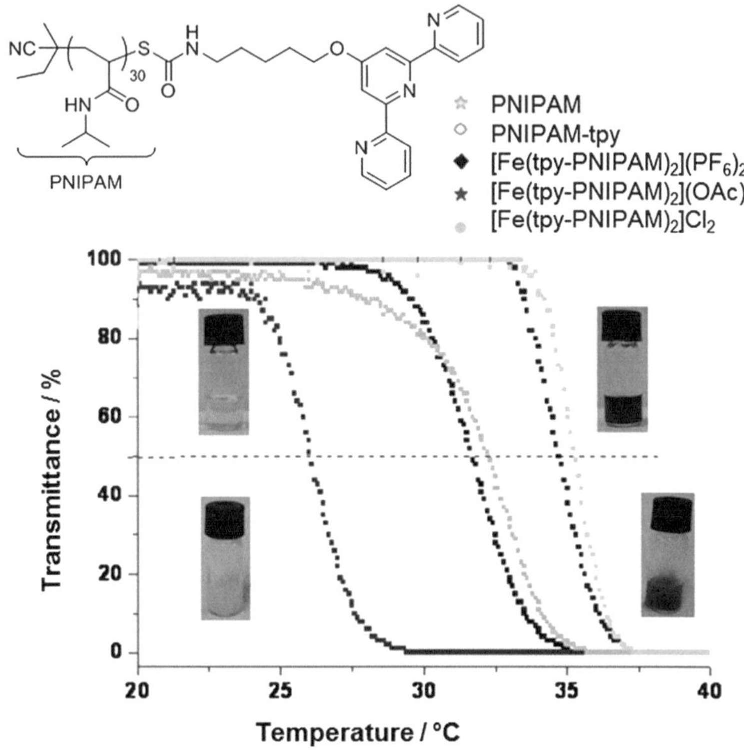

Fig. 18.10 Influence of the temperature on the transmittance of PNIPAM, PNIPAM-tpy, and [Fe(tpy-PNIPAM)$_2$]X$_2$ with various counterions (i.e., Cl^-, AcO^-, and PF_6^-) in water (5 mg mL^{-1}). An image of a vial with PNIPAM-tpy below and above the LCST (top and bottom on the left) as well as of the homopolymer with Fe^{II} below and above the LCST (top and bottom on the right). (Figure reproduced with permission from [64])

called the lower critical solution temperature (LCST), and the LCST behavior has been found to be dependent on various parameters (e.g., pH value, type and concentration of added salts, as well as the solvent composition). The authors demonstrated that the cloud point, i.e., the LCST temperature, of the PNIPAM-[M]-PNIPAM homopolymer depended on both the nature of the transition metal ion (Fe^{II} vs. Zn^{II}, attributed to the ion's radius) and of the counterion (increase with the hydrophilicity in the order Cl– > AcO– > PF_6^-).

18.4.2 Block Copolymers with One Bisterpyridine Complex: A-[M]-B Diblock Species and Beyond

Diblock copolymers with a linking bisterpyridine unit have mainly been prepared with Ru^{II} ions via a two-step procedure (see Figs. 18.1 and 18.2). Traditionally

$RuCl_3{\cdot}xH_2O$ has been used for the complexation (Ru^{III} is reduced to Ru^{II} in course of the second coordination step) [54, 85] but has recently been replaced by more reactive Ru^{II} precursor species that allow (a) to perform the coordination steps under much milder conditions and (b) to characterize the intermediate monocomplexes also by ^{1}H NMR spectroscopy (the paramagnetic Ru^{III} monocomplexes can only be investigated by SEC, MS, and UV/vis absorption spectroscopy) [15]. From the available Ru^{II} precursor complexes, $Ru(DMSO)_4Cl_2$ and $[Ru(NCCH_3)_6](BF_4)_2$ have to be named exemplarily [88, 89]. Though two-step protocols are also available for the assembly of heteroleptic Os^{II} and Ir^{III} bisterpyridine complexes, their application to the formation of A-[M]-B diblock copolymers is hampered by the required harsh reaction conditions. As mentioned in Sect. 18.2, also Ni^{II} and Co^{II} ions can be utilized for the stepwise assembly of A-[M]-B diblock copolymers; however, in the case of Co^{II}, subsequent oxidation to the +III state by adding a Ag^{I} salt was required to guarantee sufficient kinetic stability. The synthesis and SEC analysis of these kinetically stable materials are depicted in Fig. 18.11. Very recently, a concept for the selective assembly of heteroleptic Zn^{II} bisterpyridine complexes was reported and applied to obtain an A-[M]-B diblock copolymer based on this linkage (see also Sect. 18.2) [17]. This example shows how the kinetically labile nature of a complex can be balanced, to some extent, by steric hindrance and secondary non-covalent interactions (i.e., π-π stacking). However, the achieved stabilization is still not sufficient enough to enable a characterization by SEC or MS.

Phase separation is one of the characteristic properties of block copolymers that originates from the thermodynamic incompatibility between the different blocks. Due to their linkage within one macromolecular compound, either in a covalent or in supramolecular fashion, microphase separation will always be spatially limited, and the self-assembled structures feature sizes only in the range of a few times the radius of gyration of the constituent block (i.e., 10–100 nm) [90]. Often a long-range ordering of these structures throughout the bulk material can be observed, and, for example, cubic arrays of spheres or hexagonally packed cylinders are formed as nanostructures. The size and morphology of these features, in turn, are governed by the degree of incompatibility between the different blocks and the relative volume fraction and length of the blocks [225]. Besides this behavior in the solid state, microphase separation ("micellization") will also occur in solution – as long as a selective solvent is used (i.e., a good solvent for one of the blocks and a bad solvent for the other blocks). A typical block copolymer micelle consists of a micellar core and a corona, which are formed by the insoluble and the soluble block(s), respectively [2, 8, 91]. Such micellar systems have attracted significant scientific interest due to their numerous applications in various fields ranging from drug delivery to templates for nanoparticles [92, 93].

PEG and PS show distinct differences regarding their solubility and miscibility in the solid state; thus, diblock copolymers comprising these two constituent blocks are among the most intensively studied ones – this holds also true for the metallo-supramolecular variant PEG-[Ru]-PS. In pioneering work, Schubert and co-workers investigated the micellization of PEG-[Ru]-PS diblock copolymers in which the

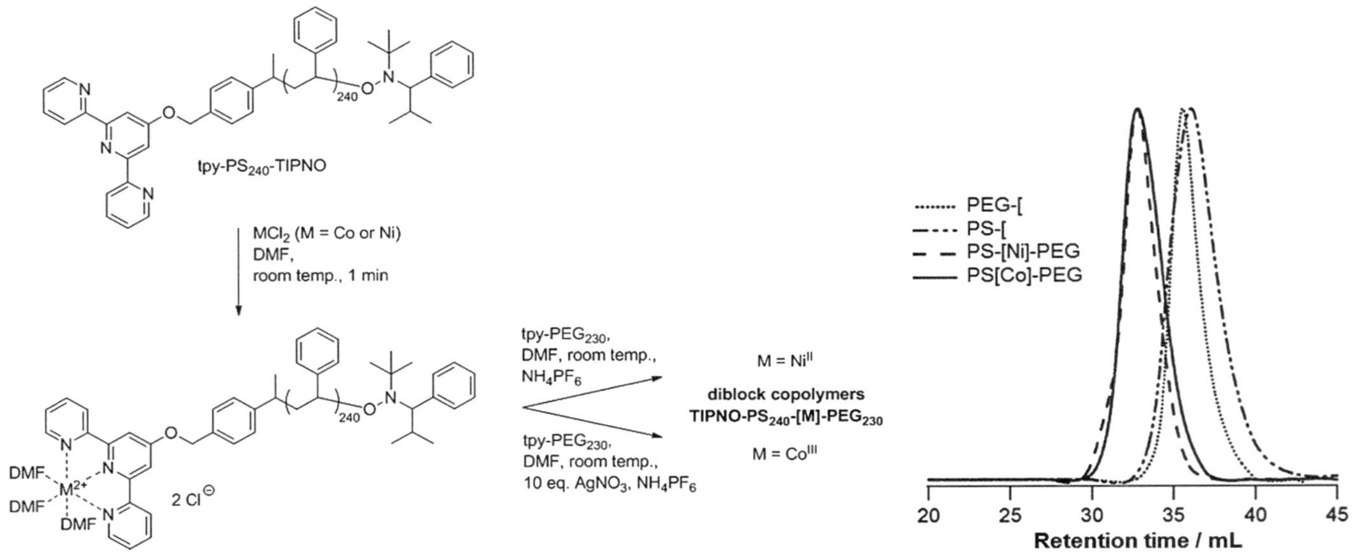

Fig. 18.11 Schematic representation of the two-step synthesis of A-[M]-B diblock copolymers with Ni^{II} or Co^{III} ions. The SEC traces of the diblock copolymers along with those of the metal-free macroligands are also shown (eluent: DMF with 5 mM NH_4PF_6). (Figure reproduced with permission from [18])

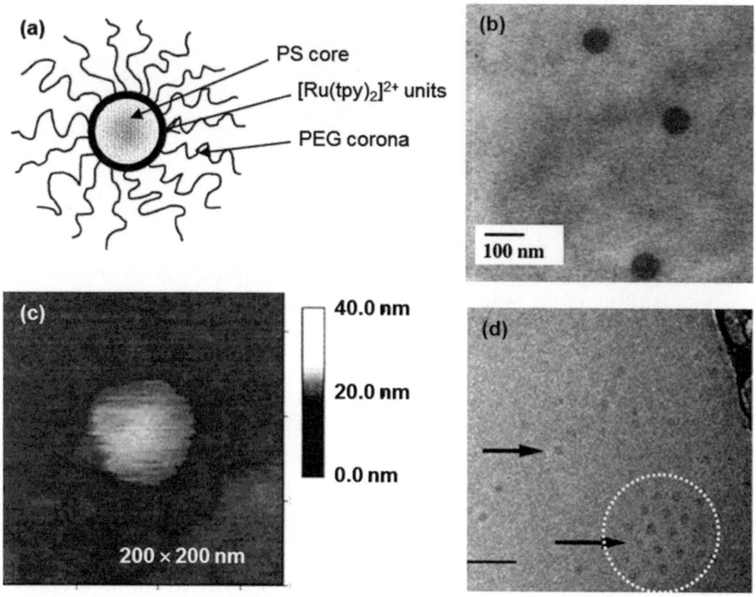

Fig. 18.12 (**a**) Schematic representation of the micellar structure of a PEG-[Ru]-PS diblock copolymer in aqueous medium. (**b**) TEM image at high magnification of an individual micelle in water (without staining). (**c**) AFM height image of an individual micelle in water. (**d**) Cryo-TEM image of micelles (the arrows indicate an individual micelle as well as a small cluster of micelles). For (**a–c**) diblock copolymers of the composition PEG_{70}-[Ru]-PS_{20} were used. (Figure reproduced with permission from [94, 97])

block lengths were varied over a wide regime [94, 95]. The micelles were formed by adding water to a solution of the metallopolymer in DMF and subsequent dialysis against water. As depicted in Fig. 18.12a, the thusly obtained spherical micelles comprised a hydrophobic PS core and a corona of the hydrophilic PEG chains; the charged Ru^{II} bisterpyridine sites were assumed to be located at the core-corona interface. For example, a material of the composition PEG_{70}-[Ru]-PS_{20} was studied in detail. Dynamic light scattering (DLS) analysis of the resulting opalescent solution indicated the presence of two different species: (a) a narrow distribution assigned to spherical micelles with an average hydrodynamic diameter (D_h) of ca. 65 nm and (b) a population with broad dispersity arising from large aggregates of individual micelles (D_h of ca. 202 nm). This aggregation behavior was found to be characteristic for the block copolymers with a charged linkage – "classical" covalent diblock copolymers of similar compositions do not prone to aggregation. The aggregation of micelles can, however, be efficiently suppressed by very cautious addition of water to the initial DMF solution via a syringe pump [96]. Additionally, various microscopy techniques were applied to visualized individual micelles (representatively, some cryogenic transmission electron (cryo-TEM), and atomic force microscopy (AFM) images are shown in Fig. 18.12b–d) [94, 97]. Also AUC could be successfully be utilized to investigate the micellization behavior; as shown by

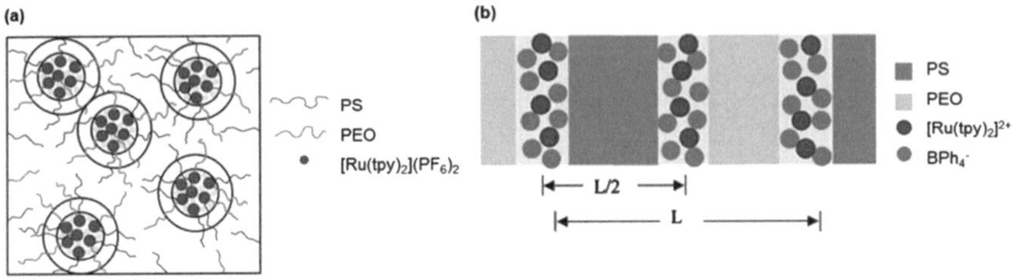

Fig. 18.13 Schematic representation of the melt morphology of the diblock copolymer PEG_{70}-[Ru]-PS_{20} with (a) PF_6^- or (b) as BPh_4^- counterions. (Figure reproduced with permission from [99, 100])

DLS, also AUC revealed the presence of unimers, micelles, and higher aggregates. For the micelles, an average molar mass of 318,000 g mol^{-1} was calculated from the hydrodynamic experiments, corresponding to an aggregation number of about 50 diblock copolymer chains per micelle [96, 98]. Basically, the same self-assembly behavior of the diblock copolymers based on Ni^{II} or Co^{III} bisterpyridine complexes, as for the above discussed Ru^{II} based ones, was found [18]. This leads to the conclusion that the nature of the transition metal ion in A-[M]-B diblock copolymers has only a negligible influence, at least when considering the micellization behavior.

Not only the behavior of the diblock copolymer PS_{20}-[Ru]-PEG_{70} in solution but also in the molten state was studied. In contrast to its "covalent" counterpart, PS_{20}-*b*-PEG70, small angle X-ray scattering (SAXS) measurements indicated a microphase separation in the melt and a profound influence of the accompanying counterion on the morphology thereof (Fig. 18.13). In detail, with PF_6^- aggregation of the charged complex sites lead to well-defined spheres (radius of ca. 1.5 nm) that were surrounded by the polymer shell (outer radius of ca. 2.4 nm) [99]. The theoretically predicted lamellar structure could, however, be generated by exchanging the counterion to the more bulky BPh_4^-: After annealing at 55 °C for 40 h, a highly ordered lamellar melt with a periodicity of 11.9 nm was obtained [100]. According to SAXS measurements, segregation of the incompatible PEG and PS blocks formed the lamellar structure with the Ru^{II} bisterpyridine complexes being confined to the lamellar interfaces. These results show that A-[M]-B diblock copolymers can, under certain experimental conditions, indeed be regarded as quasi-ABC-triblock copolymers with the bisterpyridine complex (including its counterions) acting as middle block that is highly incompatible with the two outer ones.

Lohmeijer et al. followed a combinatorial approach to prepare a (4 × 4)-library of PS_x-[Ru]-PEG_y diblock copolymers from a small set of tpy-functionalized macroligands [101]. The obtained polymers featured a wide variation of the PS/PEG ratio, and the influence of the composition on the morphologies of the thin spin-coated films was investigated. AFM imaging revealed that a broad range of structures with tunable domain size could be obtained (Fig. 18.14). In agreement with the models from polymer physics, the diblock copolymers 3, 4, 8, and 13 yielded a spherical morphology in the solid state; in the same manner, cylinders could be visualized for the block copolymers 2, 7, 9, 11, and 12.

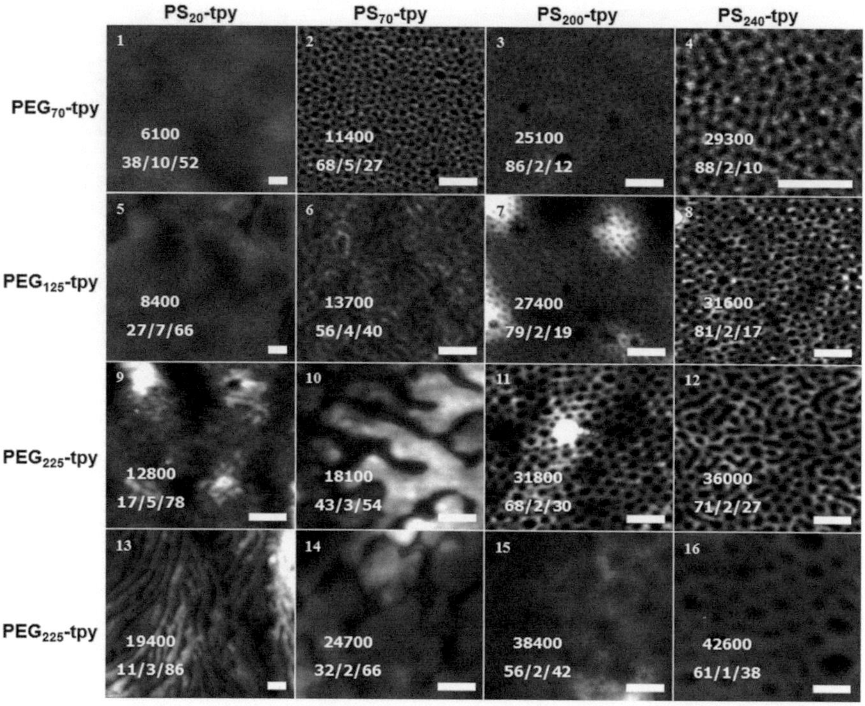

Fig. 18.14 AFM phase images of the diblock copolymer library PS_x-[Ru]-PEG_y (thin spin-casted films, no annealing). In each case, the upper number denotes the molar mass; the lower numbers correspond to the volume fractions of the three-component systems. The scale bar always represents 100 nm. (Figure reproduced with permission from [101])

It is worth mentioning in this respect that the crystalline PEG microdomains, perpendicularly oriented to the surface, fully penetrated the films. This particular phase separation behavior was applied by Gohy and co-workers for the fabrication of a nanoporous membrane [86]: Spin-coating of the PS_{375}-[Ru]-PEG_{225} diblock copolymer from a nonselective solvent onto a Si support directly yielded the targeted morphology (Fig. 18.15a). In the next step, the Ru^{II} bisterpyridine linkages were cleaved oxidatively (with Ce^{IV} under acidic conditions). The thusly released PEG minor block was rinsed off to leave a nanoporous PS membrane behind. The successful removal of the hydrophilic block was confirmed by AFM (Fig. 18.15b) and X-ray photoelectron spectroscopy (XPS) (Fig. 18.15c). The authors showed later that the degree of ordering could be increased further by annealing the spin-coated film with polar solvent vapor (e.g., THF) [102].

In one further study, the micellization of PS_x-[Ru]-PEG_y diblock copolymers in aqueous medium was studied. In total 13 different diblock copolymers were screened [103]. As revealed by AFM and TEM imaging (Fig. 18.16a shows the diblock copolymer PS_{200}-[Ru]-PEG_{375}), a contradiction to the theories of classical copolymer physics was observed – this again emphasizes that the charged

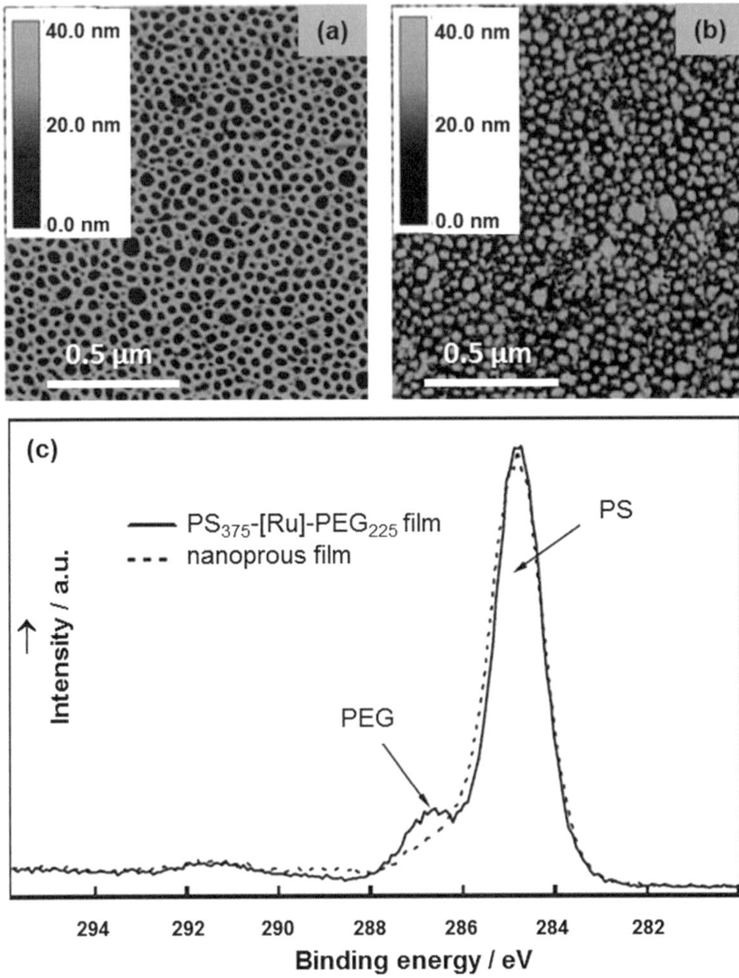

Fig. 18.15 (**a**) AFM phase image of a spin-coated film (75 nm thickness) of PS_{375}-[Ru]-PEG_{225} diblock copolymer on a Si substrate. (**b**) AFM phase image of the same film after formation of the nanoporous membrane. (**c**) XPS (C1s regime) of the film before (solid line) and after treatment with the oxidant solution. (Figure reproduced with permission from [86])

bisterpyridine complexes can, in general, not be regarded as an "innocent" linkage between the blocks but may bring special properties in that, in this example, effect the polymer physics. Both methods showed that the micelles core size did not increase linearly with the DP of the PS block. AFM imaging was carried out with the micelles being deposited on a Si substrate. As indicated in Fig. 18.16b, the flexible PEG chains of the dried micelles, i.e., the corona, are expected to lay flat on the surface, and, therefore, the height of the micelles can be related to the size of the core. In order to visualize the core by TEM, selective staining of the PS block with RuO_4 is required. Independent from the imaging method, only two core sizes could

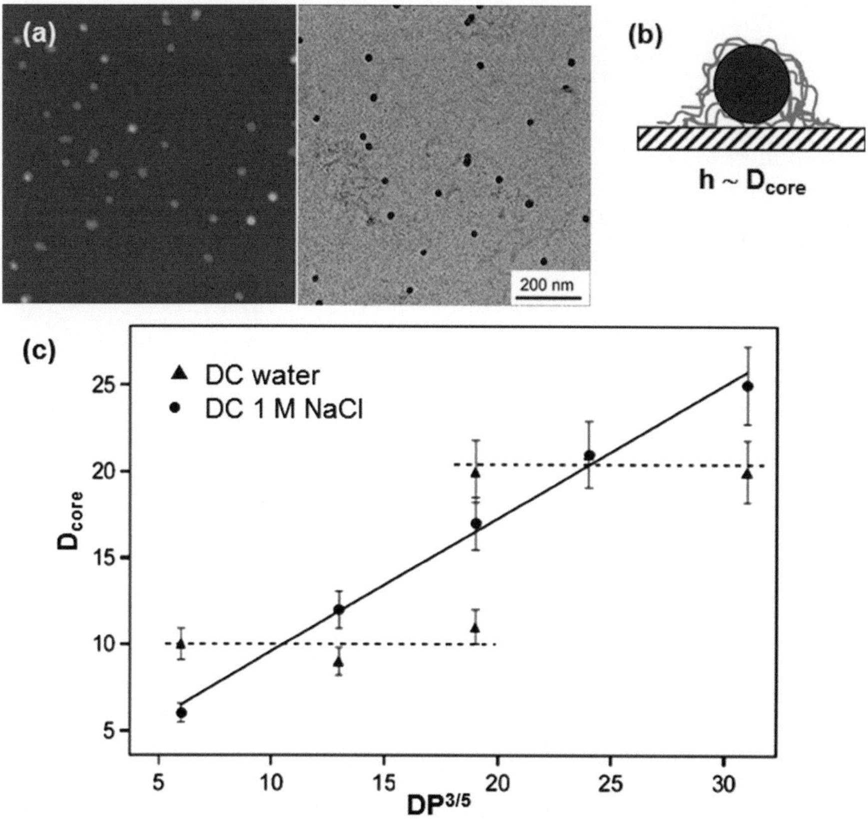

Fig. 18.16 (**a**) Microscopy imaging of the micelles prepared from PS_{200}-[Ru]-PEG_{375} in water. Left: AFM phase image (left, 1 × 1 μm^2). Right: TEM image. (**b**) Schematic representation of a dried micelle. (**c**) Relationship between the AFM-measured core size (D_{core}) the $DP^{3/5}$ of the PS block under different conditions (i.e., pure water vs. 1 M NaCl). The solid line represents the linear regression obtained from the data in 1 M NaCl; the dashed lines, evidencing the presence of only two sizes, are only a guide for the eyes. (Figure reproduced with permission from [103])

be observed: one of ca. 10 nm for DPs up to 70 and one of ca. 20 nm for DPs above 200. For DP values in between (i.e., $70 \geq \times \geq 200$), both population could be identified. Apparently, the Ru^{II} bisterpyridine linkage strongly affected the micellization due to electrostatic repulsion. It could be shown that this effect could be suppressed by performing the self-assembly in the presence of salt, e.g., NaCl. Under these conditions, the metallo-supramolecular diblock copolymers behaved as "normal" copolymer, i.e., complying the theories of copolymer physics (linear increase of the core diameter with $DP^{3/5}$; see Fig. 18.16c).

Guillet et al. addressed the influence of polyelectrolyte-type counterions, i.e., various poly(styrene sulfonate)s (PSSs), on the micellization behavior of the PS_{20}-[Ru]-PEG_{70} diblock copolymer [104]. In DMF as a nonselective solvent amphiphilic polymer, brushes of random-coil conformation could be identified (Fig. 18.17). It was shown that the morphology of the micelles could be manipulated via the length

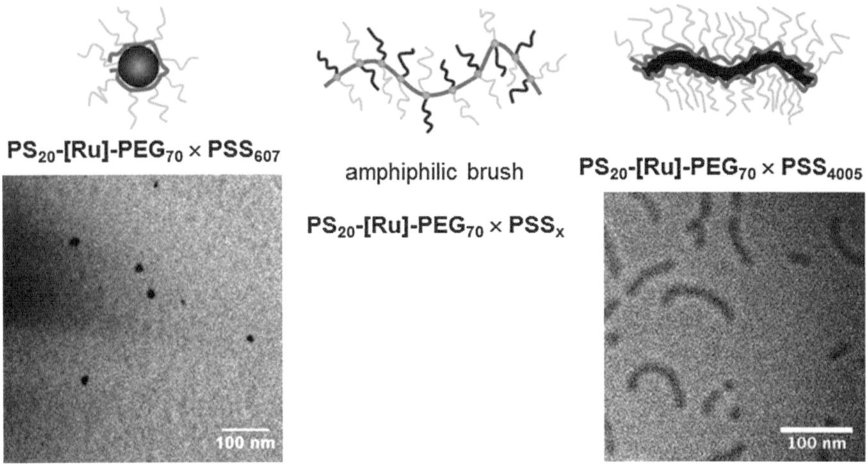

Fig. 18.17 Schematic representation of the amphiphilic brush that self-assembled into micelles in water. Depending on the length of the accompanying PSS chain, either spherical (left) or worm-like species were observed. For both cases a representative cryo-TEM image is included. (Figure reproduced with permission from [104])

of the PSS chains as well as the ratio between Ru^{II} terpyridine and sulfonate units. The micelles obtained in aqueous medium were analyzed by DLS and (cryo-)TEM measurements: The shorter PSS chains (with a DP up to 607) yielded well-defined spheres, whereas the longer polyanions assembled into worm-like species (Fig. 18.17).

Though the PS-[Ru]-PEG diblock copolymers represent the, by far, most intensively studied systems of the A-[M]-B type, some other examples, such as PDMS-[Ru]-PEG [62] and PEB_{70}-[Ru]-PEG_{70} [85, 105, 106], also have to be mentioned in this context. In aqueous medium, the latter one assembled into spherical micelles (D_h of ca. 36 nm) and clusters of micelles (D_h of ca. 115 nm); the clusters, however, could be transformed into spheres simply by dilution or by increasing the temperature [105]. Due to the glassy nature of the hydrophobic block, the analogous PS_{20}-[Ru]-PEG_{70} diblock copolymer gave only "frozen" micelles, and the reversible aggregation of second order was absent. Small PEB particles (D_h of ca. 13 nm, according to DLS and AFM measurements) were prepared from PEB_{70}-[Ru]-PEG_{70} by cleavage of the Ru^{II} complex with HEEDTA; the stability of the particles was rationalized by electrostatic repulsion by the remaining Ru^{II} ions. The groups of Manners and Winnik have introduced the concept of "crystallization-driven self-assembly" (CDSA), in which block copolymer micelles with a crystalline core grow further in a quasi-living fashion to yield cylindrical assemblies with very narrow dispersities [107]. The micelles obtained from the $PFDS_{12}$-[Ru]-PEG_{70} diblock copolymer represent one early example in this respect (PFDS: poly (ferrocenyldimethylsilane)) [63]. It was shown by differential scanning calorimetry (DSC) that the organometallic PFDS blocks already crystallized during the micellization process; the obtained cylindrical, rod-like micelles were further

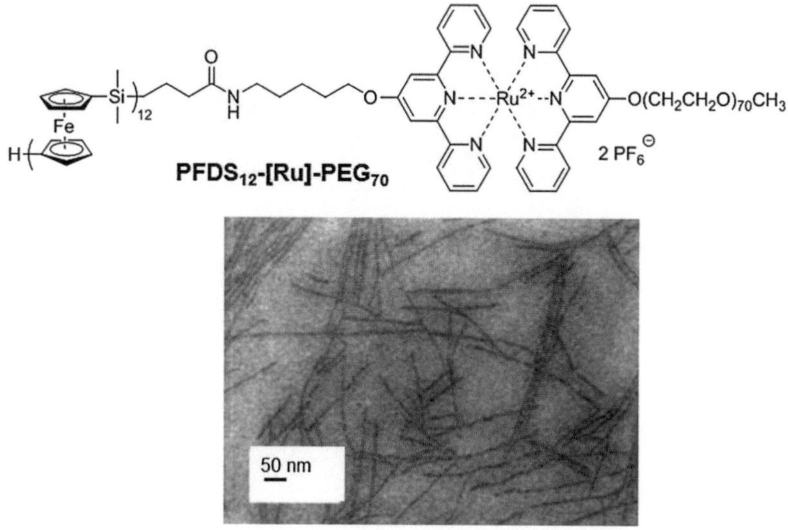

Fig. 18.18 Schematic representation of the $PFDS_{12}$-[Ru]-PEG_{70} diblock copolymer. A TEM image of the cylindrical micelles formed in water is also shown. (Figure reproduced with permission from ref. [63])

investigated by DLS as well as AFM and TEM imaging (Fig. 18.18). The micelles featured a small and constant diameter but were long and of high dispersity in length. Upon sonication, the cylindrical micelles were shorted, while their diameter was unaffected. The CDSA of micelles represents an emerging branch in today's nanoscience, and potential applications of the accessible nanostructures as semiconductors, precursor to ferromagnetic materials, and others have been envisioned [108]. Thus, diblock copolymers of the general type PFDS-[M]-B might be versatile building blocks to be investigated further in this respect.

Very recently, Chan and co-workers reported on the (P3HT)-[Zn]-PEG diblock copolymers that were formed via the selective self-assembly of the two complementary tpy ligands as shown in Fig. 18.3 (P3HT: poly(3-hexlthiophene) [17]. AFM and TEM imaging was applied to investigate the micellization in a chloroform/methanol mixture. The morphology of the structures could be tuned via the length ratio of the constituent blocks (Fig. 18.19): With increasing P3HT-content, the formation of fiber-like assemblies became favored. As in the previous example, also the P3HT-based corona showed a CDSA behavior. Due to still rather labile metallo-supramolecular linkage, facile dissociation of the block copolymer could be achieved with EDTA, thereby establishing an unshelled P3HT nanofiber network; the nanofibers could be redispersed by addition of a Zn^{II}-complexed telechelic bisterpyridine PEG macroligand.

A thermoresponsive A-[M]-A homopolymer has already been introduced in Sect. 18.4.1 [64]. Within the field of A-[M]-B diblock copolymers, more examples for this behavior can be found. Zhou and Harruna reported the synthesis of the amphiphilic

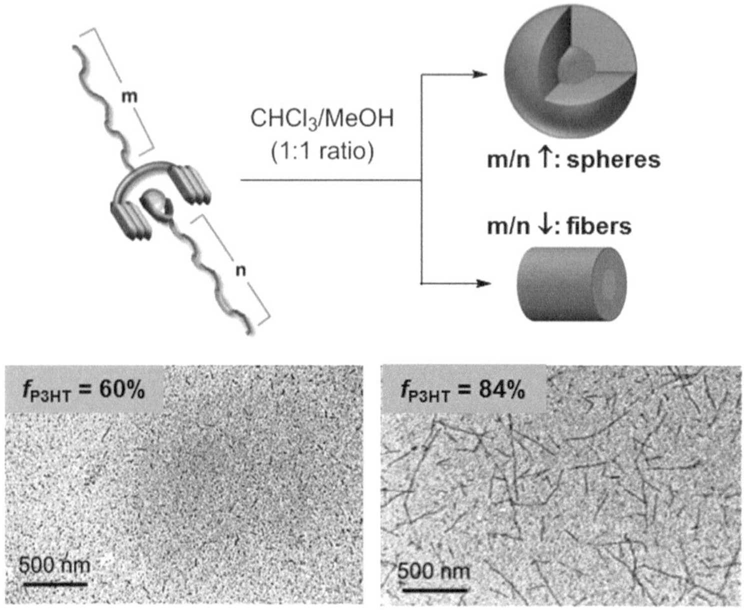

Fig. 18.19 Schematic representation of the (P3HT)-[Zn]-PEG diblock copolymer and their self-assembly into spheres and fibers as a function of m/n ratio (m, length of the PEG block; n, length of the P3HT block). Typical TEM images for both cases (i.e., low and high content of P3HT, respectively) are also shown. (Figure reproduced with permission from [17])

PNIPAM-[Ru]-PS diblock copolymer in which the PNIPAM block exhibited LCST behavior [43]. This work represents the historically first example for a temperature-responsive metallo-supramolecular block copolymer. The synthesis of cross-linked hollow nanoparticles with tpy sites within the hydrophobic nanocage is summarized in Fig. 18.20 [84]. O'Reilly et al. assembled micelles of the PS_{120}-[Ru]-PAA_{135} diblock copolymer in water and applied amidation chemistry for the subsequent cross-linking with 2,2′-(ethylenedioxy)-*bis*(ethylamine). Finally, decomplexation with HEEDTA afforded the hollow nanoparticles; the purification thereof was achieved by dialysis into aqueous THF. The hollow nature of the spheres was confirmed by ingestion experiments with a hydrophobic dye (i.e., Red Nile), and the pH responsiveness of the nanocages was also tested: The diameter could be switched between 90 and 240 nm by adjusting the pH value in a regime from 5 to 9.

The PS_{32}-*b*-$P2VP_{13}$-[Ru]-PEG_{70} triblock copolymer of the AB-[M]-C type represents one further example for a pH-responsive material [60]. This material assembled into core-shell-corona-type micelles (PS core, P2VP shell, and PEG corona). The P2PV block responded sensitively to changes in the pH value: Under acidic conditions, a D_h of 81 nm was determined, the micelles shrank upon neutralization down to a D_h of 63 nm; in agreement with the pKa value of P2VP, a sharp transition between both regimes was apparent at a pH value of 5.5. This pH responsiveness

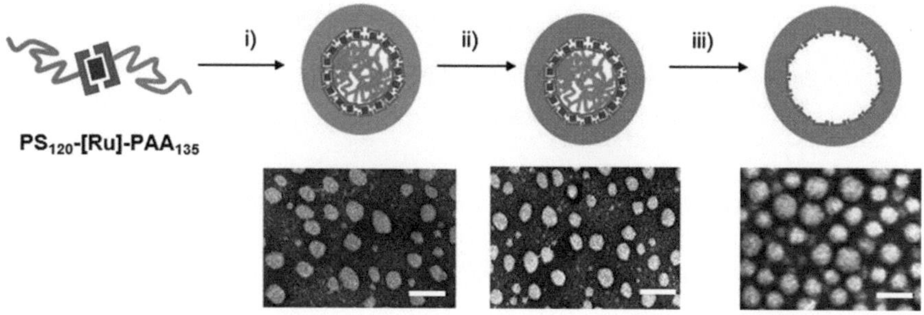

Fig. 18.20 Schematic representation of the formation of hollow nanostructures from a PS-[Ru] PAA diblock copolymer: (i) self-assembly of the copolymer into micelles in water, (ii) cross-linking of the PAA corona, and (iii) cleavage of the Ru^{II} bisterpyridine complexes and removal of the PS blocks by dialysis in THF/water mixtures. The corresponding TEM images, after staining with phosphotungstic acid, deposition onto a carbon-coated Cu grid and drying, are also depicted (the scale bar is 100 nm). (Figure reproduced with permission from [84])

could, for example, be applied for the reversible encapsulation of active species within in the P2VP shell. Such behavior has already been shown by Lodge et al. for micelles featuring a phase-separated core [109]. Gohy et al. explored the applicability of a covalent PS-*b*-P2VP-*b*-PEG triblock copolymer as "nanoreactor" for the preparation of, e.g., Au nanoparticles [110].

Amphiphilic ABC-type triblock copolymers, as in the previous example, are known to self-assembly into so-called multicompartment micelles under certain requisites [111]: Such a micelle is formed, if only one block is soluble in the used solvent and, by the same time, phase separation of the other two non-soluble blocks within the micellar core is favored. A different morphology can, however, be observed, if two segments are soluble in the solvent – in this case a mixed-arm soluble corona surrounds the core. With respect to the former one, the hydrophobic sub-domains in the micellar core must have pronounced different characters, e.g., when combining a hydrocarbon block with a fluorocarbon one. Schubert and co-workers synthesized the triblock copolymer PS_{76}-*b*-$PTFMS_{42}$-[Ru]-PEG_{70} and observed such a behavior in various alcohols (PTFMS: poly (4-trifluoromethylstyrene)) [50]: Depending on the polarity of the solvent, the solubility of the central fluorine-rich block could be modulated; thus, micelles of different morphologies were obtained: spherical and worm-like micelles as well as vesicles (Fig. 18.21). In addition to this, the thermoresponsive nature of the fluorinated block, i.e., its upper critical solution temperature (UCST), enabled a highly reversible control over both the size and morphology of the assemblies. Later, the extension of the concept even to an ABCD tetrablock copolymer was published by the Gohy and Schubert groups [49]. $PTFMS_{20}$-*b*-$PtBA_{25}$-*b*-PS_{35}-[Ru]-PEG_{70} yielded multicompartment micelles of different morphologies: Spherical micelles with a PS core and a corona comprising all three other blocks were formed in ethanol. In isopropanol, however, larger multicompartment micelles with the core comprising discrete PS and PTFMS sub-domains, linked by the PtBA block, and a

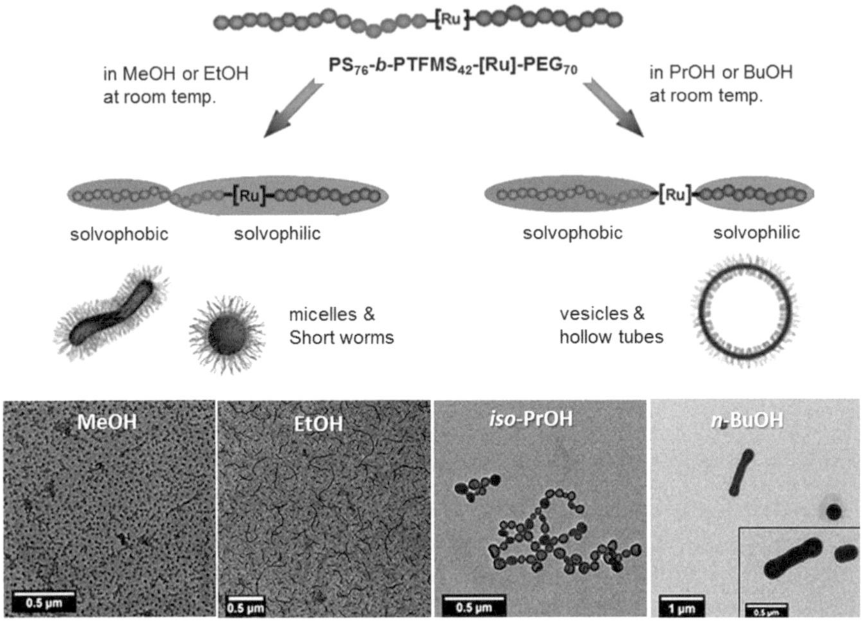

Fig. 18.21 Schematic representation of the self-assembly of the triblock copolymer PS_{76}-*b*-$PTFMS_{42}$-[Ru]-PEG_{70} in different solvents. Some representative TEM images of the assemblies in different solvents are also depicted. (Figure reproduced with permission from [50])

stabilizing PEG corona were assembled at temperatures below 45 °C (however, at higher temperatures a transition to the micelles, identical to those observed in ethanol, occurred).

Micelles of completely different architectures are formed if the order of preparation is inversed, i.e., a amphiphilic AB-[diblock copolymer with a vacant metal-binding site is assembled first (if A is the hydrophobic block, the tpy units will be located on the termini of the corona's arms), followed by the metal ion complexation. For this, two scenarios are conceivable: complexation in a highly diluted or rather concentrated regime. In the former case, *intra*-micellar complexation will be favored, whereas *inter*-micellar complexation will preferentially occur in the latter scenario. The *intra*-micellar complexation was studied by Guillet et al. to understand how complexation of the "sticky end groups" influences the overall size of the micelles [112]. For this purpose, the tpy-functionalized diblock copolymer PS_{47}-*b*-$PtBA_{55}$-tpy was assembled into spherical micelles in ethanol (R_h of 22 nm); these micelles featured exactly the morphology as described above (i.e., the tpy moieties were located at the ends of the PtBA corona chains, Fig. 18.22a). Various transition metal ions (i.e., Zn^{II}, Ni^{II}, or Fe^{II}) were added to a diluted solution of the micelles, and, at a metal/tpy ratio of 1:2, the formation of bisterpyridine complexes within the micellar coronas occurred (Fig. 18.22b). As a result of the looping of the flexible coronal chains, stable flower-shaped micelles were formed, and the R_h decreased significantly. When using an excess of transition metal ions, opening of the

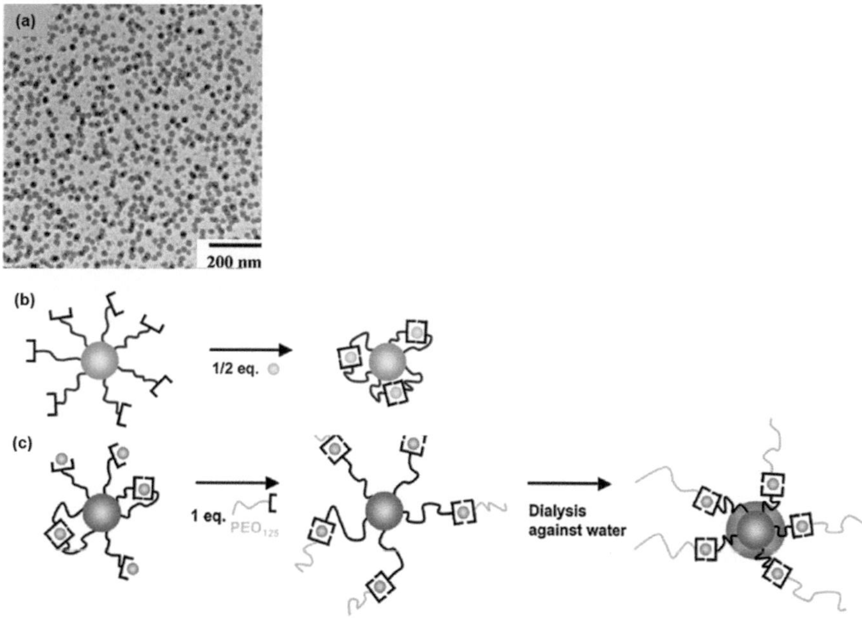

Fig. 18.22 (**a**) TEM image of the micelles obtained from the diblock copolymer PS_{47}-*b*-$PtBA_{55}$-tpy in EtOH. (**b**) Schematic representation of the formation of flower-like micelles by addition of 0.5 eq. of a transition metal ion. (**c**) Schematic representation of the grafting of PEG_{125}-tpy onto micelles of PS_{47}-*b*-$PtBA_{55}$-tpy pre-loaded with 1 eq. of Fe^{II} ions. (Figure reproduced with permission from [112])

biscomplexes occurred, and, finally, monoterpyridine complexes were generated on the termini of the coronal chains. The initial core-shell assembly could be extended to a core-shell-corona-type architecture by metallo-supramolecular grafting of appropriate tpy macroligands onto the preformed micelles. For example, PS_{47}-*b*-$PtBA_{55}$-[Fe]-PEO_{125} micelles were obtained by coordination of PEG_{125}-tpy to the preformed micelles, pre-loaded with one equivalent of Fe^{II} ions (Fig. 18.22c).

The *inter*-micellar complexation, giving rise to the formation of "micellar gels," was also in the focus of interest. Using a significantly higher concentration of micelles of PS_{80}-*b*-$PtBA_{200}$-tpy enabled the linking of individual micelles upon addition of, e.g., Zn^{II}, Ni^{II}, or Fe^{II} ions (Fig. 18.23a). For this purpose, a much higher micelle concentration was required to suppress *intra*-micellar complexation. A variety of rheological measurements was performed on the micellar gels to investigate their mechanical properties (i.e., determination of the storage (G') and loss moduli (G''), Fig. 18.23b, c). The gels exhibited a stimuli-responsive behavior – the metallo-supramolecular networks could be dissembled either chemically or mechanically. Concerning the latter one, the initial properties were recovered almost instantaneously upon stopping the extraneous mechanical trigger. This highly reversible mechanical response was found to be dependent on the nature of the linking metal ion [51].

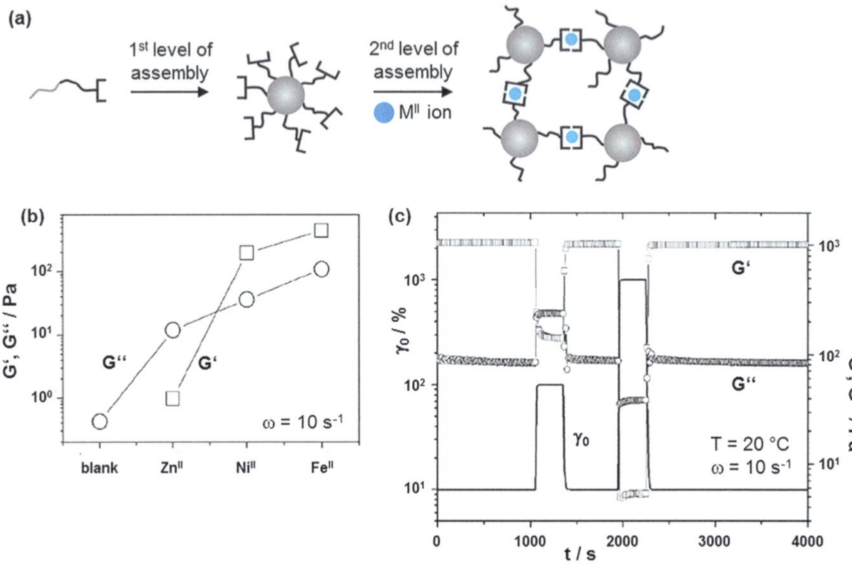

Fig. 18.23 (**a**) Schematic representation of the directed self-assembly of the PS_{80}-*b*-$PtBA_{200}$-tpy diblock copolymer into a "micellar gel." (**b**) Comparison of the storage (G′) and loss modulus (G″) at $\omega = 10\ s^{-1}$ between the initial micelles and those charged with 0.5 eq. of various transition metal ions. (**c**) Evolution of G′ and G″ with time following two successive pulses of high deformation (full lines), recorded on a "micellar gel" based on Fe^{II} ions at a copolymer concentration of 14%. (Figure reproduced with permission from [51])

18.4.3 Block Copolymers with Two Bisterpyridine Complexes: A-[M]-B-[M]-A Triblock Species and Related Systems

The A-[M]-A homopolymers and A-[B]-diblock copolymers, both comprising a single metal-bisterpyridine linkage, on the one hand as well as the multinuclear assemblies, such as the chain-extended polymers, e.g., block copolymers of the general formula A-[M]-(B-[M]-)$_n$B-[M]-A type, or the supramolecular polymers of the (A-[M]-)$_n$ type on the other hand represent the borderline cases for linear metal-containing (block co)polymers. Examples for all these structural motifs are well-documented in literature (see also Sects. 18.4.1, 18.4.2, and 18.4.4) [10, 15, 31, 113]. However, (block copolymers) with a discrete number of metal centers within the main chain are rather uncommon. One reason for this might be found in the more demanding synthesis and purification resulting from the required multistep assembly methodology – in particular when compared to the mononuclear homo- and diblock copolymers (vide supra). Representatively, the work by Möhwald et al. has to be mentioned in this context [114]. Though not dealing with polymers, the scope and limitations of preparing oligonuclear assemblies with good control over the molar mass become evident. The authors utilized a divergent stepwise assembly of a short telechelic bisterpyridine ligand with Ru^{II} ions and combined this with a convergent step to prepare a linear rod-shaped molecules featuring seven Ru^{II} centers. In each assembly step, the monodisperse nature of the compounds was retained, but the yield decreased significantly with increasing number of reaction steps.

Only a few detailed studies are known where dinuclear A-[M]-B-[M]-A triblock copolymers have been prepared and characterized seriously. Architectures of the general composition PS-[Ru]-PEG-[Ru]-PS and PS-[Ru]-PEB-[Ru]-PS have been mentioned [115, 116], but the focus of interest has been laid onto triblock assemblies with a short rigid-rod middle block. The combination of such telechelic π-conjugated bisterpyridine ligands with flexible PEG_{44}-tpy and Ru^{II} ions units enabled the synthesis of coil-rod-coil-type triblock copolymers [117, 118]. In aqueous medium, the amphiphilic triblock copolymers assembled into spherical micelles that were analyzed by DLS and cryo-TEM measurements (Fig. 18.24). The characteristic of metallo-supramolecular diblock copolymer micelles to exhibit an aggregation of second order was also found for their triblock counterparts. The micellization process of the coil-rod-coil triblock copolymers was found to be more complex than for those block copolymers only comprising coiled blocks: besides phase separation also π-π interactions between the conjugated rod-like segments or the formation of liquid-crystalline domains have to be taken into account (however, this assumption could not yet been verified experimentally). The micelles and aggregates were investigated in more detail by applying AUC, and the determined hydrodynamic radii were in good agreement with those measured by DLS and TEM (Fig. 18.25) [118]. Moreover, the hydrodynamic methods allowed to conclude on the number of triblock copolymer molecules per micelle (five molecules, micellar radius of ca. 3 nm) and aggregate (ca. 5000 molecules, average radius of 30 nm).

The Gohy and Schubert groups extended their work further toward three- and four-arm star-shaped copolymers with a rigid central blocks and flexible PEG arms [119]. The three-armed assembly yielded well-defined vesicles in pure acetone, whereas the analogous four-armed system only aggregated into spherical micelles in aqueous acetone (Fig. 18.26). The difference in the self-assembly behavior was attributed to (a) the higher overall hydrophilicity of the four-armed copolymer and (b) to the different conformations of the rigid cores (the three-armed core was planar, whereas the four-armed one was not).

18.4.4 Chain-Extended (Block Co)polymers with Bisterpyridine Complexes

In the mid-1990s, Constable and co-workers utilized bisterpyridine ligands for the synthesis of oligomers and polymers via the coordination of transition metal ions [120]. This metallo-supramolecular polymerization follows the rules of a typical polyaddition reaction, giving rise to the formation of linearly extended chains containing metal-bisterpyridine complexes as linkers (Fig. 18.27) [10, 121,

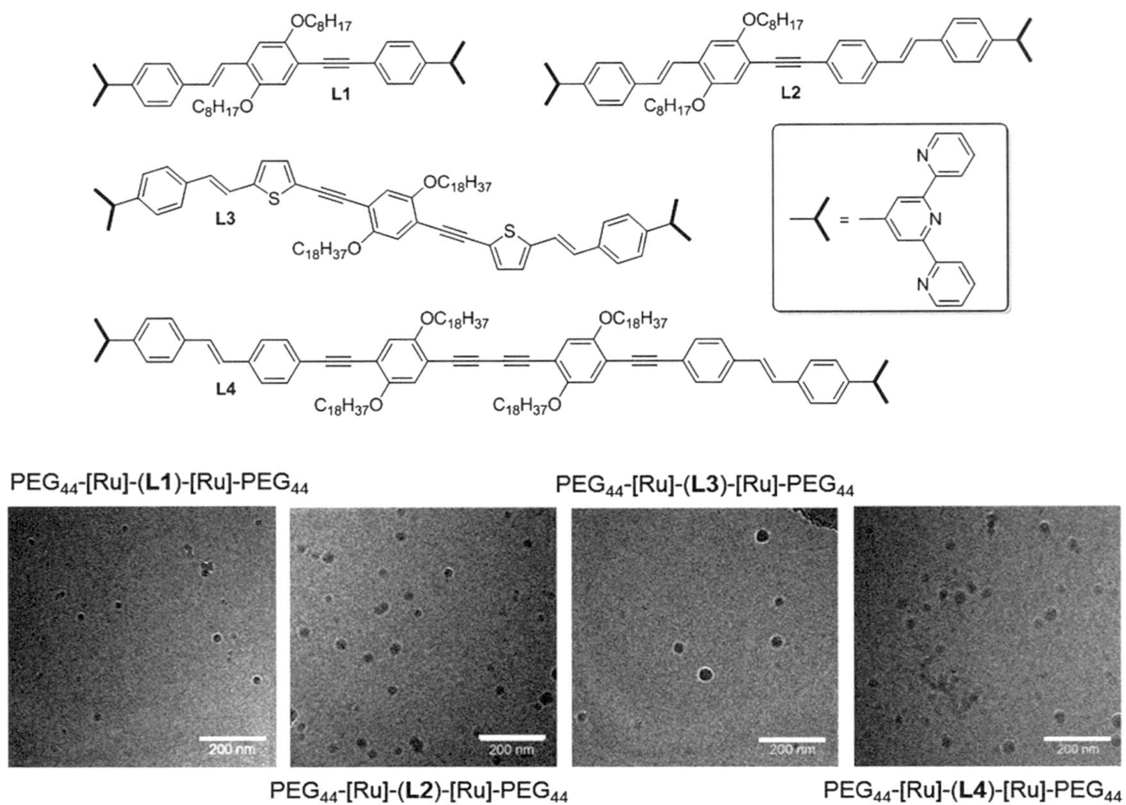

Fig. 18.24 Schematic representation of the rigid-rod bisterpyridine ligands that were used for the assembly of the coil-rod-coil triblock copolymers. The cryo-TEM images of the micelles formed by the triblock copolymers in water are also shown (the scale bar is 200 nm). (Figure reproduced with kind permission from [117])

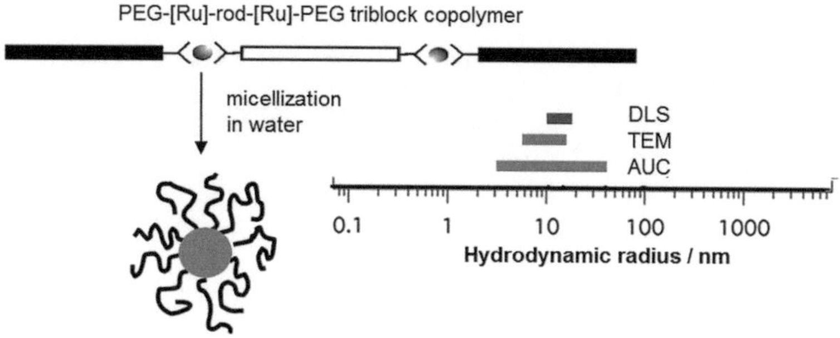

Fig. 18.25 Schematic representation of the micellization of a coil-rod-coil triblock copolymer in water. The hydrodynamic radius of the micelles as determined via three independent techniques is also depicted. (Figure reproduced with kind permission from [10])

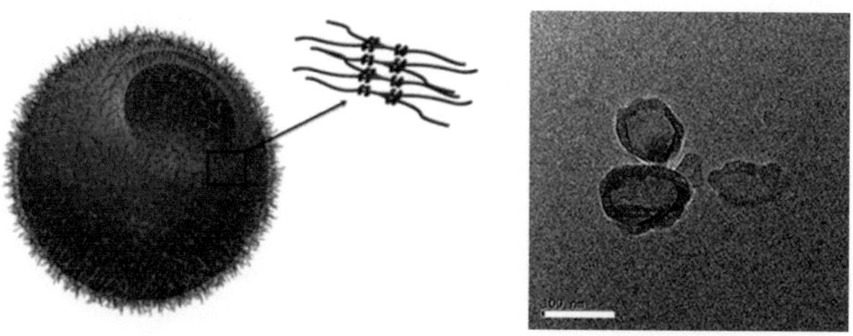

Fig. 18.26 Left: schematic representation of a vesicle formed from the three-armed copolymer in acetone. Right: cryo-TEM image of the spherical micelles formed by the four-armed copolymer in aq. acetone (scale bar: 100 nm). (Figure reproduced with permission from [119])

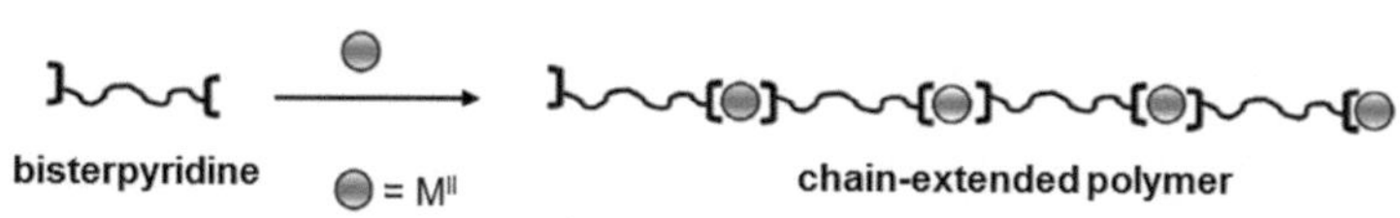

Fig. 18.27 Schematic representation of the self-assembly of bisterpyridine ligands into chain-extended polymers via metallo-supramolecular polymerization. (Figure redrawn according to [29])

122]. As detailed elsewhere, a plethora of telechelic ligands has been employed in this context, rigid-linear and flexible as well as small-molecule and polymeric ones [10, 123, 124]. In particular, when employing flexible ligands, the macrocyclization, as an entropy-driven process, represents an issue to deal with. For example, Constable et al. observed the formation of the targeted Fe^{II}-based chain-extended polymer (Fig. 18.28) besides two other species that could be identified by mass spectrometry as macrocycles [125].

In contrast to the Fe^{II}-containing assemblies, where the equilibrium between linear and cyclic structures is difficult to control, the use of Ru^{II} ions leads to kinetically stable assemblies that can be isolated and be investigated by ^{1}H NMR

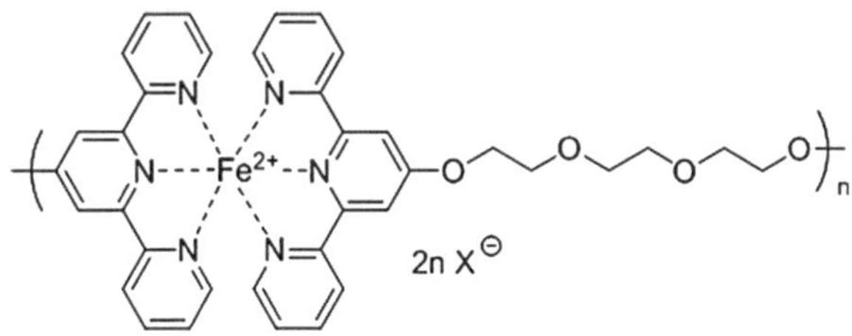

Fig. 18.28 Schematic representation of a Fe^{II}-based chain-extended polymer

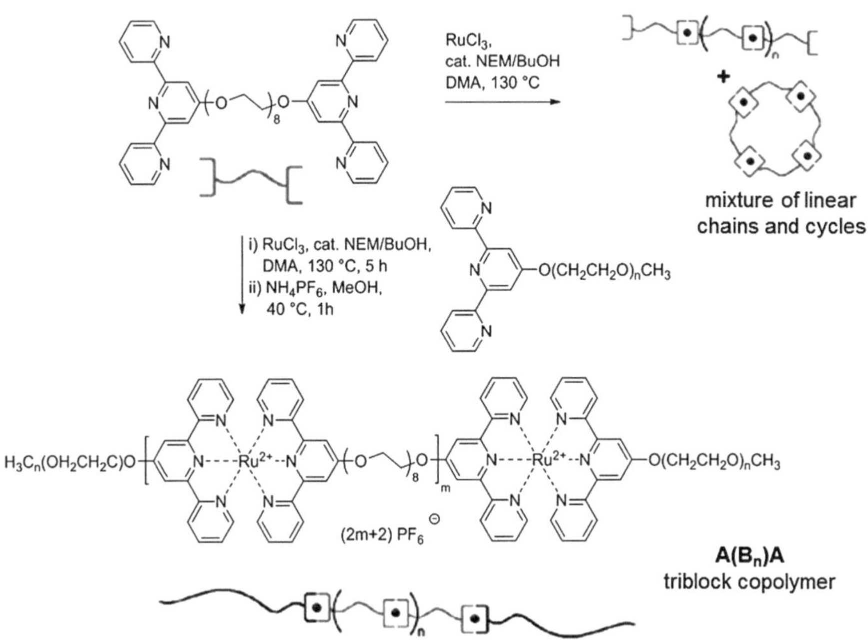

Fig. 18.29 Schematic representation of the synthesis of metallo-supramolecular species via self-assembly of a flexible telechelic bisterpyridine with Ru^{II} ions. (Figure reproduced with permission from [127])

spectroscopy, viscometry, AUC, and, most importantly, SEC [55, 56, 58, 126]. Meier et al. investigated the self-assembly of a small-molecule bisterpyridine ligand with $RuCl_3 \times xH_2O$ in a mixture of *n*-butanol/*N*-ethylmorpholine (NEM), i.e., under reducing conditions (Fig. 18.29) [127]. By increasing the monomer concentration (i.e., up to 0.2 mmol mL^{-1}), the formation of low-molar-mass species (i.e., cycles and oligomers) could be suppressed. When adding 10 mol-% of tpy-functionalized PEG as chain-stopping agent, a $A(B_n)A$ triblock copolymer could be obtained. SEC analysis of this triblock copolymer, in comparison to the homopolymer $[Ru(tpy\text{-}PEG)_2](PF_6)_2$, revealed the expected high molar mass and broad dispersity of the material (Đ of ca. 1.6). Due to the nonavailability of an

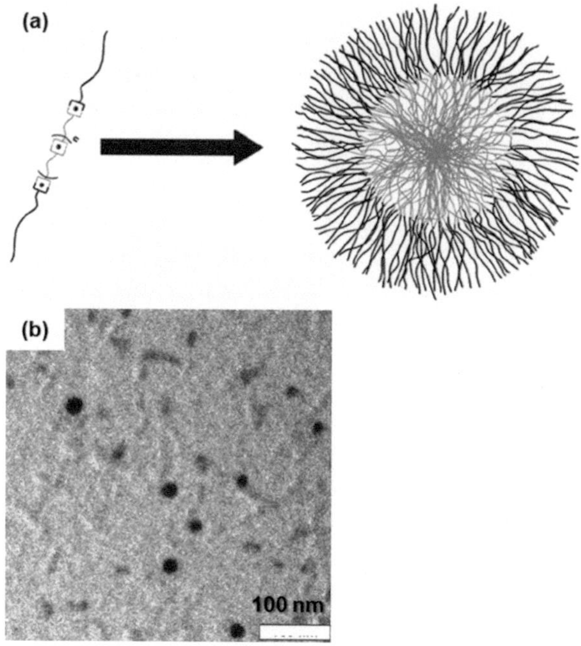

Fig. 18.30 (**a**) Schematic representation of the micellization of the A(B_n)A triblock copolymer in water. (**b**) TEM image of the micelles (without staining). (Figure reproduced with permission from [127])

appropriate calibration method, a reliable estimation of the molar mass was not feasible: Both linear PS and PMMA calibrations overestimated the M_n value, whereas the PEG calibration underestimated the molar mass of the (B_n)-block. The authors also investigated the micellization of the amphiphilic material in water (the outer PEG blocks are water-soluble; the central metal-rich (B_n)-block is hydrophobic, Fig. 18.30a). Micellization was induced by the slow addition of water to a solution of the triblock copolymer in an unselective solvent (e.g., acetone or DMSO), followed by dialysis against water. The thusly formed micelles were analyzed by DLS, AFM, and TEM imaging (Fig. 18.30b). According to these measurements, the total size (i.e., core and corona) of the hydrated micelles was estimated to be ca. 70 nm.

As already pointed out in the previous sections, the scope of SEC analysis has been extended to further bisterpyridine complexes, and, in analogy to the A(B_n)A triblock copolymer comprising Ru^{II} ions, similar materials with Ni^{II}, Co^{II}, and Fe^{II} centers have also been prepared and subjected to SEC analysis (Fig. 18.31a) [128]. However, in all three cases, the reaction conditions as well as the characterization tools had to be optimized carefully. For the Ni^{II}-containing species, the amount of the added chain-stopping agent (i.e., tpy-PEG) was varied; by this, the length of the central (B_n)-block could be adjusted over a wide range. Furthermore, the micellization of the amphiphilic materials in acetone/water was investigated, and spherical objects with radii in the range of 10–14 nm could be identified by cryo-TEM imaging (Fig. 18.31b).

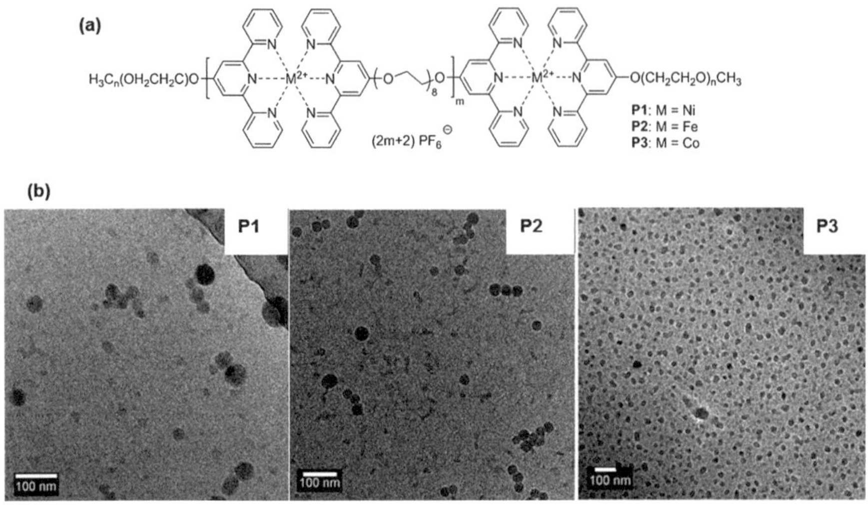

Fig. 18.31 (**a**) Schematic representation of the metallo-supramolecular $A(B_n)A$ triblock copolymers. (**b**) Cryo-TEM images of micelles of A(Bn)A triblock copolymers in acetone/water (scale bar: 100 nm). (Figure reproduced with permission from [128])

18.5 Conclusion

Can metal-bisterpyridine moieties, or metal-containing linkages in general, within copolymer architectures be regarded as a "block" on its own? From a polymer chemist's point of view, the answer to this is rather straightforward and a distinct "no." However, when not only considering length and molar mass but properties in more general, the perspective becomes different. In this contribution, we have shown that block copolymers comprising metal-bisterpyridine linkages exhibit properties that significantly differ from the covalent counterparts, i.e., having a covalent bond instead of the complex. Thus, an A-[M-]-B-type system comprises in total three "blocks" with their own characteristics that all contribute to the overall (macroscopic) behavior: the two polymer chains (named A and B) as well as the linking metal complex -[M]-. This more property-centered view on the metallo-supramolecular block copolymers does well justify this statement.

Beyond these abstract considerations, the versatility and modularity of using terpyridine-functionalized macroligands for the directed synthesis of advanced block copolymer architectures have been summarized in this chapter. This concept has been referred to as "playing LEGO™ with macromolecules" about 15 years ago and is today an established method in macromolecular chemistry and nanoscience.

Acknowledgment The authors acknowledge financial support of research by the *Deutsche Forschungsgemeinschaft* (DFG).

References

1. Yang SK, Ambade AV, Weck M (2011) Main-chain supramolecular block copolymers. Chem Soc Rev 40(1):129–137. https://doi.org/10.1039/C0CS00073F
2. Fustin C-A, Guillet P, Schubert US, Gohy J-F (2007) Metallo-supramolecular block copolymers. Adv Mater 19(13):1665–1673. https://doi.org/10.1002/adma.200602170
3. Sijbesma RP, Beijer FH, Brunsveld L, Folmer BJB, Hirschberg JHKK, Lange RFM, Lowe JKL, Meijer EW (1997) Reversible polymers formed from self-complementary monomers using quadruple hydrogen bonding. Science 278(5343):1601–1604. https://doi.org/10.1126/science.278.5343.1601
4. Binder WH, Zirbs R (2007) Hydrogen bonded polymers. Adv Polym Sci 207:7–78. https://doi.org/10.1007/12_2006_109
5. Guo D-S, Liu Y (2012) Calixarene-based supramolecular polymerization in solution. Chem Soc Rev 41(18):5907–5921. https://doi.org/10.1039/c2cS35075k
6. Harada A, Takashima Y, Yamaguchi H (2009) Cyclodextrin-based supramolecular polymers. Chem Soc Rev 38(4):875–882. https://doi.org/10.1039/b705458k
7. Rauwald U, Scherman OA (2008) Supramolecular block copolymers with cucurbit[8]uril in water. Angew Chem Int Ed 47(21):3950–3953. https://doi.org/10.1002/anie.200705591
8. Gohy J-F (2009) Metallo-supramolecular block copolymer micelles. Coord Chem Rev 253(17–18):2214–2225. https://doi.org/10.1021/ma012042t
9. Higuchi M (2016) Metallo-supramolecular polymers: design, function and device application. In: Ji S, Tsuji S (eds) Intelligent Nanosystems for energy, information and biological technologies. Springer, Japan, pp 217–248
10. Winter A, Schubert US (2016) Synthesis and characterization of metallo-supramolecular polymers. Chem Soc Rev 45(19):5311–5357. https://doi.org/10.1039/c6cS00182c
11. Morgan GT, Burstall FH (1932) Dehydrogenation of pyridine by anhydrous ferric chloride. J Chem Soc 1:20–30. https://doi.org/10.1039/JR9320000020
12. Holyer RH, Hubbard CD, Kettle SFA, Wilkins RG (1966) The kinetics of replacement reactions of complexes of the transition metals with 2,2′,2″-terpyridine. Inorg Chem 5(4):622–625. https://doi.org/10.1021/ic50038a027
13. Irving H, Willimas RJP (1948) Order of stability of metal complexes. Nature 162:746–747. https://doi.org/10.1038/162746a0
14. Meier MAR, Lohmeijer BGG, Schubert US (2003) Relative binding strength of terpyridine model complexes under matrix-assisted laser desorption/ionization mass spectrometry conditions. J Mass Spectrom 38(5):510–516. https://doi.org/10.1002/jms.463
15. Schubert US, Winter A, Newkome GR (2011) Terpyridine-based materials. Wiley-VCH, Weinheim
16. Chiper M, Meier MAR, Kranenburg JM, Schubert US (2007) New insights into nickel(II), iron (II) and cobalt(II) bis-complex-based metallo-supramolecular polymers. Macromol Chem Phys 208(7):679–689. https://doi.org/10.1002/macp.200600605
17. He Y-J, Tu T-H, Su M-K, Yang C-W, Kong KV, Chan Y-T (2017) Facile construction of metallo-supramolecular poly(3-hexylthiophene)-*block*-poly(ethylene oxide) diblock copolymers via complementary coordination and their self-assembled nanostructures. J Am Chem Soc 139(11):4218–4224. https://doi.org/10.1021/jacs.7b01010
18. Mugemana C, Guillet P, Hoeppener S, Schubert US, Fustin C-A, Gohy J-F (2010) Metallo-supramolecular diblock copolymers based on heteroleptic cobalt(III) and nickel(II)

bis-terpyridine complexes. Chem Commun 46(8):1296–1298. https://doi.org/10.1039/b923270b
19. Williams JAG, Wilkinson AJ, Whittle VL (2008) Light-emitting iridium complexes with tridentate ligands. Dalton Trans :2081–2099. https://doi.org/10.1039/b716743a
20. Schönle J, Constable EC, Housecroft CE, Prescimone A, Zampese JA (2015) Homoleptic and heteroleptic complexes of chromium(III) containing 4′-diphenylamino-2,2′:6′,2″-terpyridine ligands. Polyhedron 89:182–188. https://doi.org/10.1016/j.poly.2015.01.015
21. Wild A, Winter A, Schlütter F, Schubert US (2011) Advances in the field of π-conjugated 2,2′:6′,2″-terpyridines. Chem Soc Rev 40(3):1459–1511. https://doi.org/10.1039/c0cS00074d
22. Heller M, Schubert US (2003) Syntheses of functionalized 2,2′:6′,2″-terpyridines. Eur J Org Chem 6:947–961. https://doi.org/10.1002/ejoc.200390150
23. Heller M, Schubert US (2002) Functionalized 2,2′-bipyridines and 2,2′:6′,2″-terpyridines via Stille-type cross-coupling procedures. J Org Chem 67(23):8269–8272. https://doi.org/10.1021/jo0260600
24. Constable EC, Ward MD (1990) Synthesis and co-ordination behaviour of 6,6-bis(2-pyridyl)-2,2′:4,4″:2″,2‴-quaterpyridine; “back-to-back” 2,2′:6′,2″-terpyridine. J Chem Soc Dalton Trans (4):1405–1410. https://doi.org/10.1039/dt9900001405
25. Hetcat (2017) www.hetcat.com/People/terpy.pdf
26. Schubert US, Hofmeier H, Newkome GR (2006) Modern terpyridine chemistry. Wiley-VCH, Weinheim
27. Schubert US, Eschbaumer C (2002) Macromolecules containing bipyridine and terpyridine metal complexes: towards metallosupramolecular polymers. Angew Chem Int Ed 41(16):2892–2926. https://doi.org/10.1002/1521-3773(20020816)41:16<2892::AID-ANIE2892>3.0.CO;2-6
28. Schubert US, Eschbaumer C, Hien O, Andres PR (2001) 4′-functionalized 2,2′:6′,2″-terpyridines as building blocks for supramolecular chemistry and nanoscience. Tetrahedron Lett 42(28):4705–4707. https://doi.org/10.1016/S0040-4039(01)00796-1
29. Chiper M, Hoogenboom R, Schubert US (2009) Toward main-chain metallo-terpyridyl supramolecular polymers: the metal does the trick. Macromol Rapid Commun 30(8):565–578. https://doi.org/10.1002/marc.200800625
30. Yang L, Tan X, Wang Z, Zhang X (2015) Supramolecular polymers: historical development, preparation, characterization and functions. Chem Rev 115(15):7196–7239. https://doi.org/10.1021/cr500633b
31. Whittle GR, Hager MD, Schubert US, Manners I (2011) Functional soft materials from metallopolymers and metallosupramolecular polymers. Nat Mater 10:176–188. https://doi.org/10.1038/nmat2966
32. Andres PR, Schubert US (2004) New functional polymers and materials based on 2,2′:6′,2″-terpyridine metal complexes. Adv Mater 16(13):1043–1068. https://doi.org/10.1002/adma.200306518
33. Schroot R, Schlotthauer T, Schubert US, Jäger M (2016) Modular assembly of poly(naphthalene diimide) and Ru(II) dyes for an efficient light-induced charge separation in hierarchically controlled polymer architectures. Macromolecules 49(6):2112–2123. https://doi.org/10.1021/acs.macromol.5b02717
34. Schroot R, Schlotthauer T, Jäger M, Schubert US (2017) Hydrophilic poly(naphthalene diimide)-based acceptor-photosensitizer dyads: toward water-processible modular photoredox-active architectures. Macromol Chem Phys 218(6):1600534-n/a. https://doi.org/10.1002/macp.201600534
35. Schlotthauer T, Schroot R, Glover S, Hammarstrom L, Jager M, Schubert US (2017) A multidonor-photosensitizer-multiacceptor Ttriad for long-lived directional charge separation. Phys Chem Chem Phys. https://doi.org/10.1039/C7CP05593E
36. Schroot R, Schlotthauer T, Dietzek B, Jäger M, Schubert US (n.d.) Extending long-lived charge separation between donor and acceptor blocks in novel copolymer architectures featuring a sensitizer core. Chem Eur J:n/a-n/a. https://doi.org/10.1002/chem.201704180

37. Hoogenboom R, Schubert US (2006) The use of (metallo-)supramolecular initiators for living/controlled polymerization techniques. Chem Soc Rev 35(7):622–629. https://doi.org/10.1039/B515911C
38. Heller M, Schubert US (2001) Optically active supramolecular poly(L-lactide)s end-capped with terpyridine. Macromol Rapid Commun 22(16):1358–1363. https://doi.org/10.1002/1521-3927(20011101)22:16<1358::AID-MARC1358>3.0.CO;2-X
39. Hofmeier H, Hoogenboom R, Wouters MEL, Schubert US (2005) High molecular weight supramolecular polymers containing both terpyridine metal complexes and ureidopyrimidinone quadruple hydrogen-bonding units in the main chain. J Am Chem Soc 127(9):2913–2921. https://doi.org/10.1021/ja042919e
40. Winter A, Schubert US (2007) New polyester-based terpyridine macroligands and their blue iron(II) complexes. Macromol Chem Phys 208(18):1956–1964. https://doi.org/10.1002/macp.200700256
41. Heller M, Schubert US (2001) Terpyridines as supramolecular initiators for living polymerization methods. Macromol Symp 177(1):87–96. https://doi.org/10.1002/1521-3900(200201)177:1<87::AID-MASY87>3.0.CO;2-1
42. Pefkianakis EK, Tzanetos NP, Chochos CL, Andreopoulou AK, Kallitsis JK (2009) End-functionalization of semiconducting species with dendronized terpyridine-Ru(II)-terpyridine complexes. J Polym Sci Part A: Polym Chem 47(7):1939–1952. https://doi.org/10.1002/pola.23289
43. Zhou G-C, Harruna II (2005) Synthesis and characterization of bis(2,2′:6′,2″-terpyridine) ruthenium(II)-connected diblock polymers via RAFT polymerization. Macromolecules 38(10):4114–4123. https://doi.org/10.1021/ma047955c
44. Zhang L-W, Zhang Y-H, Chen Y-M (2006) Synthesis of bis(2,2′:6′,2″-terpyridine)-terminated telechelic polymers by RAFT polymerization and ruthenium-polymer complexation thereof. Eur Polym J 42(10):2398–2406. https://doi.org/10.1016/j.eurpolymj.2006.05.017
45. Lohmeijer BGG, Schubert US (2004) Expanding the supramolecular polymer LEGO system: nitroxide-mediated living free-radical polymerization as a tool for mono- and telechelic polystyrenes. J Polym Sci Part A: Polym Chem 42(16):4016–4027. https://doi.org/10.1002/pola.20189
46. Hawker CJ, Bosman AW, Harth E (2001) New polymer synthesis by nitroxide mediated living radical polymerizations. Chem Rev 101(12):3661–3688. https://doi.org/10.1021/cr990119u
47. Lohmeijer BGG, Schubert US (2005) The LEGO toolbox: supramolecular building blocks by nitroxide-mediated controlled radical polymerization. J Polym Sci Part A: Polym Chem 43(24):6331–6344. https://doi.org/10.1002/pola.20967
48. Ott C, Lohmeijer BGG, Wouters D, Schubert US (2006) Terpyridine-terminated homo and diblock copolymer LEGO units by nitroxide-mediated radical polymerization. Macromol Chem Phys 207(16):1439–1449. https://doi.org/10.1002/macp.200600269
49. Gohy J-F, Ott C, Hoeppener S, Schubert US (2009) Multicompartment micelles from a metallo-supramolecular tetrablock quatercopolymer. Chem Commun (40):6038–6040. https://doi.org/10.1039/b912504c
50. Ott C, Hoogenboom R, Hoeppener S, Wouters D, Gohy J-F, Schubert US (2009) Tuning the morphologies of amphiphilic metallo-supramolecular triblock terpolymers: from spherical micelles to switchable vesicles. Soft Matter 5(1):84–91. https://doi.org/10.1039/b813161a
51. Guillet P, Mugemana C, Stadler FJ, Schubert US, Fustin C-A, Bailly C, Gohy J-F (2009) Connecting micelles by metallo-supramolecular interactions: towards stimuli responsive hierarchical materials. Soft Matter 5(18):3409–3411. https://doi.org/10.1039/b910325b
52. Ott C, Hoogenboom R, Schubert US (2008) Post-modification of poly(pentafluorostyrene): a versatile "click" method to create well-defined multifunctional graft copolymers. Chem Commun (30):3516–3518. https://doi.org/10.1039/b807152g
53. Ott C, Ulbricht C, Hoogenboom R, Schubert US (2012) Metallo-supramolecular materials based on amine-grafting onto polypentafluorostyrene. Macromol Rapid Commun 33(6–7):556–561. https://doi.org/10.1002/marc.201100769

54. Lohmeijer BGG, Schubert US (2002) Supramolecular engineering with macromolecules: an alternative concept for block copolymers. Angew Chem Int Ed 41(20):3825–3829. https://doi.org/10.1002/1521-3773(20021018)41:20<3825::AID-ANIE3825>3.0.CO;2-6
55. Meier MAR, Hofmeier H, Abeln CH, Tziatzios C, Rasa M, Schubert D, Schubert US (2006) First GPC results of terpyridine based chain extended supramolecular polymers: comparison with viscosity and analytical ultracentrifugation. E-Polymers:16. https://doi.org/10.1515/epoly.2006.6.1.209
56. Lohmeijer BGG, Schubert US (2003) Water-soluble building blocks for terpyridine-containing supramolecular polymers: synthesis, complexation and pH stability studies of poly(ethylene oxide) moieties. Macromol Chem Phys 204(8):1072–1078. https://doi.org/10.1002/macp.200390078
57. Chiper M, Hoogenboom R, Schubert US (2008) Ruthenium(II) ions triggered direct supramolecular polymerization of bis-terpyridine poly(ethylene glycol): new insights on synthesis and optimization. E-Polymers:157. https://doi.org/10.1515/epoly.2008.8.1.1811
58. Hofmeier H, Schmatloch S, Wouters D, Schubert US (2003) Linear terpyridine-ruthenium(II) poly(ethylene glycol) coordination polymers. Macromol Chem Phys 204(18):2197–2203. https://doi.org/10.1002/macp.200300011
59. Lohmeijer BGG, Wouters D, Yin Z, Schubert US (2001) Block copolymer libraries using supramolecular strategies. PMSE Prepr 85:460–461
60. Gohy J-F, Lohmeijer BGG, Varshney SK, Décamps B, Leroy E, Boileau S, Schubert US (2002) Stimuli-responsive aqueous micelles from an ABC metallo-supramolecular triblock copolymer. Macromolecules 35(26):9748–9755. https://doi.org/10.1021/ma021175r
61. Chiper M, Hoogenboom R, Schubert US (2010) New terpyridine macroligands as potential synthons for supramolecular assemblies. Eur Polym J 46(2):260–269. https://doi.org/10.1016/j.eurpolymj.2009.10.021
62. Landsmann S, Winter A, Chiper M, Fustin C-A, Hoeppener S, Schubert US (2008) Poly(dimethylsiloxane)-substituted $2,2^{\prime}:6^{\prime},2^{\prime\prime}$-terpyridines: synthesis and characterization of new amphiphilic supramolecular diblock copolymers. Macromol Chem Phys 209(16):1666–1672. https://doi.org/10.1002/macp.200800219
63. Gohy J-F, Lohmeijer BGG, Alexeev AS, Wang X-S, Manners I, Winnik MA, Schubert US (2004) Cylindrical micelles from the aqueous self-assembly of an amphiphilic poly(ethylene oxide)-*b*-poly(ferrocenylsilane) (PEO-*b*-PFS) block copolymer with a metallo-supramolecular linker at the block junction. Chem Eur J 10(17):4315–4323. https://doi.org/10.1002/chem.200400222
64. Chiper M, Fournier D, Hoogenboom R, Schubert US (2008) Thermosensitive and switchable terpyridine-functionalized metallo-supramolecular poly(*N*-isopropylacrylamide). Macromol Rapid Commun 29(20):1640–1647. https://doi.org/10.1002/marc.200800339
65. Winter A, Wild A, Hoogenboom R, Fijten MWM, Hager MD, Fallahpour R-A, Schubert US (2009) Azido- and ethynyl-substituted $2,2^{\prime}:6^{\prime},2^{\prime\prime}$-terpyridines as suitable substrates for click reactions. Synthesis (9):1506–1512. https://doi.org/10.1055/s-0028-1088159
66. Xiao N, Chen Y, Shen X, Zhang C, Yano S, Gottschaldt M, Schubert US, Kakuchi T, Satoh T (2013) Synthesis of miktoarm star copolymer Ru(II) complexes by click-to-chelate approach. Polym J 45(2):216–225. https://doi.org/10.1038/pj.2012.100
67. Guerrero-Sanchez C, Lohmeijer BGG, Meier MAR, Schubert US (2005) Synthesis of terpyridine-terminated polymers by anionic polymerization. Macromolecules 38(25):10388–10396. https://doi.org/10.1021/ma051002c
68. Ott C, Kranenburg JM, Guerrero-Sanchez C, Hoeppener S, Wouters D, Schubert US (2009) Supramolecular assembly via noncovalent metal coordination chemistry: synthesis, characterization and elastic properties. Macromolecules 42(6):2177–2183. https://doi.org/10.1021/ma802298p
69. Ott C, Pavlov GM, Guerrero-Sanchez C, Schubert US (2009) Alternating terpyridine-end functionalized copolymers of styrene and diphenylethylene *via* anionic polymerization

techniques: a detailed characterization study. J Polym Sci Part A: Polym Chem 47(14):3691–3701. https://doi.org/10.1002/pola.23407
70. Pavlov GM, Amorós D, Ott C, Zaitseva II, Garcia de la Torre J, Schubert US (2009) Hydrodynamic analysis of well-defined flexible linear macromolecules of low molar mass. Macromolecules 42(19):7447–7455. https://doi.org/10.1021/ma901027u
71. Henderson IM, Hayward RC (2010) Synthesis of end-functionalized polystyrene by direct nucleophilic addition of polystyryllithium to bipyridine or terpyridine. Macromolecules 43(7):3249–3255. https://doi.org/10.1021/ma1001657
72. Schubert US, Hien O, Eschbaumer C (2000) Functionalized polymers with metal complexing segments: a simple and high-yield entry towards 2,2′:6′,2″-terpyridine-based oligomers. Macromol Rapid Commun 21(16):1156–1161. https://doi.org/10.1002/1521-3927(20001101)21:16<1156::AID-MARC1156>3.0.CO;2-O
73. Schubert US, Schmatloch S, Precup AA (2002) Access to supramolecular polymers: large scale synthesis of 4′-chloro-2,2′: 6′,2″-terpyridine and an application to poly(propylene oxide) telechelics. Des Monomers Polym 5(2–3):211–221. https://doi.org/10.1163/156855502760157935
74. Lohmeijer BGG, Schubert US (2003) Playing LEGO with macromolecules: design, synthesis and self-organization with metal complexes. J Polym Sci Part A: Polym Chem 41(10):1413–1427. https://doi.org/10.1002/pola.10685
75. Schubert US, Eschbaumer C, Andres PR, Hofmeier H, Weidl CH, Herdtweck E, Dulkheit E, Morteani A, Hecker NE, Feldmann J (2001) 2,2′:6′,2″-Terpyridine metal complexes as building blocks for extended functional metallo-supramolecular assemblies and polymers. Synth Metals 121(1–3):1249–1252. https://doi.org/10.1016/S0379-6779(00)01430-2
76. Meier MAR, Lohmeijer BGG, Schubert US (2003) Characterization of defined metal-containing supramolecular block copolymers. Macromol Rapid Commun 24(14):852–857. https://doi.org/10.1002/marc.200350031
77. Raşa M, Tziatzios C, Lohmeijer BGG, Schubert D, Schubert US (2006) Analytical ultracentrifugation studies on terpyridine-end-functionalized poly(ethylene oxide) and polystyrene systems complexed via Ru(II) ions. Progr Colloid Polym Sci 131:165–171. https://doi.org/10.1007/2882_015
78. Tziatzios C, Precup AA, Lohmeijer BGG, Börger L, Schubert US, Schubert D (2004) Dimerization of monofunctionalized poly(ethylene oxide) via metal–ligand interactions and hydrogen bonds. Progr Colloid Polym Sci 127:54–60. https://doi.org/10.1007/b98013
79. Schubert D, Tziatzios C, Schuck P, Schubert US (1999) Characterizing the solution properties of supramolecular systems by analytical ultracentrifugation. Chem Eur J 5(3):1377–1383. https://doi.org/10.1002/(SICI)1521-3765(19990503)5:5<1377::AID-CHEM1377>3.0.CO;2-H
80. Oudhoff KA, Schoenmakers PJ, Kok WT (2004) Characterization of metallo bis(terpyridine) diblock polymers by nonaqueous capillary zone electrophoresis. Chromatographia 60(7–8):475–480. https://doi.org/10.1365/s10337-004-0390-x
81. Hager MD, Greil P, Leyens C, van der Zwaag S, Schubert US (2011) Self-healing materials. Adv Mater:5424–5430. https://doi.org/10.1002/adma201003036
82. Yang Y, Urban MW (2013) Self-healing polymeric materials. Chem Soc Rev 42(17):7446–7467. https://doi.org/10.1039/C3CS60109A
83. Burnworth M, Tang L, Kumpfer JR, Duncan AJ, Beyer FL, Fiore GL, Rowan SJ, Weder C (2011) Optically healable supramolecular polymers. Nature 472(7343):334–337. https://doi.org/10.1038/nature09963
84. Ievins AD, Moughton AO, O'Reilly RK (2008) Synthesis of hollow responsive functional nanocages using a metal-ligand complexation strategy. Macromolecules 41(10):3571–3578. https://doi.org/10.1021/ma800047r
85. Gohy J-F, Lohmeijer BGG, Schubert US (2003) From supramolecular block copolymers to advanced nano-objects. Chem Eur J 9(15):3472–3479. https://doi.org/10.1002/chem.200204640

86. Fustin C-A, Lohmeijer BGG, Duwez A-S, Jonas AM, Schubert US, Gohy J-F (2005) Nanoporous thin films from self-assembled metallo-supramolecular block copolymers. Adv Mater 17(9):1162–1165. https://doi.org/10.1002/adma.200402073
87. Roy D, Brooks WLA, Sumerlin BS (2013) New directions in thermoresponsive polymers. Chem Soc Rev 42(17):7214–7243. https://doi.org/10.1039/c3CS35499g
88. Ziessel R, Grosshenny V, Hissler M, Stroh C (2004) cis-[Ru(2,2′:6′,2″-terpyridine)(DMSO) Cl_2]: useful precursor for the synthesis of heteroleptic terpyridine complexes under mild conditions. Inorg Chem 43(14):4262–4271. https://doi.org/10.1021/ic049822d
89. Underwood CC, Stadelman BS, Sleeper ML, Brumaghim JL (2013) Synthesis and electrochemical characterization of $[Ru(NCCH_3)_6]^{2+}$, tris(acetonitrile) tris(pyrazolyl)borate and tris (acetonitrile) tris(pyrazolyl)methane ruthenium(II) complexes. Inorg Chim Acta 405:470–476. https://doi.org/10.1016/j.ica.2013.02.027
90. Hamley IW (1998) The physics of block copolymers. Oxford Science Publications, Oxford
91. Gohy J-F (2005) Block copolymer micelles. Adv Polym Sci 190:65–136. https://doi.org/10.1007/12_048
92. Kataoka K, Harada A, Nagasaki Y (2012) Block copolymer micelles for drug delivery: design, characterization and biological significance. Adv Drug Deliv Rev 64:37–48. https://doi.org/10.1016/j.addr.2012.09.013
93. Riess G (2003) Micellization of block copolymers. Prog Polym Sci 28(7):1107–1170. https://doi.org/10.1016/S0079-6700(03)00015-7
94. Gohy J-F, Lohmeijer BGG, Schubert US (2002) Metallo-supramolecular block copolymer micelles. Macromolecules 35(12):4560–4563. https://doi.org/10.1021/ma012042t
95. Gohy J-F, Lohmeijer BGG, Varshney SK, Schubert US (2002) Covalent *vs.* metallo-supramolecular block copolymer micelles. Macromolecules 35(19):7427–7435. https://doi.org/10.1021/ma0204812
96. Mayer G, Vogel V, Lohmeijer BGG, Gohy J-F, van den Broek JA, Haase W, Schubert US, Schubert D (2004) Metallo-supramolecular block copolymer micelles: improved preparation and characterization. J Polym Sci Part A: Polym Chem 2(17):4458–4465. https://doi.org/10.1002/pola.20263
97. Regev O, Gohy J-F, Lohmeijer BGG, Varshney SK, Hubert DHW, Frederik PM, Schubert US (2004) Dynamic light scattering and cryogenic transmission electron microscopy investigations on metallo-supramolecular aqueous micelles: evidence of secondary aggregation. Colloid Polym Sci 282(4):407–411. https://doi.org/10.1007/s00396-003-0975-1
98. Vogel V, Gohy J-F, Lohmeijer BGG, van den Broek JA, Haase W, Schubert US, Schubert D (2003) Metallo-supramolecular micelles: studies by analytical ultracentrifugation and electron microscopy. J Polym Sci Part A: Polym Chem 41(20):3159–3168. https://doi.org/10.1002/pola.10902
99. Al-Hussein M, Lohmeijer BGG, Schubert US, de Jeu WH (2003) Melt morphology of polystyrene-poly(ethylene oxide) metallo-supramolecular diblock copolymer. Macromolecules 36(25):9281–9284. https://doi.org/10.1021/ma035183z
100. Al-Hussein M, de Jeu WH, Lohmeijer BGG, Schubert US (2005) Phase behavior of the melt of polystyrene-poly(ethylene oxide) metallo-supramolecular diblock copolymer with bulky counterions. Macromolecules 38(7):2832–2836. https://doi.org/10.1021/ma047784b
101. Lohmeijer BGG, Wouters D, Yin Z-H, Schubert US (2004) Block copolymer libraries: modular versatility of the macromolecular LEGO system. Chem Commun (24):2886–2887. https://doi.org/10.1039/b411777h
102. Fustin C-A, Guillet P, Misner MJ, Russell TP, Schubert US, Gohy J-F (2008) Self-assembly of metallo-supramolecular block copolymers in thin films. J Polym Sci Part A: Polym Chem 46(14):4719–4724. https://doi.org/10.1002/pola.22805
103. Guillet P, Fustin C-A, Lohmeijer BGG, Schubert US, Gohy J-F (2006) Study of the influence of the metal-ligand complex on the size of aqueous metallo-supramolecular micelles. Macromolecules 39(16):5484–5488. https://doi.org/10.1021/ma060929p

104. Guillet P, Fustin C-A, Wouters D, Hoeppener S, Schubert US, Gohy J-F (2009) Amphiphilic brushes from metallo-supramolecular block copolymers. Soft Matter 5(7):1460–1465. https://doi.org/10.1039/b817320f
105. Gohy J-F, Lohmeijer BGG, Schubert US (2002) Covalent *vs.* metallo-supramolecular block copolymer micelles. Macromol Rapid Commun 23(9):555–560. https://doi.org/10.1021/ma0204812
106. Gohy J-F, Lohmeijer BGG, Schubert US (2002) Reversible metallo-supramolecular block copolymer micelles containing a soft core. Macromol Rapid Commun 23(9):555–560. https://doi.org/10.1002/1521-3927(20020601)23:9<555::AID-MARC555>3.0.CO;2-K
107. Gilroy JB, Gädt T, Whittell GR, Chabanne L, Mitchels JM, Richardson RM, Winnik MA, Manners I (2010) Monodisperse cylindrical micelles by crystallization-driven living self-assembly. Nat Chem 2(7):566–570. https://doi.org/10.1038/nchem.664
108. Hailes RLN, Oliver AM, Gwyther J, Whittell GR, Manners I (2016) Polyferrocenylsilanes: synthesis, properties and applications. Chem Soc Rev 45(19):5358–5407. https://doi.org/10.1039/c6cS00155f
109. Lodge TP, Rasdal A, Li Z-B, Hillmyer MA (2005) Simultaneous, segregated storage of two agents in a multicompartment micelle. J Am Chem Soc 127(50):17608–17609. https://doi.org/10.1021/ja056841t
110. Gohy J-F, Willet N, Varshney SK, Zhang J-X, Jérôme R (2001) Core-shell-corona micelles with a responsive shell. Angw Chem Int Ed 40(71):3214–3216. https://doi.org/10.1002/1521-3773(20010903)40:17<3214::AID-ANIE3214>3.0.CO;2-F
111. Fustin C-A, Abetz V, Gohy J-F (2005) Triblock terpolymer micelles: a personal outlook. Eur Phys J E 16(3):291–302. https://doi.org/10.1140/epje/i2004-10086-0
112. Guillet P, Fustin C-A, Mugemana C, Ott C, Schubert US, Gohy J-F (2008) Tuning block copolymer micelles by metal-ligand interactions. Soft Matter 4(11):2278–2282. https://doi.org/10.1039/b808280d
113. Abd-El-Aziz AS, Shipman PO, Boden BN, NcNeil WS (2010) Synthetic methodologies and properties of organometallic and coordination macromolecules. Prog Polym Sci 35(6):714–836. https://doi.org/10.1016/j.progpolymsci.2010.01.004
114. Janini TE, Fattore JL, Mohler DL (1999) Rapid assembly of rigid rods by metal complexation of bis(terpyridyl) ligands. J Organomet Chem 578(1–2):260–263. https://doi.org/10.1016/S0022-328X(98)01110-3
115. Lohmeijer BGG (2004) Playing LEGO with macromolecules: connecting polymer chains using terpyridine metal complexes. PhD Thesis, Eindhoven University of Technology, Eindhoven
116. Ott C (2008) Advances in supramolecular polymer chemistry: well-defined terpyridine-functionalized materials. PhD Thesis, Eindhoven University of Technology, Eindhoven
117. Chiper M, Winter A, Hoogenboom R, Egbe DAM, Wouters D, Hoeppener S, Fustin C-A, Gohy J-F, Schubert US (2008) Synthesis and micellization of coil-rod-coil ruthenium (II) terpyridine assemblies. Macromolecules 41(22):8823–8831. https://doi.org/10.1021/ma801217v
118. Schlütter F, Pavlov GM, Gohy J-F, Winter A, Wild A, Hager MD, Hoeppener S, Schubert US (2011) Synthesis, characterization and micellization studies of coil-rod-coil and ABA ruthenium(II) terpyridine assemblies with π-conjugated electron acceptor systems. J Polym Sci Part A: Polym Chem 49(6):1396–1408. https://doi.org/10.1002/pola.24560
119. Gohy J-F, Chiper M, Guillet P, Fustin C-A, Hoeppener S, Winter A, Hoogenboom R, Schubert US (2009) Self-organization of rod-coil tri- and tetra-arm star metallo-supramolecular block copolymers in selective solvents. Soft Matter 5(15):2954–2961. https://doi.org/10.1039/b903111a
120. Constable EC (1995) Towards helical coordination polymers: molecular wires in chiral coats. Macromol Symp 98:503–524. https://doi.org/10.1002/masy.19950980141

121. De Greef TFA, Smulders MMJ, Wolffs M, Schenning APHJ, Sijbesma RP, Meijer EW (2009) Supramolecular polymerization. Chem Rev 109(111):5687–5754. https://doi.org/10.1021/cr900181u
122. Winter A, Hager MD, Schubert US (2012) Supramolecular polymers. In: Möller M, Matyjaszewski K (eds) Polymer science: a comprehensive reference, vol 5.13. Elsevier, Amsterdam, pp 269–310
123. Schmatloch S, González MF, Schubert US (2002) Metallo-supramolecular diethylene glycol: water-soluble reversible polymers. Macromol Rapid Commun 23(16):957–961. https://doi.org/10.1002/1521-3927(200211)23:16<957::AID-MARC957>3.0.CO;2-W
124. Schmatloch S, van den Berg AMJ, Alexeev AS, Hofmeier H, Schubert US (2003) Soluble high-molecular-mass poly(ethylene oxide)s via self-organization. Macromolecules 36(26):9943–9949. https://doi.org/10.1021/ma0350359
125. Constable EC, Housecroft CE, Smith CB (2003) Self-assembly of two discrete polynuclear iron(II) metallomacrocycles from a ligand containing two $2,2':6',2''$-terpyridine binding domains. Inorg Chem Commun 6(8):1011–1013. https://doi.org/10.1016/S1387-7003(03)00168-0
126. Raşa M, Lohmeijer BGG, Hofmeier H, Thijs HML, Schubert D, Schubert US, Tziatzios C (2006) Characterization of metallo-supramolecular block copolymers by analytical ultracentrifugation. Macromol Chem Phys 207(22):2029–2041. https://doi.org/10.1002/macp.200600235
127. Meier MAR, Wouters D, Ott C, Guillet P, Fustin C-A, Gohy J-F, Schubert US (2006) Supramolecular ABA triblock copolymers *via* a polycondensation approach: synthesis, characterization and micelle formation. Macromolecules 39(4):1569–1576. https://doi.org/10.1021/ma052045w
128. Chiper M, Meier MAR, Wouters D, Hoeppener S, Fustin C-A, Gohy J-F, Schubert US (2008) Supramolecular self-assembled Ni(II), Fe(II) and Co(II) ABA triblock copolymers. Macromolecules 41(8):2771–2777. https://doi.org/10.1021/ma0718954

Chapter 19
Three-Dimensional Coordination Polymers Composed of Luminescent Lanthanide Element Blocks

Yasuchika Hasegawa, Takayuki Nakanishi, and Yuichi Kitagawa

Abstract In this chapter, three-dimensional coordination polymers composed of luminescent lanthanide complexes (lanthanide element blocks: luminescent blocks) and organic joint parts (joint blocks) are introduced. The luminescent blocks are composed of lanthanide ions and low-vibrational frequency hfa (hexafluoroacetylacetonato)-ligand for suppression of vibrational relaxation. The joint blocks are containing phosphine oxide groups, which are attached with lanthanide ions, resulting in formation of asymmetric eight-coordination geometry of lanthanide coordination sites. The three-dimensional coordination polymers with characteristic structure (e.g., liner-typed, zigzag, two-dimensional, three-dimensional, and clustering structures) provide thermostable and strong-luminescence properties. They also show unique photophysical properties (temperature-sensitive and triboluminescence properties). The lanthanide coordination nanoparticles with good dispersibility in water and organic solvents are also applied for fabrication of optical devices. Three-dimensional coordination polymers composed of lanthanide element blocks are expected to open up frontier field of material and polymer science.

Keywords Coordination polymer · Lanthanide · Luminescence · Complex

19.1 Formation of Three-Dimensional Ln(III) Coordination Polymers

Coordination polymers composed of metal ions and organic joint parts have attracted considerable attention in the fields of coordination chemistry, inorganic chemistry, supramolecular chemistry, and polymer and material science. One-, two-, and three-

Y. Hasegawa (✉) · T. Nakanishi · Y. Kitagawa
Division of Applied Chemistry, Faculty of Engineering, Hokkaido University, Sapporo, Hokkaido, Japan
e-mail: hasegaway@eng.hokudai.ac.jp

Y. Chujo (ed.), *New Polymeric Materials Based on Element-Blocks*,
https://doi.org/10.1007/978-981-13-2889-3_19

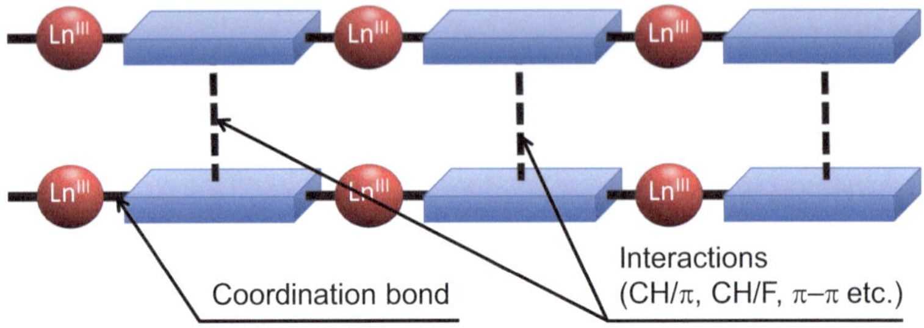

Fig. 19.1 Structural image of three-dimensional Ln(III) coordination polymer

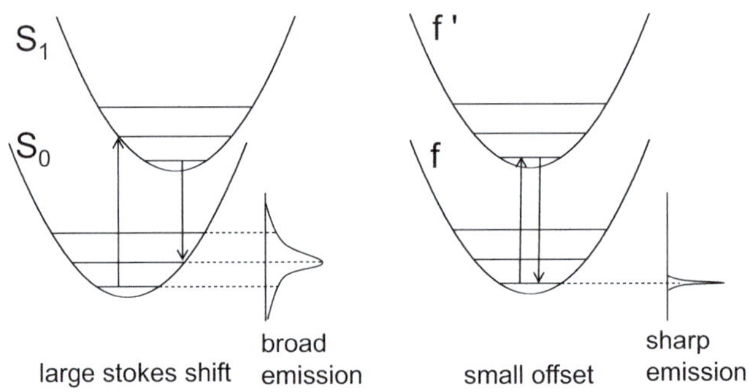

Fig. 19.2 Small offset and sharp emission spectrum of Ln(III)

dimensional alternating sequences of metal ions and organic joint parts (Fig. 19.1) provide remarkable characters as novel organic–inorganic hybrid materials with various structures and unique physical properties. In this chapter, three-dimensional coordination polymers composed of luminescent lanthanide complexes (lanthanide element blocks) and organic joint parts are introduced.

The lanthanides are composed of lanthanum (La) and 14 other elements (Ce, Pr, Nd, Pm, Sm, Eu, Gd, Tb, Dy, Ho, Er, Tm, Yb, and Lu) [1]. Lanthanides are typically represented by the symbol Ln. The lanthanides in the stable III oxidation state are simply characterized by the electronic structure in the 4f orbital. The 4f orbital is shielded by the outer $5s^2$ and $5p^2$ orbitals. In a configurational coordinate diagram, energy levels in 4f orbital appear as parallel parabolas (small offset case) because the 4f electrons are well shielded from outer filled $5s^2$ and $5p^2$ orbitals [2]. Thus, electronic transitions in absorption and emission processes show sharp spectral lines (Fig. 19.2). The sharp emission spectra of Ln(III) are useful application in photonic devices such as lasers, photo-amplified optical fibers, and displays.

The radiative emission of Ln(III) ions comes mainly from the electric dipole transition. The electric transitions in the inner 4f orbital of free ions are forbidden because they do not correspond to a change of parity. However, the transitions are partially allowed by mixing of the 4f orbital (odd parity) and 5d orbital (even parity) under the ligand field. Thus, the luminescence properties of lanthanide complexes

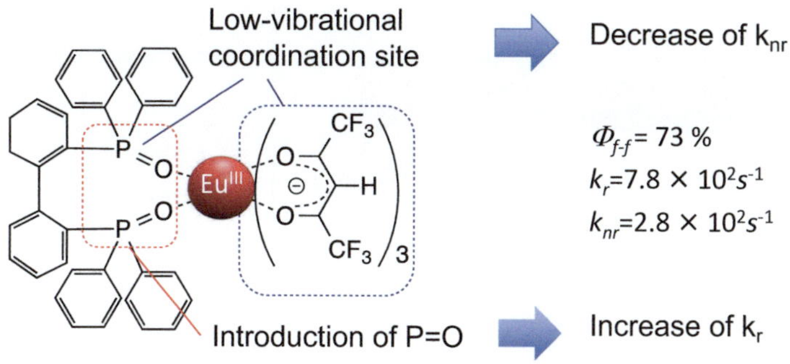

Fig. 19.3 Conceptual design of Eu(III) complex with hfa and phosphine oxide ligands

composed of Ln(III) ion and organic ligands are affected by their ligand filed, namely, coordination structure. The characteristic coordination structures and luminescence properties of lanthanide complexes are dominated by the steric and vibrational structures of the organic ligands [3]. We have reported on strong luminescent Ln(III) complexes with low-vibrational frequency hfa (hexafluoroacetylacetonato) and phosphine oxide ligands (Fig. 19.3). The coordination structure composed of hfa and phosphine oxide (P=O: 1125 cm^{-1}) ligands provides the Ln(III) complex with a high emission quantum yield and a relatively small non-radiative rate constant, k_{nr}. The coordination geometry of the Ln(III) complexes is categorized as square antiprism (8-SAP) and trigonal dodecahedron (8-TDH) without symmetry axis [4–6]. The characteristic coordination structures are composed of three hfa ligands and phosphine oxide ligands, which lead to reduction of the geometrical symmetry of the Ln(III) complex, and show relatively large radiative rate constants, k_r.

Three-dimensional ordering and assembling of strong luminescent Ln(III) complexes with hfa and phosphine oxide ligands are expected to open up a new field of material science. We here designed three-dimensional coordination polymers composed of luminescent lanthanide complexes as a lanthanide element blocks. In the next section, formation of polymer structures (liner, zigzag, two-dimensional, three-dimensional, and clustering structures) and their photophysical properties of lanthanide coordination polymers are introduced. Characteristic photo-functional properties of lanthanide coordination polymers (temperature-sensitive and triboluminescence properties) are also explained.

19.2 Polymer Structures and Photophysical Properties

19.2.1 Formation of Liner-Typed Polymer Structures

Introduction of aromatic aryl groups in the linker part of a lanthanide coordination polymer expected to provide construction of three-dimensional packing structures with intermolecular interactions such as CH/F, π–π, and CH/π interactions. Based on

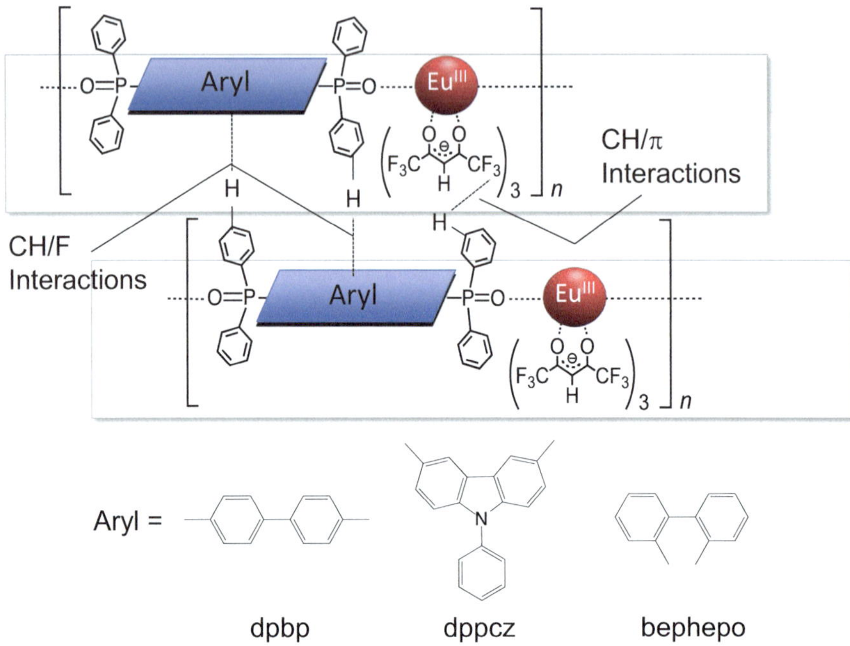

Fig. 19.4 Liner-typed Eu(III) coordination polymers

consideration of these chemical interactions, coordination polymers composed of Eu (III) and three types of liner-typed aryl units, $[Eu(hfa)_3(dpbp)]_n$, $[Eu(hfa)_3(dppcz)]_n$ (dpbp: 4,4′-bis(diphenyl-phosphoryl)biphenyl, and dppcz: 3,6-bis(diphenyl-phosphoryl)-9-phenylcarbazole), were prepared by the reaction of $[Eu(hfa)_3(H_2O)]_2$ with dpbp or dppcz in methanol under reflux (Fig. 19.4) [7].

The X-ray single-crystal analysis shows that the phosphine oxide ligand acts as a bidentate bridge between lanthanide ions in liner-typed one-dimensional polymeric chains. The coordination sites of $[Eu(hfa)_3(dpbp)]_n$ and $[Eu(hfa)_3(dppcz)]_n$ comprise three hfa ligands and two phosphine oxide units. X-ray analysis also reveals intermolecular interactions between one-dimensional polymeric chains. In one unit, two CH/F interactions and one CH/π interaction were identified for $[Eu(hfa)_3(dppcz)]_n$ and $[Eu(hfa)_3(dpbp)]_n$. These tight-binding structures are directly linked to their thermal stability. Thermogravimetric analysis (TGA) and differential scanning calorimetry (DSC) were conducted to estimate the thermal stability of the Eu(III) coordination polymers. The thermal decomposition points from the TGA thermograms were 308 and 300 °C for $[Eu(hfa)_3(dpbp)]_n$ and $[Eu(hfa)_3(dppcz)]_n$, respectively. In contrast, the thermal decomposition point of the reported Eu(III) complex, $[Eu(hfa)_3(biphepo)]_n$ (biphepo: 1,1′-biphenyl-2,2′-diylbis(diphenyl phosphine oxide), was reported to be 230 °C [5]. The high thermal stability of $[Eu(hfa)_3(dpbp)]_n$ and $[Eu(hfa)_3(dppcz)]_n$ is due to the tight-packing structure supported by a combination of CH/F and CH/π interactions. Combination of both CH/F and

CH/π interactions in coordination polymers is effective for the construction of thermostable organo Ln(III) materials.

$[Eu(hfa)_3(dpbp)]_n$ and $[Eu(hfa)_3(dppcz)]_n$ have high emission quantum yields ($\Phi_{4f\text{-}4f}$ $[Eu(hfa)_3(dpbp)]_n$ = 72%, $\Phi_{4f\text{-}4f}$ $[Eu(hfa)_3(dppcz)]_n$ = 83%). The non-radiative rate constants for the Eu(III) coordination polymers were approximately ten times smaller than that for $[Eu(hfa)_3(H_2O)_2]$. The smaller k_{nr} for the Eu (III) coordination polymers is attributed to the prevention of vibrational relaxation. Introduction of the low-vibrational frequency phosphine oxide ligand and aromatic aryl group for intermolecular interactions in the coordination polymer system is effective for the preparation of luminescent Ln(III) materials with high thermal stabilities and high emission quantum yields.

19.2.2 Formation of Zigzag Polymer Structures

Next, formation of specific intra-ligand charge transfer (ILCT) states via charge redistribution of the hfa ligands in Ln(III) coordination polymers for enhancement of the energy transfer efficiency is introduced. The ILCT states are found under the specific packing structures and are known to contribute to the photosensitized emission process in Ln(III) complexes. Eliseeva and co-workers recently reported that the formation of low-lying ILCT states contributes to the improvement of ligand-to-metal energy transfer efficiency [8, 9]. The formation of an ILCT band should affect the efficiency of energy transfer from ligands to Ln(III) ions in solid systems. A dense and tight coordination structure in crystal units may induce the formation of ILCT states, leading to high emission quantum yields. We consider that luminescent Ln(III) coordination polymers with a characteristic ILCT band can be constructed by a zigzag polymer structure.

We used hfa for ideal antenna ligands, and we designed novel thiophene-based bridging ligands for a close-packed coordination system. The novel zigzag packing system should induce more dense coordination structures for higher efficiencies of ligand-to-metal energy transfer due to the ILCT between antenna ligands. Thiophene derivatives were selected as ideal bridging ligands because of their aromaticity for rigidity, bite angles, and hetero S atom for a large dipole moment. Polar thiophene bridging ligands are expected to lead to a characteristic alternative orientation in polymer chains. Based on the molecular designs, $[Eu(hfa)_3(dpt)]_n$ and [Eu (hfa)$_3$(dpedot)]$_n$ (dpt: 2,5-bis(diphenyl-phosphoryl)thiophene) and (dpedot: 2,5-bis (diphenyl-phosphoryl)ethylenedioxythiophene) were prepared (Fig. 19.5) [10].

The crystal structures were determined to be typical eight-coordination with three hfa and two phosphine oxide ligands. The coordination $[Eu(hfa)_3(dpt)]_n$ showed highly ordered and densely packed structure with multiple intra- and intermolecular CH–F interactions. The intermolecular CH/π interactions were also found. In the case of $[Eu(hfa)_3(dpedot)]_n$, intramolecular π–π interactions were observed, and the number of intermolecular CH–F interactions was much smaller than that for [Eu (hfa)$_3$(dpt)]$_n$. The crystal densities for $[Eu(hfa)_3(dpt)]_n$ (d_{cal} = 1.77 gm^{-3}) and [Eu

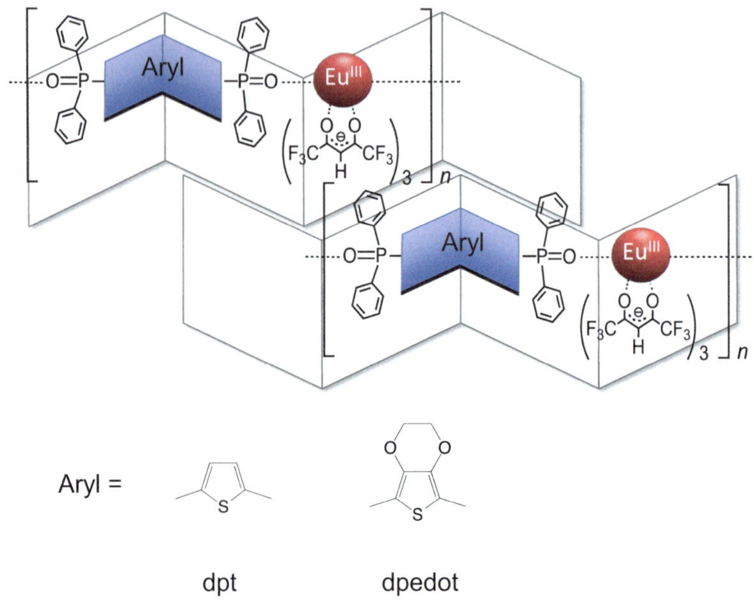

Fig. 19.5 Zigzag-typed Eu(III) coordination polymers

$(hfa)_3(dpedot)]_n$ (d_{cal} = 1.71 gm^{-3}) are larger than that of liner-typed Ln(III) coordination polymer $[Eu(hfa)_3(dpbp)]_n$ (d_{cal} = 1.66 gm^{-3}). The decomposition point(dp) of $[Eu(hfa)_3(dpt)]_n$ and $[Eu(hfa)_3(dpedot)]_n$ are estimated to be 322 and 260 °C. The coordination polymer $[Eu(hfa)_3(dpedot)]_n$ exhibits relatively low decomposition temperature as compared to the alternative orientation in single polymer chains. A small drop in the thermogravimetric curve is assumed to be responsible for degradation of the dioxane ring in bridging ligands.

The $\Phi_{4f\text{-}4f}$ values for $[Eu(hfa)_3(dpt)]_n$ and $[Eu(hfa)_3(dpedot)]_n$ were estimated to be 75% and 85%, respectively. The most remarkable point is that both polymers exhibited large $\Phi_{\pi\text{-}\pi^*}$ values (excitation at the ligand: ca. 60%) due to high photosensitized energy transfer efficiency η_{sens}. The η_{sens} of $[Eu(hfa)_3(dpt)]_n$ and that of $[Eu(hfa)_3(dpedot)]_n$ were estimated to be 80% and 66%. The high η_{sens} might be due to the formation of ILCT states induced by the densely packed zigzag orientation. To confirm the formation of ILCT states, an absorption spectrum in methanol and diffuse reflectance spectra in solid state were measured. In 10^{-5} M methanol solution, a π–π* transition band of hfa was observed at around 300 nm. In the solid state, both polymers exhibited large π–π* absorption bands at 330 nm. We also found characteristic low-lying bands at around 400 nm, which might be assigned to ILCT states caused by the densely packed coordination structure. The η_{sens} is considered to be linked to ILCT in solid state.

Along with the conventional molecular design of ligand fields around Ln(III) ions, we have provided guidelines for a densely packed assembly of luminescent Ln (III) coordination polymers. The reported strategy of constructing a zigzag coordination structure would be advantageous for efficient energy transfer and strong luminescence in the solid state.

19.2.3 Formation of Two-Dimensional Stacked Polymer Structures

The Eu(III) coordination polymers with triangular tridentate phosphine oxides are expected to provide two-dimensional close-packed structures and effective photosensitized luminescence such as zigzag coordination structure. Here, two-dimensional stacked coordination polymers composed of luminescent Eu $(hfa)_3$ and novel tridentate phosphine oxide ligands, tris(4-diphenylphosphorylphenyl)benzene (tppb) and tris(4-diphenyl-phosphorylphenyl)-1,3,5-triazine (tpptz), were prepared as a two-dimensional joint ligands (Fig. 19.6) [11].

Based on their elemental analyses, the ratio of $Eu(hfa)_3$ and tridentate phosphine oxides was estimated to be 3:2. The polymeric structures of $[Eu_3(hfa)_9(tppb)_2]_n$ and $[Eu_3(hfa)_9(tpptz)_2]_n$ were characterized using XRD and ESI-MS. The thermal stability of $[Eu_3(hfa)_9(tppb)_2]_n$ and $[Eu_3(hfa)_9(tpptz)_2]_n$ was evaluated using TGA. The decomposition temperatures for $[Eu_3(hfa)_9(tppb)_2]_n$ and $[Eu_3(hfa)_9(tpptz)_2]_n$ were estimated to be 354 and 355 °C, respectively. These decomposition temperatures are higher than those of corresponding previous liner-typed $[Eu(hfa)_3(dpbp)]_n$ (dp = 308 °C) and zigzag-typed dpt (dp = 322 °C). The increase in the decomposition temperatures of $[Eu_3(hfa)_9(tppb)_2]_n$ and $[Eu_3(hfa)_9(tpptz)_2]_n$ is caused by tight-packing structures of Eu(III) coordination polymers.

Remarkable absorption shoulder bands of solid-state $[Eu_3(hfa)_9(tppb)_2]_n$ at around 400 nm were observed. The characteristic absorption shoulder band could

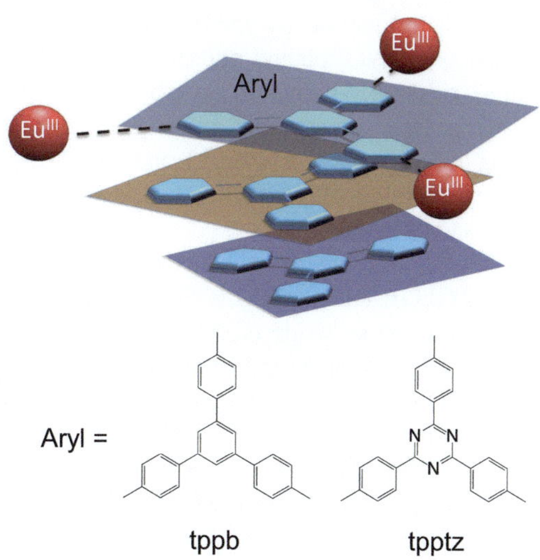

Fig. 19.6 Two-dimensional stacked Ln(III) coordination polymers

be assigned to the intra-ligand charge transfer (ILCT) via charge redistribution of the hfa ligands. The ILCT state via charge redistribution is formed in close-packing crystals. Observation of the characteristic bands at around 350 nm in solid is significant for the formation of close-packing structures in $[Eu_3(hfa)_9(tppb)_2]_n$ crystals.

The $\Phi_{4f\text{-}4f}$ of $[Eu_3(hfa)_9(tppb)_2]_n$ was estimated to be 82% in solid state, which is one of the best luminescence performances among Eu(III) complexes and coordination polymers. The $\Phi_{4f\text{-}4f}$ of $[Eu_3(hfa)_9(tpptz)_2]_n$ was found to be 78%. These high emission quantum yields are caused by small k_{nr} values of Eu(III) coordination polymers with triangular tridentate phosphine oxides. The non-radiative rate constant k_{nr} is generally related to the thermal relaxation from the excited state, which is dependent on the vibrational structure of the luminescent materials. The remarkably small k_{nr} constants of Eu(III) coordination polymers with triangular spacers may be induced by the close-stacking structure of the two-dimensional tppb and tpptz ligands in the crystals. The close-packing structure of $[Eu_3(hfa)_9(tpptz)_2]_n$ might also produce an asymmetric coordination structure at around the Eu(III) ions with molecular strain. In particular, the molecular strain in $[Eu_3(hfa)_9(tppb)_2]_n$ forms an ILCT band and creates effective photosensitized energy efficiency ($[Eu_3(hfa)_9(tppb)_2]_n$: $\eta_{sen} = 78\%$, $[Eu_3(hfa)_9(tpptz)_2]_n$:$\eta_{sen} = 48\%$]. The synergetic effect on the tight-packing structure of Eu(III) coordination polymers with rigid triangular spacers leads to an enhancement in structural stability, high emission quantum yield, and effective energy transfer efficiency.

19.2.4 Formation of Three-Dimensional Networked Polymer Structures

Eu(III) coordination polymers with hfa and carboxyphosphine oxide (CPO: 4-carboxyphenyl diphenyl phosphine oxide/TCPO: 4,4′,4″-tricarboxyphenyl phosphine oxide) were prepared. Coordination of the phosphine oxide parts in CPO and TCPO as three-dimensional joint ligands provides a low-vibrational frequency for strong luminescence. The CPO and TCPO ligands are also designed to include carboxy groups for construction of the thermostable Ln-MOF structure reported by Hong and co-workers [12].

The three-dimensional networked Ln(III) coordination polymers $Eu(hfa)_x(TCPO)_y$ and $Eu(hfa)_x(CPO)_y$ were synthesized by the complexation of the carboxyphosphine oxide (CPO or TCPO) with $Eu(hfa)_3(H_2O)_2$ in methanol under reflux. The phosphine oxide parts ($P = O$) and the carboxy groups (COO^-) in CPO and TCPO ligands effectively promote the formation of polymeric structures. The x and y in formulas in $Eu(hfa)_x(CPO)_y$ and $Eu(hfa)_x(TCPO)_y$ are defined $0 < x < 1$ and $0 < y < 3$. We estimated $x = 0.38$, $y = 2.12$ in $Eu(hfa)_x(CPO)_y$ and $x = 0.03$, $y = 1.92$ in $Eu(hfa)_x(TCPO)_y$ using EDX data [13].

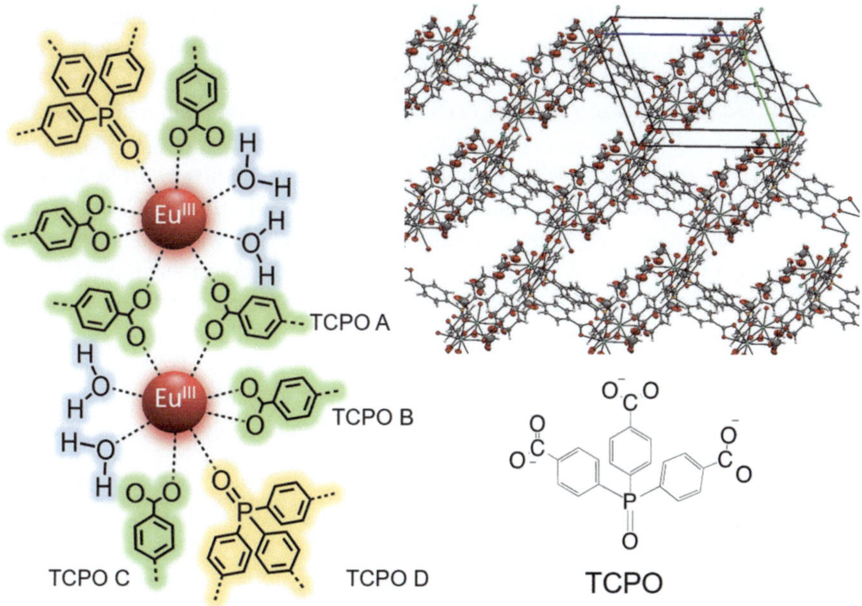

Fig. 19.7 Three-dimensional networked Ln(III) coordination polymer

In order to identify the structure of Eu(hfa)$_x$(TCPO)$_y$, we tried to measure by single-crystal X-ray structure analysis. The structure was determined to be eight-coordinated structure with two water molecules and five TCPO ligands. The two TCPO ligands show bidentate bridged connection between two Eu(III) ions (TCPO A in Fig. 19.7). Two TCPO ligands show bidentate (TCPO B) and monodentate (TCPO C) connection in one Eu(III) ion. Final TCPO ligand is attached to one Eu (III) ion by P=O group (TCPO D). The Eu(hfa)$_x$(TCPO)$_y$ crystal provides three-dimensional network structure.

This single crystal is including four methanol molecules in one unit. The two waters and four methanol molecules in one unit are removed after heat treatment (90 °C, 2 h, under reduced pressure). The fragment peaks of Eu(hfa)$_x$(CPO)$_y$ and Eu (hfa)$_x$(TCPO)$_y$ in the FAB-MS spectra agree with those calculated for $[Eu_2(hfa)_3(CPO)_2]^+$ and $[Eu(hfa)_2(TCPO)]\cdot 5H_2O]^+$ fragments, respectively. The EDX measurements indicated the percentage of hfa moieties in Eu(hfa)$_x$(CPO)$_y$ and Eu(hfa)$_x$(TCPO)$_y$ were 10.8% and 0.89%, respectively. We propose that the small amount of hfa molecules attached on the crystal surface. The hfa molecules on the surface were successfully detected by ionized-fragment information using FAB-MS spectrum. The decomposition temperature of Eu(hfa)$_x$(TCPO)$_y$ was 450 °C. We cannot observe the elimination of solvent from the material. This result indicates that Eu(hfa)$_x$(TCPO)$_y$ have no solvent in the structure after heat treatment. Therefore, XRD measurements of Eu(hfa)$_x$(TCPO)$_y$ were kept under 450 °C. The decomposition temperature of Eu(hfa)$_x$(TCPO)$_y$ is the highest among the organo-Eu

(III) luminophores with photosensitized hfa moieties. Thus, a three-dimensional Eu (III) coordination polymer with extra-high thermostability was successfully synthesized.

Bright red luminescence $Eu(hfa)_x(TCPO)_y$ heated on a hot plate under UV light irradiation ($\lambda = 365$ nm) was successfully observed from $Eu(hfa)_x(TCPO)_y$ under 400 °C. Thus, $Eu(hfa)_x(TCPO)_y$ exhibits both effective photosensitized luminescence and thermostability. The emission quantum yield of $Eu(hfa)_x(TCPO)_y$ excited at 355 nm ($\Phi_{\pi\text{-}\pi^*}$: $\pi - \pi^*$ transition band of hfa moieties) was also measured to calculate the energy transfer efficiency (η_{sens}), which was determined as 34%. The energy transfer efficiency of $Eu(hfa)_x(TCPO)_y$ ($\eta_{sens} = 59\%$) is larger than that of the previously reported thermostable liner-typed Eu(III) coordination polymers, $[Eu(hfa)_3(dpbp)]_n$ ($\eta_{sens} = 40\%$). The luminescence of $Eu(hfa)_x(TCPO)_y$ is due to photosensitized energy transfer from hfa moieties to Eu(III) ions. The three-dimensional networked structure in Ln(III) coordination polymer is an effective hyper-thermostability and effective photosensitized luminescence.

19.2.5 Formation of Metal-Jointed Structure

In the previous study, organic molecules were mainly used for bridge ligands of the lanthanide coordination polymer. We here have attempted to introduce the transition metal complex as bridge ligands for development of novel lanthanide coordination polymer. Here, luminescent Eu(III) coordination polymer $[Eu(hfa)_3(dppy)_2ZnCl_2]n$ (dppy: 4-pyridyldiphenylphosphane oxide) was synthesized by the complexation of $Eu(hfa)_3(H_2O)_2$ with $ZnCl_2(dppy)_2$ (Fig. 19.8) [14].

The photophysical properties were estimated using the emission spectra, the emission lifetimes, and the emission quantum yields. The emission quantum yields are larger than that of precursor $Eu(hfa)_3(H_2O)_2$. The emission quantum yield of Eu–Zn was found to be 59%. Luminescent lanthanide coordination polymers cross-linked with metal complexes are expected to open up new coordination polymers in the frontier field of material science.

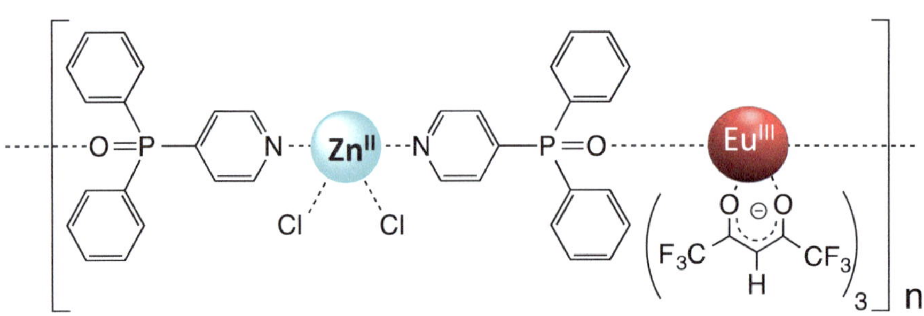

Fig. 19.8 Metal-joined Ln(III) coordination polymer

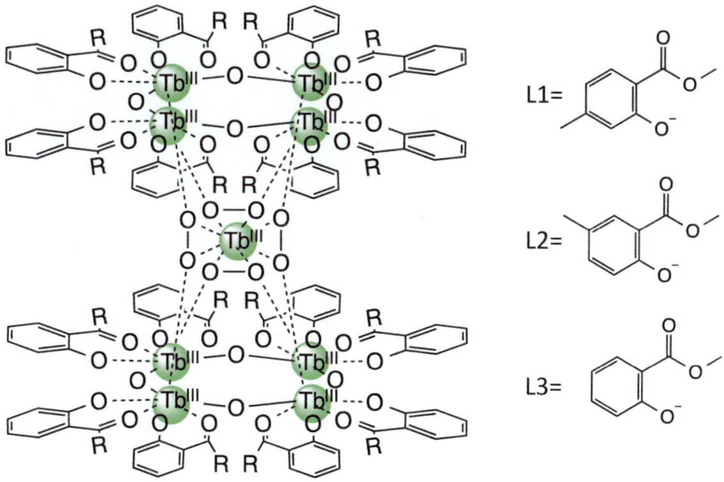

Fig. 19.9 Nona-nuclear Tb(III) coordination clusters

19.2.6 Formation of Clustering Structure

Polynuclear Ln(III) cluster composed of nine Ln(III) ions and organic joint ligands is regarded as a three-dimensional clustering-structured Ln(III) coordination polymer. Nona-nuclear Tb(III) cluster with a rigid Tb–O–Tb lattice center containing 16 methyl salicylate ligands was prepared by the reaction of terbium nitrate with salicylates in methanol (Fig. 19.9).

The nona-nuclear Tb(III) clusters showed characteristic sandglass-shaped structures composed of 9 Tb(III) ions, 16 salicylate ligands, 8 bridged μ_3-OH^- parts, and 2 μ_4-OH^- parts. All Tb(III) ions are surrounded by eight oxygen, which is typical coordination number of lanthanide(III) complexes.

Optical Faraday effect of nona-nuclear Tb(III) cluster was observed in the visible region. The optical Faraday effect rotates the plane of polarized light in linear proportion to the component of the magnetic field in the direction of propagation. The Faraday effect is important for the construction of optical isolators for fiber-optic telecommunication systems. The magneto-optical properties are induced in cubic Ln–X lattices. The Ln–O ordering bonds in polynuclear lanthanide clusters provide effective optical Faraday effect. This is the first observation of optical Faraday rotation in Ln(III) coordination polymer. The nona-nuclear Tb(III) clusters are much larger than those of previous Tb(III) oxide glass [15].

The chiral nona-nuclear Tb(III) clusters Tb-sal-(R) and (S) (sal-(R/S) = (R/S)-2-butyl salicylate) were also found to exhibit a unique magneto-optical property: the Faraday effect (Fig. 19.10). The clusters were composed of 9 Tb(III) ions bridged by 10 μ-OHs and 16 chiral salicylic acid esters. The Faraday rotation angle of Tb-sal-(R) was greater than that of Tb-sal-(S) indicating that the Faraday effect was affected by the chirality of the Tb(III) clusters. The chiroptical properties of the

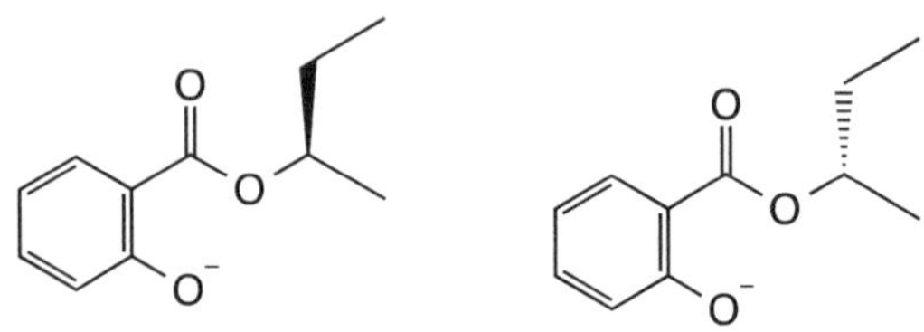

Fig. 19.10 Chiral ligands for chiral Tb(III) clusters

Tb(III) clusters were estimated using circular dichroism and circularly polarized luminescence [16, 17].

The nona-nuclear Tb(III) clusters also show green luminescence under irradiation at salicylate joint ligands. The emission quantum yield of Tb-L_1 ($\Phi_{\pi-\pi^*} = 31\%$) was 13 times larger than that of Tb-L_2 ($\Phi_{\pi-\pi^*} = 2.4\%$) and approximately 5 times larger than that of Tb-L_3 ($\Phi_{\pi-\pi^*} = 6.7\%$) [18]. Tb-L_2 showed a more than 4-fold increase in lifetime and 13-fold increase in quantum yield compared with Tb-L_1 and a 3-fold increase in lifetime and 4-fold increase in quantum yield compared with Tb-L_3. DFT calculations revealed that the position of the methyl group is directly linked to the HOMO and LUMO of the ligands. The introduction of the methyl group and variation of its position on the aromatic ring can significantly alter the photophysical properties of the nona-nuclear Tb(III) cluster, particularly the luminescence efficiency.

19.3 Characteristic Photo-Functional Properties

19.3.1 Temperature-Sensitive Property

In 2003, Amao and co-workers reported the first temperature-sensitive dye that employed an Eu(III) complex in a polymer film [19]. Khalil et al. demonstrated the high performance of an Eu(III) complex for a temperature-sensitive paint (temperature sensitivity: $4.42\%^{\circ}C^{-1}$) [20]. We have reported a Tb(III) complex, $Tb(hfa)_3(H_2O)_2$, that is suitable as a temperature-sensing probe since it exhibits effective energy back transfer (BEnT) from the emitting level of the Tb(III) ion to the excited triplet state of the hfa ligand [21]. Since BEnT depends on the energy barrier of the process, the emission intensity varies with temperature.

Here, we consider that introducing Tb(III) ion and hfa ligands to coordination polymer frameworks will produce a Tb(III) coordination polymer that can be used as a temperature-sensing probe. The triplet state of hfa (22,000 cm^{-1}) is very close to the emitting level of the Tb(III) ion (20,500 cm^{-1}), resulting in effective EnT1 and BEnT and thus high-performance thermo-sensing dyes (Fig. 19.9). Second, we attempted to impart ratio metric temperature sensing by using luminescent Eu(III) and Tb(III) ions in the frameworks of the coordination polymer to realize a high thermo-sensing ability. Two independent emission bands are expected to enable more accurate thermal measurements than a previous single lanthanide complex [22–26]. Their emission intensities based on the energy transfer from Tb(III) to Eu

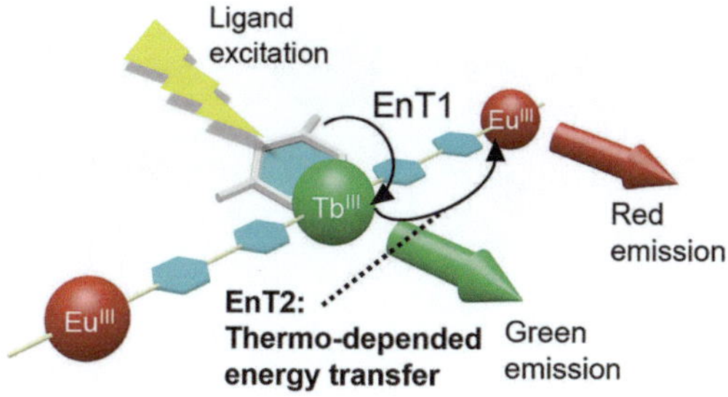

Fig. 19.11 Temperature-sensitive Tb(III)/Eu(III) coordination polymer

(III), EnT2 (Fig. 19.11), may also be dependent on the temperature. Based on the photophysical concepts, thermo-sensing dyes that consist of a lanthanide coordination polymer, $[Tb(hfa)_3(dpbp)]_n$ and $[Tb_{0.99}Eu_{0.01}(hfa)_3(dpbp)]_n$ (dpbp: 4,4′-bis (diphenyl-phosphoryl) biphenyl, Ln=Eu, Tb), were prepared [27].

$[Tb_{0.99}Eu_{0.01}(hfa)_3(dpbp)]_n$ has a higher-temperature sensitivity (0.83% C^{-1}) than $[Tb(hfa)_3(dpbp)]_n$ (0.64% C^{-1}). This result indicates that energy is transferred to both the excited triplet state of the hfa ligands (BEnT) and to the Eu(III) ion from the emitting level of the Tb(III) ion. The energy transfer efficiency $\eta_{Tb\text{-}Eu}$ between Tb(III) and Eu(III) ion incoordination polymer can be estimated using the emission lifetime decay profiles. The values of $\eta_{Tb\text{-}Eu}$ at 200, 250, 275, and 300 K are estimated to be 1%, 19%, 26%, and 38%, respectively. $\eta_{Tb\text{-}Eu}$ increases with increasing temperature. The energy transfer occurs to both the excited triplet state of hfa ligands and to the Eu(III) ion.

We also demonstrated that the luminescence reversibly undergo repeated thermo-cycles. The reversible changes of emission intensity ratio of Eu(III) and Tb(III) (I_{Eu}/I_{Tb}) of $[Tb_{0.99}Eu_{0.01}(hfa)_3(dpbp)]_n$ are observed by the alternative thermo-cycles in the range of 300 and 400 K. The changes in the emission intensity ratio are stably repeated between the I_{Eu}/I_{Tb} values of 1.0 (300 K, yellow emission) and 8.7 (400 K, red emission). $[Tb_{0.99}Eu_{0.01}(hfa)_3(dpbp)]_n$ also exhibits red emission under UV irradiation, even at 500 K. We thus successfully synthesized an effective luminophore with a wide temperature-sensing range of 200–500 K. In the future, the chameleon luminophores $[Tb_{0.99}Eu_{0.01}(hfa)_3(dpbp)]_n$ are expected to be promising candidates for temperature-sensitive dyes, which are used for temperature distribution measurements of material surfaces such as planes, cars, chemical plans, etc. Such lanthanide coordination polymers with thermo-sensing properties have the potential to open up new fields in supramolecular chemistry, polymer science, and molecular engineering [27, 28].

19.3.2 Glass Transition Property

We recently succeeded in synthesizing luminescent and amorphous Ln(III) coordination polymer composed of Eu(III) ions, light-harvesting hfa ligands, and 120-°-angled bridging ligands with ethynyl groups (Fig. 19.12, m-dpeb) [29]. Quantum

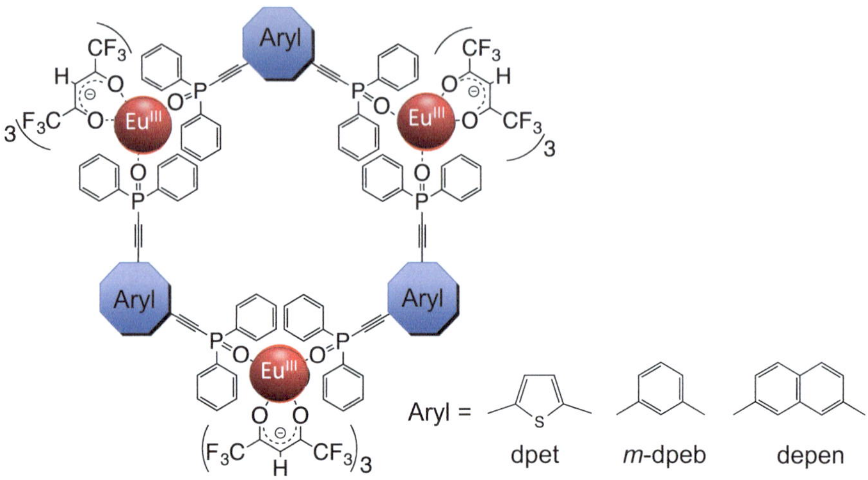

Fig. 19.12 Ln(III) coordination polymers with glass-transition property

calculations and mass spectrometry revealed the formation of trimer structure with pseudo-C_3 axis. The glass formability should be caused by the thermodynamically non-equilibrium states that arise from multiple quasi-stable states, suppressing easy crystal packing to form amorphous solid without polymer matrices. Systematic construction of Ln(III) coordination compounds with glass-forming ability by simply altering the aryl cores of organic bridging ligands to broaden the range of molecular designs has been demonstrated. We also prepared three bidentate phosphine oxide ligands with thienyl and naphthyl cores (Fig. 19.12: dpet, depen) [30].

The glass-transition temperatures were clearly identified for all compounds, which were characteristic of amorphous molecular materials. The $[Eu(hfa)_3(dpen)]_3$ showed the highest T_g ($T_g = 87$ °C) among these compounds. We consider this to be responsible for the largest molecular weight and availability of aromatic surfaces for π-stacking of the bridging ligand, dpen. The ethynyl groups in bridging ligands may also play an important role in amorphous formation by suppressing tight-binding interaction and crystallization of assembled Eu(III) coordination compounds. The reported thiophene-linked Eu(III) coordination compound without ethynyl groups $[Eu(hfa)_3(dpt)]_n$ (dpt: 2,5-bis(diphenyl-phosphoryl)thiophene) [10] exhibits no glass-transition point. The glass-forming ability was thus systematically introduced by bent-angled bridging ligands with ethynyl groups, which made more irregular-shaped molecules harder to crystallize and more likely to form amorphous glass. We consider that trimers do not form longer-range bonding or ordering out of each structure. Since the present Ln(III) coordination compounds exhibit broken C_3 symmetrical structures with lack of intermolecular-linking, the stability at high temperature may be caused by stable coordination bonds in discrete trimer complexes. The $\Phi_{4f\text{-}4f}$ of $[Eu(hfa)_3(dpen)]_3$ and $[Eu(hfa)_3(m\text{-}dpeb)]_3$ were estimated to be 46% and 72%, respectively. Unfortunately, $[Eu(hfa)_3(dpet)]_3$ exhibited weak red luminescence along with the luminescence from bridging ligands in the visible region.

We also prepared a Ln(III)-mixed coordination glass [Tb,Eu(hfa)$_3$(m-dpeb)]$_3$ (Tb/Eu = 50) to introduce the temperature-responsive emission property in glassy state. The emission spectra were recorded in the range of 100–400 K, and corresponding green, yellow, orange, and red photoluminescence of amorphous glass was observed. We estimated the emission intensity ratios of Tb(III) and Eu (III) emission bands (543 nm and 613 nm, respectively). The Ln(III)-mixed coordination glass was found to show higher-temperature sensitivity (0.92% K^{-1}) than that of reported [Tb,Eu(hfa)$_3$(dpbp)]$_3$ (0.83% K^{-1}). We consider this to be responsible for the lower-energy barrier in linker-mediated energy transfer from Tb(III) to Eu (III) ions [31].

19.3.3 Triboluminescence I

Triboluminescence is the emission of light originating from mechanical stress on bulk solid materials [32, 33]. To date, a number of different types of materials that exhibit triboluminescence, such as organic crystals, polymers, and metal complexes, have been studied. While there have been extensive discussions on the origin of triboluminescence, some studies have indicated a contribution of the piezoelectric effect upon breaking non-centrosymmetric bulk crystals [34, 35]. Some lanthanide complexes have also been studied for their triboluminescence in the crystalline state. Strongly luminescent lanthanide coordination polymer crystals with non-centrosymmetric structure are expected to show efficient triboluminescence because of the generation of opposite electric charges on opposing faces of cracks perpendicular to their polar axis. Brilliant triboluminescence from lanthanide coordination polymer crystals with an ideal non-centrosymmetric structural network, [Eu (hfa)$_3$(bypypo)]$_n$ (bypypo: 3,3 -bis(diphenyl-phosphoryl)-2,2 -bipyridine), was observed (Fig. 19.13) [36].

The packing structure of [Eu(hfa)$_3$(bypypo)]$_n$ using X-ray single-crystal analysis is non-centrosymmetric *Cc*. This space group is suitable for piezoelectricity and triboluminescence. The intense triboluminescence seems to originate from the non-centrosymmetric structure along with the polymer-like structure.

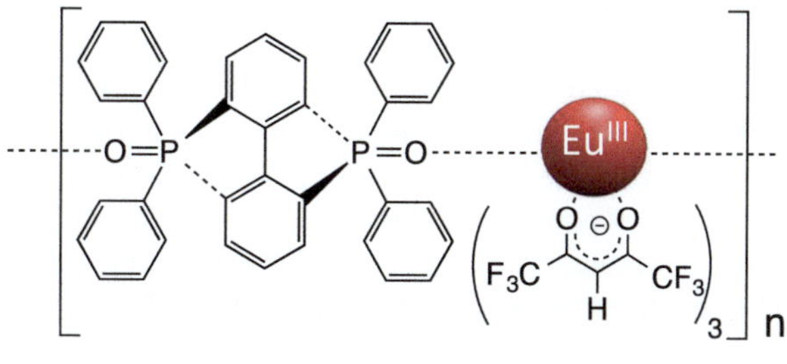

Fig. 19.13 Chemical structure of [Eu (hfa)$_3$(bypypo)]$_n$

The photoluminescence quantum yield of the $[Eu(hfa)_3(bypypo)]_n$ crystals excited at 380 nm is as high as 61%. The novel lanthanide coordination polymer, with a specific coordination structure of low-vibrational frequency and poly-Eu-BIPYPO coordination networks, exhibits intense triboluminescence upon breaking, which is clearly observed even at ambient temperature in daylight.

Previous introduced two-dimensional Eu(III) coordination polymer $[Eu_3(hfa)_9(tppb)_2]_n$ also shows effective triboluminescence [11]. The spectral shape of the electric dipole transition band (613 nm: $^5D_0$-$^7F_2$ transition) of the triboluminescence spectrum was different from that of photoluminescence (excited at 365 nm: UV-LED). We consider that the coordination symmetry of the luminescent Eu(III) center upon the breaking crystals by the rubbing process (triboluminescence) is different from that of photoluminescence of normal bulk crystals. To estimate the TL efficiency under an argon atmosphere, we introduced of shockwave irradiation using a nanosecond laser pulse (Nd: YAG, l=1064 nm, fwhm = 5 nm, pulse energy = 0.1 mJ) for high-speed crushing of the crystals. The TL lifetime is calculated to be 0.57 ms. The TL efficiency of $[Eu_3(hfa)_9(tppb)_2]_n$ using lifetime and emission spectra was calculated to be 49%. The smaller k_r value for TL compared to photoluminescence indicates that the coordination symmetry of the triboluminescent Eu(III) center is different from that of the luminescent Eu(III) center [6]. The k_{nr} for triboluminescence is four times larger than that for photoluminescence in $[Eu_3(hfa)_9(tppb)_2]_n$ crystals. These results indicate that mechanical crushing of molecular crystals leads to enhancement of the non-radiative transition in the quantum process.

19.3.4 Triboluminescence II

We consider that strong TL Ln(III) coordination polymers can be systematically constructed by introducing disordered arrangement of substituents between single polymer chains in the solid state. Novel Ln(III) coordination polymers composed of Ln(III) ions, hfa, and furan-based bridging ligands (2,5-bis(diphenyl-phosphoryl) furan = dpf) are introduced to provide the disordered face-to-face arrangement of CF_3 substituents. The polar character and small aromatic ring of the furyl bridging ligand prevent the polymer chains from forming highly ordered alternate packing structures (Fig. 19.14). Based on the molecular design, $[Eu(hfa)_3(dpf)]_n$ was prepared [37].

A large spectral difference between TL and PL is confirmed in Tb(III)/Eu(III) mixed polymers using a charge-coupled detector (CCD) system. The overall emission quantum yield excited by π–π* bands of hfa ligands ($\Phi_{\pi-\pi^*} = 64\%$) and energy transfer efficiency ($\eta_{sens} = 88\%$) of $[Eu(hfa)_3(dpf)]_n$ was higher than those of previously reported solid-state Eu(III) coordination polymers [10]. The intrinsic emission quantum yield ($\Phi_{4f\text{-}4f} = 73\%$) was comparable to the previous compounds. The enhanced radiative rate constant ($kr = 1.0 \times 10^3\ s^{-1}$) in $[Eu(hfa)_3(dpf)]_n$ was

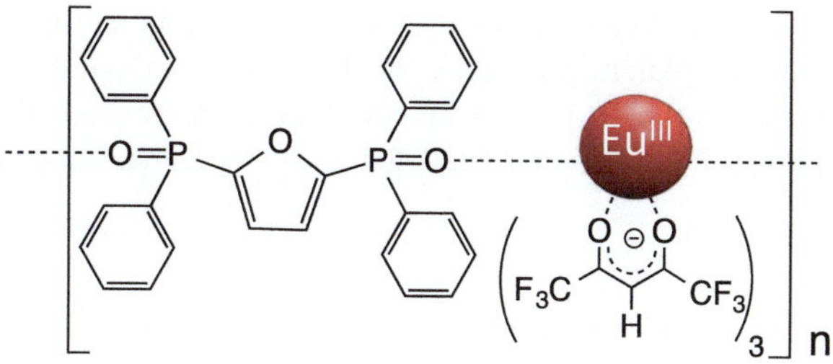

Fig. 19.14 Chemical structure of $[Eu(hfa)_3(dpf)]_n$

due to the polar character of bridging ligands and asymmetric coordination geometry without an inversion center around Eu(III) ions.

The coordination polymers $[Tb/Eu(hfa)_3(dpf)]_n$ (Tb/Eu = 1, 10) were also prepared, and the obtained compounds exhibited strong TL and PL under UV irradiation and grinding. The emission intensity of Eu(III) ions is higher than that of Tb(III) ions in the Tb/Eu = 1 compound under UV irradiation, resulting in reddish-orange PL. The PL colors of Tb(III)/Eu(III) mixed coordination polymers are dominated by Eu(III)-centered emission, since the Tb(III)-centered emission is affected by both the BET process and excitation energy transfer from Tb(III) to Eu(III) ions at room temperature [27].

Interestingly, the observed TL colors were clearly different from those of PL. The Tb/Eu = 1 compound exhibited yellow TL and reddish-orange PL, while the Tb/Eu = 10 compound exhibited green TL and greenish-yellow PL. These spectral differences between TL and PL were the most remarkable among previously reported TL- and PL-active compounds. The TL colors of these coordination polymers might not be explained by the simple excitation mechanisms in PL. Because the TL colors correspond to the Tb/Eu mixture ratios, TL would be dominated by direct excitation of Ln(III) ions; however, the possibility of antenna ligand (hfa)-mediated emission cannot be ruled out. The hfa-centered blue TL and PL of $[Gd(hfa)_3(dpf)]_n$ were also observed in the range from 450 to 550 nm with identical spectral shapes, indicating the formation of excited states of hfa ligands under grinding. In the case of $[Ln(hfa)_3(dpf)]_n$ (Ln = Tb, Eu), hfa-centered emission is quenched through efficient ligand-to-Ln(III) energy transfer, resulting in intense Ln(III)-centered emission.

The excitation process of TL in Ln(III) complexes has been considered as ligand excitation or direct Ln(III) excitation. Sweeting and Rheingold reported that the charge separation upon cleavage excited the antenna ligands, followed by the formation of Ln(III) excited states through an intramolecular energy transfer (ligand excitation) [38]. Bourhill and co-workers also described Ln(III) excitation by electron bombardment (direct Ln(III) excitation) [39]. The TL phenomenon is based on an electric origin, which might be understood by comparison of electroluminescence (EL) spectra of the compounds.

The strategy for constructing strong TL-active compounds would be a key to revealing the ambiguous relationships between luminescence and mechanical stress. These compounds with Tb(III) and Eu(III) ions in particular are also expected to be useful as optical sensors in the field of fluid dynamics and aeronautical engineering, being sensitive to impact, pressure, and temperature to visualize damage and fluid flow on material surfaces. In addition, compounds with varied emission colors depending on mechanical stress and photo-irradiation are highly attractive as advanced security materials for future identification cards.

19.3.5 Photo-Energy Conversion Property

Remarkable improvements of environmental durability and energy conversion efficiency in silicon solar cells were achieved using ethylene vinyl acetate (EVA) protection film containing poly(methyl methacrylate) (PMMA) polymer beads with Eu(III) coordination polymer $[Eu(hfa)_3(dpbp)]_n$, PB-$[Eu(hfa)_3(dpbp)]_n$.

The average diameter of the beads was estimated to be 46.7 nm. The emission quantum yield and lifetime of PB-$[Eu(hfa)_3(dpbp)]_n$ in solid are the same as those of $[Eu(hfa)_3(dpbp)]_n$ powder. The durability was evaluated using photophysical measurements after degradation test (85 °C 85% RH). The durability of $[Eu(hfa)_3(dpbp)]_n$ in EVA film was estimated to be 25 years, which is much longer than those of previous lanthanide complexes. The temperature dependency of emission spectra of $[Eu(hfa)_3(dpbp)]_n$ confirmed that the luminescence properties are maintained even at high temperatures up to 120 °C. Increased value of the solar cell short-circuit current efficiency using EVA protection film containing $[Eu(hfa)_3(dpbp)]_n$ was found to be 1.1%, which is a drastic increase as a photovoltaic solar system.

The EVA protection films with $[Eu(hfa)_3(dpbp)]_n$ are effective for enhancing the performance of silicon solar cells [40].

19.4 Coordination Nanoparticles

19.4.1 Micelle Method

In general, characteristic tight-stacking structures of Ln(III) coordination polymers lead to the formation of insoluble compounds, microsized particles, in water and organic solvents. These insoluble microsized particles prevent the preparation of transparent materials for optical use because of multiple light scattering in the UV-vis region. The nanosized particles of Ln(III) coordination polymers without multiple light scattering may provide future optical and luminescent materials.

We here focused on characteristic formation of organo-nanoparticles using micelles in liquid media, such as preparation of nanoscaled organic compounds. The preparation of polystyrene nanoparticles using micelle techniques has been reported in the field of polymer science [41]. The micelle sizes are also controlled by concentration and molecular structure of organo-surfactants in water media [42, 43]. The luminescent nanoparticles $[Eu(hfa)_3(dpbp)]_n$ were obtained by the

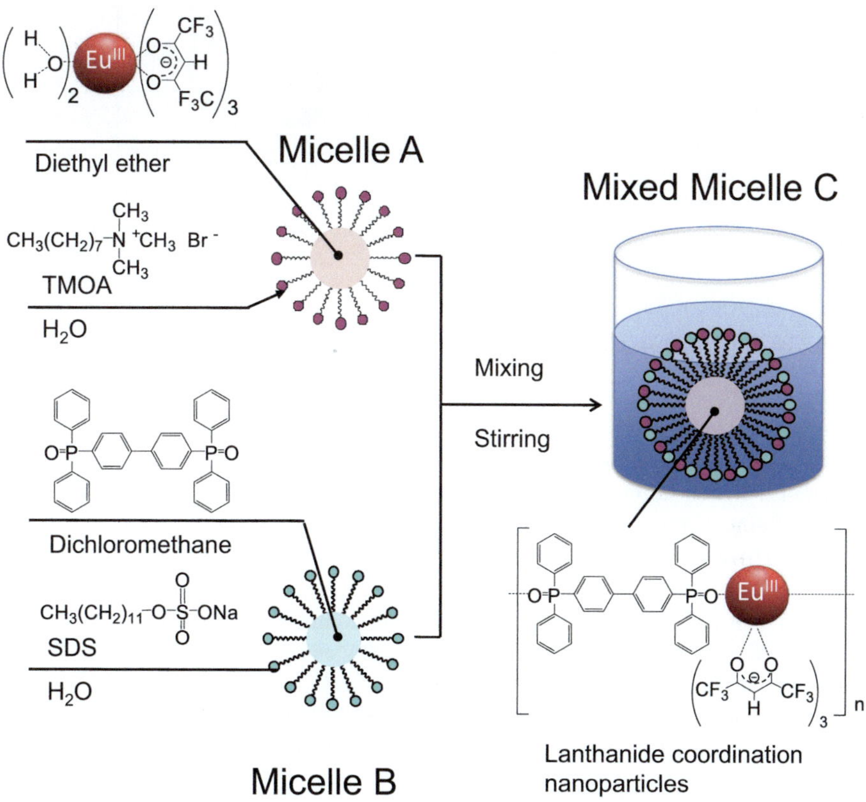

Fig. 19.15 Eu(III) coordination nanoparticles prepared using micelle techniques

polymerization of [Eu(hfa)$_3$(H_2O)$_2$] with dpbp in micelles under water [44]. The particle size is controlled using SDS (sodium dodecyl sulfate) and TMOA (trimethyl (octyl)ammonium bromide) in water (Fig. 19.15).

The prepared particle size using SEM measurements is estimated to be approximately 70 nm in diameter. Observed signals at 7.5, 8.8, 9.4, 10.2, 20.1, and 21.4° are attributed to the geometric structures of [Eu(hfa)$_3$(dpbp)]$_n$. We found that the signals of nanoparticles agree with those of previously reported [Eu(hfa)$_3$(dpbp)]$_n$. The decomposition temperature of Ln(III) coordination nanoparticles was estimated to be 301 °C and agreed well with that of bulk powders. We found that nanoparticles composed of [Eu(hfa)$_3$(dpbp)]$_n$ show effective thermal stability based on the characteristic rigid structure of [Eu(hfa)$_3$(dpbp)]$_n$ with CH-π and CH-F interactions. The time-resolved emission profiles of the Ln(III) coordination polymers also revealed single exponential decays with lifetimes in the millisecond timescale. The emission lifetimes were determined from the slopes of the logarithmic plots of the decay profiles. Emission lifetimes of nanoparticles were found to be 0.91 ms, which is the same as that of bulk powders (0.91 ms). We consider that the emission properties of

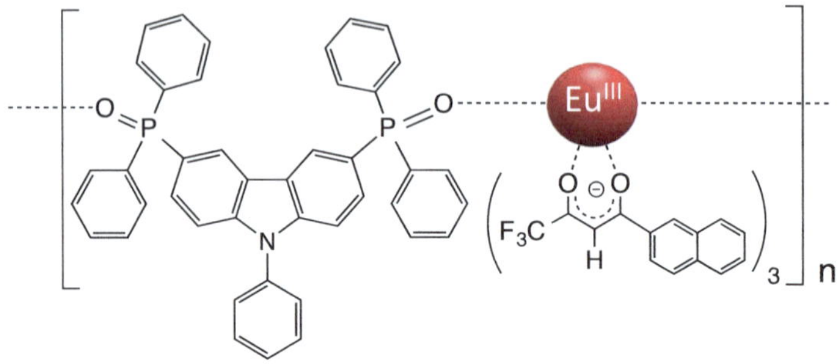

Fig. 19.16 Chemical structure of $[Eu(ntfa)_3(dppcz)]_n$

nanoparticles composed of Ln(III) coordination polymers are the same as those of bulk Ln(III) coordination polymers.

The Ln(III) coordination polymers $[Eu(ntfa)_3(dppcz)]_n$ [ntfa = 3-(2-naphthoyl)-1,1,1-trifluoroacetone, dppcz = 3,6-bis(diphenyl-phosphoryl)-9-phenylcarbazole] (Fig. 19.16) were also obtained by the polymerization of Eu(III) complexes ($[Eu(hfa)_3(H_2O)]_2$ and $[Eu(ntfa)_3(H_2O)]_2$ with shared dpbp and dppcz ligands in micellar capsules in water [45]. Their particle sizes were controlled using ferrocenyl surfactants (FcPEG=1,1-ferrocenylundecyl polyoxyethylene glycol [46]) in water. Using micelle reactions and electrochemical deposition techniques, characterized luminescent thin films composed of micelle-encapsulated Eu(III) coordination polymers on indium tin oxide (ITO) glass electrodes were successfully prepared.

19.4.2 Reaction Control Method

We also attempted to introduce terminator ligands with vinyl groups on the surface of the lanthanide coordination nanoparticles. The vinyl groups of terminator ligands, diphenyl-(*p*-vinyl phenyl) phosphine oxide (SDPO) [47], can promote polymerization for surface protection under acidic conditions (Figs. 19.17 and 19.18). The size-controlled nanoparticles are obtained by the reaction of $Eu(hfa)_3(H_2O)_2$ with joint dpbp ligands and terminator ligands (SDPO) ($[Eu(hfa)_3(dpbp)_x(SDPO)_y]_n$).

The acid-protected Eu(III) coordination nanoparticles are prepared by the polymerization of $[Eu(hfa)_3(dpbp)_x(SDPO)_y]_n$ with styrene monomers under 80 °C ($[Eu(hfa)_3(dpbp)_x(SDPO)_y]_n$-PS (PS: polystyrene).

The particle size distribution of $[Eu(hfa)_3(dpbp)_x(SDPO)_y]_n$ and $[Eu(hfa)_3(dpbp)_x(SDPO)_y]_n$-PS was characterized using DLS measurements. The maximum distributions of $[Eu(hfa)_3(dpbp)_x(SDPO)_y]_n$ and $[Eu(hfa)_3(dpbp)_x(SDPO)_y]_n$-PS are estimated to be 156.7 nm and 272.9 nm, respectively. We observed an increase of the particle size of $[Eu(hfa)_3(dpbp)_x(SDPO)_y]_n$-PS for the formation of the amorphous polystyrene thin layers on the $[Eu(hfa)_3(dpbp)_x(SDPO)_y]_n$ surface.

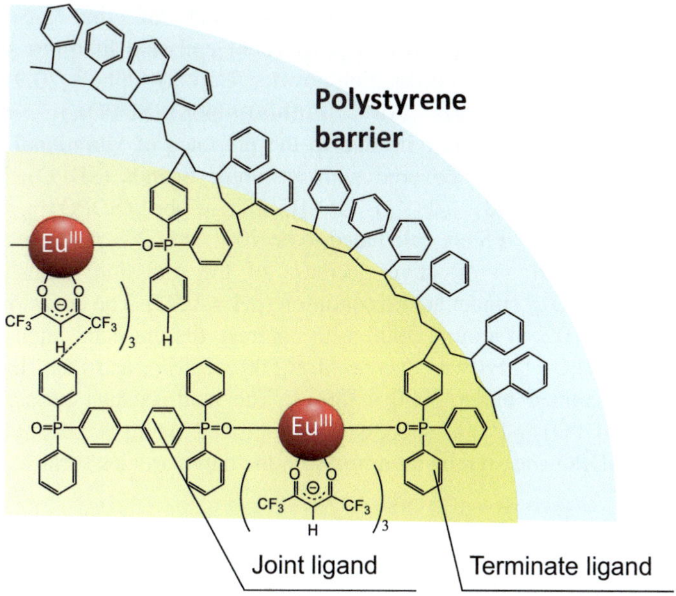

Fig. 19.17 Structural image of Eu(III) coordination nanoparticles covered with polystyrene

Fig. 19.18 Chemical structure of [Eu(hfa)$_3$(dpbp)$_x$(SDPO)$_y$]$_n$ and [Eu(hfa)$_3$(dpbp)$_x$(SDPO)$_y$]$_n$-PS

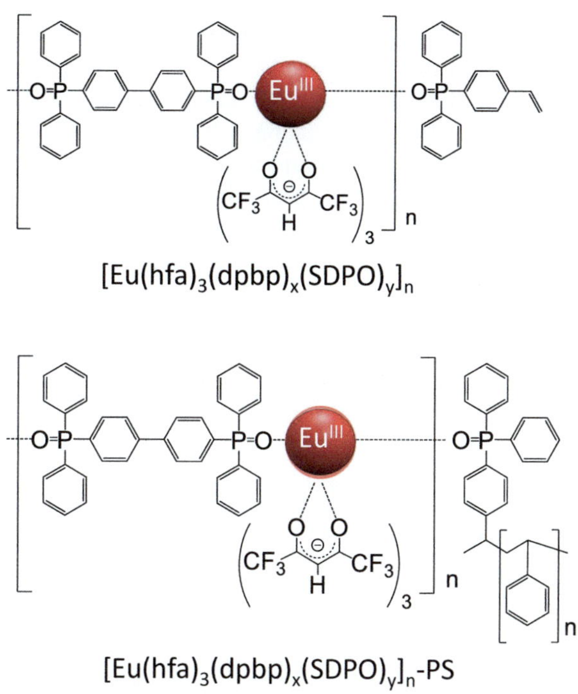

The emission lifetimes are similar to that of the previous [Eu$(hfa)_3(dpbp)_{0.20}(TPPO)_{0.80}]_n$ (0.66 ms) [48]. These emission lifetimes are smaller than that of the bulk Eu(III) coordination powder [Eu$(hfa)_3(dpbp)]_n$ (0.91 ms). The smaller emission lifetimes of [Eu$(hfa)_3(dpbp)_x(SDPO)_y]_n$ and [Eu$(hfa)_3(dpbp)_x(SDPO)_y]_n$-PS may be due to the presence of vibrational relaxation on the nanoparticle surface covered with terminate ligands (SDPO). The 4f–4f emission quantum yields ($\Phi_{4f\text{-}4f}$) of [Eu$(hfa)_3(dpbp)_x(SDPO)_y]_n$ and [Eu$(hfa)_3(dpbp)_x(SDPO)_y]_n$-PS are calculated to be 70% and 63%, respectively.

We also observed an effective decrease of the emission intensity of [Eu$(hfa)_3(dpbp)_x(SDPO)_y]_n$ under acidic condition (pH = 0.92). The emission intensity was found to be 60% at around 3000 s. In contrast, the emission intensity of [Eu$(hfa)_3(dpbp)_x(SDPO)_y]_n$-PS was decreased at 500 s (88%), and then the emission intensity was retained under 3000 s (80%). The acid-resistant property of [Eu$(hfa)_3(dpbp)_x(SDPO)_y]_n$-PS is successfully improved using a polystyrene layer composed of SDPO and styrene monomers on the nanoparticles surface.

19.5 Summary of Three-Dimensional Ln(III) Coordination Polymer

In this chapter, characteristic structures and photophysical and photo-functional properties of three-dimensional coordination polymers composed of joint organic parts and luminescent Ln(III) complexes as lanthanide element blocks were explained. Ln(III) coordination polymers show excellent luminescence properties and thermal stability. From these photophysical and structural findings, Ln(III) coordination polymer is a promising polymer material for future photonic devices, display, and sensing applications [49, 50]. The Ln(III) coordination polymer is also expected to open up a frontier field of coordination chemistry and material and polymer science.

Acknowledgment This work was partly supported by Grants-in-Aid for Scientific Research on Innovative Areas of "New Polymeric Materials Based on Element-Blocks (No. 2401)" (24102012) of the Ministry of Education, Culture, Sports, Science and Technology (MEXT) of Japan.

References

1. Shriver DF, Atkins PW (1999) Inorganic chemistry, 3rd edn. Oxford University Press, Oxford, p 320
2. Hasegawa Y, Wada Y et al (2004) Strategies for the design of luminescent lanthanide(III) complexes and their photonic applications. J Photochem Photobiol C 5:183–202
3. Hasegawa Y (2014) Photofunctional lanthanoid complexes, coordination polymers, and nanocrystals for future photonic applications. Bull Chem Soc Jpn 87:1029–1057

4. Hasegawa Y, Yamamuro M et al (2003) Luminescent polymer containing the Eu(III) complex having fast radiation rate and high emission quantum efficiency. J Phys Chem A 107:1697–1702
5. Nakamura K, Hasegawa Y et al (2007) Enhanced lasing properties of dissymmetric Eu(III) complex with bidentate phosphine ligands. J Phys Chem A 111:3029–3037
6. Miyata K, Hasegawa Y et al (2011) Remarkable luminescence properties of lanthanide complexes with asymmetric dodecahedron structures. Chem Eur J 17:521–528
7. Miyata K, Hasegawa Y et al (2012) Thermostable organo-phosphor: low-vibrational coordination polymers that exhibit different intermolecular interactions. ChemPlusChem 77:277–280
8. Eliseeva SV, Kuzmina NP et al (2008) Role of the ancillary ligand N,N-Dimethylaminoethanol in the sensitization of Eu and Tb luminescence in dimeric β-Diketonates. J Phys Chem A 112:3614–3626
9. Trivedi ER, Pecoraro VL et al (2014) Highly emitting near-infrared lanthanide "encapsulated sandwich" metallacrown complexes with excitation shifted toward lower energy. J Am Chem Soc 136:1526–1534
10. Hirai Y, Hasegawa Y et al (2016) Luminescent europium(III) coordination zippers linked with thiophene-based bridges. Angew Chem Int Ed 55:12059–12062
11. Hasegawa Y, Tateno S et al (2017) Effective photo- and triboluminescent europium(III) coordination polymers with rigid triangular spacer ligands. Chem Eur J 23:2666–2672
12. Lee WR et al (2010) Microporous lanthanide-organic frameworks with open metal sites: unexpected sorption propensity and multifunctional properties. Inorg Chem 49:4723–4725
13. Nakajima A, Hasegawa Y et al (2016) Hyper-stable organo-EuIII luminophore under high temperature for photo-industrial application. Sci Rep 6:24458
14. Yamamoto M, Hasegawa Y et al (2016) Luminescent Eu(III) coordination polymer cross-linked with Zn(II) complexes. Mater Lett 167:183–187
15. Nakanishi T, Hasegawa Y et al (2014) Enhancement of optical Faraday effect of nonanuclear Tb (III) complexes. Inorg Chem 53:7635–7641
16. Wada S, Hasegawa Y et al (2016) The relationship between magneto-optical properties and molecular chirality. NPG Asia Mater 8:e251
17. Wada S, Hasegawa Y et al (2015) Chiroptical properties of nonanuclear Tb(III) clusters with chiral champhor derivative ligands. e-J Surf Sci 13:31–34
18. Omagari S, Nakanishi T et al (2015) Effective photosensitized energy transfer of Nonanuclear terbium clusters using methyl salicylate derivatives. J Phys Chem A 119:1943–1947
19. Mitsuishi M, Amao Y et al (2003) Characterization of an ultrathin polymer optode and its application to temperature sensors based on luminescent europium complexes. J Mater Chem 13:2875–2879
20. Khalil GE, Dalton LR et al (2004) Europium Beta-diketonate temperature sensors: effects of ligands, matrix, and concentration. Rev Sci Instrum 75:192
21. Katagiri S, Hasegawa Y et al (2004) Thermo-sensitive luminescence based on the Back energy transfer in terbium(III) complexes. Chem Lett 33:1438–1439
22. Liu Y, Wang M et al (2005) Temperature-dependent luminescent properties of Eu–Tb complexes synthesized in situ in gel glass. Appl Phys Lett 286:071907
23. Tremblay MS, Sames D et al (2007) Cocktails of Tb^{3+} and Eu^{3+} complexes: a general platform for the design of ratiometric optical probes. J Am Chem Soc 129:7570–7577
24. Kerbellec N, Guillou OD et al (2009) An unprecedented family of lanthanide-containing coordination polymers with highly tunable emission properties. Inorg Chem 48:2837–2843
25. Xiao Y, Yuan J et al (2012) A Ratiometric luminescence probe for highly reactive oxygen species based on lanthanide complexes. Inorg Chem 51:2940–2946
26. Comby S, Gunnlaugsson T et al (2012) New trick for an old ligand! The sensing of Zn(II) using a lanthanide based ternary Yb(III)-cyclen-8-hydroxyquinoline system as a dual emissive probe for displacement assay. Inorg Chem 51:10158–10168
27. Miyata K, Hasegawa Y et al (2013) Chameleon luminophore for sensing temperatures: control of metal-to-metal and energy back transfer in lanthanide coordination polymers. Angew Chem Int Ed 52:6413–6416

28. Hirai Y, Hasegawa Y et al (2014) Thermo-sensitive luminescent materials composed of Tb(III) and Eu(III) complexes. Mater Lett 130:91–93
29. Hirai Y, Hasegawa Y et al (2014) Luminescent coordination glass: remarkable morphological strategy for assembled Eu(III) complexes. Inorg Chem 54:4364–4370
30. Hirai Y, Hasegawa Y et al (2017) Glass transition properties and temperature-sensitive luminescence of lanthanide coordination glasses linked by thienyl, naphthyl, and phenyl bridges with ethynyl group. Bull Chem Soc Jpn 90:322–326
31. Hatanaka M, Morokuma K et al (2017) Organic linkers control the thermosensitivity of the emission intensities from Tb(III) and Eu(III) in a chameleon polymer. Chem Sci 8:423–429
32. Wiedemann G, Schmidt F (1895) Ann Phys (Leipzig) 54:604–625
33. Walton AJ (1977) Triboluminescence. Adv Phys 26:887–948
34. Chen XF, You XZ et al (1999) Triboluminescence and crystal structures of non-ionic europium complexes. J Mater Chem 9:2919–2922
35. Feffrey GA (1973) Intramolecular hydrogen-bonding in carbohydrate crystal-structures. Carbohydr Res 28:233–241
36. Hasegawa Y, Kawai T et al (2011) Brilliant triboluminescence of a lanthanide coordination polymer with low-vibrational-frequency and non-centrosymmetric structural networks. Eur J Inorg Chem 32:4978–4984
37. Hirai Y, Hasegawa Y et al (2017) Triboluminescence of lanthanide coordination polymers with face-to-face arranged substituents. Angew Chem Int Ed 56:7171–7175
38. Sweeting LM, Rheingold AL (1987) Crystal disorder and triboluminescence: triethylammonium tetrakis (dibenzoylmethanato)europate. J Am Chem Soc 109:2652–2658
39. Sage I, Bourhill G (2001) Triboluminescent materials for structural damage monitoring. J Mater Chem 11:231–245
40. Kataoka H, Hasegawa Y et al (2016) Drastically improved durability and efficiency of silicon solar cells using hyper-stable lanthanide coordination polymer beads. Bull Chem Soc Jpn 89:103–109
41. Atik SS, Thomas JK (1981) Polymerized microemulsions. J Am Chem Soc 103:4279–4280
42. Antonietti M, Niel CV (1992) Polymerization in microemulsion. 2. Surface control and functionalization of microparticles. Macromolecules 25:1139–1143
43. Boutonnet M, Maire G (1982) The preparation of monodisperse colloidal metal particles from microemulsions. Colloids Surf 5:209–225
44. Onodera H, Hasegawa Y et al (2014) Thermo-stable lanthanoid coordination nanoparticles composed of luminescent Eu(III) complexes and organic joint ligands using micelle techniques in water. Bull Chem Soc Jpn 87:1386–1390
45. Hasegawa Y, Fushimi K et al (2016) Luminescent thin films composed of nano-sized europium coordination polymers on glass electrodes. ChemPlusChem 81:187–193
46. Saji T, Goto M (1991) Formation of organic thin films by electrolysis of surfactants with the ferrocenyl moiety. J Am Chem Soc 113:450–456
47. Onodera H, Hasegawa Y et al (2016) Acid-protected Eu(III) coordination nanoparticles covered with polystyrene. J Mater Chem C 4:75–81
48. Onodera H, Hasegawa Y et al (2015) Thermostable nano Luminophores composed of europium ions and organic ligands. e-J Surf Sci Nanotech 13:219–222
49. Hasegawa Y, Nakanishi T (2015) Luminescent lanthanide coordination polymers for photonic applications. RSC Adv 5:338–353
50. Hirai Y, Hasegawa Y et al (2016) Organo-lanthanide luminophores bridged by phosphine oxide ligands. J Lumin 170:801–807

Chapter 20
Organic Hybrid Thermoelectric Materials Containing Nano-dispersed Poly(nickel 1,1,2,2-ethenetetrathiolate) as an Element-Block

Naoki Toshima, Keisuke Oshima, and Yukihide Shiraishi

Abstract Thermoelectric technology is expected to convert waste heat to electrical energy. Organic hybrid thermoelectric materials are some of the candidates applicable to this technology to obtain electrical energy from waste heat below 150 °C. Nanoparticles of poly(nickel 1,1,2,2-ethenetetrathiolate) (abbreviated as nPETTs) were prepared from 1,3,4,6-tetrathiapentalene-2,5-dione using a surfactant, dodecyltrimethylammonium bromide (DTAB), as an additive. The nPETT, recognized as an "element-block" composed of Ni, S, and C, can be dispersed in N-methylpyrrolidone (NMP), although solid PETT cannot be dispersed in NMP. The nPETTs can help to produce the dispersions of carbon nanotubes (CNTs) in NMP and play a role as the building block to connect the independent CNTs to each other by strong d-π interactions to form dispersed nPETT/CNT hybrids. Thus, the organic hybrid films, prepared by drop-casting the dispersed solution of the nPETTs, CNTs, and poly(vinyl chloride) (PVC) in NMP on the substrates and recognized as element-block polymeric materials, showed a good thermoelectric performance. A series of thermoelectric materials were also produced using other kinds of CNTs and polymers.

Keywords Poly(nickel 1,1,2,2-ethenetetrathiolate) · Organic hybrid thermoelectric materials · Nanoparticles · Carbon nanotube

N. Toshima (✉) · K. Oshima · Y. Shiraishi
Tokyo University of Science Yamaguchi, Sanyo-Onoda, Yamaguchi, Japan
e-mail: toshima@rs.tus.ac.jp

Y. Chujo (ed.), *New Polymeric Materials Based on Element-Blocks*,
https://doi.org/10.1007/978-981-13-2889-3_20

20.1 Introduction

Energy and environment are the most important global issues which require a long time and a significant effort to be solved for humans. Many researchers are making an effort to solve these global problems from various viewpoints. One of the main viewpoints to solve the energy problem is to minimize the consumption of fossil fuels. In fact, we still use a lot of fossil fuels, i.e., coal, petroleum, and natural gas. Figure 20.1 shows the energy flow in Japan. Most of the energy, actually about 90% of the primary energy in Japan, comes from fossil fuels. They are consumed for industries, consumers, and transportation, and finally transformed to motive power, electric energy, heat, light, etc. However, the most important point in this energy flow is that about 60% of the primary energy is lost as waste heat without being utilized in any form. This means that we use a lot of fossil fuels, more than half of which is excreted as waste heat that causes a temperature increase in the atmosphere on a global scale. This may result in the recent abnormal weather around the world.

Thus, the reduction of the waste heat emission and creation of the electrical energy from the waste heat are two of the urgent issues for humans. The thermoelectric technology can provide the energy conversion from heat (temperature difference) to electrical energy and vice versa. The technology is mainly based on thermoelectric inorganic semiconducting materials, because their energy conversion efficiency is high at high temperature [1]. At low temperature below 150°C, however, the efficiency is usually low. Because most of the waste heat is generated at temperatures below 150 °C, the thermoelectric energy conversion using organic hybrid thermoelectric materials, which is usually applicable below 150°C, is expected to be achieved in practice.

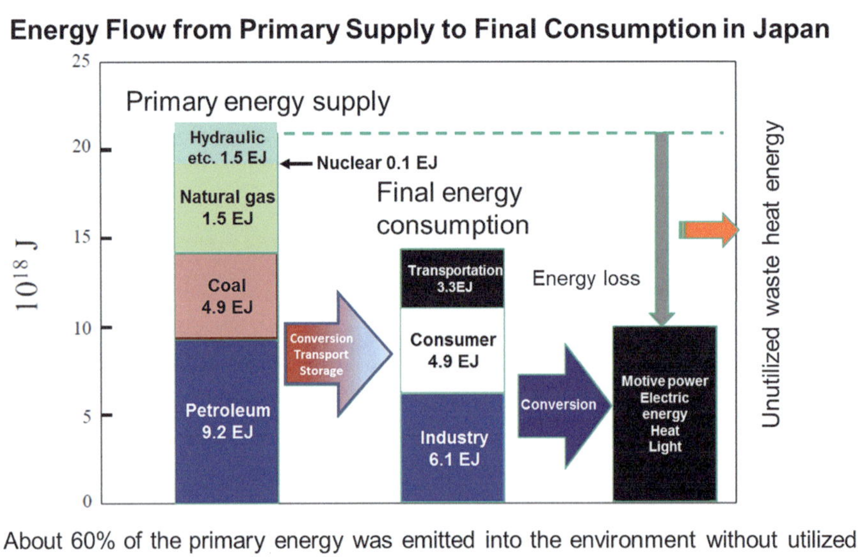

Fig. 20.1 Energy flow from primary supply to final consumption. (Prepared on the basis of the data of energy demand and supply in 2012 from the Agency for Natural Resources and Energy, Japan)

In contrast to the inorganic thermoelectric materials, the organic hybrid thermoelectric materials have a short research history and may have the possibility to show a high thermoelectric energy conversion efficiency at relatively low temperatures. In the last two decades, in fact, tremendous improvement in the thermoelectric efficiency has already been reported [2]. In addition, the organic hybrid thermoelectric materials have many advantages, which have been described in a review paper [3]. The most important advantage of the organic hybrid thermoelectric materials is based on the easy processing to manufacture the thermoelectric modules or devices, which can reduce the cost of the thermoelectric energy conversion to increase the efficiency.

The "element-block" proposed by Professor Yoshiki Chujo [4] is an effective concept to produce a new type of organic or polymeric hybrid material. The element-block can be constructed of one or a few elements and work as a building block to produce the polymeric materials with various functions. Thus, the concept of the element-block has really provided many functional polymeric materials, some of which are reported in this book.

In this chapter, we report the preparation and application of the element-block which is important for the construction of organic hybrid thermoelectric materials. This new type of element-block is "nano-dispersed poly(nickel 1,1,2,2-ethenetetrathiolate)" (abbreviated as "nPETT"). The nPETTs can be dispersed in N-methylpyrrolidone (NMP) and form the carbon nanotubes (CNTs) dispersed in NMP. The organic hybrid thermoelectric materials, composed of nPETTs, CNTs, and poly(vinyl chloride) (PVC), were prepared using this characteristic property of the nPETTs. Similar hybrid materials can also be produced using other kinds of CNTs and polymers and show high thermoelectric performances.

The thermoelectric performance of the materials can be evaluated by the dimensionless thermoelectric figure of merit, ZT, which can be calculated by Eq. (20.1):

$$ZT = \left(S\sigma^2/\kappa\right)T \tag{20.1}$$

where S, σ, κ, and T denote the Seebeck coefficient, electrical conductivity, thermal conductivity of the materials, and the absolute temperature of the measurement condition, respectively [5]. In the case of organic thermoelectric thin films, the Seebeck coefficient and electrical conductivity are usually measured in the direction of the plane of the films. However, the accurate measurement of the thermal conductivity in the plane is not easy, especially for thin films. Thus, the thermoelectric power factor PF ($= S\sigma^2$) is often used instead of the ZT value. Nevertheless, the apparent ZT values (shown as ZT_a hereafter) are often calculated using the Seebeck coefficient and electrical conductivity in the plane and the thermal conductivity through the plane. These calculated ZT_a values cannot provide the correct properties, but have traditionally been used, especially for the inorganic semiconducting thermoelectric materials. In this chapter, we carefully distinguish the ZT and ZT_a values.

20.2 Nano-dispersed Poly(nickel 1,1,2,2-ethenetetrathiolate), nPETT

Poly(nickel 1,1,2,2-ethenetetrathiolate) (abbreviated as "PETT," Fig. 20.2a) was reported by the group of Professor Daoben Zhu [6] as a new type of organic thermoelectric material having a negative Seebeck coefficient. However, the solid PETT cannot be dispersed in any solvent, which makes its application difficult for use in organic thermoelectric devices by the solution process.

We have succeeded in preparing nanoparticles of PETT (abbreviated as nPETTs) by adding a surfactant, dodecyltrimethylammonium bromide (DTAB), to the reaction mixtures of 1,3,4,6-tetrathiapentalene-2,5-dione, nickel(II) chloride, and sodium (or potassium) methoxide (the nPETTs prepared by addition of sodium and potassium methoxide are abbreviated as nPETT(Na) and nPETT(K), respectively, in order to distinguish them) in methanol [7]. A photograph of the NMP dispersion of nPETT in a glass bottle, shown in Fig. 20.3a, reveals the homogeneity of the dispersion. The

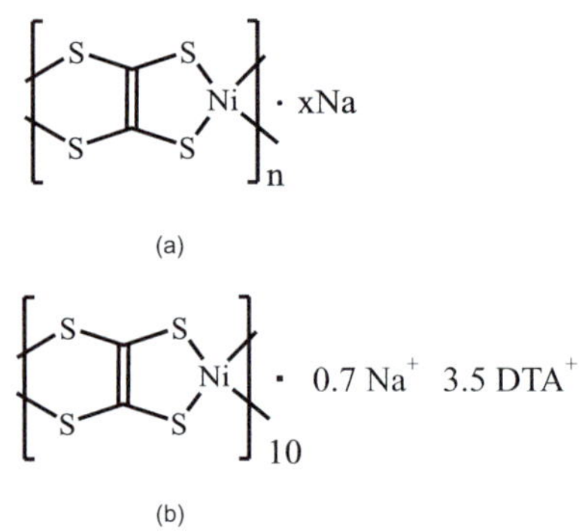

Fig. 20.2 (**a**) The chemical structure of PETT. (**b**) The possible chemical structure of nPETT, in which the degree of polymerization was estimated as 10

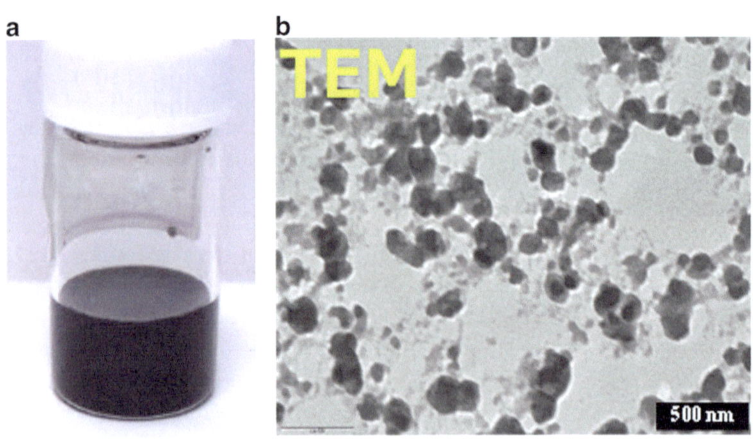

Fig. 20.3 (**a**) Photograph of the NMP dispersion of nPETT in a glass bottle and (**b**) TEM image of nPETT ((b): cited from Fig. 2b in [7] under the permission of the Chemical Society of Japan)

Table 20.1 Elemental analysis (wt.%) of PETT and nPETT

	Ni	Na	S	C	N	H
	22.82	1.11	48.33	13.01	0	1.02
nPETT	15.25	0.48	38.56	30.76	1.49	4.38

Cited from Table S1 in supporting information of [7] under the permission of the Chemical Society of Japan

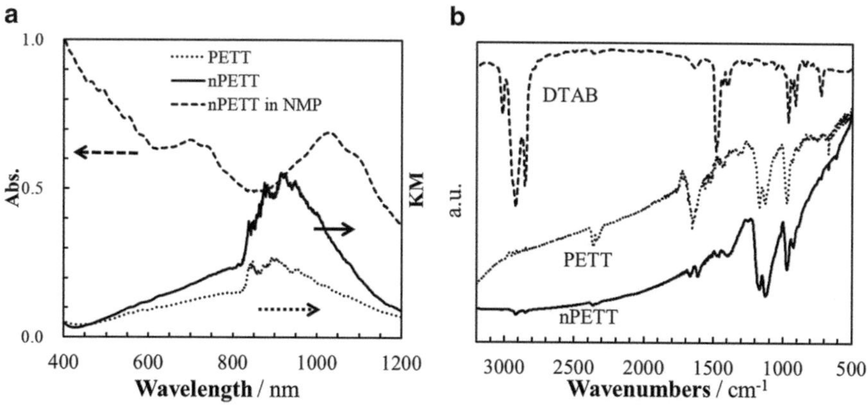

Fig. 20.4 (**a**) UV-vis reflection spectra of solid PETT and nPETT and the absorption spectrum of nPETT in NMP. (**b**) FTIR spectra of nPETT, PETT, and DTAB. (Cited from Fig. 1 in [7] under the permission of the Chemical Society of Japan)

transmission electron micrograph (TEM) image, shown in Fig. 20.3b, indicates that the average diameter of the nPETT is 38 ± 12 nm.

The elemental analysis data of nPETT(Na) are shown in Table 20.1, revealing that a small amount of the DTAB surfactant (0.38 mole of the surfactant versus the central Ni atom of PETT) is included in the nPETT. Ultraviolet-visible (UV-vis) reflection and absorption spectra as well as Fourier transform infrared (FTIR) spectra, shown in Fig. 20.4, indicate that nPETT maintains the same chemical structure as the PETT except for the coexistence of a small amount of DTAB. Based on these data, we proposed the chemical structure of nPETT(Na) as shown in Fig. 20.2b.

The thermoelectric properties of nPETT, measured by an ULVAC ZEM-3 instrument for a pellet prepared by cold-pressing of the nPETT powders, revealed that the Seebeck coefficient, electrical conductivity, and power factor at 345 K were -42 ± 1 μV K^{-1}, 0.01 ± 0.001 S cm^{-1}, and $(2.2 \pm 0.4) \times 10^{-3}$ μW $m^{-1}K^{-2}$, respectively. Although the Seebeck coefficient of the pressed pellet of the nPETT was at a level similar to that of PETT, the electrical conductivity, thus also the power factor of nPETT was inferior to that of PETT. Nevertheless, the nPETT, prepared by subdividing PETT, can play an important role as an element-block as described in the following sections.

20.3 Interaction of nPETTs with CNTs

CNTs have recently received much attention as a building block to construct organic hybrid thermoelectric materials with a high performance. The group at Texas A&M University found that the composites of the CNTs and conducting poly (3,4-ethylenedioxythiophene)-poly(styrene sulfonate) (PEDOT-PSS) showed a high thermoelectric performance [8] and, in addition, that the thin composite films, laminated with polyaniline, graphene, and CNTs by a layer-by-layer method, showed a much higher performance with $PF = 2710\ \mu W\ m^{-1}K^{-2}$ [9].

Although the highly conducting CNTs are very interesting materials as a building block of hybrid thermoelectric materials, they strongly bind to each other to form bundles, which are difficult to disperse in any solvent. Thus, the preparation of homogeneous dispersions of the CNTs is an important technology to obtain homogeneous thermoelectric films with a high performance by a solution process.

The new type of element-block nPETT, mentioned above, has been discovered to have a strong ability to disperse CNTs in NMP. Figure 20.5 shows photographs of the NMP dispersion (concentration: 1 mg/L) of supergrowth CNTs (abbreviated as SGCNTs), which is one of the least expensive single-walled CNTs, with (a and c) and without (b and d) the addition of nPETT immediately after strong sonication for 5 min (a and b) and after storing for a day under ambient air conditions (c and d). The comparison of these photographs reveals that the SGCNTs can be dispersed in NMP by strong sonication without nPETT, but after 1 day, the SGCNTs formed aggregates. In contrast, the addition of nPETT to the SGCNTs did stabilize the NMP dispersion of the SGCNTs and kept the dispersion stable even after 1 day.

Figure 20.6 shows photographs of the NMP dispersions of SGCNT (concentration: 1 g/L) alone (a), SGCNT with poly(vinyl chloride) (abbreviated as PVC) (b), and SGCNT with PVC and nPETT (c), placed between two glass plates immediately after a 5-min strong sonication. The photographs, taken by an optical microscope and shown in Fig. 20.6a, b, reveal the nonhomogeneous dispersions of the black SGCNTs even immediately after the strong sonication. In contrast, the addition of nPETT to the dispersion of the SGCNTs and PVC provided a homogeneous dispersion even under the microscope as shown in Fig. 20.6c.

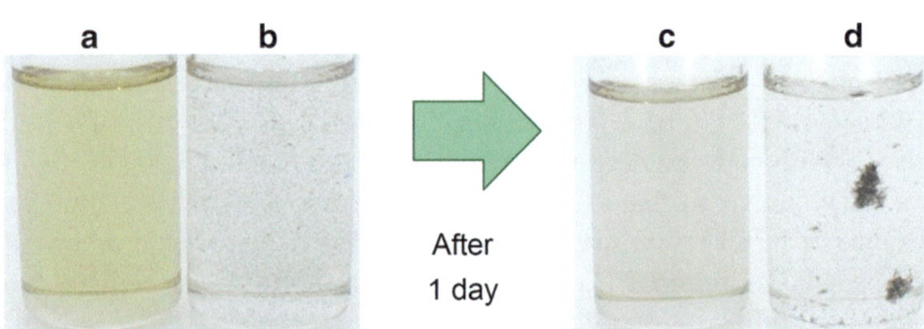

Fig. 20.5 Photographs of the dispersions of SGCNTs in NMP with (**a**, **c**) and without (**b**, **d**) the help of nPETT immediately after the sonication (**a**, **b**) and after storing 1 day after the sonication (**c**, **d**)

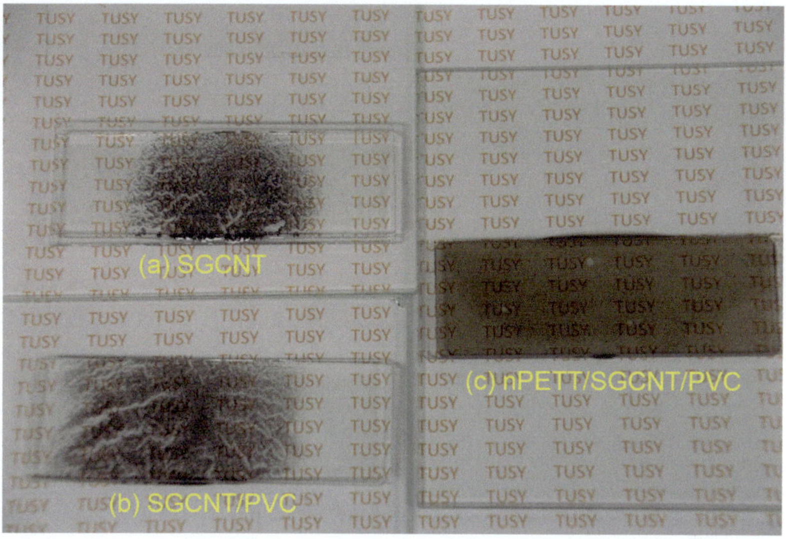

Fig. 20.6 Photographs of the dispersions of SGCNT alone (**a**), SGCNT with PVC (**b**), and SGCNT with PVC and nPETT (**c**), placed between two glass plates immediately after a 5-min strong sonication

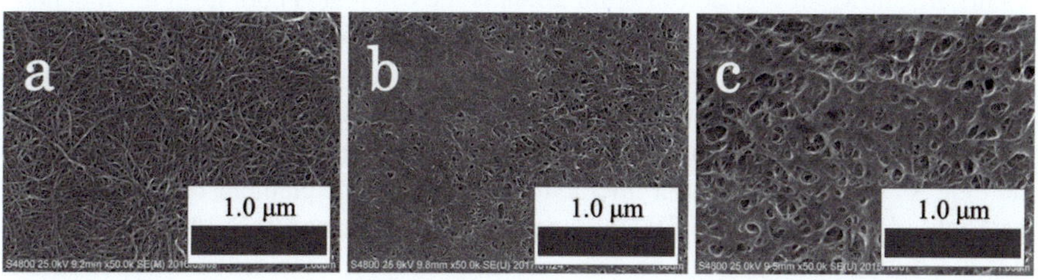

Fig. 20.7 SEM photographs of SGCNT-containing films. (**a**) SGCNT buckypaper prepared by the help of a surfactant. (**b**) SGCNT/PVC (40/60 in wt.) cast film. (**c**) nPETT/SGCNT/PVC (45/40/15 in wt.) cast hybrid film ((c): cited from Fig. 6 of [14] published by Springer)

The cast films prepared from the dispersion were observed by scanning electron microscopy (SEM). The SEM images are shown in Fig. 20.7. The SEM photographs of the cast films of SGCNT/PVC (b) and nPETT/SGCNT/PVC (c) reveal the presence of separated CNTs, which are covered by smooth PVC film. For the SGCNT film, in contrast, the SEM photograph of a buckypaper of SGCNT (a), prepared by the help of a surfactant, is presented instead of the cast film, because the dispersion of only the SGCNTs could not be cast to prepare a smooth film. The SEM observation suggests that the polymer PVC works to cover and stabilize the cast films of the CNTs.

The strong dispersing ability of nPETT has suggested the presence of an interaction of nPETT with the SGCNTs. In order to clarify the interaction, the change in

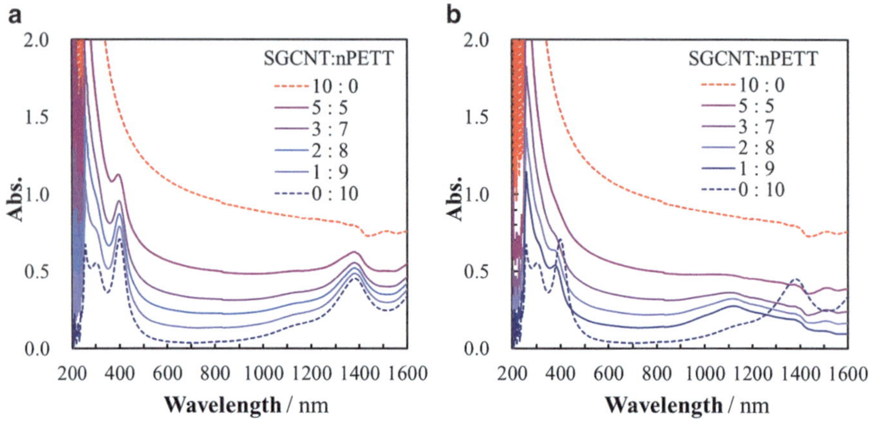

Fig. 20.8 The change in UV-vis absorption spectra of the mixtures of SGCNT and nPETT depending on the concentration ratio between SGCNT and nPETT in NMP. (**a**) The theoretical spectra calculated from the absorption spectra of SGCNT and nPETT. (**b**) The observed spectra measured for the mixed solutions

the UV-vis absorption spectra of the mixtures of the SGCNT and nPETT depending on the concentration ratio between the SGCNT and nPETT was measured. The results are shown in Fig. 20.8. Figure 20.8a is the calculated data from the independent spectra of SGCNT and nPETT, and Fig. 20.8b shows the measured data, which is quite different from the calculated ones, and has the small maximum peak observed near 1100 nm. This suggests the presence of a strong interaction between SGCNT and nPETT. In other words, the nPETT element-blocks are in contact with the SGCNTs possibly by a d-π interaction to produce the element-block polymeric materials [4]. Although SGCNTs cannot be well dispersed in NMP, the nPETT/SGCNT hybrid materials can be dispersed in NMP as the element-block polymeric materials. The addition of PVC to the nPETT/SGCNT is postulated to stabilize the dispersion of nPETT/SGCNT by covering the element-block polymers to form stable organic hybrid materials [10].

20.4 Thermoelectric Properties of Hybrid Materials of nPETT/CNT/PVC

The formation of the stable organic hybrid materials of nPETT, CNT, and PVC as the element-block polymeric materials has encouraged us to measure their thermoelectric properties. We initially used an arc-discharged single-walled CNT (abbreviated as Arc-CNT), which is expensive but has a high electrical conductivity, thus expected to show a high thermoelectric performance. The dispersion of Arc-CNT, nPETT(Na), and PVC (10:8:3 in wt.) in NMP was cast to prepare smooth organic

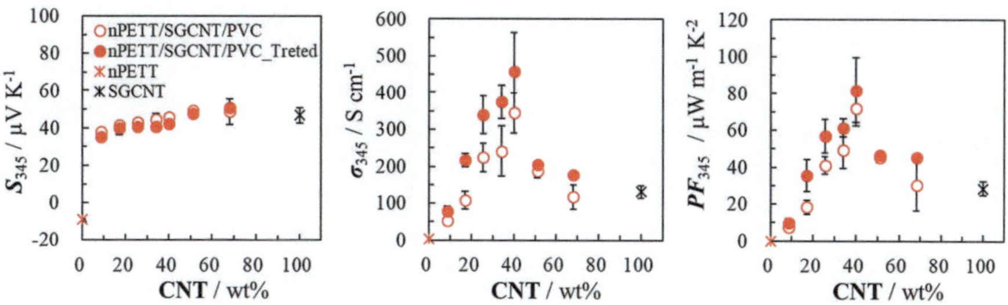

Fig. 20.9 Seebeck coefficient S, electrical conductivity σ, and power factor PF of ternary hybrid films of nPETT(K)/SGCNT/PVC at 345 K before (open red circle) and after (closed red circle) the methanol treatment depending on the CNT concentration

hybrid films. The thermoelectric properties of the films were measured after being treated with methanol. The Seebeck coefficient, electrical conductivity, power factor, and thermal conductivity through the plane at 345 K were 30.5 ± 0.4 μV K^{-1}, $630 \pm$ S cm^{-1}, 58.6 ± 1.2 μW m^{-1}K^{-2}, and 0.07 W m^{-1}K^{-1}, respectively. Based on these data, the apparent figure of merit, ZT_a, was calculated to be 0.29 [11]. These data were near the same level of those of the highest in organic thermoelectric thin films composed of PEDOT-tosylate (p-toluenesulfonate) ($ZT = 0.25$, [12]) and PEDOT-PSS ($ZT = 0.42$, [13]).

As already mentioned, the Arc-CNT is expensive, while the SGCNT is much less expensive. However, the buckypaper of SGCNT has a lower electrical conductivity than that of Arc-CNT, which may result in the lower thermoelectric performance. This is because the SGCNT has more defects in the nanotubes. In other words, the SGCNT has more active spots and is more porous than the Arc-CNT. Thus, we tried to use SGCNT instead of Arc-CNT for the preparation of the ternary hybrid thermoelectric materials of nPETT/CNT/PVC. The thermoelectric properties of the ternary hybrid films of nPETT(K)/SGCNT/PVC depending on the CNT concentration are shown in Fig. 20.9. The dependence of the power factor on the CNT concentration revealed an interesting result. The PF has a maximum at the CNT concentration at about 40 wt.%, which means that the real hybrid effect is observed in this case, while the property of the hybrids usually linearly changes with the concentration of the component. The highest power factor of the film after methanol treatment was 81.9 ± 18 μW m^{-1}K^{-2} at 345 K.

In order to calculate the apparent thermoelectric figure of merit, ZT_a, the thermal conductivities of the hybrid cast films through the plane were measured before and after the methanol treatment. The results are shown in Fig. 20.10. The thermal conductivities of the cast films significantly decreased after the methanol treatment. This is attributed to the decrease in the density or the increase in the porosity of the films by the methanol treatment. The porous structure of the films after the methanol treatment is also observed in the SEM photographs as shown in Fig. 20.11 [14]. This is attributed to the removal of PVC and nPETT from the films by the methanol

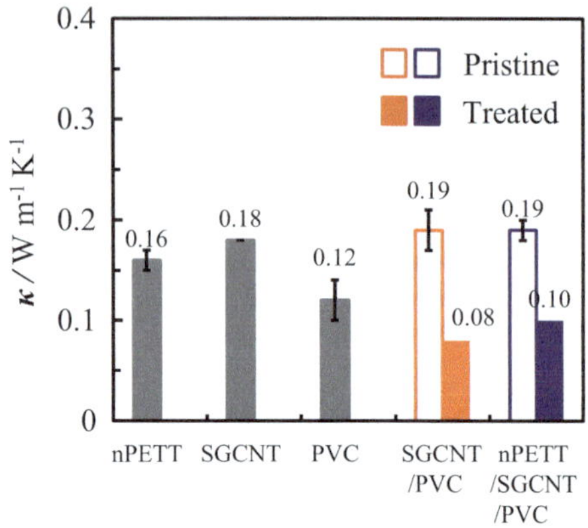

Fig. 20.10 The thermal conductivity of the samples of nPETT(Na) (cold-pressed pellet), SGCNT (buckypaper), and PVC (cast film) and the films of SGCNT/PVC binary and nPETT/SGCNT/PVC ternary hybrid cast films before (open bars) and after (closed bars) methanol treatment

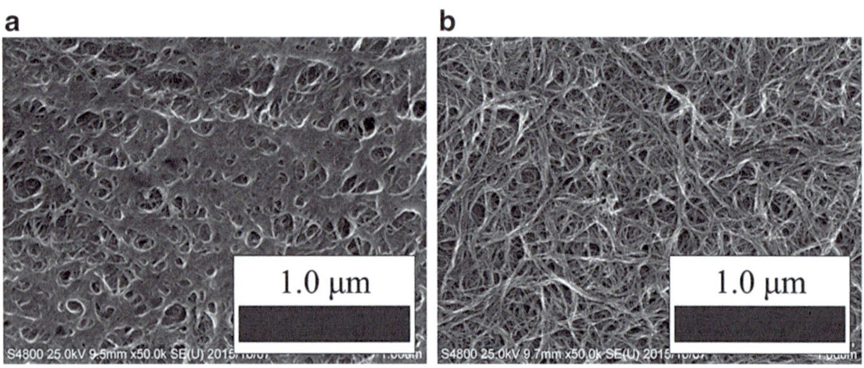

Fig. 20.11 SEM images of ternary hybrid cast films of nPETT(Na)/SGCNT/PVC before (**a**) and after (**b**) the methanol treatment. (Cited from Fig. 6 of [14] published by Springer)

Table 20.2 Thermoelectric properties of CNT hybrid films at 345 K

	S_{345}	σ_{345}	PF_{345}	κ	ZT_{a345}
	μV/K	S/cm	μW/(m K^2)	W/(m K)	–
Arc-CNT	31.9 ± 0.4	690 ± 16	100 ± 1.9	–	–
SGCNT	44.1 ± 2.2	335 ± 15	64.9 ± 3.7	0.18	0.12
nPETT(Na)/Arc-CNT/PVC (MeOH-treated)	30.5 ± 0.4	630 ± 23	58.6 ± 1.2	0.07	0.29
nPETT(K)/SGCNT/PVC (MeOH-treated)	42.3 ± 0.2	459 ± 104	81.9 ± 18	0.10	0.28

treatment, which is evidenced by the elemental analyses of the films before and after the methanol treatment.

The thermoelectric properties of the ternary films with different CNTs are compared in Table 20.2. Since the thermal conductivity of the ternary hybrid film of nPETT/SGCNT/PVC through the plane was 0.10 W $m^{-1}K^{-1}$, the apparent thermoelectric figure of merit, ZT_a, was calculated to be 0.28. This is at the same level as that of the previously reported nPETT/Arc-CNT/PVC ternary film, even though the less expensive SGCNT is used in this system.

20.5 Conclusions

Nano-dispersed conducting polymer complex nPETTs were prepared as a kind of element-block composed of Ni, S, and C in addition to a small amount of the DTAB surfactant. The nPETTs have a negative Seebeck coefficient but a low electrical conductivity. Thus, they cannot be used as effective thermoelectric materials by themselves. However, the nPETTs have a strong ability to disperse the CNTs, which are interesting as a building block of hybrid thermoelectric materials because of their high electrical conductivity. This ability of the nPETTs is attributed to their strong interaction with the CNTs. In other words, the nPETTs can be bound to each other by the CNTs as illustrated in Fig. 20.12 to form element-block polymeric materials. In the ternary organic hybrid thermoelectric materials of nPETT/CNT/PVC, the vinyl polymer PVC stabilizes the nPETT/CNT element-block polymers by covering them.

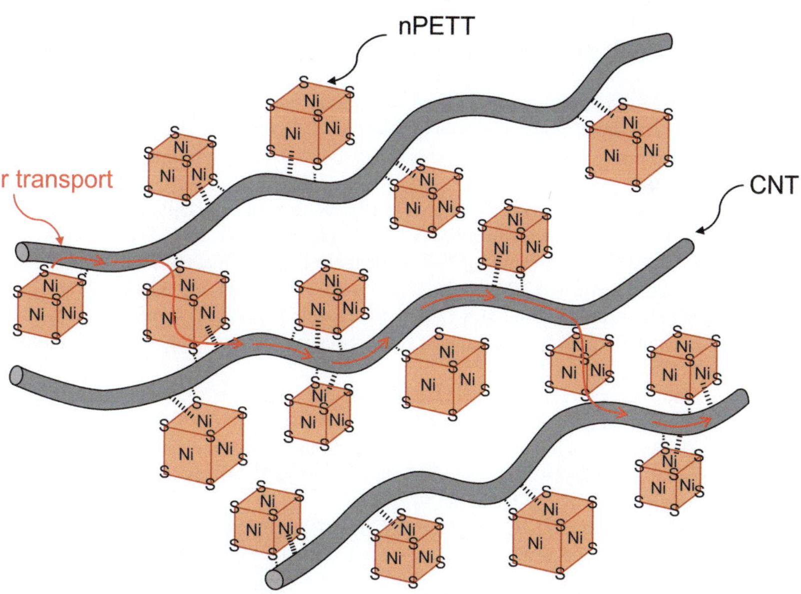

Fig. 20.12 Illustration of element-block polymeric materials of nPETT/CNT. Many nPETTs as element blocks (shown in brown dices) combine with long CNTs (shown in gray strings) to produce element-block polymeric materials. The carrier transport is expected to easily occur from one CNT to another CNT through the nPETTs

The films of the ternary element-block polymeric materials of nPETT/CNT/PVC showed a sufficiently high thermoelectric performance with the apparent thermoelectric figure of merit of $ZT_a \sim 0.3$. The films are tough and flexible, thus having the possibility to be used as p-type thermoelectric materials. In addition, it should be emphasized that we have succeeded in producing ternary hybrid thermoelectric materials based on the concept of element-block polymeric materials by using SGCNTs less expensive than the Arc-CNTs with a high electrical conductivity.

The films of the ternary organic hybrid thermoelectric materials with a high thermoelectric performance can be easily manufactured by drop-casting and drying the corresponding dispersions. The films are stable and maintain their properties under ambient conditions for a long time. Thus, they may be expected to be used for practical applications in the near future.

Acknowledgments This research was supported by the Ministry of Education, Culture, Sports, Science and Technology (MEXT) Japan (Yamaguchi Green Materials Cluster, 2009–2014, and KAKENHI), the New Energy and Industrial Technology Development Organization (NEDO) Japan (2015–2017), and the Nippon ZEON Corporation, Japan.

References

1. Koumoto K, Mori T (eds) (2013) Thermoelectric nanomaterials: materials design and application. Springer, Heidelberg
2. Toshima N (2017) Recent progress of organic and hybrid thermoelectric materials. Synth Met 225:3–21
3. Toshima N (2015) Making an advance of organic thermoelectric materials: acceleration of energy recovery from waste heat with hybridization. Gendai Kagaku 532:42–46
4. Chujo Y, Tanaka K (2015) New polymeric materials based on element-blocks. Bull Chem Soc Jpn 88:633–643
5. Terasaki I (2008) Introduction to thermoelectric conversion. In: Kajikawa T (ed) Handbook of thermoelectric conversion technology. NTS, Tokyo, pp 17–20
6. Sun YM, Sheng P, Di CA et al (2012) Organic thermoelectric materials and devices based on p- and n-type poly(metal 1,1,2,2-ethenetetrathiolate)s. Adv Mater 24:932–937
7. Oshima K, Shiraishi Y, Toshima N (2015) Novel nanodispersed polymer complex, poly(nickel 1,1,2,2-ethenetetrathiolate): preparation and hybridization for n-type of organic thermoelectric materials. Chem Lett 44:1185–1187
8. Yu C, Choi K, Yin L et al (2011) Light-weight flexible carbon nanotube based organic composites with large thermoelectric power factors. ACS Nano 5:7885–7892
9. Cho C, Wallace KL, Tzeng P et al (2016) Outstanding low temperature thermoelectric power factor from completely organic thin films enabled by multidimensional conjugated nanomaterials. Adv Energy Mater 6:1502168
10. Oshima K, Yanagawa Y, Asano H et al (2017b) Improvement of stability of n-type super growth CNTs by hybridization with polymer for organic hybrid thermoelectrics. Synth Met 225:81–85
11. Toshima N, Oshima K, Anno H et al (2015) Novel hybrid thermoelectric materials: three-component hybrid films consisting of a nanoparticle polymer complex, carbon nanotubes, and vinyl polymer. Adv Mater 27:2246–2251

12. Bubnova O, Khan ZU, Malti A et al (2011) Optimization of the thermoelectric figure of merit in the conducting polymer poly(3,4-ethylenedioxythiophene). Nat Mater 10:429–433
13. Kim GH, Shao L, Zhang K et al (2013) Engineering doping of organic semiconductors for enhanced thermoelectric efficiency. Nat Mater 12:719–723
14. Oshima K, Inoue J, Sadakata et al (2017a) Hybrid-type organic thermoelectric materials containing nanoparticles as a carrier transport promoter. J Electron Mater 46(5):3207–3214

Chapter 21
Crystallization Kinetics-Induced Self-Assembly of Inorganic Element Blocks and the Surface-Enhanced Raman Scattering Based on Ag Hierarchical Structures

Jinguang Cai and Akira Watanabe

Abstract In this chapter, on basis of the concept of element blocks which can be extended to inorganic element blocks consisting of inorganic nano-cores such as inorganic nanoclusters, nanoparticles, and nanocrystals. The self-assembly behaviors and strategies of inorganic element blocks, as well as novel properties of assembly structures applied in high-performance surface-enhanced Raman scattering (SERS) were first briefly introduced. Crystallization behaviors of symmetric and asymmetric polyhedral oligomeric silsesquioxane (POSS) molecules were applied to the formation of unique fractal hierarchical structures of POSS/Ag hybrid films by using a facile spin-coating method, where POSS molecules and Ag nanoparticles have comparable sizes. A mechanism of crystallization kinetics-induced self-assembly of inorganic element blocks was proposed and demonstrated to be a novel universal method for various kinds of POSS molecules, other inorganic nanoparticles, and arbitrary substrates to form a fractal hierarchical structure. Ag fractal hierarchical structures obtained by heat treatment of a POSS/Ag hybrid film showed excellent SERS activity and performed as an effective in situ SERS-active media due to their stable adhesion to the substrate. In addition, an ultra-stable in situ SERS sensor based on Ag hybrid nanoassembly and an in situ SERS detection using a smartphone were also demonstrated. It is expected that the mechanism of crystallization kinetics-induced self-assembly could be extended to other material systems, such as small organic molecules, polymers, salts, atomic clusters, biomolecules, as well as relatively large colloidal particles.

J. Cai
Institute of Multidisciplinary Research for Advanced Materials, Tohoku University, Sendai, Japan

Institute of Materials, China Academy of Engineering Physics, Jiangyou, Sichuan, China

A. Watanabe (✉)
Institute of Multidisciplinary Research for Advanced Materials, Tohoku University, Sendai, Japan
e-mail: watanabe@tagen.tohoku.ac.jp

Y. Chujo (ed.), *New Polymeric Materials Based on Element-Blocks*,
https://doi.org/10.1007/978-981-13-2889-3_21

Keywords Crystallization kinetics-induced self-assembly · Inorganic element blocks · Surface-enhanced Raman scattering · In situ SERS sensor · Smartphone sensing

21.1 Introduction

An “element block” is a structural unit composed of different elements or elemental groups [1]. The new element blocks with typical properties and functions can be designed and synthesized and act as basic building blocks to construct higher-level hierarchical structures by polymerization or self-assembly to obtain expected multifunctions such as chemical, electrical, optoelectronic, optical, magnetic, and mechanical properties. Such polymerized materials with different types of element blocks can be easily processed by molding, showing high flexibility. It is definite that this novel concept “element block polymers” can be employed to exploit and develop lots of innovative smart polymers with various properties and functions.

Besides such smart element blocks based on chemical reactions of organic groups, there are plenty of inorganic building blocks such as nanoclusters, nanoparticles, and nanocrystals, which possess various novel optical, electrical, optoelectronic, magnetic, and catalytic properties, resulting from the typical effects caused by the size reduction [2]. These inorganic nanomaterials with typical properties and functions can be recognized as generalized inorganic element blocks of inorganic nano-cores with surface-capped molecules. Such inorganic element blocks can act as building blocks to assemble and construct higher-order hierarchical structures with multifunctions by employing the bottom-up methods. The assembly structures can exhibit not only the intrinsic properties of the original element blocks but also the novel collective properties resulting from coupling or synergetic effects [3, 4]. Generally, the self-assembly of nanocrystals is a thermodynamic process, resulting in ordered assembly structures with planar or formal shapes. However, there is another type of hierarchical structures called fractals, which are existing widely in nature. Fractals show self-similarity where a structure or dynamic pattern is repeated over multiple spatial or time scales, which are important in aesthetics, mathematics, science, and engineering. Fractal hierarchical structures assembled by molecules or nanoparticles have showed some interesting properties in sensing, electronics, optoelectronics, and optics.

In this chapter, we will first briefly introduce inorganic element blocks and their self-assembly, fractal hierarchical structures, and the widely studied applications in surface-enhanced Raman scattering (SERS) of hierarchical self-assembly structures based on metal nanoparticles. Crystallization kinetics-induced self-assembly for POSS/Ag nanoparticle fractals is proposed based on their crystallization behaviors, and self-assembly mechanism and its universality are emphasized for various POSS molecules and metal nanoparticles. Finally, the SERS performance of Ag fractal hierarchical structures prepared by crystallization kinetics-induced self-assembly is discussed, and the application as SERS-active media for a smartphone sensing is demonstrated. The summary and outlook are also discussed in the last part.

21.1.1 Inorganic Element Blocks

Generally, inorganic building blocks represent nanocrystals, nanoparticles, or nanoclusters of metal, semiconductor, or insulator, which include inorganic nanocores recognized as generalized inorganic element blocks with surface capping molecules. The sizes of them are small enough to bring novel chemical, optical, optoelectronic, electrical, and magnetic properties that are different from those of corresponding bulk materials [2]. For example, nanoparticles with an ultra-small size, which possess a very high surface area, would bring much more reaction active sites on the surface due to the high portions of the atoms with high surface energy, so they can exhibit much higher catalytic and reactive activities. Metal nanoparticles would exhibit strong surface plasmonic properties when their size locates in the region where the surface electrons can resonate periodically with the incident photons with matched wavelength. Such nanoparticle plasmonics can manipulate the incident lights elaborately by tuning the resonant light wavelength through controlling the size and shape of metal nanoparticles [5, 6], which can find many promising applications in bio-imaging, photothermal therapy, photocatalysis, optoelectronics, surface-enhanced Raman scattering, and photovoltaics [7]. Semiconductor nanoparticles show tunable energy band and quantum confinement effect depending on the sizes and are called quantum dots. The energy band of a semiconductor nanoparticle becomes discrete energy level when the size of a semiconductor nanoparticle is reduced to less than its Bohr radius [8]. Quantum dots can be used as emitter materials due to their narrow and symmetric emission spectra and have been successfully commercialized as luminescent materials in bio-imaging and quantum dots LED displays [9]. Besides, quantum dots can also be used as effective absorbents in quantum dots solar cells [10]. Magnetic nanoparticles with small sizes are widely studied for high-contrast MRI imaging, effective therapy, and high-capacity magnetic data storage [11]. Figure 21.1 shows the summary of the materials, structures, novel features, properties, and applications of inorganic building blocks.

21.1.2 Fractal Hierarchical Structures

To create practical multifunctional materials or devices, it is critical to organize nanomaterials with pre-programmed functionality and self-assembly instructions through bottom-up methods. Different from the covalent bonds interactions in polymerized element blocks or atomic crystals, the inorganic building blocks are generally assembled together through weak forces (e.g., van der Waals, hydrogen bonding), hard-particle (e.g., excluded volume) interactions, or molecular linkers. A

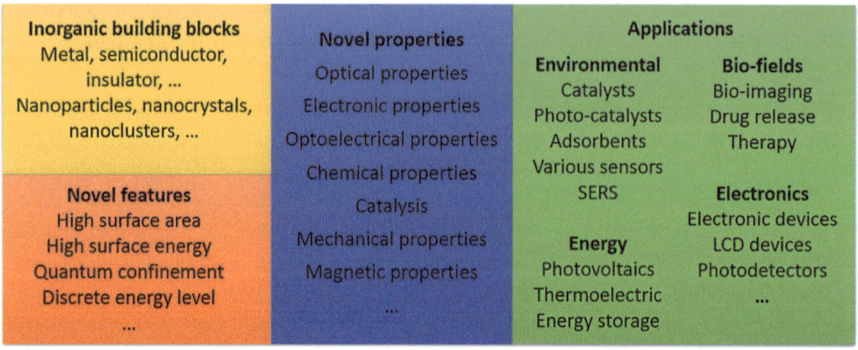

Fig. 21.1 Summary of the materials, structures, novel features, properties, and applications of inorganic building blocks

schematic self-assembly strategy and main influence parameters were summarized by Talapin et al. as shown in Fig. 21.2 [12]. From inner to outer, there are three levels of parameters that influence the self-assembly process largely. First is the inorganic nano-core, i.e., generalized inorganic element block, the size and shape of which (e.g., nanospheres, nanopolyhedra, nanorods, nanosheets, nanopods, etc.) show large influences on the final assembly structures through space interaction, especially nonspherical nanoparticles. The second level of influence parameters is the surface-bond ligands. By modifying the surface with ions, surfactants, polymers, and even DNA molecules, the inorganic element blocks will show different assembly behaviors and can even be directed by the interaction of surface ligands, forming programmed assembly structures. In addition, generalized environments such as solvents, temperature, depletion, confinement, patterning, and outer forces also play important roles in the self-assembly process and the final structures [12]. Generally, self-assembly of inorganic element blocks is controlled by the comprehensive effect of several above parameters. Most of such self-assembly strategies are thermodynamically stable processes, which could produce well-crystallized ordered self-assembly structures with planar or formal shapes. But it is difficult to produce very complicated structures, such as fractals.

Fractal hierarchical structures are characterized with self-similarity where a structure or dynamic behavior is repeated over a broad range of spatial, time, or other scales. Fractals are complicated yet fascinating patterns and widely existing in nature, such as cauliflower, ferns, peacock feathers, Koch snowflake, lightning, and waterfall, inspiring many creations in aesthetics, mathematics, science, and engineering. There are some other well-known fractals including Brownian motion, polymer networks, aggregation growth phenomena, porous media, glasses, brain networks, structural details of genomes, and complex dynamics in human physiology [13]. Fractal structures have shown advantages in many application fields. For example, silicon detectors on transparent electrodes designed with Hilbert and Peano fractal geometries show broadband enhanced responsivity in the 450–750 nm range without any polarization dependence different from the case of the grating and

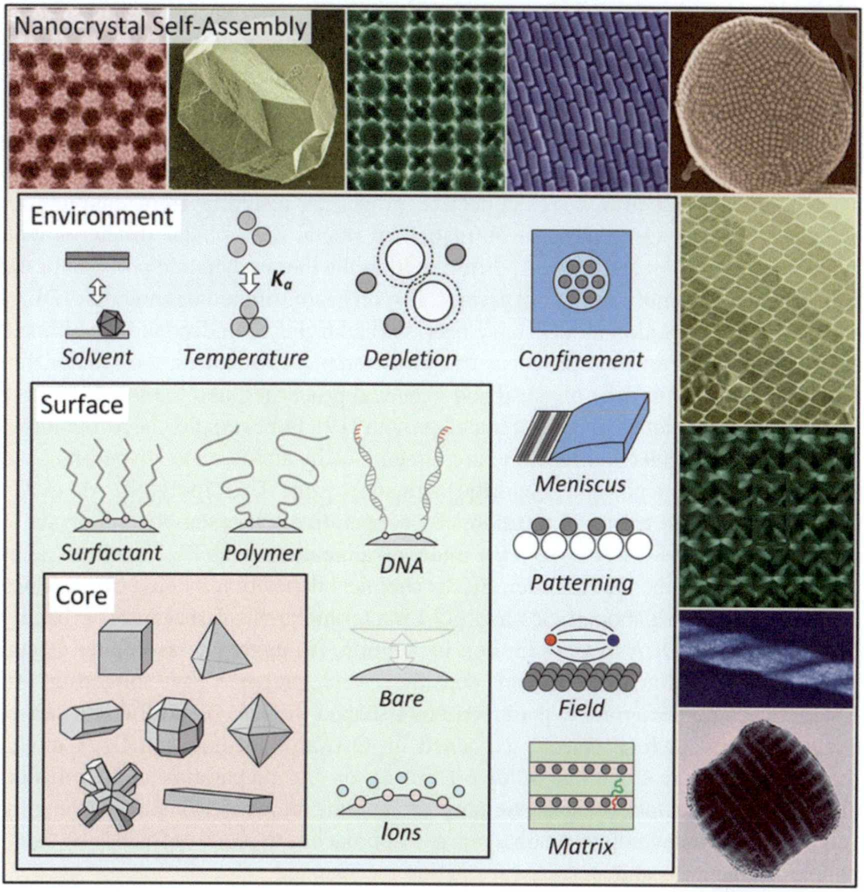

Fig. 21.2 Self-assembly process of inorganic building blocks involves many parameters ranging from the inorganic nano-core, i.e., inorganic element block, the surface capping ligands, to the environments and outer forces (Reprinted with permission from [12])

grids [14]. Fan et al. demonstrated the fractal design concepts for stretchable electronics by using Peano, Greek cross, Vicsek, and other fractal constructs in space-filling structures of electronic materials, including monocrystalline silicon, for electrophysiological sensors, precision monitors and actuators, and radio-frequency antennas [15]. Fractal antennas with multiband, broadband, compact, and light-weight properties are widely studied for wireless communication in mobile devices [16]. Fractal geometries allow large area, conformal electrodes which are suitable for delivering cardiac electrical stimulation and for sensing cardiac electrical activity [17]. A three-dimensional fractal nanobiointerface was constructed for efficient cancer cell capture by growing a hierarchical assembled ITO nanowire array with both horizontal and vertical nanowire branches on a substrate [17]. In addition,

Gottheim et al. demonstrated that the fractal Cayley tree nanostructures exhibit multimodal plasmon spectrum with the fractal order increasing, indicating that fractals with large N would show broad, multipeaked spectra from plasmons with large degeneracy numbers [18].

Fractals constructed by inorganic element blocks would be interesting hierarchical structures because they not only exhibit the intrinsic properties of original element blocks but also novel collective properties resulting from coupling or synergetic effects. Generally, the formation of fractal hierarchical structures with nanomaterials is a kinetic process, different from the thermodynamic process for the common self-assembly of nanomaterials. The diffusion-limited aggregation (DLA) and ballistic aggregation models have been studied trying to understand the ultimate aggregation dynamics that produces fractal hierarchical structures, the relationship of this fractality with their physical and chemical properties, and the most effective methods and techniques to control fractal growth [19]. For example, the formation of highly ordered fractal dendritic structures assembled by amido-ended hyperbranched polyester [20], fractal and dendritic structure from CdTe/poly(NIPAM-AAc) microgels [21], and fractal-like patterns by cooperative self-assembly and crystallization of $(PNIPAM\text{-}b\text{-}PAA)_2\text{-}(PVP)_2$ under alkaline conditions [22] has been demonstrated obeying the DLA model. Electrochemical deposition method can produce silver fractals with nanoparticles and gold fractal hierarchical structures, probably controlled by the DLA model [23, 24]. In addition, Au-aggregate-assembled fractal patterns with tailored sizes and densities were prepared by sputtering Au nanoparticles on hexagonally patterned bowl-shaped-dimples on Al foil and subsequent annealing, which can be explained by Ostwald ripening and DLA model [25]. Though there are some studies focusing on the preparation and formation mechanism of fractals by self-assembly of building blocks, it is still challenging and important to develop new strategies and approaches to produce fractals by using bottom-up methods and study the formation mechanism.

21.1.3 Surface-Enhanced Raman Scattering

One of the characteristic features of self-assembly hierarchical structures of metal nanoparticles is the extremely significant enhancement of Raman spectra of the surface molecules under matched exited light irradiation, in comparison with isolated nanoparticles. In this section, we will briefly introduce surface-enhanced Raman scattering (SERS) by hierarchical structures constructed by metal nanoparticles and its enhancement mechanism.

When a metal nanoparticle or nanocrystal is irradiated by light, the oscillating electric field causes the conduction electrons to oscillate coherently [5]. This oscillation is called surface plasmon resonance (SPR), inducing a strong enhancement of absorption and scattering of electromagnetic radiation in resonance with SPR frequency. Four factors determine the oscillation frequency, i.e., the density of electrons, the effective electron mass, and the shape and size of the charge distribution. SPR of metal structures with enhanced electromagnetic (EM) fields on the surfaces has been widely studied in many applications, such as sensing, imaging, catalysis, nanoantenna, and energy fields [4]. This SPR enhancement is significantly favored for Raman spectroscopy because of the very small cross sections of Raman scattering. Raman scattering by molecules adsorbed on metal surfaces is enhanced through the EM and chemical enhancement mechanisms based on the excitation of localized surface plasmons and the formation of charge-transfer complexes, respectively, which is called surface-enhanced Raman scattering (SERS). SERS intensity is proportional to the fourth power of the localized electric field, leading to an enhancement factor (EF) up to more than 10^{10} through EM mechanism mode, while the EFs based on chemical enhancement mode are about 10–10^4. However, the calculated electromagnetic EFs for isolated particles, either single spheres or single polygons, are not very large, which are typically on the order of only 10^3. Very high SERS intensity results from the extraordinarily strong EM fields generated on rough metal surfaces, at sharp tips of nanostructures or plasmonically enhanced "hot spots" formed in the nanoscale gaps between closely packed nanostructures (Fig. 21.3), which are meaningful for high EF SERS sensors, even single-molecule SERS [26–28].

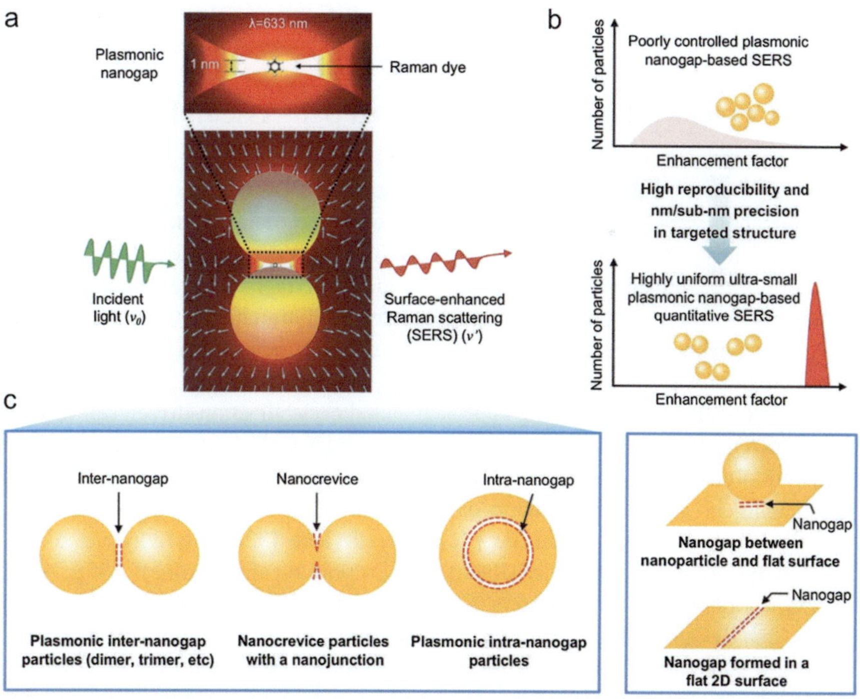

Fig. 21.3 Plasmonic nanogap structures for surface-enhanced Raman scattering. (**a**) Electromagnetic (EM) field distribution in the 1 nm gap of a dimer, where the SERS is highly enhanced. (**b**) Controlling of narrower nanogap distributions with larger SERS enhancement factors. (**c**) Different types of plasmonic nanogaps generated in particles (left) and 2D film (right) (Reprinted with permission from [28]. Copyright (2016) American Chemical Society)

Theoretical calculation indicates that the maximum EF in the hot spot decreases by three orders of magnitude when the gap size is increased from 1 nm to 5.5 nm, which can also be demonstrated by the experiments. The enhancement effect becomes rather low when the gap size is larger than 15 nm. Therefore, rational design and preparation of self-assembly hierarchical structures with controlled nanosize gaps less than 10 nm is beneficial for the SERS with high EFs. Till now, various nanostructures with nanogaps have been prepared by the direct growth method or self-assembly approaches [28]. Generally, the nanogaps consist of two types, i.e., intra-particle nanogaps and inter-particle nanogaps (Fig. 21.3). The intra-particle nanogaps are always formed in the core-shell or yolk-shell metal nanoparticles or metal nanourchin structures, which can be fabricated by direct growth or controlled stepwise growth with or without an etching process [26]. The inter-particle nanogaps are generally constructed by controlling the assembly of metal building blocks or the templated deposition or growth [27]. For example, the hierarchical structures such as multimeric nanostructures including dimers and trimers and superlattices with controlled distance can be prepared by molecular linker directed self-assembly and evaporating-induced self-assembly, respectively [29], while the nanostructure arrays and nanowire arrays can be fabricated by nanosphere crystal lithography and Al_2O_3 templated growth, separately [30].

21.2 Crystallization Kinetics-Induced Self-Assembly

Many approaches, such as evaporation-induced self-assembly of nanoparticles, DNA- or polymer-directed organization of building blocks, and environments-guided assembly of nanoparticles, have been used to prepare ordered hierarchical structures with controlled size, morphology, and even inter-particle distance. A few studies have been reported for preparation of fractal hierarchical structures obeying DLA mechanism. However, it still needs to develop new strategies to construct fractal hierarchical structures assembled by inorganic building blocks. There is another interesting and promising approach called crystallization-driven self-assembly (CDSA) which has been developed to construct well-defined structures from block copolymers. To realize the crystallization process, there should be some polymer blocks with crystallization capability incorporated in the polymer chain, typically, such as polyacrylonitrile [31], polyethylene [32], poly(ethylene oxide) [33], poly(ferrocenyl dimethylsilane) [34], poly(ε-caprolactone) [35], poly (3-hexylthiophene) [36], or stereoregular polylactides [37]. At the same time, typical groups can be grafted on the crystallized bones to realize multi-functionality. However, such block polymers almost only form specific structures, such as one-dimensional cylindrical or wormlike crystalline-core micelles [38]. In this part, we will discuss the other type of crystallized species called polyhedral oligomeric silsesquioxane (POSS) and its crystallization kinetics-induced self-assembly of nanoparticles to fractal hierarchical structures.

21.2.1 POSS Family and Their Crystallization Behaviors

POSS is a type of polymeric structures of highly ordered discrete molecular species with a general formula $(RSiO_{3/2})_n$ where n is commonly 6, 8, or 10. In this chapter, we focus on the POSS compounds based on the $(RSiO_{3/2})_8$ formula, which show some important features of POSS molecules such as a nanosize siloxane cage core and outer eight pendant arms distributed surrounding the cube in a three-dimensional arrangement. It is interesting that such cubic siloxane cages could be substituted by different groups or chains, respectively, to realize expected properties and functionalities; thus they have become popular as nanometer-scale building blocks to achieve typical functionality in a wide range of polymeric materials. For example, POSS species are always incorporated in polymeric materials or nanocomposites to modify the properties and performances. The flammability and the viscoelastic, thermal, optical, fluidic, and mechanical properties can be tuned by incoporating POSS species. POSS can also be introduced into catalysts, biomedical materials, cosmetics, and even act as building blocks for preparing zeolite-like materials [39].

At the molecular level, POSS species contains hydrophobic cubic siloxane cores with crystallization capability, to which other various functional groups or chains can be attached, thus abundant self-assembly behaviors can be achieved in bulk, solution, and thin films, resulting in many interesting assembly structures such as periodical nanostructures, microphase-separated nanostructures, spherical core-shell micelles, cylinders, vesicles, and thin films with ordered structures [40–50]. The CDSA mechanism has also been used to design and prepare hierarchical structures from POSS-based molecules. For example, the square nanosheets with tunable size have been successfully prepared by using living crystallization-driven two-dimensional self-assembly of hyperbranched poly(ether amine) capped with heptaisobutyl-POSS [38]. However, the preparation of hybrid fractal hierarchical structures through co-assembly of inorganic building blocks and organic element blocks by crystallization-driven self-assembly is still a big challenge.

POSS-based molecules could act as effective building blocks due to the tunable size and surface wettability, crystallization capability, and modification possibility. First, the crystallization behaviors of POSS molecules should be investigated. In this chapter, two types of POSS molecules are involved, i.e., symmetric POSS and asymmetric POSS, which are defined according to the difference of substitute groups. Here, the behaviors of two typical molecules representing symmetric POSS (octavinyl-silsesquioxane, OV-POSS) and asymmetric POSS (allyl-heptaisobutyl-silsesquioxane, AHI-POSS), respectively, are studied by drop-casting their toluene solutions on the glass or Si substrates and allowing thermodynamic crystallization by gradual evaporation at room temperature in air. The structures formed on the substrates are then characterized by optical microscopy (OM) and X-ray diffraction (XRD).

The OM image and XRD patterns of OV-POSS structures formed on the substrates by drop-casting the OV-POSS solution in toluene and allowing evaporation at room temperature in air are shown in Fig. 21.4. The OM image indicates that lots of particles

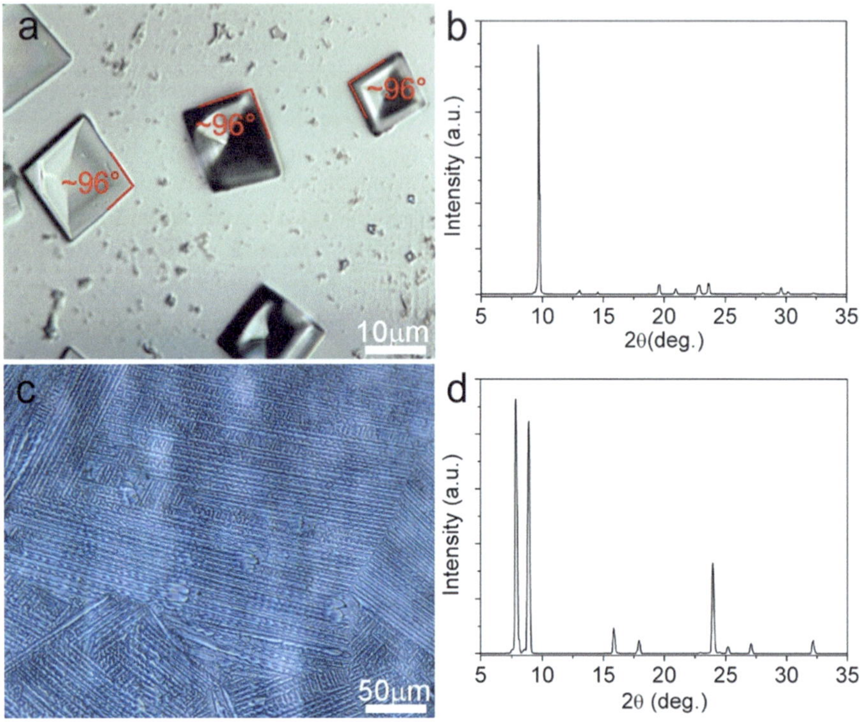

Fig. 21.4 (**a**) Optical microscopy image of OV-POSS crystals formed by drop casting of OV-POSS solution. The crystals of parallelepiped shape show an identical angle around 96°. (Adapted with permission from [51]. Copyright (2015) American Chemical Society.) (**b**) XRD pattern of OV-POSS structures formed on glass substrate. (**c**) Optical microscopy image of AHI-POSS crystals formed by drop casting of AHI-POSS solution. (**d**) XRD pattern of AHI-POSS structures formed on glass substrate

with parallelepiped shape and size ranging from several micrometers to tens of micrometers (Fig. 21.4a). Interestingly, such parallelepiped particles show a coincident angle of about 96°, which may be dominated by the crystal lattice parameters. XRD pattern in Fig. 21.4b shows a strong X-ray diffraction intensity, indicating a highly crystallization property. The analysis of the XRD pattern indicates that OV-POSS crystal belongs to monoclinic crystal system, and the lattice parameter is $a = 9.16$ Å, $b = 6.83$ Å, $c = 9.10$ Å, $\alpha = 90°$, $\beta = 96.33°$, and $\gamma = 90°$. The specific angle of the micrometer-size parallelepipeds in Fig. 21.4a may be determined by the lattice parameter β. The crystal structures may be eventually dominated by the molecular structure of OV-POSS. Such crystallization behavior can be essentially recognized as a molecular self-assembly process. Therefore, the crystallization behaviors of asymmetric POSS may be different from symmetric POSS. The OM image of AHI-POSS structures formed on the substrate is shown in Fig. 21.4c, indicating a very interesting one-dimensional branched structure, which is quite different from OV-POSS structures. The XRD pattern of AHI-POSS structures suggests the high crystallization property and the difference from that of OV-POSS (Fig. 21.4d). Therefore, the molecular structures dominate the crystallization behaviors, which can be used to tune self-assembly structures.

21.2.2 Solvent Evaporation-Induced Crystallization Kinetics

The molecules with POSS cores always have sizes from about 1 nm to several nanometers depending on substituted groups, which is comparable to the size of metal nanoparticles of about several nanometers. Therefore, it is expected that POSS-based molecules mixed with nanoparticles of comparable size may show interesting assembly behaviors. Here, Ag nanoparticles with a size of 4.5 nm and the typical OV-POSS of about 1 nm in toluene were used to study and discuss the crystallization and self-assembly behaviors.

The typical concentrations of Ag nanoparticle dispersion in toluene (ULVAC Ag1T) and OV-POSS solution in toluene are 25 wt% and 5 wt%, respectively. First, 20 μL of Ag nanoparticle dispersion and 100 μL of OV-POSS solution were mixed thoroughly and then drop-casted on the substrate, allowing the evaporation of solvent in air. Figure 21.5 shows the OM image of the resultant structures formed on the substrate by drop-casting method. It is apparent that many parallelepiped-shape particles were formed, which are the same shape as pure OV-POSS particles formed by drop-casting method. Other colored parts surrounding OV-POSS particles are recognized as Ag nanoparticles. This suggests that the OV-POSS and Ag nanoparticles are crystallized and self-assembled separately, and no co-assembly behaviors took place. The separate assembly behavior may be caused by the slow evaporation process of the solvent, which allows enough time for the diffusion of

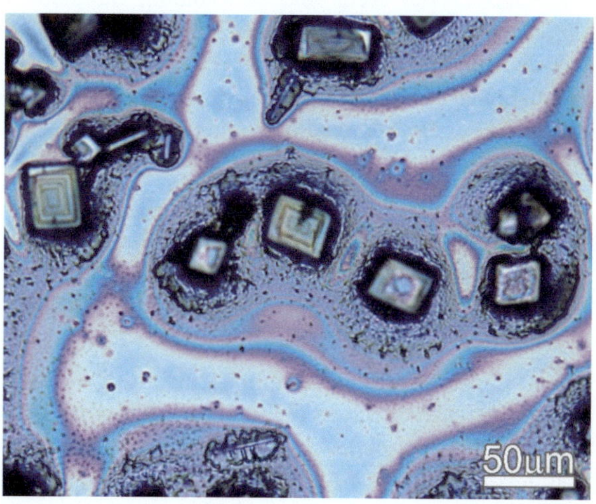

Fig. 21.5 OM image of the structures formed on the substrate by drop-casting the mixture of Ag nanoparticle and OV-POSS in toluene

OV-POSS molecules to form thermodynamically stable crystals in the equilibrium crystallization.

Accelerating solvent evaporation process may induce fast crystallization within a very short time, largely reducing the diffusion time of OV-POSS molecules in the solution for crystallization, thus different co-assembly structures may be formed (Fig. 21.6a). Spin-coating method can provide such non-equilibrium condition, because fast rotation can throw out redundant liquid, leaving a thin layer of the solution, which can be evaporated in a very short time [51]. The mixture solution of Ag nanoparticles and OV-POSS was spin-coated on the substrate at 2000 rpm for 30 s with an accelerating time of 5 s from 0 to 2000 rpm. The OV-POSS was removed by heat treatment on a hotplate at 250 °C. Figure 21.6 shows as-prepared OV-POSS/Ag structures and Ag structures after heat treatment. OM image in Fig. 21.6b indicates that Ag/OV-POSS hybrid fractal hierarchical structures formed on the substrate have uniform cross-star shapes with a scale of about 30–60 μm over a large area. The enlarged OM image in Fig. 21.6c clearly shows sub-branches with the same direction with length decreasing gradually from cross center to tip in the branches of cross-star structures. This is demonstrated by the SEM image in Fig. 21.6d, which also shows different brightness between center and other parts, suggesting the different distribution of Ag nanoparticles and OV-POSS crystals. A high-magnification SEM image in Fig. 21.6e shows the ordered arrangement of Ag nanoparticles in bright part in Fig. 21.6d, even on curved surfaces. The SEM image of the structures after heat treatment in Fig. 21.6f suggests that the evaporated center parts are OV-POSS skeletons surrounded by Ag nanoparticles. Enlarged SEM image in Fig. 21.6g indicates the fusion of Ag nanoparticles during heat treatment process. OV-POSS hybrid fractal hierarchical structures and remaining Ag nanostructures after heat treatment were also demonstrated by corresponding FT-IR spectra and XRD patterns. Therefore, fast evaporation of solvent can control the crystallization kinetics of POSS molecules to induce self-assembly behaviors of Ag nanoparticles. Novel cross-star Ag/OV-POSS hybrid fractal hierarchical structures with inner OV-POSS crystal cores and surrounding inorganic nanoparticles can be easily fabricated within tens of seconds by using a facile spin-coating method. Besides, hierarchical cross-star Ag fractals consisting of interconnected larger Ag nanoparticles can also be easily obtained by a simple heat treatment, where POSS crystals are easily removed by sublimation.

21.2.3 Effect of Parameters on the Self-Assembly and the Self-Assembly Mechanism

There are many parameters influencing the OV-POSS crystallization-driven self-assembly process, such as the rotating rate during spin-coating process, concentrations of Ag nanoparticles and OV-POSS solution. As we have known, spin coating from mixed solution of Ag nanoparticles and OV-POSS can produce quite different structures from drop-casting, which means rotating plays a critical role in the

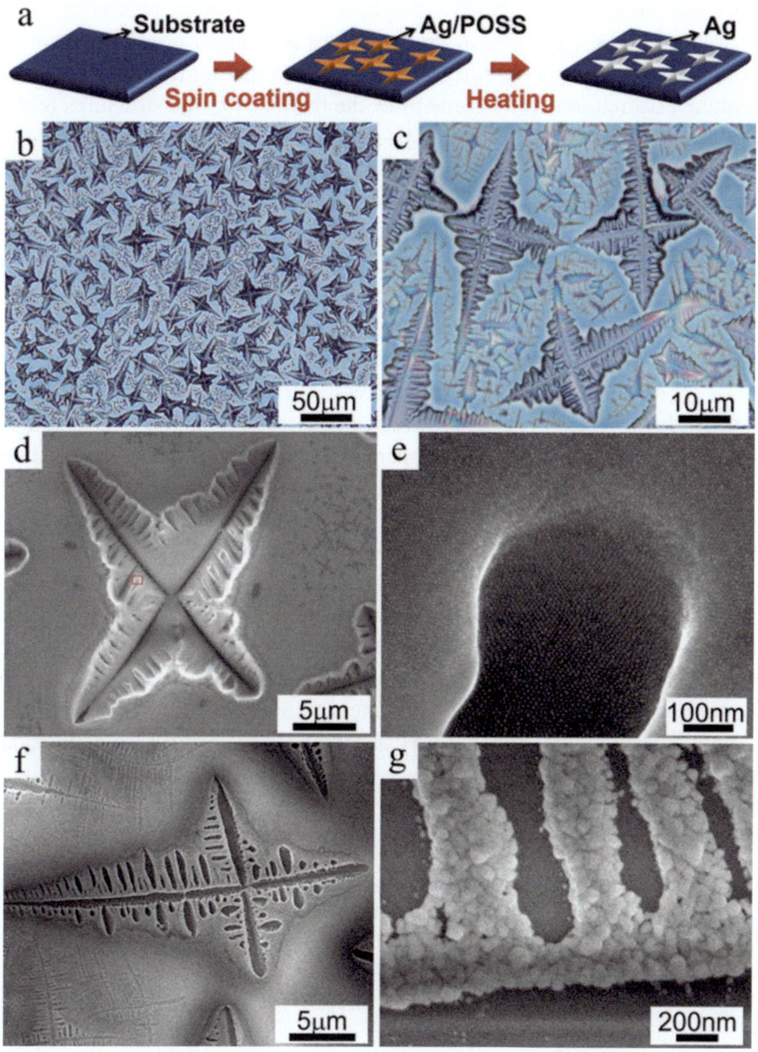

Fig. 21.6 (**a**) Schematic illustration for preparation of Ag/OV-POSS fractal hierarchical structures and the evaporation of OV-POSS by heating treatment. (**b**, **c**) Optical microscopy images of Ag/OV-POSS fractal hierarchical structures prepared by spin coating. (**d**, **e**) SEM images of Ag/OV-POSS fractal hierarchical structures prepared by spin coating. (**f**, **g**) Corresponding SEM images after heat treatment. (Reprinted with permission from [51]. Copyright (2015) American Chemical Society)

formation of fractal hierarchical structures. Therefore, rotating rate must be an important parameter to control resultant hybrid structures. To demonstrate the influence of rotating rates on the hierarchical structures more effectively, the solution was directly spin-coated on the substrate at rotating rates ranging from 500 to 3000 rpm without accelerating time. As the rotating rate was changed from 500 to

3000 rpm, the cross-star Ag/OV-POSS hybrid fractal hierarchical structures can still be prepared. However, the size of the structures is reduced gradually from around 30–60 μm at 500 rpm to 5–15 μm at 3000 rpm, and the number of sub-branches is also gradually decreased. At the same time, the thickness of the structures is changed from around 1.4–2 μm at 500 rpm to 300–500 nm. Therefore, rotation is not only one key factor in the formation of cross-star Ag/OV-POSS hybrid fractal hierarchical structures but also an important factor to tune the size and thickness of the structures.

The presence and concentration of Ag nanoparticles must have an important influence on the formation of fractal hierarchical structures, because there are no fractal hierarchical structures of OV-POSS under the absence of Ag nanoparticles. The original concentration of OV-POSS solution was kept constant at 5 wt%, and the concentration of original Ag nanoparticle dispersion was changed. As the concentration of original Ag nanoparticle dispersion is decreased gradually from 62.5 to 2.5 wt%, the hybrid fractal hierarchical structures with sub-branches are gradually changed to a large crystal particle surrounded with Ag nanoparticles and sub-branches gradually disappear (Fig. 21.7). Therefore, the presence of Ag nanoparticles is another key point to get fractal hierarchical structures. As the crystallization material, the concentration of OV-POSS also influences the assembly structures. As the concentration of OV-POSS solution decreases, the size and sub-branches of OV-POSS crystallization cores gradually decreases, until forming nearly uniform film of Ag nanoparticles. Synchronous decrease of the concentrations of Ag nanoparticles and OV-POSS can also produce hybrid fractal hierarchical structures with similar morphology but lower size and thickness. In addition, uniform long band-like structures with length of tens of micrometers and thickness of 200–400 nm was also obtained by doctor-blading method with the same mixture solution.

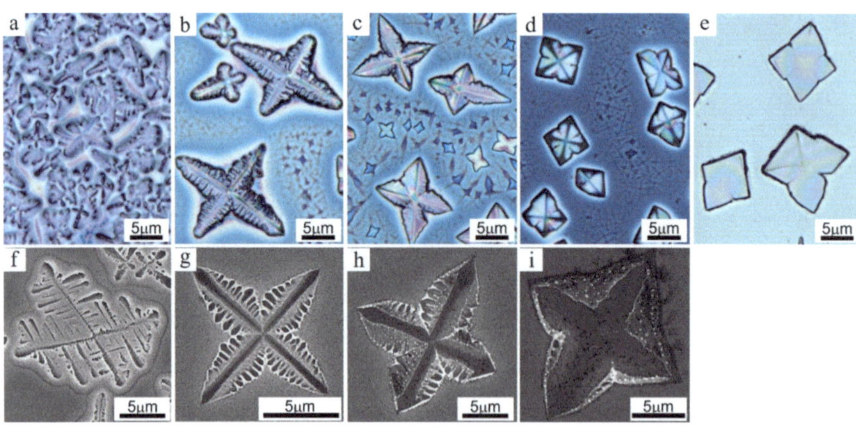

Fig. 21.7 Optical microscopy images of as-prepared Ag/OV-POSS hybrid structures at different Ag concentrations (**a**–**e**) and SEM images of corresponding samples after heat treatment (**f**–**i**). Ag concentrations: (**a**, **f**) 62.5 wt%, (b, g) 12.5 wt%, (c, h) 6.25 wt%, (**d**, **i**) 2.5 wt%, and (**e**) 0 wt% (Reprinted with permission from [51]. Copyright (2015) American Chemical Society)

The formation mechanism was proposed based on the above investigation (Fig. 21.8). Basically, the crystallization behavior is controlled by the diffusion of OV-POSS molecules. All the studied parameters, such as rotating rates, Ag nanoparticles, and concentrations of each species in spin-coating method, as well as doctor-blading method, influence the resultant structures by influencing the diffusion and available quantity of the OV-POSS molecules. In the drop-casting case, OV-POSS molecules in the solution have enough time for diffusion to form large crystal particles (equilibrium crystallization). In the spin-coating case (non-equilibrium crystallization), only one thin layer of solution remains on the substrate after high speed rotating, inducing a fast evaporation of the solvent. The very fast evaporation process allows a very short time for OV-POSS diffusion, meanwhile the presence of Ag nanoparticles with comparable size obstacles the diffusion path of OV-POSS, thus formig fractal hierarchical OV-POSS/Ag hybrid structures. The increase of rotating speed can reduce the thickness of remaining solution layer, leading to a shorter evaporation time and less OV-POSS molecules, thus producing hybrid fractal hierarchical structures with smaller size and thickness. The increase of Ag nanoparticle concentration enhances the space resistance to OV-POSS diffusion, thus resulting in hybrid fractal hierarchical structures with increasing sub-branches. In the doctor-blading case, the evaporation of solvent from the thin layer dispersion lasts a little long time, and the evaporation direction of the thin layer is not uniform but arbitrary, so long-band hybrid branched hierarchical structures with arbitrary directions instead of cross-star branched hierarchical are formed.

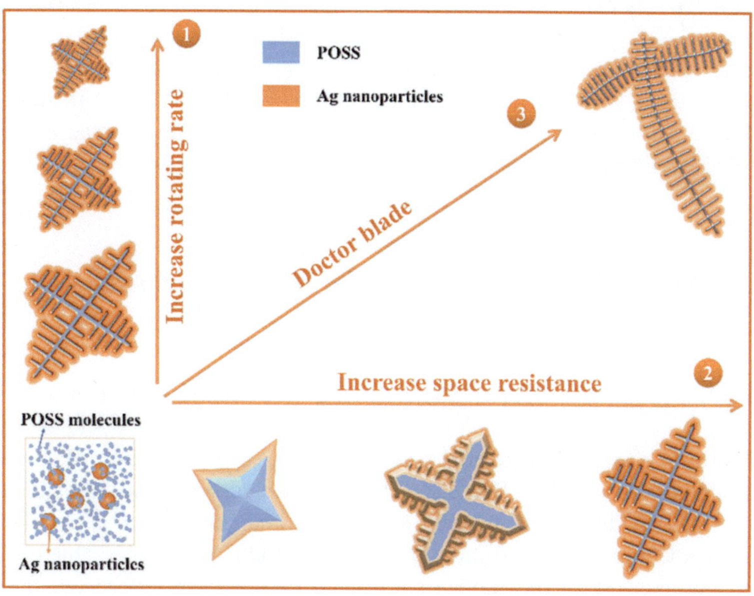

Fig. 21.8 Schematic illustration of the formation mechanism of Ag/OV-POSS fractal hierarchical structures controlled by crystallization kinetics and space resistance (Reprinted with permission from [51]. Copyright (2015) American Chemical Society)

21.2.4 Universal Methods

The formation mechanism of crystallization kinetics-driven self-assembly is a universal method for the preparation of hybrid fractal hierarchical structures in at least three aspects. First, OV-POSS can be extended to other symmetric POSS molecules such as octacyclohexyl-POSS (OCH-POSS) to prepare similar hybrid fractal hierarchical structures by tuning the concentration ratio, because the OCH-POSS possesses the similar crystallization behavior and comparable size. Second, the mechanism can be extended to other nanoparticles with comparable size and surface modification. For example, nanoparticle dispersions of Au, Cu, or InSn alloy can achieve in similar fractal hierarchical assembly structures as shown in Fig. 21.9, because they can play a similar role in hindering the diffusion path of POSS molecules. In addition, the preparation of Ag/OV-POSS hybrid fractal hierarchical structures doesn't depend on the substrates, and similar composite structures can be achieved on arbitrary substrates, such as glass, silicon wafer, Al foil, Cu foil, hydrophobic polytetrafluoroethylene, and even thin flexible polyimide film.

21.3 Surface-Enhanced Raman Scattering Based on Ag Structures

As stated in the first section, the hierarchical structures of self-assembled novel metal nanoparticles may exhibit a high enhancement factor in the SERS measurement due to the numerous presence of gaps between building blocks which can act as "hotspots." Here, pure Ag fractal hierarchical structures comprised of interconnected Ag nanoparticles without surface capping agents obtained after heat treatment are expected to act as effective SERS substrates for the detection of trace number of molecules adsorbed on the surface.

21.3.1 SERS Performance of Ag Fractal Hierarchical Structures

The *p*-aminothiophenol (PATP) was used as a probe molecule to evaluate SERS performance of Ag fractal hierarchical structures. SERS signals of PATP molecules adsorbed on Ag fractal hierarchical structures by Ag-S bonds were picked up, and Raman signals of solid PATP and PATP molecules adsorbed on Ag nanoparticle film obtained by heat-treating pure Ag nanoparticle films were also taken for comparison. Compared with Raman spectrum of solid PATP, remarkable enhancement in the intensity of the Raman shift bands can be clearly observed from the SERS spectra of PATP adsorbed on Ag fractal hierarchical structures, which is much higher than the flat part with Ag nanoparticles instead of Ag fractal hierarchical

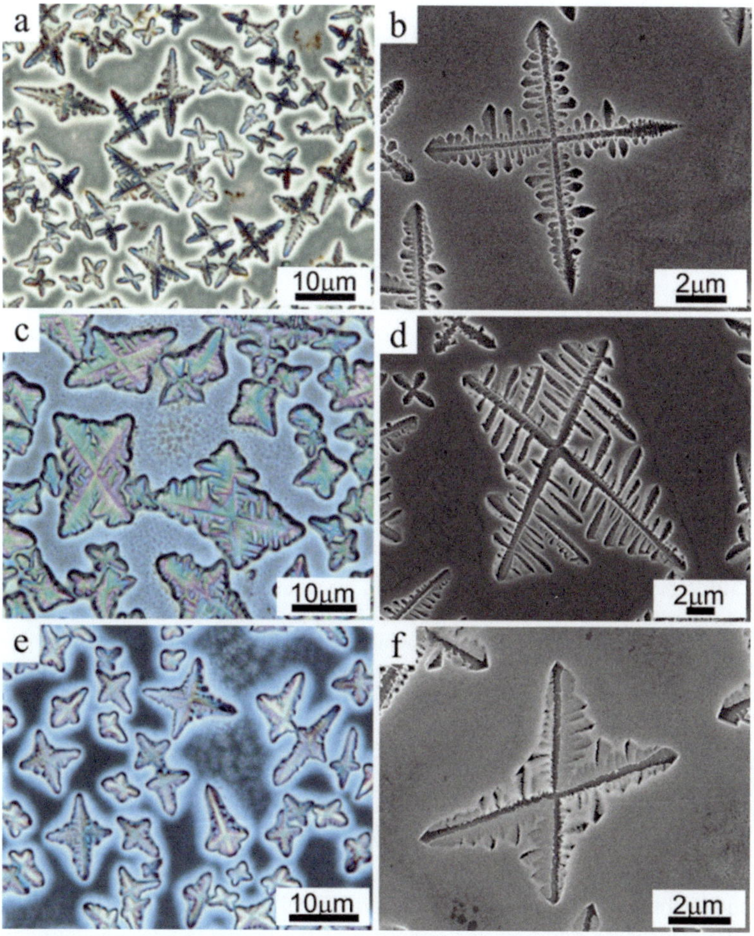

Fig. 21.9 OM images of different inorganic nanoparticles and OV-POSS composites obtained by spin-coating method (**a**, **c**, **e**) and corresponding SEM images after heat treatment (**b**, **d**, **f**). (**a**, **b**) Au, (**c**, **d**) Cu, and (**e**, **f**) InSn nanoparticles

structures and heat-treated pure Ag nanoparticle film (Fig. 21.10). The enhancement effect on the Raman signals here may be dominated by electromagnetic (EM) model instead of charge-transfer (CT) model [52–54]. The calculated EF for the Ag fractal hierarchical structures at two typical Raman shift peaks 1582 cm^{-1} and 1444 cm^{-1} of a_1 vibration mode and b_2 vibration mode, respectively, were 4.11×10^6 and 2.59×10^7. The high EF values could be attributed to the coupling of SPR between the neighboring branches and the inter-nanoparticle "hotspots," resulting in a detection limit as low as 1 nM for PATP molecules. In addition, the Ag fractal hierarchical structures almost exhibit relatively uniform and highest SERS intensity in every part of the cross-star branched structures, much higher than other flat parts.

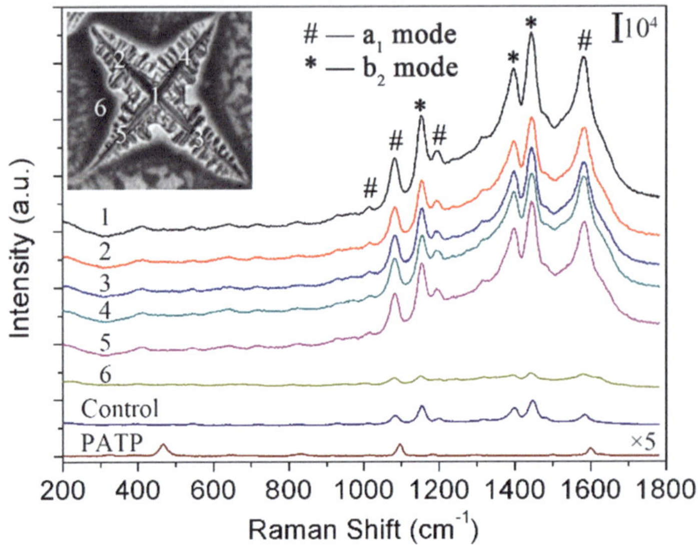

Fig. 21.10 Raman spectrum of solid PATP, SERS spectrum of PATP molecules adsorbed on pure Ag nanoparticle film after heat treatment (Control line), and SERS spectra of PATP molecules at different parts of the substrates with Ag fractal hierarchical structures obtained by heat-treating the typical Ag/OV-POSS structures (1–6). Inset is corresponding single Ag fractal structures in SERS measurement (Reprinted with permission from [51]. Copyright (2015) American Chemical Society)

21.3.2 In Situ SERS Sensor Based on Ag Fractal Hierarchical Structures

It is reasonably expected that the excellent SERS performance and substrate-independent property make Ag fractal hierarchical structures act as an effective in situ SERS sensor. In addition, the adhesion of Ag fractal hierarchical structures to the substrate is strong enough for practical use due to the heat treatment which may make Ag fractal structures tightly contacted to the substrate. The PATP used as mimic contamination was drop-casted on the apple skin. No apparent Raman signals appear from the apple skin without the contact of Ag fractal hierarchical structures, but Raman signals were largely enhanced after the substrate with Ag fractal hierarchical structures contacted to the apple skin due to the excellent SERS performance of Ag fractal hierarchical structures (Fig. 21.11). Therefore, the substrate with Ag fractal hierarchical structures may act as an effective SERS sensor for in situ detection of contamination, pesticide, and biomolecules.

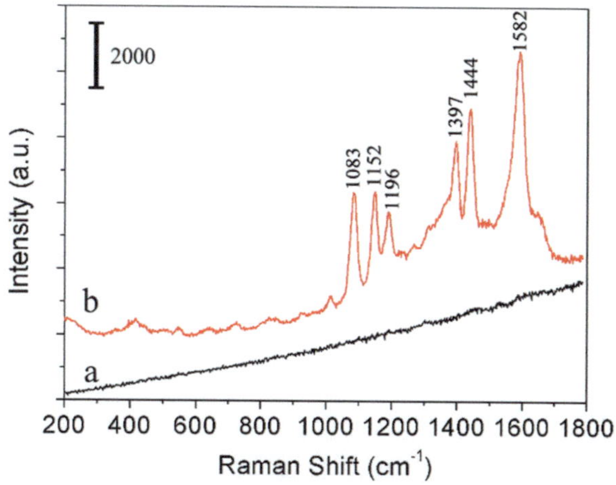

Fig. 21.11 Raman spectra of PATP molecules casted on the apple skin before (**a**) and after Ag fractal hierarchical structures contacted (**b**) (Reprinted with permission from [51]. Copyright (2015) American Chemical Society)

21.3.3 Ultra-stable In Situ SERS Sensor Based on Ag Hybrid Nano-Assembly

Generally, assembled metal structures formed on substrates by evaporation method have a relatively weak interaction with the substrates, which makes the in situ SERS detection difficult due to the stability, because the intimate contact between metal structures and the surface to be measured is necessary to receive SERS signals. In the above Ag fractal hierarchical structures, heat treatment enhanced the stability of Ag structures without damaging fractal structures, but it still caused the fusion of primary Ag nanoparticles which may reduce the SERS performance. Here, another strategy by using chemical anchoring method is introduced, which may perfectly keep the structures of primary metal nanoparticles. The marine mussel strategy of adhesion in aqueous environments inspires the preparation of a stable adhesion layer between metal nanoparticles and substrates. A catechol-functionalized polysiloxane (CFPS) was first synthesized, exhibiting a good film forming ability on various organic and inorganic substrates using a facile dip-coating method with a surface roughness of 4.6 nm [55]. Ag nanoparticles can be easily anchored onto CFPS-modified substrates by using another dip-coating process with homogeneous distribution (Fig. 21.12), and the surface number density and average interspace of Ag nanoparticles can be tuned easily by controlling the concentration of Ag nanoparticle dispersions. The enhancement factor of a substrate with high-density Ag nanoparticles can reach as high as 7.89×10^7, and an ultra-low detection limitation can be lowered to 10^{-10} M, which may be attributed to "hotspots" formed between Ag nanoparticles controlled by CFPS-induced self-assembly. The strong hydrogen bonding interactions provided by the high surface-density catechol units of CFPS make Ag nanoparticles anchored to be extremely stable on substrates. Such stable structures were demonstrated to be an effective in situ SERS sensor for the detection of contamination molecules on an apple peel with no damage to the structures.

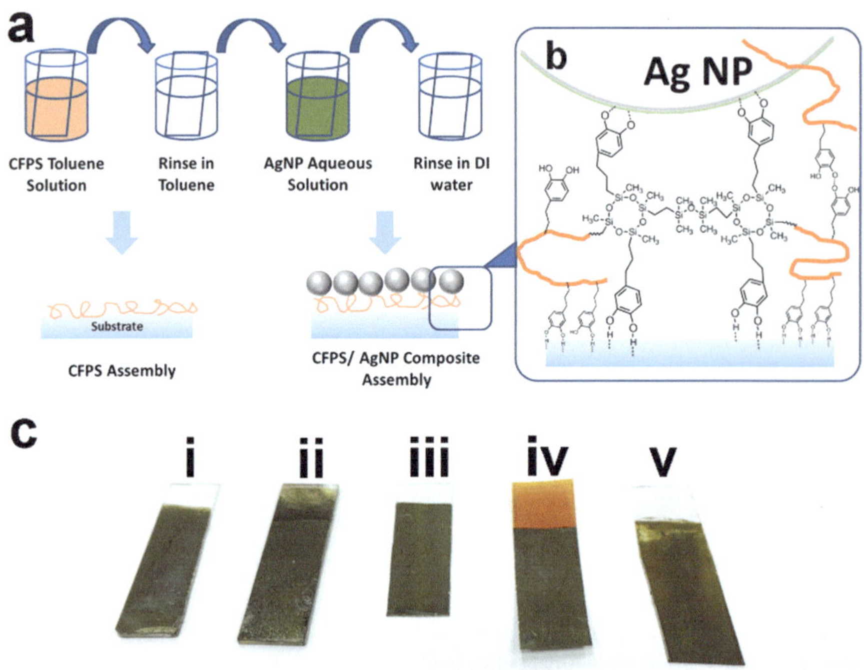

Fig. 21.12 (**a**) Scheme of the preparation process for the CFPS/Ag nanoparticles nanoassembly film, (**b**) the proposed mechanism of CFPS self-crosslinking and serve as a polymer bonder, and (**c**) digital photographs of CFPS/Ag nanoparticles nanoassembly films prepared on different substrates: (i) glass, (ii) quartz, (iii) PMMA, (iv) PI, and (v) PET (Reproduced from [55] with permission from The Royal Society of Chemistry)

21.3.4 In Situ SERS Detection Using a Smartphone

The significant SERS enhancement effect of Ag fractal hierarchical structures can be employed to develop an in situ SERS detection system in combination with a portable machine. Portable laboratory devices bring a breakthrough for field work on environmental science and so on. One of the problems in portable analytical instruments is the lower sensitivity compared to those in laboratory use. The SERS detection based on Ag fractal hierarchical structure is a promising method to realize a high-sensitive portable analytical instrument. A sensor unit and data recording, processing, and displaying units are basic components for a portable analytical system, where the instrument size is determined by the latter units. A smartphone has a large possibility for the application to portable analytical system because of the small size, light weight, and its popularization in the current society. The additional advantage of a smartphone is the compatibility with the Internet of things (IoT) technology. Researchers in field work can connect to a big data platform using a portable analytical system based on a smartphone. We have preliminary studied the application of the SERS sensing media based on Ag fractal hierarchical structures to in situ SERS detection using a smartphone.

In the in situ SERS measurements using a smartphone as a detector, SERS signals were selectively introduced to a camera unit of a smartphone through a Raman longpass edge filter (Semrock LP03-532RU-25, laser wavelength 532 nm, T_{avg} >93% 538.9–1200 nm) and a single-band bandpass filter (Semrock FF01-579/34-25, center wavelength 579 nm, FWHM 43.5 nm), which were adjusted to the spectral features of the SERS signals from PATP molecules used as mimic contamination. The optical features of the filters in the Raman spectroscopy using 532 nm laser source are shown in Fig. 21.13 with the SERS spectrum of PATP. By using the longpass edge and single-band bandpass filters, the SERS signals of PATP can be selectively observed by a camera unit. In the in situ measurements, a 200× zoom clip-on microscope lens (WONBSDOM MCL-001) was attached to an iPhone camera unit to correct effectively SERS signals from the 532 nm laser irradiated area. Figure 21.14 shows the clip-on microscope, a camera adaptor, and a Raman

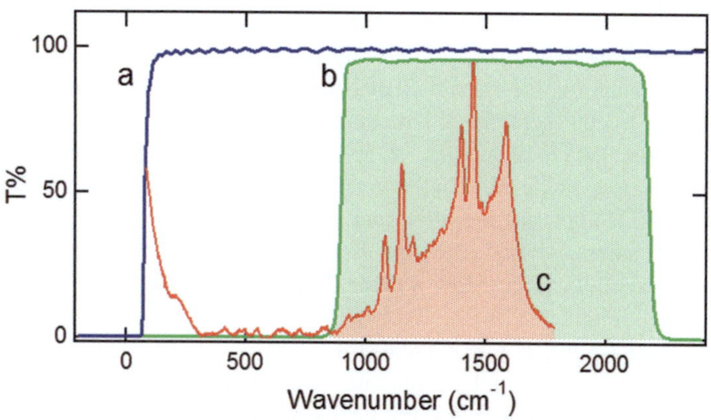

Fig. 21.13 Spectral features of (**a**) longpass edge filter, (**b**) single-band bandpass filter in the Raman spectroscopy using 532-nm laser source, and (**c**) SERS spectrum of PATP

Fig. 21.14 Clip-on microscope (**a**), a camera adaptor (**b**), and Raman filter holder (**c**) units for the in situ SERS detection using a smartphone

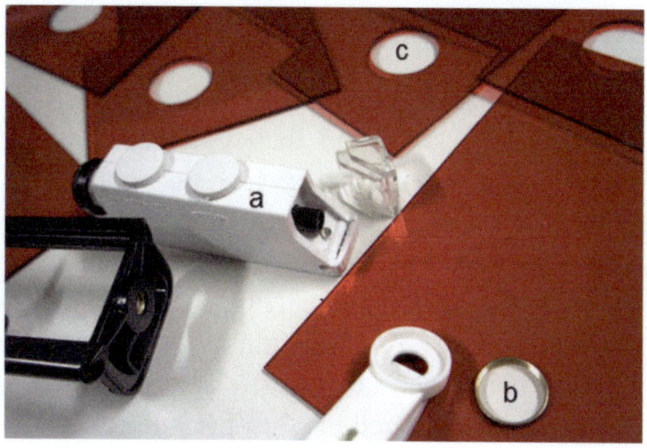

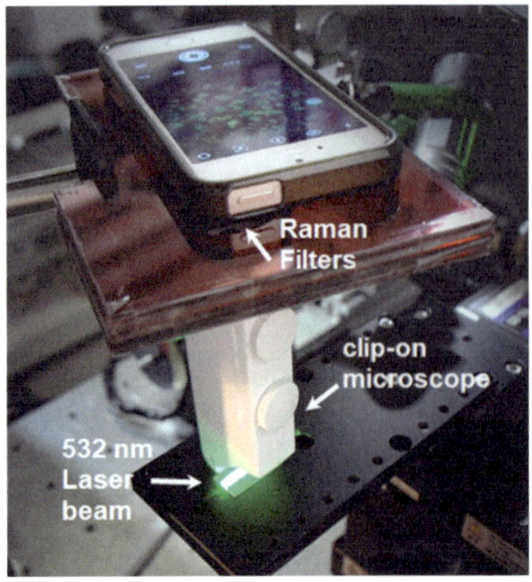

Fig. 21.15 Experimental setup for the in situ SERS detection using a smartphone

filter holder unit, where red acrylic plates (2 mm thickness) were used to eliminate the 532 nm laser scattering to the camera unit, and the longpass edge and single-band bandpass filters are held in a 26 mm diameter hole of red acrylic plate. The experimental setup for the in situ SERS detection using an iPhone is shown in Fig. 21.15. Ag fractal hierarchical structures were formed on a glass plate and then contaminated by PATP molecules. A 532 nm laser beam was irradiated on the surface with a low angle of incidence. A 2D SERS image through the clip-on microscope lens, longpass edge filter, and single-band bandpass filter was detected by the iPhone camera, where SERS signals were accumulated by long-time exposure using an iPhone application (Slow Shutter Cam) [56] and the 2D SERS image was captured on the iPhone.

The image of Ag fractal hierarchical structures was observed by the iPhone clip-on microscope under white light illumination as shown in Fig. 21.16a. Although the cross-star shapes of Ag fractal hierarchical structures were observed under white light illumination, the image was not clear due to the scattering from all surface area. On the other hand, the image of the cross-star shapes became remarkably clear by 532-nm laser irradiation (Fig. 21.16b), which can be assigned to the 2D SERS image of PTAP adsorbed on the Ag fractal hierarchical structures. The 2D SERS image was obtained by accumulation for 10 s using long-time exposure mode of iPhone application. The background scattering was eliminated by Raman longpass edge and single-band bandpass filters. The 2D SERS image was transferred to an Internet server from iPhone and then received by an Android tablet Pad for the image processing and analysis by Google application (IJ_Mobile) which is based on a popular software known as ImageJ [57, 58]. Figure 21.17 shows the images on an Android tablet Pad during the image processing and analysis, where the SERS intensity in a selected area (Fig. 21.17a) was plotted against the distance (pixel) as shown in Fig. 21.17b. The numerical intensity of the 2D SERS image also can be measured on the portable analysis system, and then it can be connected to a big data at the same time. Such in situ SERS detection of contamination is possible using the

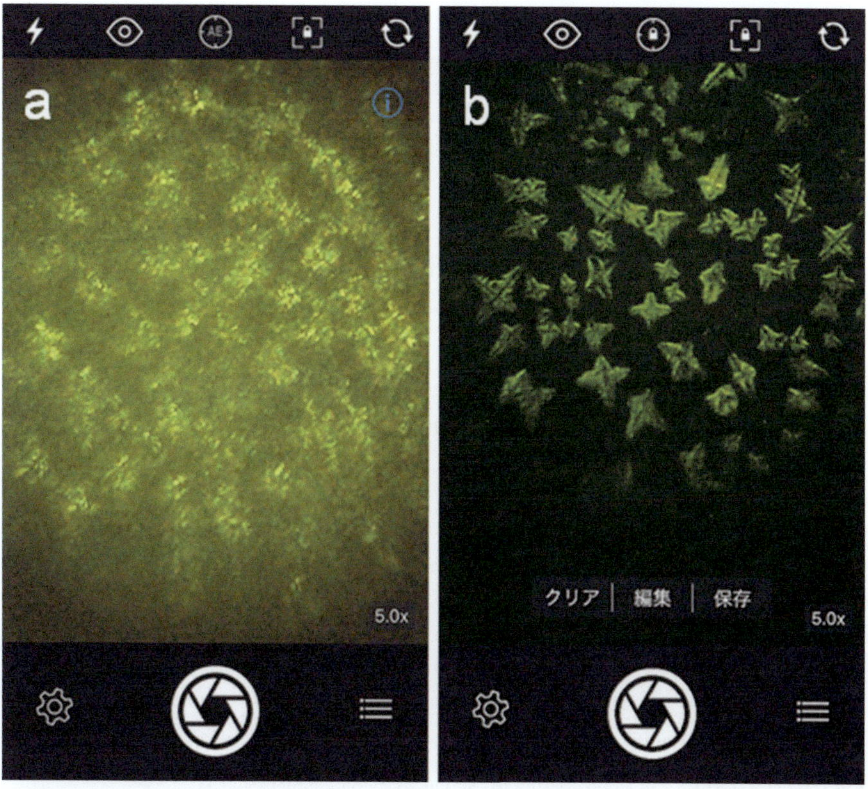

Fig. 21.16 Images of Ag fractal hierarchical structures observed by iPhone clip-on microscope (**a**) under white light illumination and (**b**) 532 nm laser beam irradiation. The 2D SERS image (**b**) was obtained by accumulation for 10 s

SERS sensing media based on Ag fractal hierarchical structures and a smartphone, where the sensitivity limitation of a smartphone can be overcome by the large SERS enhancement factor of the Ag fractal hierarchical structures.

21.4 Conclusion

In this chapter, we briefly introduced the properties of the inorganic building blocks consisting of inorganic nano-cores such as nanoclusters, nanoparticles, and nanocrystals, which can be recognized as generalized inorganic element blocks, and their self-assembly behaviors and self-assembly strategies, especially the fractal structures through diffusion-limited aggregation, as well as the application of metal assembly structures in high-performance SERS and corresponding mechanism. The control on the crystallization kinetic process of symmetric POSS molecules could be

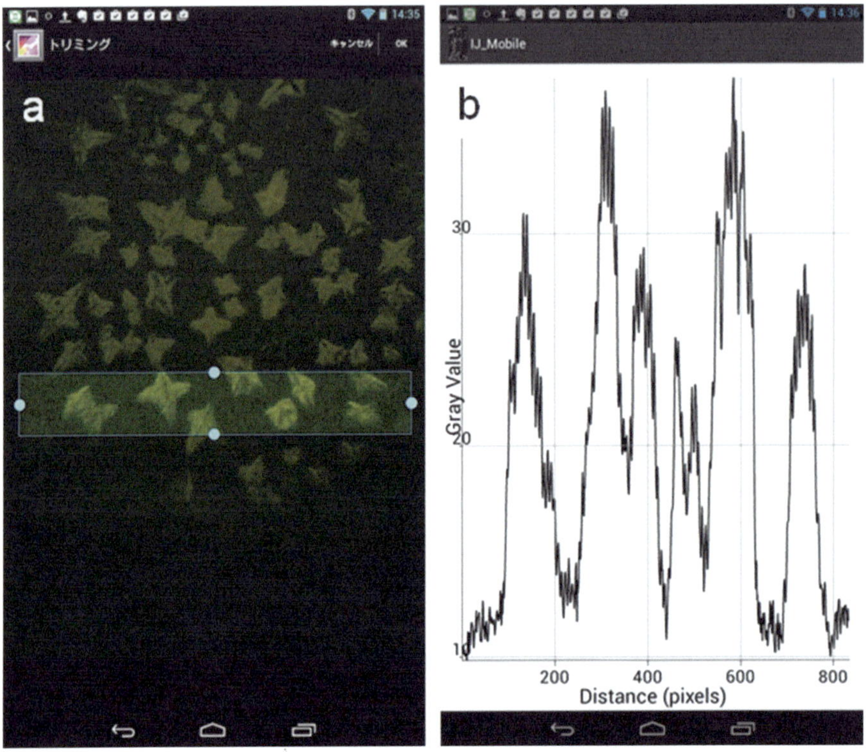

Fig. 21.17 2D SERS image processing (**a**) and analysis on a tablet Pad (**b**)

employed to construct POSS/nanoparticles hybrid fractal hierarchical structures by using a facile spin-coating method to fasten the evaporation inducing the non-equilibrium condition. The parameters such as rotating rates and concentrations of the components influence the crystallization kinetic process, thus the final fractal structures. The formation mechanism is clearly explained and demonstrated to be a universal method in other symmetric POSS molecules, other inorganic element blocks, and arbitrary substrates. The Ag fractal hierarchical structures obtained after heat treatment showed excellent SERS performance and have been demonstrated to act as an effective in situ SERS sensor. This crystallization kinetics-driven self-assembly method has potential to be applied to other building blocks and crystallization materials, such as inorganic salts with different crystal phases, polymers, and atomic nanoclusters. However, the synthesis of novel inorganic building blocks and the new self-assembly strategies and mechanism, especially the application of the assembly structures, need further study in the future.

References

1. Chujo Y, Tanaka K (2015) New polymeric materials based on element-blocks. B Chem Soc Jpn 88(5):633–643. https://doi.org/10.1246/bcsj.20150081
2. Talapin DV, Lee J-S, Kovalenko MV, Shevchenko EV (2010) Prospects of colloidal nanocrystals for electronic and optoelectronic applications. Chem Rev 110(1):389–458. https://doi.org/10.1021/cr900137k
3. Nie Z, Petukhova A, Kumacheva E (2010) Properties and emerging applications of self-assembled structures made from inorganic nanoparticles. Nat Nanotechnol 5(1):15–25
4. Gwo S, Chen H-Y, Lin M-H, Sun L, Li X (2016) Nanomanipulation and controlled self-assembly of metal nanoparticles and nanocrystals for plasmonics. Chem Soc Rev 45(20):5672–5716. https://doi.org/10.1039/C6CS00450D
5. Kelly KL, Coronado E, Zhao LL, Schatz GC (2003) The optical properties of metal nanoparticles: the influence of size, shape, and dielectric environment. J Phys Chem B 107(3):668–677. https://doi.org/10.1021/jp026731y
6. Murphy CJ, Sau TK, Gole AM, Orendorff CJ, Gao J, Gou L, Hunyadi SE, Li T (2005) Anisotropic metal nanoparticles: synthesis, assembly, and optical applications. J Phys Chem B 109(29):13857–13870. https://doi.org/10.1021/jp0516846
7. Polavarapu L, Mourdikoudis S, Pastoriza-Santos I, Perez-Juste J (2015) Nanocrystal engineering of noble metals and metal chalcogenides: controlling the morphology, composition and crystallinity. CrystEngComm 17(20):3727–3762. https://doi.org/10.1039/C5CE00112A
8. Li L-s, Hu J, Yang W, Alivisatos AP (2001) Band gap variation of size- and shape-controlled colloidal CdSe quantum rods. Nano Lett 1(7):349–351. https://doi.org/10.1021/nl015559r
9. Zhou J, Yang Y, Zhang C-Y (2015) Toward biocompatible semiconductor quantum dots: from biosynthesis and bioconjugation to biomedical application. Chem Rev 115(21):11669–11717. https://doi.org/10.1021/acs.chemrev.5b00049
10. Kamat PV (2008) Quantum dot solar cells. Semiconductor nanocrystals as light harvesters. J Phys Chem C 112(48):18737–18753. https://doi.org/10.1021/jp806791s
11. Hao R, Xing R, Xu Z, Hou Y, Gao S, Sun S (2010) Synthesis, functionalization, and biomedical applications of multifunctional magnetic nanoparticles. Adv Mater 22(25):2729–2742. https://doi.org/10.1002/adma.201000260
12. Boles MA, Engel M, Talapin DV (2016) Self-assembly of colloidal nanocrystals: from intricate structures to functional materials. Chem Rev 116(18):11220–11289. https://doi.org/10.1021/acs.chemrev.6b00196
13. Hashemi SM, Jagodič U, Mozaffari MR, Ejtehadi MR, Muševič I, Ravnik M (2017) Fractal nematic colloids. Nat Commun 8:14026. https://doi.org/10.1038/ncomms14026
14. Afshinmanesh F, Curto AG, Milaninia KM, van Hulst NF, Brongersma ML (2014) Transparent metallic fractal electrodes for semiconductor devices. Nano Lett 14(9):5068–5074. https://doi.org/10.1021/nl501738b
15. Fan JA, Yeo W-H, Su Y, Hattori Y, Lee W, Jung S-Y, Zhang Y, Liu Z, Cheng H, Falgout L, Bajema M, Coleman T, Gregoire D, Larsen RJ, Huang Y, Rogers JA (2014) Fractal design concepts for stretchable electronics. Nat Commun 5:3266. https://doi.org/10.1038/ncomms4266
16. Kumar Y, Singh S (2015) A compact multiband hybrid fractal antenna for multistandard Mobile wireless applications. Wirel Pers Commun 84(1):57–67. https://doi.org/10.1007/s11277-015-2593-x
17. Xu L, Gutbrod SR, Ma Y, Petrossians A, Liu Y, Webb RC, Fan JA, Yang Z, Xu R, Whalen JJ, Weiland JD, Huang Y, Efimov IR, Rogers JA (2015) Materials and fractal designs for 3D multifunctional integumentary membranes with capabilities in cardiac electrotherapy. Adv Mater 27(10):1731–1737. https://doi.org/10.1002/adma.201405017
18. Gottheim S, Zhang H, Govorov AO, Halas NJ (2015) Fractal nanoparticle Plasmonics: the Cayley tree. ACS Nano 9(3):3284–3292. https://doi.org/10.1021/acsnano.5b00412

19. Nicolás-Carlock JR, Carrillo-Estrada JL, Dossetti V (2016) Fractality à la carte: a general particle aggregation model. Sci Rep 6:19505. https://doi.org/10.1038/srep19505
20. Zhang D, Xu Z, Li J, Chen S, Cheng J, Zhang A, Chen S, Miao M (2014) Self-assembly of Amido-ended Hyperbranched polyester films with a highly ordered dendritic structure. ACS Appl Mater Interfaces 6(18):16375–16383. https://doi.org/10.1021/am504705c
21. Li J, Liu B, Li J (2006) Controllable self-assembly of CdTe/poly (N-isopropylacrylamide–acrylic acid) microgels in response to pH stimuli. Langmuir 22(2):528–531. https://doi.org/10.1021/la052519k
22. Sun S, Xu S, Zhang W, Wu P, Zhang W, Zhu X (2013) Cooperative self-assembly and crystallization into fractal patterns by PNIPAM-based nonlinear multihydrophilic block copolymers under alkaline conditions. Polym Chem 4(24):5800–5809. https://doi.org/10.1039/C3PY00682D
23. Dong J, Qu S, Zheng H, Zhang Z, Li J, Huo Y, Li G (2014) Simultaneous SEF and SERRS from silver fractal-like nanostructure. Sensors Actuators B Chem 191:595–599. https://doi.org/10.1016/j.snb.2013.09.088
24. Zhang P, Chen L, Xu T, Liu H, Liu X, Meng J, Yang G, Jiang L, Wang S (2013) Programmable fractal nanostructured interfaces for specific recognition and electrochemical release of cancer cells. Adv Mater 25(26):3566–3570. https://doi.org/10.1002/adma.201300888
25. Hou C, Meng G, Huang Q, Zhu C, Huang Z, Chen B, Sun K (2014) Ag-nanoparticle-decorated Au-fractal patterns on bowl-like-dimple arrays on Al foil as an effective SERS substrate for the rapid detection of PCBs. Chem Commun 50(5):569–571. https://doi.org/10.1039/C3CC46878J
26. Li J-F, Zhang Y-J, Ding S-Y, Panneerselvam R, Tian Z-Q (2017) Core–shell nanoparticle-enhanced Raman spectroscopy. Chem Rev 117(7):5002–5069. https://doi.org/10.1021/acs.chemrev.6b00596
27. Schlücker S (2014) Surface-enhanced Raman spectroscopy: concepts and chemical applications. Angew Chem Int Ed 53(19):4756–4795. https://doi.org/10.1002/anie.201205748
28. Nam J-M, Oh J-W, Lee H, Suh YD (2016) Plasmonic Nanogap-enhanced Raman scattering with nanoparticles. Acc Chem Res 49(12):2746–2755. https://doi.org/10.1021/acs.accounts.6b00409
29. Wustholz KL, Henry A-I, McMahon JM, Freeman RG, Valley N, Piotti ME, Natan MJ, Schatz GC, Duyne RPV (2010) Structure–activity relationships in gold nanoparticle dimers and trimers for surface-enhanced Raman spectroscopy. J Am Chem Soc 132(31):10903–10910. https://doi.org/10.1021/ja104174m
30. Wu D-Y, Li J-F, Ren B, Tian Z-Q (2008) Electrochemical surface-enhanced Raman spectroscopy of nanostructures. Chem Soc Rev 37(5):1025–1041. https://doi.org/10.1039/B707872M
31. Lazzari M, Scalarone D, Hoppe CE, Vazquez-Vazquez C, Lòpez-Quintela MA (2007) Tunable Polyacrylonitrile-based micellar aggregates as a potential tool for the fabrication of carbon nanofibers. Chem Mater 19(24):5818–5820. https://doi.org/10.1021/cm7019894
32. Schmelz J, Karg M, Hellweg T, Schmalz H (2011) General pathway toward crystalline-Core micelles with tunable morphology and Corona segregation. ACS Nano 5(12):9523–9534. https://doi.org/10.1021/nn202638t
33. Mihut AM, Crassous JJ, Schmalz H, Drechsler M, Ballauff M (2012) Self-assembly of crystalline-coil diblock copolymers in solution: experimental phase map. Soft Matter 8(11):3163–3173. https://doi.org/10.1039/C2SM06359J
34. McGrath N, Schacher FH, Qiu H, Mann S, Winnik MA, Manners I (2014) Synthesis and crystallization-driven solution self-assembly of polyferrocenylsilane diblock copolymers with polymethacrylate corona-forming blocks. Polym Chem 5(6):1923–1929. https://doi.org/10.1039/C3PY01383A
35. He W-N, Zhou B, Xu J-T, Du B-Y, Fan Z-Q (2012) Two growth modes of semicrystalline cylindrical poly(ε-caprolactone)-b-poly(ethylene oxide) micelles. Macromolecules 45(24):9768–9778. https://doi.org/10.1021/ma301267k
36. Patra SK, Ahmed R, Whittell GR, Lunn DJ, Dunphy EL, Winnik MA, Manners I (2011) Cylindrical micelles of controlled length with a π-conjugated polythiophene core via

crystallization-driven self-assembly. J Am Chem Soc 133(23):8842–8845. https://doi.org/10.1021/ja202408w

37. Sun L, Petzetakis N, Pitto-Barry A, Schiller TL, Kirby N, Keddie DJ, Boyd BJ, O'Reilly RK, Dove AP (2013) Tuning the size of cylindrical micelles from poly(l-lactide)-b-poly(acrylic acid) diblock copolymers based on crystallization-driven self-assembly. Macromolecules 46(22):9074–9082. https://doi.org/10.1021/ma401634s
38. Yu B, Jiang X, Yin J (2014) Size-tunable nanosheets by the crystallization-driven 2D self-assembly of Hyperbranched poly(ether amine) (hPEA). Macromolecules 47(14):4761–4768. https://doi.org/10.1021/ma500845e
39. Cordes DB, Lickiss PD, Rataboul F (2010) Recent developments in the chemistry of cubic polyhedral oligosilsesquioxanes. Chem Rev 110(4):2081–2173. https://doi.org/10.1021/cr900201r
40. Zhang W, Müller AHE (2013) Architecture, self-assembly and properties of well-defined hybrid polymers based on polyhedral oligomeric silsequioxane (POSS). Prog Polym Sci 38(8):1121–1162. https://doi.org/10.1016/j.progpolymsci.2013.03.002
41. Hirai T, Leolukman M, Jin S, Goseki R, Ishida Y, Kakimoto M-A, Hayakawa T, Ree M, Gopalan P (2009) Hierarchical self-assembled structures from POSS-containing block copolymers synthesized by living anionic polymerization. Macromolecules 42(22):8835–8843. https://doi.org/10.1021/ma9018944
42. Raftopoulos KN, Koutsoumpis S, Jancia M, Lewicki JP, Kyriakos K, Mason HE, Harley SJ, Hebda E, Papadakis CM, Pielichowski K, Pissis P (2015) Reduced phase separation and slowing of dynamics in polyurethanes with three-dimensional POSS-based cross-linking moieties. Macromolecules 48(5):1429–1441. https://doi.org/10.1021/ma5023132
43. Peng J, Xing Y, Xu K, Lin W, Wu J, Yu Z, Zhang Y, Chen M (2015) Fabrication of silsesquioxane-based nano-wrinkled structures by coupling the polymeric surface onto rigid templates assembled from unique deca-silsesquioxane. J Mater Chem C 3(12):2897–2908. https://doi.org/10.1039/C4TC02720E
44. Liu F, Zhang Y, Xu L, Zhang W (2015) Morphology-controlled self-assembly of an organic/inorganic hybrid porphyrin derivative containing polyhedral oligomeric silsesquioxane (POSS). Chem-Eur J 21(14):5540–5547. https://doi.org/10.1002/chem.201405334
45. Li D, Niu Y, Yang Y, Wang X, Yang F, Shen H, Wu D (2015) Synthesis and self-assembly behavior of POSS-embedded hyperbranched polymers. Chem Commun 51:8296–8299. https://doi.org/10.1039/C5CC01338K
46. Liao W-H, Yang S-Y, Hsiao S-T, Wang Y-S, Li S-M, Ma C-CM, Tien H-W, Zeng S-J (2014) Effect of octa(aminophenyl) polyhedral oligomeric silsesquioxane functionalized graphene oxide on the mechanical and dielectric properties of polyimide composites. ACS Appl Mater Interfaces 6(18):15802–15812. https://doi.org/10.1021/am504342j
47. Li Z, Wu D, Liang Y, Fu R, Matyjaszewski K (2014) Synthesis of well-defined microporous carbons by molecular-scale templating with polyhedral oligomeric silsesquioxane moieties. J Am Chem Soc 136(13):4805–4808. https://doi.org/10.1021/ja412192v
48. Li J, Zhou Z, Ma L, Chen G, Li Q (2014) Hierarchical assembly of amphiphilic POSS-Cyclodextrin molecules and azobenzene end-capped polymers. Macromolecules 47(16):5739–5748. https://doi.org/10.1021/ma501100r
49. Hou H, Gan Y, Yin J, Jiang X (2014) Multifunctional POSS-based nano-photo-initiator for overcoming the oxygen inhibition of photo-polymerization and for creating self-wrinkled patterns. Adv Mater Interfaces 1(9):1400385. https://doi.org/10.1002/admi.201400385
50. Gan Y, Jiang X, Yin J (2012) Self-wrinkling patterned surface of photocuring coating induced by the fluorinated POSS containing thiol groups (F-POSS-SH) as the reactive nanoadditive. Macromolecules 45(18):7520–7526. https://doi.org/10.1021/ma301439g
51. Cai J, Lv C, Watanabe A (2015) Facile preparation of hierarchical structures using crystallization-kinetics driven self-assembly. ACS Appl Mater Interfaces 7(33):18697–18706. https://doi.org/10.1021/acsami.5b05177

52. Wang Y, Chen H, Dong S, Wang E (2006) Surface-enhanced Raman scattering of silver-gold bimetallic nanostructures with hollow interiors. J Chem Phys 125(4):044710. https://doi.org/10.1063/1.2216694
53. Zhou Q, Zhao G, Chao Y, Li Y, Wu Y, Zheng J (2007) Charge-transfer induced surface-enhanced Raman scattering in silver nanoparticle assemblies. J Phys Chem C 111(5):1951–1954. https://doi.org/10.1021/jp067045s
54. Osawa M, Matsuda N, Yoshii K, Uchida I (1994) Charge transfer resonance Raman process in surface-enhanced Raman scattering from p-aminothiophenol adsorbed on silver: Herzberg-Teller contribution. J Phys Chem 98(48):12702–12707. https://doi.org/10.1021/j100099a038
55. Liu Y, Demirci A, Zhu H, Cai J, Yamamoto S, Watanabe A, Miyashita T, Mitsuishi M (2016) A versatile platform of catechol-functionalized polysiloxanes for hybrid nanoassembly and in situ surface enhanced Raman scattering applications. J Mater Chem C 4(38):8903–8910. https://doi.org/10.1039/C6TC02963A
56. Cogitap Software (2017) Slow Shutter Cam, https://itunes.apple.com/us/app/slow-shutter-cam/id357404131?mt=8. Accessed 23 Jul 2017
57. ImageJ developers (2009) Image J, https://imagej.net/Contributors, https://imagej.nih.gov/ij/. Accessed 23 Jul 2017
58. Michael Steptoe (2017) IJ_Mobile, https://play.google.com/store/apps/details?id=comij_mobile&hl=en. Accessed 23 July2017

Chapter 22
Air-Stable Optoelectronic Devices with Metal Oxide Cathodes

Makoto Takada and Hiroyoshi Naito

Abstract Air-stable organic light-emitting diodes (OLEDs) and organic photovoltaic cells (OPVs) are fabricated by using metal oxide instead of low-work function metals such as Ca as a cathode. The optoelectronic devices are called inverted OLEDs and inverted OPVs and are important for flexible devices. Application of element-block polymers to these inverted devices is demonstrated.

Keywords Organic light-emitting diodes · Organic photovoltaic cells · Metal oxide cathode · Polyethyleneimine

22.1 Introduction

Element-block polymers, consisting of various groups of elements [1], are promising materials for optoelectronic application such as organic light-emitting diodes (OLEDs) and organic photovoltaic cells (OPVs) [2, 3]. This stems from the fact that both electronic transport properties and optical properties are controlled by tuning various groups of elements. OLEDs have been successfully applied to OLED TV, OLED lightings, and smartphones. Other potential applications of the devices are flexible displays, light sources and detectors for biosensors, and power sources for Internet of Things (IoT) devices.

It is well known that conventional OLEDs and OPVs are sensitive to oxygen and water in the atmosphere. This is due to the fact that low-work function cathodes are

---

M. Takada
Department of Physics and Electronics, Osaka Prefecture University, Sakai, Japan
e-mail: m-takada@pe.osakafu-u.ac.jp

H. Naito (✉)
Department of Physics and Electronics, Osaka Prefecture University, Sakai, Japan

The Research Institute for Molecular Electronic Devices, Osaka Prefecture University, Sakai, Japan
e-mail: naito@pe.osakafu-u.ac.jp

Y. Chujo (ed.), *New Polymeric Materials Based on Element-Blocks*,
https://doi.org/10.1007/978-981-13-2889-3_22

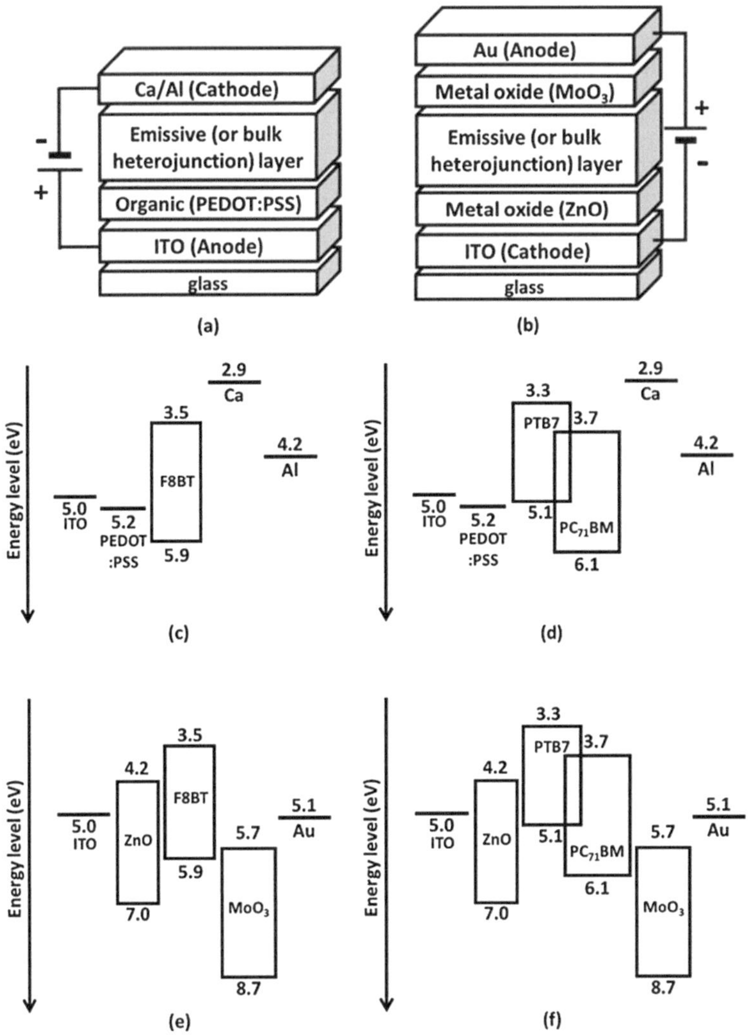

Fig. 22.1 Device structures of (**a**) conventional OLEDs (and OPVs) and (**b**) inverted OLEDs (and OPVs) and band diagrams of (**c**) conventional OLEDs, (**d**) conventional OPVs, (**e**) inverted OLEDs, and (**f**) inverted OPVs

air-sensitive materials such as magnesium:silver (Mg:Ag), lithium fluoride (LiF)/aluminum (Al), and calcium (Ca)/aluminum (Al), and hence rigorous encapsulation is necessary to protect conventional OLEDs and OPVs (Fig. 22.1a) from moisture and oxygen.

Inverted OLEDs and OPVs (Fig. 22.1b) [4, 5] with an air-stabile bottom metal oxide cathode have recently attracted considerable attention as substitutes for conventional OLEDs and OPVs. The air-stable inverted OLEDs and OPVs are suitable for flexible light-emitting devices and saving production cost because of allowing the preparation of devices without rigorous encapsulation and using flexible substrates such as a plastic film. Generally, flexible substrates suffer from poor gas and moisture barrier performance, and the water vapor transmission rate (WVTR) of commercially available flexible films for organic devices is 10^{-5}–10^{-2} g m^{-2} day^{-1} (the WVTR of conventional glass substrates is 10^{-8} g m^{-2} day^{-1}). It is widely known that conventional OLEDs and OPVs require barrier films with a WVTR of less than 10^{-6} g m^{-2} day^{-1}.

The configurations of inverted OLEDs and OPVs are inverted with respect to that of conventional OLEDs and OPVs: the representative device structures and the band diagrams of inverted OLEDs and OPVs are shown in Fig. 22.1. In this chapter, we show the architectures and the basic properties of inverted OLEDs and OPVs, based on simple element-block polymers. The inverted devices are suitable application of element-block polymers, and the development of highly efficient inverted OLEDs and OPVs will be possible by using newly synthesized element-block polymers.

22.2 Fabrication of Inverted Devices

As examples of the fabrication of inverted OLEDs and OPVs, we show in this section the typical fabrication processes of our inverted devices.

The device configuration of the inverted OLEDs was Al-doped zinc oxide (AZO)/polyethyleneimine (PEI) [6]/poly(9,9-dioctylfluorene-alt-benzothiadiazole) (F8BT) [7]/MoO_3/Al. A patterned AZO glass (Geomatec) as a cathode was cleaned using acetone, 2-propanol, and UV-ozone. Subsequently, a thin layer of PEI (1–5-nm) (Fig. 22.2a), an electron injection layer, was spun onto the surface of the AZO glass from an ethanol solution (0.05 wt%, 2000 rpm, 30 s). The substrate was then

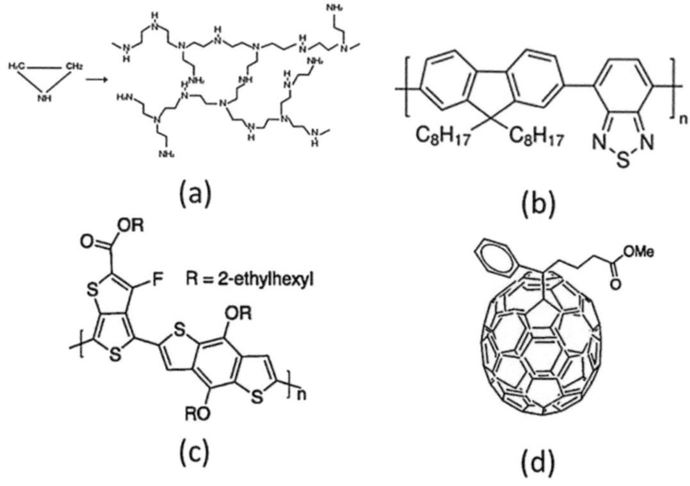

Fig. 22.2 Chemical structures of (**a**) PEI, (**b**) F8BT, (**c**) PTB7, and (**d**) $PC_{71}BM$

annealed in ambient atmosphere (5 min, 150 °C). The processes mentioned below were carried out in a nitrogen-filled glove box (dew point: −80 °C). A layer of 130-nm F8BT (Fig. 22.2b), a green light-emitting polymer, was spun onto the PEI layer from a chlorobenzene solution as an emissive layer (0.75 wt%, 800 rpm, 60 s). After the F8BT emissive layer deposition, the substrates were dried at 80 °C for 20 min. 10-nm MoO_3 and 50-nm Al layer were successively thermally evaporated onto the F8BT emissive layer in a vacuum chamber at a base pressure of 10^{-3} Pa. Finally, the inverted OLEDs were encapsulated with epoxy.

The device configuration of the inverted OPVs was Ga-doped zinc oxide (GZO)/polythieno[3,4-b]-thiophene-co-benzodithiophene (PTB7): [6,6]-phenyl-C_{71}-butyric acid methyl ester ($PC_{71}BM$)/MoO_3/Al. A layer of 80-nm PTB7:$PC_{71}BM$ (Fig. 22.2c, d) was spun onto the PEI-precoated GZO layer from a chlorobenzene solution as a bulk heterojunction layer (1:1.5, 2 wt%, 800 rpm, 60 s). After the PTB7:$PC_{71}BM$ layer deposition, the substrates were dried at room temperature for 60 min in vacuum. The processes of the deposition of 10-nm MoO_3 and 50-nm Al layer and encapsulation were carried out as the inverted OLEDs.

The conventional OLEDs and OPVs were fabricated to compare the device characteristics between conventional and inverted structures. The device structures of conventional OLEDs (OPVs) were ITO/poly(3, 4-ethylenedioxythiophene): poly (styrenesulfonate) (PEDOT:PSS)/F8BT/Ca/Al and ITO/PEDOT:PSS/PTB7:$PC_{71}BM$/Ca/Al, respectively. The active area of all devices was 2 mm^2.

Current density-voltage (*J-V*) characteristics of OLEDs and OPVs were recorded with a source measure unit (Keithley 2411). A light source was a solar simulator (Asahi Spectra, HAL-320). Luminance of OLEDs was measured with a luminance meter (Konica Minolta CS-200). All measurements were performed in laboratory atmosphere.

22.3 Inverted OLEDs

Polymer OLEDs [8] have been fabricated by direct printing such as inkjet printing, offset printing, and screen printing, which are low-energy processes compared to conventional vacuum deposition processes used for small-molecule OLEDs. The first demonstration of low-cost, encapsulation-free OLEDs (inverted OLEDs with F8BT) was reported by Morii et al., and the device performance of their inverted OLEDs whose device structure was FTO/TiO_2/F8BT/MoO_3/Au was the turn-on voltages of 2.5 V and the luminance of 500 cd/m^2 at 4.8 V in air [4]. F8BT and Super Yellow [9] have been used as light-emitting polymers for basic studies of prototypical inverted OLEDs.

The *J-V* and the luminescence-voltage (*L-V*) characteristics of the inverted OLEDs are shown in Fig. 22.3a. The luminance at 4 V is 16,000 cd m^{-2} (at 190 mA cm^{-2}). The current efficiency-current density characteristics of the inverted OLEDs obtained from Fig. 22.3a are shown in Fig. 22.3b. The current efficiency at 100 mA cm^{-2} is 9.9 cd A^{-1}, which shows higher performance than

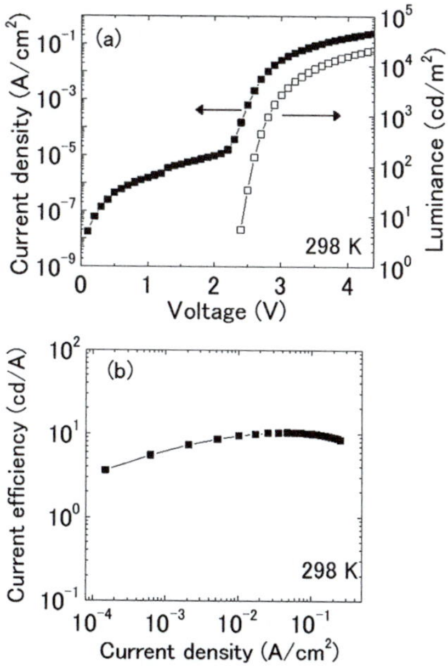

Fig. 22.3 (**a**) Current density-luminance-voltage characteristics and (**b**) current efficiency-current density characteristic of inverted OLEDs with F8BT

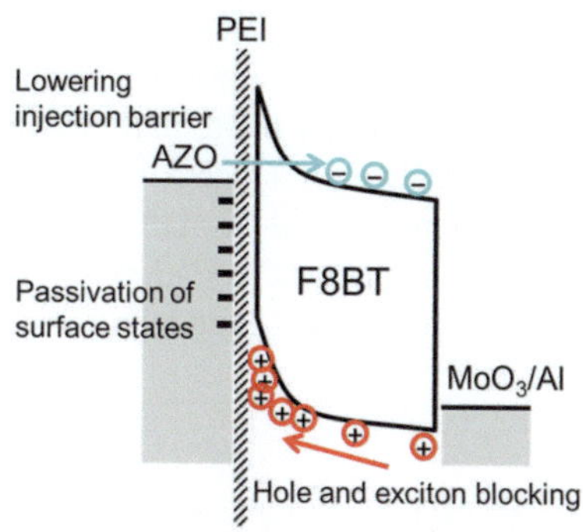

Fig. 22.4 Roles of PEI layer: electron injection barrier lowering, the passivation of the surface states of AZO, the exciton blocking, and the hole accumulation at the AZO/F8BT interface

F8BT-based single-layer inverted OLEDs in literature [10]. We examined the current efficiency of the conventional OLEDs. The maximum current efficiency of the conventional OLEDs was 3.6 cd/A at 360 mA/cm^2. The current efficiency of the inverted OLEDs is higher than that of the conventional OLEDs because the PEI layer enhances electron injection from metal oxide cathode, AZO.

The PEI layer in the inverted OLEDs can play important roles: the electron injection barrier lowering, the passivation of the surface states of AZO, the exciton blocking, and the hole accumulation at the AZO/F8BT interface, as illustrated in Fig. 22.4. Electron injection of the inverted OLEDs is greatly enhanced by the hole accumulation at the AZO (PEI)/F8BT interface rather than the electron injection barrier lowering between the AZO cathode and the F8BT emissive layer. The enhanced electron injection is due to the electric field crowding at the AZO (PEI)/

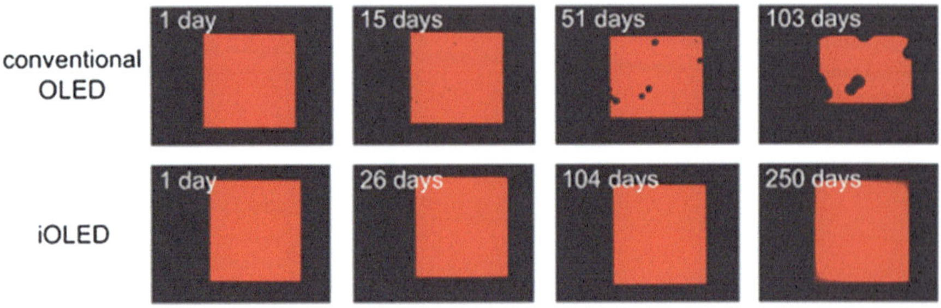

Fig. 22.5 Images of light-emitting areas of OLEDs as a function of storage time [12] (Copyright 2014, The Japan Society of Applied Physics)

F8BT interface caused by hole accumulation, and such electric field crowding greatly enhances the Fowler-Nordheim-type tunneling electron injection from the AZO cathode to the F8BT emissive layer [11].

Recently, interesting approaches of inverted OLEDs have been reported for the demonstration of high efficiency and high air stability. Inverted OLEDs with phosphorescent emitters (ITO/ZnO/PEI/ Bis(10-hydroxybenzo[h]quinolinato)beryllium ($Bebq_2$): tris[1-phenylisoquinolinato-C^2,N]iridium(III) ($Ir(piq)_3$)/N,N′-Di (1-naphthyl)-N,N′-diphenylbenzidine (α-NPD)/MoO_3/Au) exhibited high external quantum efficiency (EQE) of over 15% and no dark spot formation after 250 days (the barrier films used for the encapsulation has a WVTR of 3×10^{-4} gm^{-2} day^{-1}) [12]. On the other hand, dark spot formation was clearly observed in the conventional OLED after 15 days of exposure to laboratory atmosphere, and the emitting area decreased by about half after 103 days. The observed dark spot formation may originate from the degradation/oxidization of lithium fluoride and/or aluminum, both of which are widely used in conventional OLEDs (Fig. 22.5).

22.4 Inverted OPVs

Polymer OPVs have been also fabricated by direct printing processes [13]. An important difference in device structures between OLEDs and OPVs is that the active layer of OPVs is the disordered blend of electron acceptor and donor materials, known as bulk heterojunction. Recently, the power conversion efficiency (PCE) of OPVs has been improved steadily because of the advances made in device structures [14–16], interfacial engineering [17], morphological control [18], and synthesis of acceptor and donor materials [19, 20].

The highest PCE of over 11% has been reported in OPVs based on conjugated polymers and fullerene derivatives [21], and nonfullerene acceptors provide a more promising way for improving device performance, (3,9-bis(2-methylene-(3-(1,1-dicyanomethylene)-indanone))-5,5,11,11-tetrakis(4-hexyl phenyl)-dithieno [2,3-d:2′,3′-d′]-s-indaceno[1,2-b:5,6-b′] dithiophene) (ITIC), and perylene diimide-based electron acceptors are prototypical nonfullerene acceptors for replacing fullerene acceptors [22–29]. In particular, the PCE of OPVs using ITIC-core-based nonfullerene acceptors has been reported to be 12.05% [23].

Generally, inverted OPVs exhibit higher PCE than conventional OPVs [13, 21]. Unlike OLEDs, most of donor polymers in bulk heterojunction are polycrystalline, and backbone orientation of the donor polymers such as edge-on and face-on orientation is spontaneously optimized in inverted OPVs to increase charge carrier mobilities and reduce bimolecular recombination constants [13, 21].

In the last several years, OPVs with a specific donor polymer family based on fluorinated thieno[3,4-b]thiophene, for example, PTB7, and a specific fullerene, $PC_{71}BM$, show relatively high performance [17, 30, 31]. PTB7: $PC_{71}BM$ OPVs are thus good examples to examine the performance of inverted OPVs. Indeed, such studies have been performed and showed that the PCE of the inverted OPV (ITO/ZnO/PTB7:$PC_{71}BM$/MoO_3/Ag) is 7.2% while that of the conventional OPV (ITO/PEDOT:PSS/PTB7:$PC_{71}BM$/Ca/Ag) is 6.7%. The main origin of the improvement of the PCE results from the reduction in the series resistance of the inverted OPVs [32]. We found the improvement similar to the results in [32].

PEI also plays important roles in inverted OPVs. We show further improvement of the performance of inverted OPVs using PEIs. Figure 22.6 shows the *J-V* characteristics of inverted OPVs with and without PEI (GZO/PEI/PTB7:$PC_{71}BM$/MoO_3/Al and GZO/PTB7:$PC_{71}BM$/MoO_3/Al). The PCE of the inverted OPV with PEI (7.5%) is higher than that of the inverted OPV without PEI (6.3%). The fill factor is also improved by depositing PEI layer on GZO in the inverted OPVs.

Simple equivalent circuit analysis based on a single-diode model [33] was carried out in the inverted OPVs with and without PEI and showed that the series resistance of the inverted OPV with PEI is reduced to almost half value of the series resistance of the inverted OPV without PEI. In addition to the reduction, the increase in the open-circuit photovoltage in the inverted OPV with PEI leads to overall improvement of the PCE, and the changes caused by PEI layer are consistent with the roles of PEI found in inverted OLEDs shown in Fig. 22.4.

The inverted optoelectronic devices described above are important for low-cost flexible and IoT devices. Synthesis of element-block polymers for highly efficient

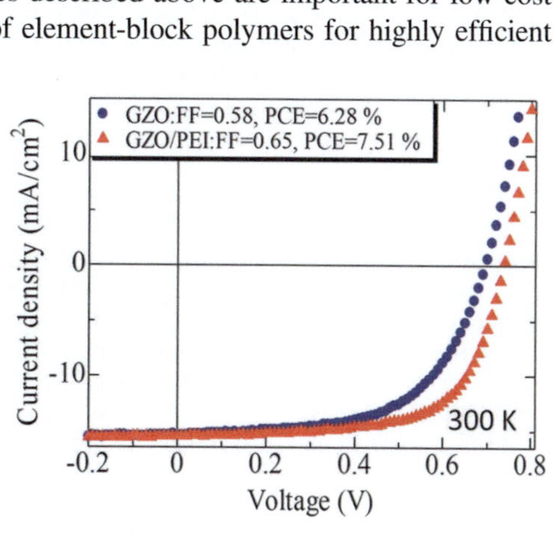

Fig. 22.6 Current density-voltage characteristics of inverted OPVs without and with PEI, under the illumination of AM 1.5G, 100 mWcm^{-2}

inverted devices is thus a critical challenge. For instance, the synthesis of element-block polymers with a group of elements exhibiting thermally activated delayed fluorescence [34] will be an interesting approach for the fabrication of highly efficient inverted OLEDs.

22.5 Conclusions

We showed that the device performance of inverted devices is better than that of conventional devices. Inverted OLEDs and OLEDs are air stable and are promising as flexible and IoT devices. We examined the basic properties of inverted OLEDs with the structure of AZO/PEI/F8BT/MoO_3/Al and of inverted OPVs with that of GZO/PEI/PTB7:PC_{71}BM/MoO_3/Al. F8BT and PTB7 can be regarded as simple prototypical element-block polymers, and further improvement of device performance of inverted devices (for instance, inverted OLEDs with high EQE, inverted OPVs with high PCE, and full-color inverted OLEDs) is expected by newly synthesized element-block polymers.

References

1. Chujo Y, Tanaka K (2015) New polymeric materials based on element-blocks. Bull Chem Soc Jpn 88:633–643. https://doi.org/10.1246/bcsj.20150081
2. Müllen K, Scherf U (eds) (2006) Organic light emitting devices: synthesis, properties and applications. Wiley-VCH, Weinheim
3. Brabec C, Scherf U, Dyakonov V (eds) (2014) Organic photovoltaics: materials, device physics, and manufacturing technologies, 2nd edn. Wiley-VCH, Weinheim
4. Morii K, Ishida M, Takashima T et al (2006) Encapsulation-free hybrid organic-inorganic light-emitting diodes. Appl Phys Lett 89:183510-1–183510-3. https://doi.org/10.1063/1.2374812
5. Waldauf C, Morana M, Denk P et al (2006) Highly efficient inverted organic photovoltaics using solution based titanium oxide as electron selective contact. Appl Phys Lett 89:233517-1–233517-3. https://doi.org/10.1063/1.2402890
6. Zhou Y, Fuentes-Hernandez C, Shim J et al (2012) A universal method to produce low-work function electrodes for organic electronics. Science 336:327–332. https://doi.org/10.1126/science.1218829
7. He Y, Gong S, Hattori R, Kanicki J (1999) High performance organic polymer light-emitting heterostructure devices. Appl Phys Lett 74:2265–2267. https://doi.org/10.1063/1.123862
8. Friend RH, Gymer RW, Holmes AB et al (1999) Electroluminescence in conjugated polymers. Nature 397:121–128. https://doi.org/10.1038/16393
9. Swensen J, Moses D, Heeger AJ et al (2005) Light emission in the channel region of a polymer thin-film transistor fabricated with gold and aluminum for the source and drain electrodes. Synth Met 153:53–56. https://doi.org/10.1016/j.synthmet.2005.07.258
10. Bolink HJ, Brine H, Coronado E, Sessolo M (2010) Hybrid organic-inorganic light emitting diodes: effect of the metal oxide. J Mater Chem 20:4047–4049. https://doi.org/10.1039/b927408a

11. Takada M, Nagase T, Kobayashi T, Naito H (2017) Electron injection in inverted organic light-emitting diodes with poly(ethyleneimine) electron injection layers. Org Electron 50:290–295. https://doi.org/10.1016/j.orgel.2017.07.049
12. Fukagawa H, Morii K, Hasegawa M et al (2014) Highly efficient and air-stable inverted organic light-emitting diode composed of inert materials. Appl Phys Express 7:082104-1–082104-4. https://doi.org/10.7567/APEX.7.082104
13. Zhang F, Inganäs O, Zhou Y, Vandewal K (2016) Development of polymer-fullerene solar cells. Natl Sci Rev 3:222–239. https://doi.org/10.1093/NSR/NWW020
14. Yu G, Gao J, Hummelen JC et al (1995) Polymer photovoltaic cells: enhanced efficiencies via a network of internal donor-acceptor heterojunctions. Science 270:1789–1791. https://doi.org/10.1126/science.270.5243.1789
15. Kim JY, Lee K, Coates NE et al (2007) Efficient tandem polymer solar cells fabricated by all-solution processing. Science 317:222–225. https://doi.org/10.1126/science.1141711
16. Vohra V, Kawashima K, Kakara T et al (2015) Efficient inverted polymer solar cells employing favourable molecular orientation. Nat Photon 9:403–408. https://doi.org/10.1038/nphoton.2015.84
17. He Z, Zhong C, Su S et al (2012) Enhanced power-conversion efficiency in polymer solar cells using an inverted device structure. Nat Photon 6:591–595. https://doi.org/10.1038/nphoton.2012.190
18. Li G, Shrotriya V, Huang J et al (2005) High-efficiency solution processable polymer photovoltaic cells by self-organization of polymer blends. Nat Mater 4:864–868. https://doi.org/10.1038/nmat1500
19. Li Y (2012) Molecular design of photovoltaic materials for polymer solar cells: toward suitable electronic energy levels and broad absorption. Acc Chem Res 45:723–733. https://doi.org/10.1021/ar2002446
20. Saito M, Osaka I, Suzuki Y et al (2015) Highly efficient and stable solar cells based on thiazolothiazole and naphthobisthiadiazole copolymers. Sci Rep 5:14202-1–14202-9. https://doi.org/10.1038/srep14202
21. Liu Y, Zhao J, Li Z et al (2014) Aggregation and morphology control enables multiple cases of high-efficiency polymer solar cells. Nat Commun 5:5293-1–5293-8. https://doi.org/10.1038/ncomms6293
22. Tang CW (1986) Two-layer organic photovoltaic cell. Appl Phys Lett 48:183–185. https://doi.org/10.1063/1.96937
23. Li S, Ye L, Zhao W et al (2016) Energy-level modulation of small-molecule electron acceptors to achieve over 12% efficiency in polymer solar cells. Adv Mater 28:9423–9429. https://doi.org/10.1002/adma.201602776
24. Lin Y, Zhao F, He Q et al (2016) High-performance electron acceptor with thienyl side chains for organic photovoltaics. J Am Chem Soc 138:4955–4961. https://doi.org/10.1021/jacs.6b02004
25. Liu J, Chen S, Qian D et al (2016) Fast charge separation in a non-fullerene organic solar cell with a small driving force. Nat Energy 1:16089-1–16089-7. https://doi.org/10.1038/nenergy.2016.89
26. Zhao W, Qian D, Zhang S et al (2016) Fullerene-free polymer solar cells with over 11% efficiency and excellent thermal stability. Adv Mater 28:4734–4739. https://doi.org/10.1002/adma.201600281
27. Bin H, Zhang ZG, Gao L et al (2016) Non-fullerene polymer solar cells based on alkylthio and fluorine substituted 2D-conjugated polymers reach 9.5% efficiency. J Am Chem Soc 138:4657–4664. https://doi.org/10.1021/jacs.6b01744
28. Lin Y, Wang J, Zhang Z-G et al (2015) An electron acceptor challenging fullerenes for efficient polymer solar cells. Adv Mater 27:1170–1174. https://doi.org/10.1002/adma.201404317
29. Bin H, Gao L, Zhang Z-G et al (2016) 11.4% efficiency non-fullerene polymer solar cells with trialkylsilyl substituted 2D-conjugated polymer as donor. Nat Commun 7:13651-1–13651-11. https://doi.org/10.1038/ncomms13651

30. Liao SH, Jhuo HJ, Cheng YS, Chen SA (2013) Fullerene derivative-doped zinc oxide nanofilm as the cathode of inverted polymer solar cells with low-bandgap polymer (PTB7-Th) for high performance. Adv Mater 25:4766–4771. https://doi.org/10.1002/adma.201301476
31. Ye L, Zhang S, Zhao W et al (2014) Highly efficient 2D-conjugated benzodithiophene-based photovoltaic polymer with linear alkylthio side chain. Chem Mater 26:3603–3605. https://doi.org/10.1021/cm501513n
32. Etxebarria I, Guerrero A, Albero J, Garcia-belmonte G (2014) Inverted vs standard PTB7:PC70BM organic photovoltaic devices. The benefit of highly selective and extracting contacts in device performance. Org Electron 15:2756–2762. https://doi.org/10.1016/j.orgel.2014.08.008
33. Nishida K, Oka M, Hase H, Naito H (2010) Determination of physical parameter in organic bulk heterojunction solar cells using a genetic algorithm. IEEJ Trans EIS 130:1–5. https://doi.org/10.1541/ieejeiss.131.283
34. Adachi C (2014) Third-generation organic electroluminescence materials. Jpn J Appl Phys 53:060101-1–060101-11. https://doi.org/10.7567/JJAP.53.060101

Chapter 23
Design of Multifunctional Soft Biomaterials: Based on the Intermediate Water Concept

Masaru Tanaka

Abstract There are numerous parameters of polymeric biomaterials that can affect the protein adsorption and cell adhesion. The mechanisms responsible for the polymer/protein/cell interactions at the molecular level have not been clearly demonstrated, although many experimental and theoretical efforts have been made to understand these mechanisms. Water interactions have been recognized as fundamental for the protein and cell response to contact with polymers. This chapter focuses on the interfacial water at the polymer/protein/cell interfaces and specific water structure in hydrated biopolymers and bio-inspired water in hydrated synthetic polymers. Additionally, it highlights recent developments in the use of biocompatible polymeric biomaterials for medical devices and provides an overview of the progress made in the design of multifunctional element-block polymers by controlling the bio-inspired water structure through precision polymer synthesis.

Keywords Biocompatibility · Blood compatibility · Water structure · Protein adsorption · Cell adhesion

23.1 Introduction

The variety of synthetic polymeric biomaterials with distinct chemical structures, with the precise control of the molecular architecture and assembly, rationalize the numerous uses of polymers in the industry over the past few decades [1]. The synthetic polymers are used as biomedical devices, artificial organs, drug delivery system, and tissue-engineering scaffold. However, the nature and biological fate of the polymers depend on their biocompatibility.

M. Tanaka (✉)
Institute for Materials Chemistry and Engineering, Kyushu University, Fukuoka, Japan

Frontier Center for Organic Materials, Yamagata University, Yamagata, Japan
e-mail: tanaka@yz.yamagata-u.ac.jp; masaru_tanaka@ms.ifoc.kyushu-u.ac.jp

Y. Chujo (ed.), *New Polymeric Materials Based on Element-Blocks*,
https://doi.org/10.1007/978-981-13-2889-3_23

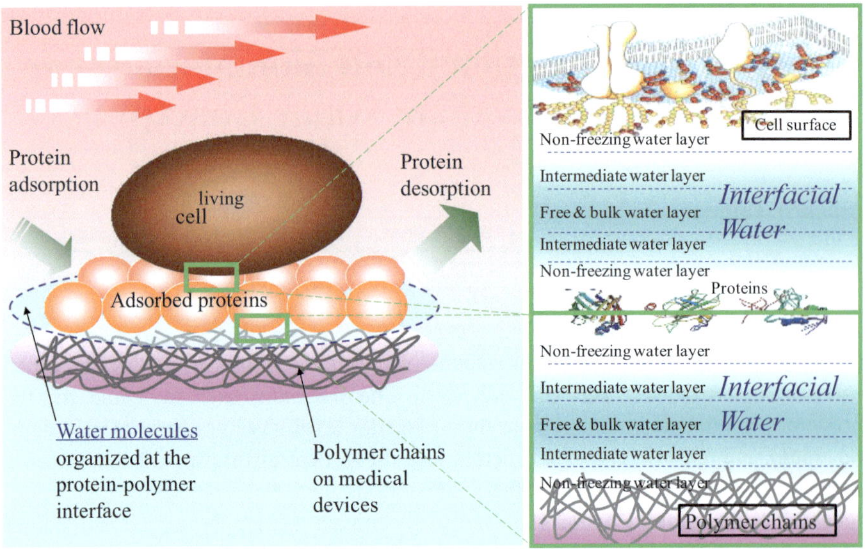

Fig. 23.1 Schematic representation of the polymer/protein/cell interfaces. When the biomaterials come into contact with the blood, the water molecules present in the blood immediately adsorb onto the surface of the polymers. This is followed, in order, by protein adsorption, restructuring and denaturation of the adsorbed proteins, and finally cell adhesion

In biomedical technology, there are continuous efforts to enhance methods, materials, and devices. The recent development of novel biomaterials and their applications to medical problems have improved the treatment and diagnosis of many diseases. Although a various types of materials in biomedicine have been used widely, most biomaterials lack the desired functional properties to interface with biological systems and have not been engineered for optimum performance. Therefore, there is an increasing demand to develop novel materials to address these problems in biomedical applications.

The mechanisms responsible for the biocompatibility of polymers at the molecular level have not been clearly demonstrated, although many experimental and theoretical efforts have been made to understand these mechanisms [1]. Water interactions have been recognized as fundamental for the blood response to contact with polymers [1]. When the biomaterials come into contact with the human body, water molecules immediately adsorb onto the surface of the materials. This is followed by protein adsorption, protein restructuring, denaturation of the adsorbed protein, and finally cell adhesions (Fig. 23.1). We have proposed the "intermediate water" concept [2] and hypothesized that intermediate water, which prevents the proteins and blood cells from directly contacting the polymer surface on the polymer surface, plays an important role in the blood compatibility of polymers.

Here, we describe the recent synthetic strategies for the design of biocompatible polymeric biomaterials, which include characteristic properties of biomaterials for various applications in the surfaces of polymeric biomaterials. In addition, we describe an overview of the recent experimental progress of the screening of biocompatible polymers based on the bio-interfacial water structure through precision polymer synthesis.

23.2 Biocompatible Polymers for Medical Device

Polymeric materials for the medical devices that may come in contact with the human blood should have capacity to resist protein adsorption and blood cell adhesion, thus triggering the organism's defense systems. Some biocompatible polymer surfaces have been developed, and they fall into the following three categories: (a) hydrophilic surfaces [3], (b) surfaces with micro-phase-separated domains [4], and (c) biomembrane-like surfaces [5], including zwitterionic groups [6, 7]. Physicochemical properties of polymers such as wettability, surface charge, surface free energy, stiffness, topography, and the presence of specific chemical functionalities, surface bound water appears to bear an essential role in the biological response induced by the polymers. A new generation polymer, poly(2-methoxyethyl acrylate) (PMEA), showed excellent biocompatibility and has been approved for medical use by the FDA [8]. For instance, PMEA-coated circuits and tubes exhibit significantly reduced blood cell activation when used in cardiopulmonary bypass and catheters. PMEA has the largest market share in the world as an antithrombogenic coating agent for artificial oxygenators. It has been maintained that PMEA's compatibility with platelets, white and red blood cells, and complement and coagulation systems has been dictated by the presence of the intermediate water [2, 9, 10].

It should be noted that the word "biocompatibility" is used in general as the term evaluating properties of materials which do not cause adverse effect when the materials come into contact with living organisms, such as proteins, biological cells, and tissues. This presentation primarily deals with "biocompatibility" of polymers against various biological elements in blood flow system.

23.3 Protein Adsorption on Polymeric Biomaterials

In order to clarify the reasons for excellent biocompatibility of PMEA, the amount and kinetics in the early stage of plasma protein adsorbed onto PMEA and the secondary structure of the protein were investigated. The amount of protein adsorbed onto PMEA was very small, and the quantity was similar to that adsorbed onto poly (2-hydroxyethyl methacrylate) (PHEMA) [11, 12]. Circular dichroism (CD) spectroscopy revealed a significant conformation change in proteins adsorbed onto PHEMA, whereas the conformational change of the proteins adsorbed on PMEA was very small [12]. Using the quartz crystal microbalance (QCM) measurement, we investigated the adsorption/desorption behavior of proteins on the PMEA surface in terms of their binding constants and association and dissociation rates. The CD and QCM results suggested that the excellent biocompatibility of PMEA is related to the low denaturation and the high dissociation rates of the proteins attached onto PMEA [12].

The adhesion force between PMEA and fibrinogen as well as PMEA and bovine serum albumin (BSA) was measured by atomic force microscopy (AFM). The PMEA surface showed almost no adhesion to native protein molecules [13, 14]. The denaturation of the adsorbed protein could lead to platelet activation and subsequent thrombus formation. In other words, when the protein molecule which adsorbs onto a polymer surface retains its native conformation, platelets cannot adhere to the surface.

23.4 Relationship Between Water Structure and Biocompatibility

Although considerable theoretical and experimental efforts have been devoted to clarifying this issue in the past few decades, the factors responsible for biocompatibility of polymers have not been elucidated. Water molecules serve as a medium for adhesion and play a role in cell morphology and other cellular functions (Fig. 23.1). Water is thought to be a fundamental factor in the biological response induced by artificial materials. Many researchers have insisted that water structure on a polymer surface is one of the key factors for its biocompatibility as mentioned above. However, the proposed structures and/or the functions of water are different in many cases, and there is little consistency among structures and dynamics. Detailed studies on the dynamics and structure of hydrated polymers are required to clarify the mechanism underlying the biocompatibility of polymers.

23.5 There States of Water Structure in Hydrated Polymers

It is well known that proteins and cells have hydration shells composed of non-freezing, intermediate, and free water [15–18], and the shells keep the blood components stable. In addition, the hydrated water in synthetic polymer can be classified into three types: free water (or freezing water), freezing bound water (or intermediate water), and non-freezing water (or non-freezing bound water) (Table 23.1).

The hydrated PMEA possessed a unique water structure, observed as cold crystallization of water in differential scanning calorimetry (DSC) [2, 9, 10]. Cold crystallization is interpreted as ice formation at low temperature below 0 °C, an attribute of intermediate water in PMEA. It should be noted that an exothermic peak of cold crystallization in the DSC heating process as well as the cooling process can

Table 23.1 Classification of water in hydrated polymer

Water structure (mode of binding)	Tightly bound water/non-freezing bound water/non-freezing water	Loosely bound water/ freezing bound water/ intermediate water	Scarcely bound water/freezing water/free water
Freezability	Not freezable below 0 °C	Freezable/melting below 0 °C	Freezable/melting at 0 °C
NMR correlation time (t_c)(s)	10^{-8}–10^{-6}	10^{-10}–10^{-9}	10^{-12}–10^{-11}
In situ ATR-IR O-H stretching region (cm^{-1})	3600	3400	3200
Binding constant to polymer	Strong	Medium	Weak

The hydrated water in polymer can be classified into three types, (1) non-freezing water (tightly bound water/non-freezing bound water), (2) intermediate water (loosely bound water/freezing bound water), and (3) free water (scarcely bound water/freezing water) on the basis of the equilibrium water content and the enthalpy changes due to the phase transition observed by differential scanning calorimetry (DSC) analysis as well as ATR-IR and solid-state NMR. The intermediate water was only found in hydrated biopolymers (proteins, polysaccharides, and nucleic acid; DNA and RNA) and hydrated biocompatible synthetic polymers. The intermediate water behaves differently from bulk water and acts as a physical barrier against protein adsorption and cell adhesion. The intermediate water, which prevents the proteins and cells from directly contacting the polymer surface, plays an important role in the compatibility of the blood with polymers

show the presence of intermediate water by a peak of crystallization below 0 °C, which is different from the free water. Also, in the heating thermograms, an endothermic peak with a shoulder on the lower temperature region below 0 °C is often considered as the effect of the melting of ice from intermediate water.

The presence of three types of water in PMEA is supported by the results of attenuated total reflection infrared (in situ ATR-IR) spectroscopy, a strong peak at 3400 cm^{-1} in a time-resolved ATR-IR spectrum, and the localized hydration structure consisting of the three hydrated water in PMEA [19, 20]. In O-H stretching region, the band for non-freezing water (3600 cm^{-1}) appears immediately within water contact with PMEA film, and C=O stretching band that assigned to hydrogen-bonded C=O (1730 cm^{-1}) increases over time. The band for free water (3200 cm^{-1}) arises in the late stage after 1 min. Consequently, the non-freezing water has a C=O...H-O type of hydrogen-bonding interaction with the carbonyl group of PMEA, and the intermediate water could interact with the methoxy moiety (ether moiety) in the PMEA side-chain terminal, forming small water cluster.

The intermediate water was also detected by solid-state NMR as a higher mobility of water as well as polymer chains [21]. Tightly bound (non-freezing) water has a NMR correlation time (τ_c) value of 10^{-8}–10^{-6} s and does not crystallize even at −100 °C. Loosely bound (intermediate) water has a τ_c value of 10^{-10}–10^{-9} s and exhibits melting/crystallization at temperatures below 0 °C. Free water has τ_c value of 10^{-12}–10^{-11} s and melting at 0 °C. The intermediate water interacts weakly with the methoxy group of PMEA. While investigating the main factor responsible for the biocompatibility of PMEA, it is important to reveal the intermediate water structure on the polymer surface.

23.6 Intermediate Water in Bio- and Synthetic Polymers

The hypothesis is supported by several reports that demonstrate the formation of intermediate water in well-known biocompatible synthetic polymers like poly(ethylene glycol) (PEG), polyvinylpyrrolidone (PVP), polyoxazoline, polyphosphazene, poly(methylvinyl ether) (PMVE), poly(2-methacryloyloxyethyl phosphorylcholine) (PMPC) including zwitterionic polymers (betine polymers), poly(tetrahydrofurfuryl acrylate) (PTHFA), poly(2-(2-ethoxyethoxy)ethyl acrylate), polyoxazoline, and PMEA analogous polymers possessing a lower critical solution temperature (LCST) and other biocompatible poly(meth)acrylates; poly poly(meth)acrylamide; biopolymers; nucleic acid, DNA and RNA; proteins, gelatin, albumin, and cytochrome C; and various polysaccharides, including hyaluronan, heparin, alginate, and gum [22–25]. The intermediate water contents were dependent on their sequence. On the other hand, cold crystallization of water was not observed in hydrated PMEA analogous polymers (non-biocompatible), which do not show excellent biocompatibility. In addition, the protein adsorption and the exposure degree of platelet adhesion sites showed decrement trends as affected by increasing the amount of intermediate water. The results of this study demonstrate that the amount of intermediate water should play a key role in expressing the blood compatibility of polymeric materials [23–25]. Based on these findings, the intermediate water, which prevents the biocomponents from directly contacting the polymer surface or non-freezing water on the polymer surface, must play an important role in the excellent biocompatibility of polymers. Moreover, the intermediate water content in polymers influenced protein adsorption behavior [26]. It is demonstrated that protein adsorption is suppressed when intermediate water content increased. Thus, it is expected that more favorite synthetic polymers for various cell culture can be designed by controlling intermediate water contents [24, 25, 27–40]. Additionally, the mechanism of the intermediate water formation in biopolymers to provide more information on the role of water molecules in biological response needs to be clarified.

23.7 Control of Intermediate Water Content

The molecular structure and dynamics which affect the water structure can be controlled by interactions between the backbone and side chain of the polymer. Therefore, in theory, by altering the structure of a polymer (e.g., by changing its length, the functional group in its side chain, or whether its backbone is acrylate or methacrylate), we could regulate its molecular structure and dynamics to control the intermediate water content. Using principles of intermediate water, which is

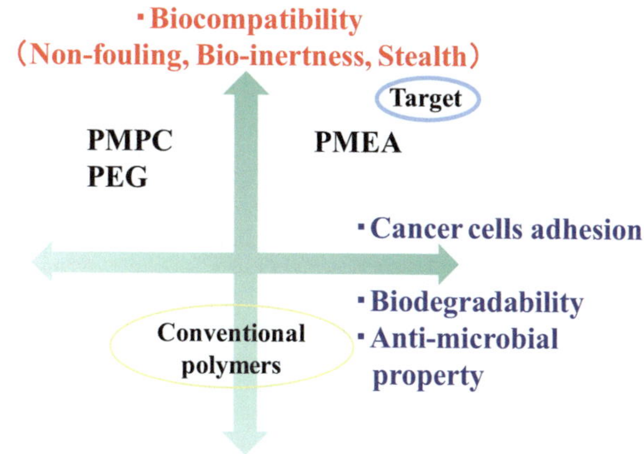

Fig. 23.2 New target and strategy to design smart soft materials. Intermediate water contents affect the results

common in hydrated biopolymers and only biocompatible synthetic polymers, we hope to find the synthetic methodology to create novel biocompatible polymers moves toward a more high-throughput way. Indeed, new designed aliphatic carbonyl polymers have ester or carbonate linkages facilitating the breakdown of their monomers and thus their degradation. We found that a higher amount of intermediate water in aliphatic carbonyl polymers is related to the blood compatibility/biodegradability and the presence of the ether bonds in the main chain of the aliphatic carbonyl polymers that are involved in the hydration and formation of intermediate water [41, 42]. Such well-defined polymeric biomaterials could find application in the age of personalized medicine (Fig. 23.2).

23.8 Conclusions

We propose intermediate water concept for directional design of functional biomaterials. The interaction of polymers with blood components is managed by cooperation of primary structure including backbone and functional groups at the side chain to tune the chemical functions and higher-order structure forming specific shape to restrict or expand the chemical function as biological system generally adopts. Using principles of intermediate water, which is common in hydrated biopolymers and in only biocompatible synthetic polymers, the synthetic methodology to create novel biocompatible polymers moves toward a more high-throughput way. Such well-defined polymeric biomaterials could find application in the age of personalized medicine.

Acknowledgments The authors are very grateful to Professor Emeritus Teiji Tsuruta (University of Tokyo) for his valuable advice. The author also would like to thank all members of the Tsuruta Forum for their helpful comments.

References

1. a). Hoffman AS (2002) Hydrogels for biomedical applications. Adv Drug Deliv Rev 43:3–13. b). Ratner BD, Hoffman AS, Schoen FJ, Lemons JE (2004) Biomaterials science; an introduction to materials in medicine. Academic, London. c) Special Issue of Prof. T. Tsuruta (2010) J Biomater Sci Polym Ed 21:1827–1970
2. Tanaka M, Mochizuki A, Ishii N, Motomura T, Hatakeyama T (2002) Study on blood compatibility of poly(2-methoxyethylacrylate). Relationship between water structure and platelet compatibility in poly(2-methoxylethylacrylate-co-2-hydroxyethylmethacrylate). Biomacromolecules 3:36–41
3. Peppas NA (1987) Hydrogel in medicine and pharmacy, vol 2. CRC Press, Boca Raton
4. Okano T, Nishiyama S, Shinohara I, Akaike T, Sakurai Y, Kataoka K, Tsuruta T (1981) Effect of hydrophilic and hydrophobic microdomains on mode of interaction between block copolymer and blob platelets. J Biomed Mater Res 15:393–403
5. Ishihara K, Nomura H, Mihara T, Kurita K, Iwasaki Y, Nakabayashi N (1998) Why do phospholipid polymers reduce protein adsorption? J Biomed Mater Res 39:323–330
6. Holmlin RE, Chen X, Chapman RG, Takayama S, Whitesides GM (2001) Zwitterionic SAMs that resist nonspecific adsorption of protein from aqueous buffer. Langmuir 17:2841–2850
7. Kitano H, Tada S, Mori T, Takaha K, Gemmei-Ide M, Tanaka M, Fukuda M, Yokoyama Y (2005) Correlation between the structure of water in the vicinity of carboxybetaine polymers and their blood-compatibility. Langmuir 21:11932–11940
8. Tanaka M, Motomura T, Kawada M, Anzai T, Kasori Y, Shiroya T, Shimura K, Onishi M, Mochizuki A (2000) Blood compatible aspects of poly(2-methoxyethylacrylate) (PMEA)–relationship between protein adsorption and platelet adhesion on PMEA surface. Biomaterials 21(14):1471–1481
9. Tanaka M, Motomura T, Ishii N, Shimura K, Onishi M, Mochizuki A, Hatakeyama T (2000) Cold crystallization of water in hydrated poly(2-methoxyethyl acrylate) (PMEA). Polym Int 49:1709–1713
10. Tanaka M, Mochizuki A (2004) Effect of water structure on blood compatibility: thermal analysis of water in poly(meth)acrylate. J Biomed Mater Res 68A:684–695
11. Tanaka M, Mochizuki A, Motomura T, Shimura K, Onishi M, Okahata Y (2001) In situ studies on protein adsorption onto a poly(2-methoxyethyl acrylate) surface by a quartz crystal microbalance. Colloids Surf A Physicochem Eng Asp 193:145–152
12. Tanaka M, Mochizuki A, Shiroya T, Motomura T, Shimura K, Onishi M, Okahata Y (2002) Study on kinetics of early stage protein adsorption and desorption on poly(2-methoxyethyl acrylate) (PMEA) surface. Colloids Surf A Physicochem Eng Asp 203:195–204
13. Hayashi T, Tanaka Y, Koide Y, Tanaka M, Hara M (2012) Mechanism underlying bioinertness of self-assembled monolayers of Oligo(ethyleneglycol)-terminated alkanethiols on gold: protein adsorption, platelet adhesion, and surface forces. Phys Chem Chem Phys 14:10194–10206
14. Sekine T, Tanaka Y, Sato C, Tanaka M, Hayashi T (2015) Evaluation of factors to determine platelet compatibility by using self-assembled monolayers with a chemical gradient. Langmuir 31:7100–7105
15. Hazlewood CF, Nichols BL, Chamberlain NF (1969) Evidence for the existence of a minimum of two phases of ordered water in skeletal muscle. Nature 222:747
16. Kuntz ID Jr, Brassfield TS, Law GD, Purcell GV (1969) Hydration of macromolecules. Science 163(3873):1329–1331
17. Uedaira H (1980) In: Pullman B, Yagi K (eds) Water and metal cations in biological systems. Japan Scientific Societies Press, Tokyo, p 47
18. Pal SK, Peon J, Zewail AH (2002) Biological water at the protein surface: dynamical solvation probed directly with femtosecond resolution. Proc Natl Acad Sci U S A 99:1763
19. Morita S, Tanaka M, Ozaki Y (2007) Time-resolved in-situ ATRIR observations of the process of water into a poly(2-methoxyethyl acrylate) (PMEA) film. Langmuir 23:3750–3761

20. Morita S, Tanaka M (2014) Effect of sodium chloride on hydration structures of PMEA and P (MPC-r-BMA). Langmuir 30:10698–10703
21. Miwa Y, Ishida H, Saitô H, Tanaka M, Mochizuki A (2009) Network structures and dynamics of dry and swollen poly(acrylate)s. Polymer 50:6091–6099
22. Hatakayama T, Tanaka M, Hatakayama H (2010) Studies on bound water restrained by poly(2-methacryloyloxyethyl phosphorylcholine) (PMPC): comparison of the polysaccharides-water systems. Acta Biomater 6:2077–2082
23. Tanaka M, Hayashi T, Morita S (2013) The roles of water molecules at the biointerface of medical polymers. Polym J 45:701–710
24. Sato K, Kobayashi S, Kusakari M, Watahiki S, Oikawa M, Hoshiba T, Tanaka M (2015) The relationship between water structure and blood compatibility in poly (2-methoxyethyl Acrylate) (PMEA) analogues. Macromol Biosci 15:1296–1303
25. Tanaka M, Sato K, Kitakami E, Kobayashi S, Hoshiba T, Fukushima K (2015) Design of biocompatible and biodegradable polymers based on intermediate water concept. Polym J 47:114–121
26. Hoshiba T, Nemoto E, Sato K, Orui T, Otaki T, Yoshihiro A, Tanaka M (2015) Regulation of the contribution of integrin to cell attachment on poly(2-Methoxyethyl Acrylate) (PMEA) analogous polymers for attachment-based cell enrichment. PLoS One 10:e0136066
27. Kobayashi S, Fukuda K, Kataoka M, Tanaka M (2016) Regioselective ring-opening metathesis polymerization of 3-substituted cyclooctenes with ether side chains. Macromolecules 49:2493–2501
28. Osawa K, Kobayashi S, Tanaka M (2016) Synthesis of sequence-specific polymers with amide side chains via regio-/stereoselective ring-opening metathesis polymerization of 3-substituted*cis*-cyclooctene. Macromolecules 49:8154–8161
29. Sato K, Kobayashi S, Sekishita A, Wakui M, Tanaka M (2017) Synthesis and thrombogenicity evaluation evaluation of poly(3-methoxypropionic acid vinyl ester): a candidate for blood compatible polymer. Biomacromolecules 18:1609–1616
30. Hoshiba T, Nikaido M, Tanaka M (2014) Characterization of the attachment mechanisms of tissue-derived cell lines to blood-compatible polymers. Adv Healthc Mater 3:775–784
31. Choi H, Tanaka M, Hiragun T, Hide M, Sugimoto K (2014) Non-tumor mast cells cultured in vitro on a honeycomb-like structured film proliferate with multinucleated formation. Nanomedicine 10:313–319
32. Hirata T, Matsuno H, Kawaguchi D, Hirai T, Yamada N, Tanaka M, Tanaka K (2015) Effect of local chain dynamics on a bio-inert interface. Langmuir 31:3661–3667
33. Hoshiba T, Otaki T, Nemoto E, Maruyama H, Tanaka M (2015) Blood com- patible polymer for hepatocyte culture with high hepatocyte-specific functions toward bioartificial liver development. ACS Appl Mater Interfaces 7:18096–18103
34. Khan F, Tanaka M, Ahmad SR, Mater J (2015) Fabrication of polymeric biomaterials: a strategy for tissue engineering and medical devices. Chem B 3:8224–8249
35. Sato C, Aoki M, Tanaka M (2016) Blood-compatible poly(2-methoxyethyl acrylate) for the adhesion and proliferation of endothelial and smooth muscle cells. Colloids Surf B: Biointerfaces 145:586–596
36. Hoshiba T, Nikaido M, Yagi S, Konno I, Yoshihiro A, Tanaka M, Bioact J (2016) Blood compatible poly(2-methoxyethyl acrylate) (PMEA) for the adhesion and proliferation of lung cancer cells toward the isolation and analysis of circulating tumor cells. Compat Polym 31:361–372
37. Kono K, Hiruma H, Kobayashi S, Sato Y, Tanaka M, Sawada R, Niimi S (2016) In vitro endothelialization test of biomaterials using immortalized endothelial cells. PLoS One 11: e01582898
38. Hoshiba T, Orui T, Endo C, Sato K, Yoshihiro A, Minagawa Y, Tanaka M (2016) Adhesion-based simple capture and recovery of circulating tumor cells using a blood-compatible and thermo-responsive polymer-coated substrate. RSC Adv 6:89103–89112

39. Murakami D, Kobayashi S, Tanaka M (2016) Interfacial structures and fibrinogen adsorption at blood-compatible polymer/water interfaces. ACS Bimater Sci Eng 2(12):2122–2126
40. Hoshiba T, Nemoto E, Sato K, Maruyama H, Endo C, Tanaka M (2016) Promotion of adipogenesis of 3T3-L1 cells on protein adsorption-suppressing poly(2-methoxyethyl acrylate) analogs. Biomacromolecules 17:3808–3815
41. Fukushima K, Tsai M, Ota T, Haga Y, Matsuzaki K, Inoue Y, Tanaka M (2015) Evaluation of haemocompatibility of hydrated biodegradable aliphatic carbonyl polymers with a subtle difference in a backbone structure on the basis of intermediate water concept and surface hydration. Polym J 47:469–473
42. Basterretxea A, Haga Y, Sanchez-Sanchez A, Isik M, Irusta L, Tanaka M, Fukushima K, Sardon H (2016) Biocompatibility and hemocompatibility evaluation of polyether urethanes synthesized using DBU organocatalyst. Eur Polym J 84:750–758

Chapter 24
Synthesis of Calcium-Phosphate-Based Nanoparticles as Biocompatible and Biofunctional Element Blocks

Ayako Oyane and Maki Nakamura

Abstract Calcium phosphate (CaP)-based nanoparticles containing functional substances such as DNA and Ag are biocompatible and biofunctional element blocks that are useful as agents for drug and gene delivery and as building blocks for higher-order biomaterials. Such CaP-based nanoparticles can be synthesized via precipitation from labile supersaturated CaP solutions supplemented with functional substances. In this chapter, conventional and laser-assisted precipitation processes for the synthesis of CaP-based nanoparticles are described with a focus on our recent studies. Both precipitation methods are simple (one-pot), rapid (nanoparticle formation occurs within a few tens of minutes), free of harmful additives, and capable of controlling the physicochemical and biological properties of the CaP-based nanoparticles. These characteristics represent advantages for future in vitro and in vivo applications of these precipitation processes and the resulting CaP-based nanoparticles.

Keywords Calcium phosphate · Precipitation · Supersaturated solution · Nanoparticle · Laser

24.1 Introduction

The human cortical bone is an organic–inorganic hybrid composed of calcium phosphate (CaP)-based element blocks, i.e., nanosized CaP (low-crystalline apatite and its precursors) precipitated on collagen nanofibers. Cortical bones are naturally fabricated in the body (in vivo) via the precipitation of nanosized CaP (a phenomenon known as biomineralization) on collagen nanofibers, which are assembled and hierarchically textured into a higher-order structure. This in vivo

A. Oyane (✉) · M. Nakamura
Nanomaterials Research Institute, National Institute of Advanced Industrial Science and Technology (AIST), Tsukuba, Ibaraki, Japan
e-mail: a-oyane@aist.go.jp; ma-ki-nakamura@aist.go.jp

Y. Chujo (ed.), *New Polymeric Materials Based on Element-Blocks*,
https://doi.org/10.1007/978-981-13-2889-3_24

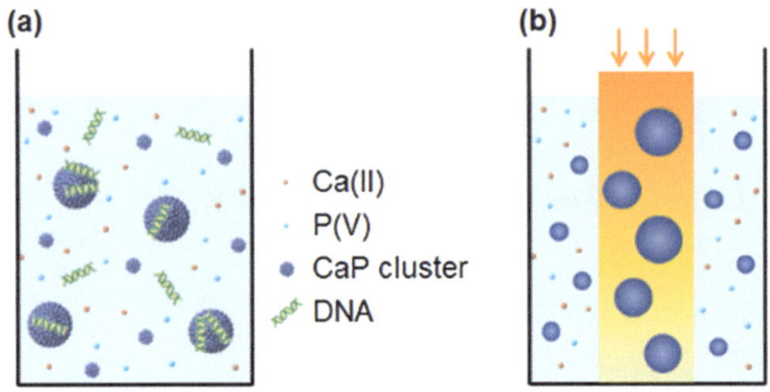

Fig. 24.1 Precipitation process for the synthesis of CaP-based nanoparticles from labile supersaturated CaP solutions: (**a**) conventional precipitation and (**b**) laser-assisted precipitation

CaP precipitation process can be reproduced outside the body (in vitro) with acellular, metastable supersaturated CaP solutions mimicking the body fluid [1]. By tuning the ionic concentrations, pH, and temperature of the CaP solution, which control the degree of supersaturation, it is possible to prepare a labile supersaturated CaP solution that induces homogeneous precipitation in solution, even without collagen nanofibers or another nucleating agent. In such a homogeneous precipitation process from a labile supersaturated CaP solution, a wide variety of nanoparticles composed of CaP matrix (hereafter referred to as CaP-based nanoparticles) can be synthesized.

In this chapter, we describe synthetic techniques for CaP-based nanoparticles as biocompatible element blocks that are useful in drug and gene delivery. First, we describe the fundamentals of CaP precipitation (Sect. 24.2.1) along with the coprecipitation of CaP and functional substances (Sect. 24.2.2) from labile supersaturated CaP solutions. Second, we describe two different precipitation processes for the synthesis of CaP-based nanoparticles, conventional (Sect. 24.3) and laser-assisted (Sect. 24.4), with a focus on our recent studies. Using the conventional process (Sect. 24.3.2), we synthesized DNA–CaP composite nanoparticles via homogeneous precipitation in labile supersaturated CaP solutions (Fig. 24.1a). Although homogeneous precipitation has long been used to synthesize DNA–CaP composite nanoparticles for gene delivery, we first reported the use of infusion fluids for the synthesis of DNA–CaP composite nanoparticles [2–4]. Laser-assisted precipitation (Sect. 24.4) is our newly developed process in which pulsed laser irradiation is applied to a labile supersaturated CaP solution during homogeneous precipitation (Fig. 24.1b) [5]. We describe our laser-assisted process for the synthesis of three types of CaP-based nanoparticles: Fe–CaP composite (Sect. 24.4.1) [6], iron oxide (IO)–CaP composite (Sect. 24.4.2) [7], and Ag–CaP composite (Sect. 24.4.3) [8] nanoparticles.

24.2 Fundamentals

24.2.1 *CaP Precipitation from Supersaturated CaP Solutions*

The process of CaP precipitation from supersaturated CaP solutions can be described by a nonclassical nucleation pathway [9]. In certain metastable, supersaturated CaP solutions with neutral pH [e.g., simulated body fluid (SBF) [1]], nanosized prenucleation CaP clusters can be stable [10, 11]. These CaP clusters are solutes with a molecular character in the solution. The prenucleation CaP clusters densify heterogeneously at extrinsic interfaces (nucleating agents or CaP seeds) to form post-critical, amorphous nuclei (i.e., heterogeneous nucleation) [12]. In more highly supersaturated and labile CaP solutions (e.g., five times concentrated SBF [13]), the prenucleation CaP clusters cannot remain stable for long periods of time; they gather spontaneously to form post-critical amorphous nuclei (homogeneous nucleation) via cluster–cluster aggregation.

The amorphous CaP nucleated at the extrinsic interface or dispersed throughout the solution grows by incorporating CaP nanoclusters, cluster aggregates, and ions in the supersaturated solution while crystallizing into more stable CaP phases with time. After aging in the solution for a certain period of time, the final CaP phase is generally low-crystalline apatite because apatite has the lowest solubility and is the most stable among all CaP phases in neutral solutions [14].

24.2.2 *Coprecipitation of CaP and Functional Substances in Supersaturated CaP Solutions*

The CaP precipitates (apatite and its precursors) synthesized from supersaturated solutions generally show good biocompatibility along with biodegradability; they partially degrade into plasma ions in the body fluid. These CaP precipitates can immobilize a variety of functional substances, including trace elements, proteins, DNAs, antibodies, and antibacterial agents, through coprecipitation if these substances supplement the supersaturated CaP solutions and have sufficient affinity for CaP [15, 16]. Even thermally non-durable, unstable biological substances (e.g., proteins and DNAs) can coprecipitate with CaP without complete denaturalization or deactivation because of the mild and biomimetic reaction condition in supersaturated CaP solutions. The resulting nanocomposites comprise a CaP matrix and functional substances dispersed [17]. The nanocomposites allow the controlled release of the functional substances, most likely because of the partial dissolution of the CaP matrix [18, 19]. The released functional substances retain their biological activity, thereby fulfilling various biomedical functions in vitro and in vivo.

24.3 Conventional Precipitation

24.3.1 Synthesis of CaP-Based Nanoparticles

As described in Sect. 24.2.1, homogeneous CaP nucleation is induced in labile supersaturated CaP solutions with relatively high degrees of supersaturation with respect to CaP. In certain labile supersaturated CaP solutions, the nucleated CaP spontaneously grows into CaP nanoparticles. When functional substances with sufficient affinity for CaP (e.g., DNA, as illustrated in Fig. 24.1a) are added to the solution, these substances attach to the CaP nanoparticles during their growth and become immobilized within the CaP nanoparticle matrix (Fig. 24.1a). The resulting CaP-based nanoparticles with immobilized functional substances have a potential as agents for drug and gene delivery. This is because the nanoparticles can penetrate into tiny interstices in the body, be injected intravenously and intra-arterially, circulate in the bloodstream, and penetrate cell membranes and blood vessel walls when they possess the appropriate size and surface chemistry. In the next section, we describe our recent study on the synthesis of DNA–CaP composite nanoparticles for gene delivery applications.

24.3.2 Use of Infusion Fluids for the Synthesis of DNA–CaP Composite Nanoparticles

Gene delivery techniques using viral and non-viral systems are useful in controlling cell properties and behavior (e.g., proliferation and differentiation). Among the various gene delivery agents, DNA–CaP composite nanoparticles are advantageous because of their safety, biocompatibility, bioresorbability, and cost efficiency [20]. The DNA–CaP composite nanoparticles can be synthesized via homogeneous CaP nucleation and the coprecipitation of DNA and CaP from labile supersaturated CaP solutions (Fig. 24.1a). When DNA–CaP composite CaP nanoparticles synthesized in this way are added to a cell culture, they are uptaken by cells via endocytosis, thereby inducing specific gene expression in the cells.

DNA–CaP composite nanoparticles have been conventionally synthesized from labile supersaturated CaP solutions prepared from chemical reagents. Recently, we have synthesized DNA–CaP composite nanoparticles from labile supersaturated CaP solutions prepared from clinically approved infusion fluids rather than chemical reagents [2–4]. We followed the protocol proposed by Sogo et al. and Mutsuzaki et al.; they used infusion fluids for the synthesis of protein–apatite composite layers [21, 22]. In contrast to chemical reagents, infusion fluids are intravenously injectable medicinal solutions that have been approved by regulatory agencies. Thus, DNA–CaP composite nanoparticles derived from infusion fluids are free of pathogens and endotoxins, and have a high degree of biological safety. This represents a significant advantage for future in vivo studies and clinical applications.

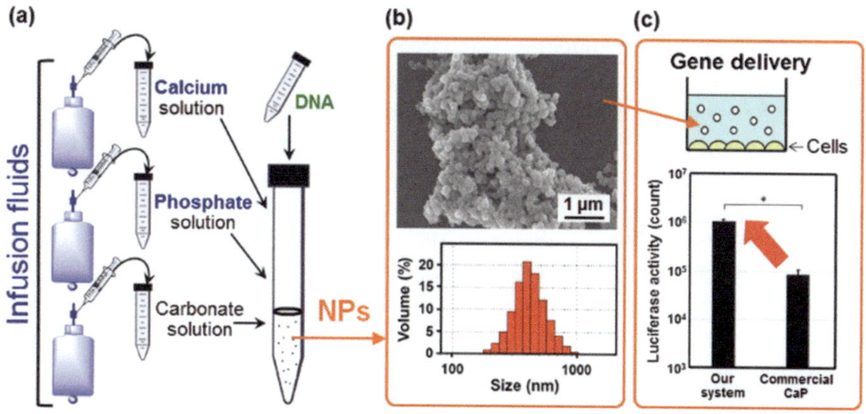

Fig. 24.2 DNA–CaP composite nanoparticles. (Reproduced from [4] with permission.) (**a**) Precipitation using infusion fluids to synthesize the nanoparticles. (**b**) Scanning electron microscopy (SEM) image and particle size distribution (analyzed by dynamic light scattering) of the nanoparticles. (**c**) Gene delivery system using the nanoparticles (upper image) and their gene delivery efficiency to CHO-K1 cells (luciferase activity of the cells) in comparison to that of commercial DNA–CaP nanoparticles (lower plot)

We prepared supersaturated CaP solutions by mixing a plasmid (cDNA of luciferase) and several infusion fluids (calcium-containing, phosphate-containing, and carbonate-containing infusion fluids along with water for injection) (Fig. 24.2a) [2–4]. The Ca and P concentrations of the solution were systematically varied according to the previous report of Bodhak et al. [23]. The thus prepared labile supersaturated CaP solutions (X1.0, X1.2, and X1.4 solutions in [3]) induced the homogeneous nucleation of amorphous CaP within 10 min after preparation. The nucleated CaP remained amorphous for up to 3 h and grew three-dimensionally into spherical nanoparticles because of the absence of crystalline anisotropy. During this precipitation process, a large majority of the DNA molecules (~90%) added to the solution was immobilized within the CaP nanoparticles, yielding nanoparticles with large negative zeta potentials [3, 4]. Further spontaneous growth of these nanoparticles and their aggregation were restricted as a result of the surface attachment of DNA molecules and charge repulsion between nanoparticles, respectively, leading to the formation of size-regulated and well-dispersed DNA–CaP composite nanoparticles (Fig. 24.2b) [3, 4]. The size-regulating effect of the DNA molecules was verified by a control experiment with a DNA-free supersaturated solution; the CaP nanoparticles quickly grew into microparticles in this solution because of the lack of DNA [3].

Over the past few decades, the gene delivery efficiency of DNA–CaP composite nanoparticles has been improved using various additives, including surfactants, cell adhesion proteins, lipids, and ligands. In our study, the efficiency was increased simply by varying the process parameters (e.g., precipitation temperature and time along with the Ca and P concentrations in the CaP solution) without using additives other than infusion fluids and plasmids [4]. These parameters affected the gene delivery efficiency of the resulting DNA–CaP composite nanoparticles by affecting the kinetics of CaP precipitation and DNA immobilization in the solution. By optimizing these parameters, we achieved a higher gene delivery efficiency compared to the commercial DNA–CaP composite nanoparticles for both hard-to-transfect pluripotent stem cells (C3H10T1/2) and easy-to-transfect CHO-K1 cells (Fig. 24.2c). MTT and protein assays confirmed that our DNA–CaP composite nanoparticles were not cytotoxic to either type of cell under the tested condition.

Our DNA-immobilized CaP nanoparticles have the advantages of biological safety because of the injectable source materials, low cytotoxicity, and relatively high and controllable gene delivery efficiency depending on the process parameters; thus, they have a potential for use in various in vitro and in vivo gene delivery applications.

24.4 Laser-Assisted Precipitation

24.4.1 Synthesis of Fe–CaP Composite Nanoparticles

In the conventional precipitation process from labile supersaturated CaP solutions, it is difficult to control the size and dispersion of the nanoparticles because, once nucleated, the CaP nanoparticles grow continuously into larger particles under supersaturated conditions. Therefore, specific additives such as DNA molecules and surfactants have been used in the precipitation process to chemically modify the particle surfaces and control their size and dispersion. However, some additives may create biological safety issues for biomedical applications. The use of physical stimulation such as laser irradiation, microwave irradiation, and ultrasonic irradiation is another effective approach in controlling the size and dispersion of nanoparticles [24–27].

Recently, we have developed an additive-free, rapid, and one-pot physicochemical process for the synthesis of CaP-based nanoparticles by combining a physical laser process with a chemical precipitation process [6–8]. For the physical laser process, we followed the process of pulsed laser melting in liquid reported by Wang et al. [28]. In their work, weak pulsed laser irradiation was applied without focusing to nanopowders dispersed in a liquid solvent. After irradiation for 10–30 min, the nanopowders with irregular shapes changed into size-regulated spherical nanoparticles via laser absorption and selective heating to form intermediate melted spherical droplets in the solvent. This process enabled the additive-free synthesis of spherical nanoparticles of various metals and metal oxides. However, spherical CaP nanoparticles were not obtained using this process under the tested conditions because of the insufficient laser absorption of CaP nanopowders (with the exception of carbon-integrated CaP nanopowders [29]).

We attempted to induce coprecipitation of CaP and light-absorbing agents from labile supersaturated CaP solutions to synthesize nanoprecipitates with increased light absorption in situ and simultaneously achieve pulsed laser melting in the same

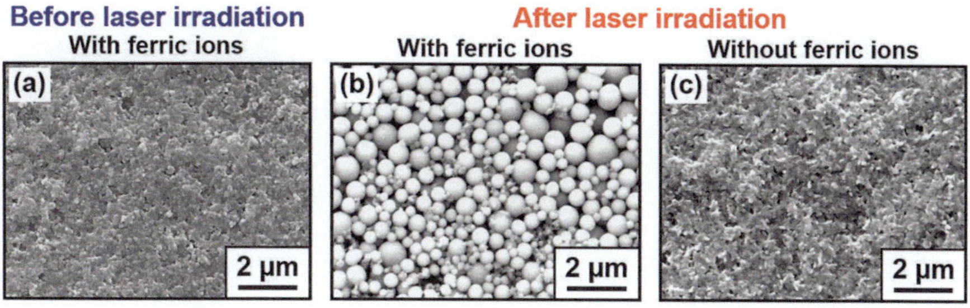

Fig. 24.3 Fe–CaP composite nanoparticles. (Modified from [6].) (**a**) SEM image of the initial precipitates (before irradiation) as-formed in a labile supersaturated CaP solution supplemented with ferric ions (5 mM). (**b**) SEM image of the Fe–CaP composite nanoparticles synthesized by the laser irradiation (200 mJ/pulse/cm^2, 30 min) of the same solution. (**c**) SEM image of the product obtained after the same laser irradiation of a labile supersaturated CaP solution without ferric ions

solution (Fig. 24.1b). As a light-absorbing agent, we first selected ferric (Fe^{3+}) ions and applied unfocused Nd:YAG pulsed laser irradiation (355 nm, 30 Hz, 200 mJ/pulse/cm^2) to a labile supersaturated CaP solution supplemented with ferric ions (5 mM) [6]. This solution formed Fe-containing nanoprecipitates (Fig. 24.3a) via homogeneous nucleation soon after preparation (i.e., upon the mixing of the source ion solutions). The thus-formed nanoprecipitates underwent pulsed laser melting and spheroidizing in the supersaturated solution because of the increased light absorption, leading to the formation of spherical Fe–CaP composite nanoparticles after 20–30 min of laser irradiation (Fig. 24.3b). As the melted spherical droplets were quenched in the surrounding aqueous solution during the pulse interval, the Fe–CaP composite nanoparticles were solidified while retaining their amorphous structure. Irregularly shaped non-spherical precipitates were obtained from the labile supersaturated CaP solution without ferric ions (Fig. 24.3c), demonstrating the significant effect of ferric ions as a light-absorbing agent.

In our laser-assisted precipitation process, the size and chemical composition (i.e., iron content) of the Fe–CaP composite nanoparticles are controllable to some extent by tuning the process parameters. Nanoparticle spheroidization and growth were enhanced by increasing the concentration of ferric ions in the labile supersaturated CaP solution (from 0 to 5 mM), laser fluence (from 67 to 200 mJ/pulse/cm^2), and irradiation time (from 0 to 30 min). These effects of the parameters were attributed to the effects on iron immobilization in the CaP nanoprecipitates and iron-mediated light–material interactions in the CaP solution.

24.4.2 *Synthesis of IO–CaP Composite Nanoparticles*

The Fe–CaP composite nanoparticles derived from ferric ions were amorphous, as described in the preceding section. We aimed to crystallize magnetic IO nanocrystals within Fe–CaP composite nanoparticles to provide the nanoparticles with magnetic

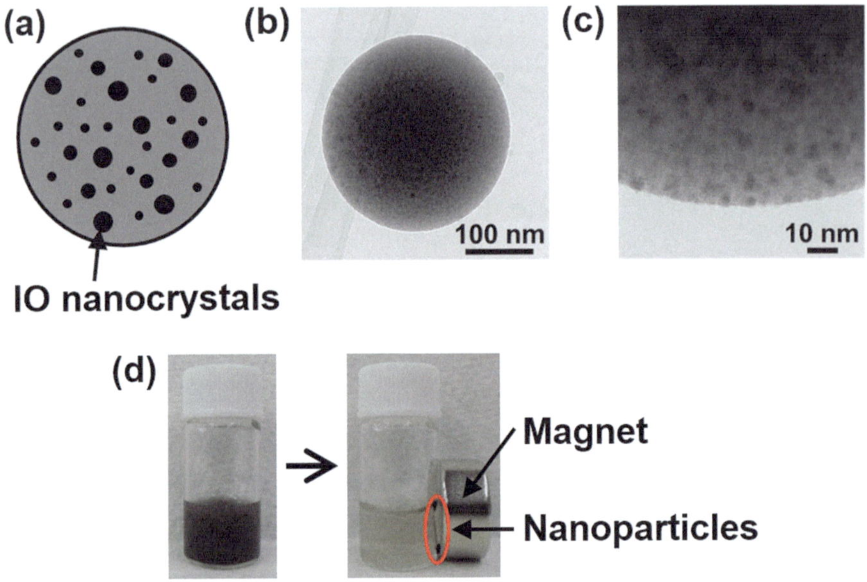

Fig. 24.4 IO–CaP composite nanoparticles. (Modified from [7].) (**a**) Schematic of the nanoparticles. (**b**) Low-magnification and (**c**) high-magnification transmission electron microscopy (TEM) images of the nanoparticles. (**d**) Dispersed nanoparticles (left) and those (right) collected by the magnet outside the bottle

properties. Magnetite (Fe_3O_4, ferrous–ferric oxide) is a major magnetic IO and is known to form in basic solutions containing both ferrous (Fe^{2+}) and ferric ions at appropriate concentrations. We hypothesized that this ferrous–ferric coexistence that is favorable for magnetite crystallization would be naturally produced by the partial oxidation of ferrous ions in the CaP solution.

Thus, we employed ferrous ions instead of ferric ions as a light-absorbing agent and applied pulsed laser irradiation to a labile supersaturated CaP solution supplemented with ferrous ions (2–20 mM) [7]. However, the synthesized spherical Fe–CaP composite nanoparticles were amorphous and did not contain IO crystals. We attributed this failure to the acidic conditions of the CaP solution, which are unfavorable for magnetite precipitation. To increase solution pH to above 7, we added sodium hydroxide (NaOH) to the CaP solution supplemented with ferrous ions (20 mM). From this CaP solution with the elevated pH, the CaP nanoparticles containing magnetite (and wüstite) nanocrystals were successfully synthesized after laser irradiation for 20 min (Fig. 24.4a–c). Laser-assisted heating was thought to facilitate the rapid crystallization of IO within the nanoparticles because the nanoprecipitates aged in the same solution for 20 min without irradiation contained a negligibly small amount of crystalline IO. The resulting IO–CaP composite nanoparticles exhibited magnetic properties, allowing the noncontact manipulation of the nanoparticles using a magnet (Fig. 24.4d). Hence, these magnetic nanoparticles are potentially applicable in magnetic resonance imaging and hyperthermia as well as in magnetic targeting, although further in vitro and in vivo studies are needed.

The synthesis of IO–CaP composite nanoparticles via conventional precipitation requires a long aging time (~1 day) [30, 31]. As our laser-assisted precipitation process is completed within a few tens of minutes in one-pot, it represents a new practical method for synthesizing IO–CaP composite nanoparticles.

24.4.3 Synthesis of Ag–CaP Composite Nanoparticles

We employed silver ions (Ag^+) as a light-absorbing agent and synthesized Ag–CaP composite nanoparticles using our laser-assisted precipitation process [8]. A labile supersaturated CaP solution supplemented with silver ions (5 mM) was irradiated with a pulsed laser. The silver ions coprecipitated with CaP in the CaP solution and enhanced laser absorption, thereby inducing the melting and spheroidization of the nanoprecipitates within 20 min. The resulting nanoparticles were spherical and consisted of amorphous CaP matrix with metallic silver (Ag^0) nanocrystals dispersed throughout the matrix (Fig. 24.5a, b). Note that metallic silver nanocrystals did not exist in the initial precipitates that were formed in the same CaP solution before laser irradiation. Thus, metalic silver was formed durig laser irradiation as a result of photoreduction of silver ions in the precipitates. Such photoreduction into metallic iron (Fe^0) was not observed in the case of ferric and ferrous ions (Sects. 24.4.1 and 24.4.2). This difference can be explained by the difference in standard redox potential (i.e., $Ag^+/Ag^0 = +0.80$ eV, $Fe^{3+}/Fe^0 = -0.04$ eV, and $Fe^{2+}/Fe^0 = -0.44$ eV) [32], indicating that silver ions are more likely to be reduced than ferric and ferrous ions.

The synthesized Ag–CaP composite nanoparticles exhibited antibacterial activity against major pathogenic oral bacteria (i.e., *Streptococcus mutans*, *Aggregatibacter actinomycetemcomitans*, and *Porphyromonas gingivalis*) [8]. As shown in Fig. 24.5c, the turbidity of the bacterial (*Streptococcus mutans*) suspension, which corresponds to the amount of bacteria, decreased with nanoparticle dose. The antibacterial activity of the nanoparticles is attributed to the silver nanocrystals immobilized in the CaP matrix. In addition, the nanoparticles may support the remineralization of tooth lesions by dissolving and neutralizing acids. In general, oral bacteria decrease ambient pH and enhance tooth demineralization [33]. We confirmed that the decrease in pH of the bacterial suspension was suppressed by the nanoparticles (Fig. 24.5d). These results suggest that the Ag–CaP composite nanoparticles have a potential as therapeutic agents in dental care because they can possibly eliminate oral bacteria and support the remineralization of tooth lesions.

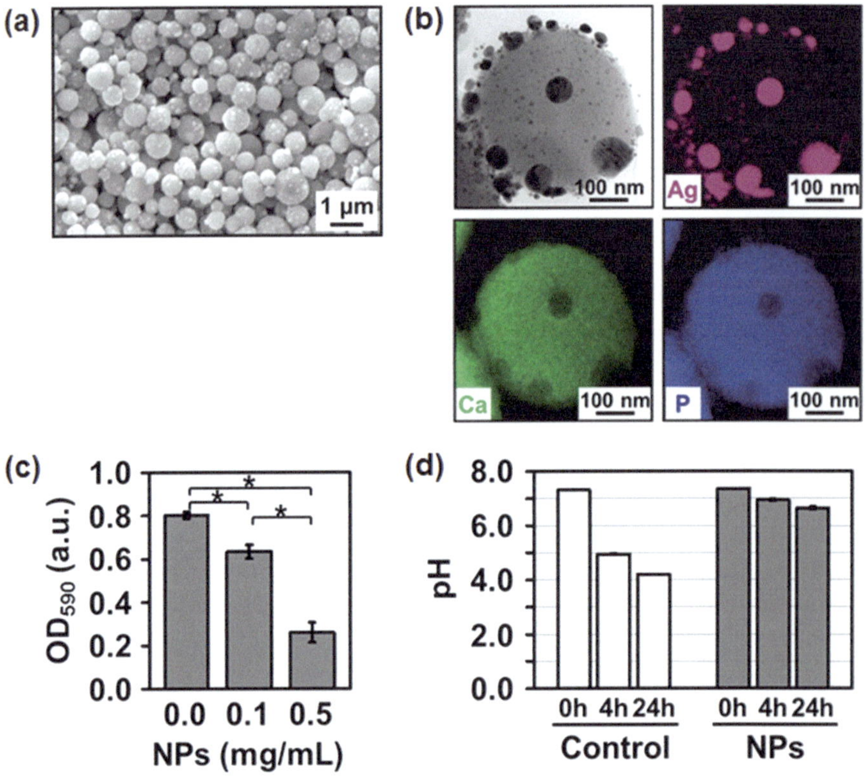

Fig. 24.5 Ag–CaP composite nanoparticles (NPs). (Modified from [8].) (**a**) SEM image. (**b**) Cross-sectional TEM image of the NPs and elemental maps of silver, calcium, and phosphorus from the region shown in the TEM image. (**c**) Turbidity of a suspension of *Streptococcus mutans* incubated for 24 h in the absence and presence of the NPs (dose of NPs: 0.1 and 0.5 mg/mL). (**d**) pH of suspensions of *Streptococcus mutans* incubated for 0, 4, and 24 h in the absence (control) and presence (0.5 mg/mL) of the NPs

24.5 Summary

CaP-based nanoparticles with various biomedical functions can be synthesized from labile supersaturated CaP solutions via homogeneous CaP nucleation and coprecipitation with functional substances. The application of pulsed laser irradiation to the supersaturated solutions enables the rapid spheroidization of CaP-based nanoparticles and dispersive crystallization of magnetite and silver within the nanoparticles through laser-mediated thermal and photoreactions (Sects. 24.4.2 and 24.4.3).

The precipitation processes both with and without laser irradiation are simple (one-pot), rapid (nanoparticle formation occurs within a few tens of minutes), free of harmful additives, and capable of controlling the physicochemical and biological properties of the CaP-based nanoparticles. The resulting CaP-based nanoparticles are useful as biocompatible carriers for drugs and genes, as demonstrated above.

These nanoparticles are also useful as building blocks for higher-order biofunctional materials. For example, DNA–CaP composite nanoparticles can be assembled onto polymeric and ceramic substrates utilizing the precipitation process described in Sect. 24.3.2 [2, 3]. Such composite materials are useful as gene-stimulating cell scaffolds for tissue engineering applications as they have biocompatible surfaces to support cell adhesion and growth and can stimulate cells via surface-mediated gene delivery [16].

Acknowledgments Our research was supported by JSPS KAKENHI grant numbers JP16H03831, JP26560250, and JP15F15030, Japan; the Magnetic Health Science Foundation, Japan; and the Amada Foundation, Japan. We would like to thank Dr. Quazi T. H. Shubhra, Ms. Hiroko Araki, Ms. Ikuko Sakamaki, Dr. Yoshiki Shimizu, Dr. Kenji Koga, Dr. Alexander Pyatenko, and Dr. Atsuo Ito from AIST, Dr. Hirofumi Miyaji and Dr. Naoto Koshizaki from Hokkaido University, and Dr. Hideo Tsurushima from the University of Tsukuba for their contributions to our research.

References

1. Kokubo T, Takadama H (2006) How useful is SBF in predicting in vivo bone bioactivity? Biomaterials 27:2907–2915
2. Oyane A, Araki H, Sogo Y, Ito A, Tsurushima H (2013) Spontaneous assembly of DNA–amorphous calcium phosphate nanocomposite spheres for surface-mediated gene transfer. CrystEngComm 15:4994–4997
3. Oyane A, Araki H, Nakamura M, Shimizu Y, Shubhra QTH, Ito A, Tsurushima H (2016) Controlled superficial assembly of DNA–amorphous calcium phosphate nanocomposite spheres for surface-mediated gene delivery. Coll Surf B, Biointerface 141:519–527
4. Shubhra QTH, Oyane A, Araki H, Nakamura M, Tsurushima H (2017) Calcium phosphate nanoparticles prepared from infusion fluids for stem cell transfection: process optimization and cytotoxicity analysis. Biomater Sci 5:972–981
5. Nakamura M, Oyane A (2016) Physicochemical fabrication of calcium phosphate-based thin layers and nanospheres using laser processing in solutions. J Mater Chem B 4:6289–6301
6. Nakamura M, Oyane A, Sakamaki I, Ishikawa Y, Shimizu Y, Koga K, Kawaguchi K, Koshizaki N (2014) A physicochemical process for fabricating submicrometre calcium iron phosphate spheres. RSC Adv 4:38442–38445
7. Nakamura M, Oyane A, Sakamaki I, Ishikawa Y, Shimizu Y, Kawaguchi K (2015) Laser-assisted one-pot fabrication of calcium phosphate-based submicrospheres with internally crystallized magnetite nanoparticles through chemical precipitation. Phys Chem Chem Phys 17:8836–8842
8. Nakamura M, Oyane A, Shimizu Y, Miyata S, Saeki A, Miyaji H (2016) Physicochemical fabrication of antibacterial calcium phosphate submicrospheres with dispersed silver nanoparticles *via* coprecipitation and photoreduction under laser irradiation. Acta Biomater 46:299–307
9. Gebauer D, Cölfen H (2011) Prenucleation clusters and non-classical nucleation. Nano Today 6:564–584
10. Onuma K, Ito A (1998) Cluster growth model for hydroxyapatite. Chem Mater 10:3346–3351
11. Oyane A, Onuma K, Kokubo T, Ito A (1999) Clustering of calcium phosphate in the system $CaCl_2-H_3PO_4-KCl-H_2O$. J Phys Chem B 103:8230–8235
12. Dey A, Bomans PHH, Müller FA, Will J, Frederik PM, de With G, Sommerdijk NAJM (2010) The role of prenucleation clusters in surface-induced calcium phosphate crystallization. Nat Mater 9:1010–1014

13. Barrère F, Layrolle P, van Blitterswijk CA, de Groot K (2000) Fast formation of biomimetic Ca-P coatings on Ti6Al4V. Mater Res Soc Symp Proc 599:135–140
14. Wang L, Nancollas GH (2008) Calcium orthophosphates: crystallization and dissolution. Chem Rev 108:4628–4669
15. Wang X, Ito A, Li X, Sogo Y, Oyane A (2011) Signal molecules-calcium phosphate coprecipitation and its biomedical application as a functional coating. Biofabrication 3:022001
16. Oyane A, Wang X, Sogo Y, Ito A, Tsurushima H (2012) Calcium phosphate composite layers for surface-mediated gene transfer. Acta Biomater 8:2034–2046
17. Oyane A, Uchida M, Onuma K, Ito A (2006) Spontaneous growth of a laminin-apatite nanocomposite in a metastable calcium phosphate solution. Biomaterials 27:167–175
18. Yazaki Y, Oyane A, Tsurushima H, Araki H, Sogo Y, Ito A, Yamazaki A (2014) Coprecipitation of DNA-lipid complexes with apatite and comparison with superficial adsorption for gene transfer applications. J Biomater Appl 28:937–945
19. Yazaki Y, Oyane A, Sogo Y, Ito A, Yamazaki A, Tsurushima H (2011) Control of gene transfer on a DNA–fibronectin–apatite composite layer by the incorporation of carbonate and fluoride ions. Biomaterials 32:4896–4902
20. Xie Y, Chen Y, Sun M, Ping Q (2013) A mini review of biodegradable calcium phosphate nanoparticles for gene delivery. Curr Pharm Biotechnol 14:918–925
21. Sogo Y, Ito A, Fukasawa K, Kondo N, Ishikawa Y, Ichinose N, Yamazaki A (2005) Coprecipitation of cytochrome C with calcium phosphate on hydroxyapatite ceramic. Curr Appl Phys 5:526–530
22. Mutsuzaki H, Ito A, Sakane M, Sogo Y, Oyane A, Ochiai N (2008) Fibroblast growth factor-2-apatite composite layers on titanium screw to reduce pin tract infection rate. J Biomed Mater Res B Appl Biomater 86:365–374
23. Bodhak S, Kikuchi M, Sogo Y, Tsurushima H, Ito A, Oyane A (2013) Calcium phosphate coating on a bioresorbable hydroxyapatite/collagen nanocomposite for surface functionalization. Chem Lett 42:1029–1031
24. Zhou H, Bhaduri S (2012) Novel microwave synthesis of amorphous calcium phosphate nanospheres. J Biomed Mater Res B Appl Biomater 100:1142–1150
25. Qi C, Zhu YJ, Zhao XY, Lu BQ, Tang QL, Zhao J, Chen F (2013) Highly stable amorphous calcium phosphate porous nanospheres: microwave-assisted rapid synthesis using ATP as phosphorus source and stabilizer, and their application in anticancer drug delivery. Chem Eur J 19:981–987
26. Zhao J, Zhu YJ, Zheng JQ, Chen F, Wu J (2013) Microwave-assisted hydrothermal preparation using adenosine 5′-triphosphate disodium salt as a phosphate source and characterization of zinc-doped amorphous calcium phosphate mesoporous microspheres. Microporous Mesoporous Mater 180:79–85
27. Rouhani P, Taghavinia N, Rouhani S (2010) Rapid growth of hydroxyapatite nanoparticles using ultrasonic irradiation. Ultrason Sonochem 17:853–856
28. Wang H, Pyatenko A, Kawaguchi K, Li X, Swiatkowska-Warkocka Z, Koshizaki N (2010) Selective pulsed heating for the synthesis of semiconductor and metal submicrometer spheres. Angew Chem Int Ed 49:6361–6364
29. Nakamura M, Oyane A, Sakamaki I, Shimizu Y, Koga K, Koshizaki N (2015) A physicochemical process for fabricating submicrometer hollow fluorescent spheres of Tb^{3+}-incorporated calcium phosphate. RSC Adv 5:22620–22624
30. Wu HC, Wang TW, Sun JS, Wang WH, Lin FH (2007) A novel biomagnetic nanoparticle based on hydroxyapatite. Nanotechnology 18:165601
31. Ansar EB, Ajeesh M, Yokogawa Y, Wunderlich W, Varma H (2012) Synthesis and characterization of iron oxide embedded hydroxyapatite bioceramics. J Am Ceram Soc 95:2695–2699
32. Atkins PW (1998) Physical chemistry, 6th edn. Oxford University Press, Oxford, p 936
33. Selwitz RH, Ismail AI, Pitts NB (2007) Dental caries. Lancet 369:51–59

MIX
Papier aus verantwortungsvollen Quellen
Paper from responsible sources
FSC® C105338

If you have any concerns about our products,
you can contact us on
ProductSafety@springernature.com

In case Publisher is established outside the EU,
the EU authorized representative is:
Springer Nature Customer Service Center GmbH
Europaplatz 3, 69115 Heidelberg, Germany

Printed by Libri Plureos GmbH
in Hamburg, Germany